Binary systems of stars are as common as single stars. Stars evolve primarily by nuclear reactions in their interiors, but a star with a binary companion can also have its evolution influenced by the companion. Multiple star systems can exist stably for millions of years, but can ultimately become unstable as one star grows in radius until it engulfs another.

This volume discusses the statistics of binary stars; the evolution of single stars; and several of the most important kinds of interaction between two (and even three or more) stars. Some of the interactions discussed are Roche-lobe overflow, tidal friction, gravitational radiation, magnetic activity driven by rapid rotation, stellar winds, magnetic braking and the influence of a distant third body on a close binary orbit. A series of mathematical appendices gives a concise but full account of the mathematics of these processes.

PETER EGGLETON is a physicist at the Lawrence Livermore National Laboratory in California. Following his education in Edinburgh, he obtained his Ph.D. in Astrophysics from the University of Cambridge in 1965. He lectured for a short period at York University before returning to the University of Cambridge to conduct research from 1967 to 2000 as a Fellow of Corpus Christi College. In 2000, he took up his current position at LLNL. He is well known throughout the community as one of the most knowledgeable experts in binary star evolution.

Cambridge Astrophysics Series

Series editors

Andrew King, Douglas Lin, Stephen Maran, Jim Pringle and Martin Ward

EVOLUTIONARY PROCESSES IN BINARY AND MULTIPLE STARS

PETER EGGLETON

Lawrence Livermore National Laboratory, California

CAMBRIDGE
UNIVERSITY PRESS

CAMBRIDGE UNIVERSITY PRESS
Cambridge, New York, Melbourne, Madrid, Cape Town,
Singapore, São Paulo, Delhi, Tokyo, Mexico City

Cambridge University Press
The Edinburgh Building, Cambridge CB2 8RU, UK

Published in the United States of America by Cambridge University Press, New York

www.cambridge.org
Information on this title: www.cambridge.org/9781107403420

First published 2006
First paperback edition 2011

A catalogue record for this publication is available from the British Library

ISBN 978-0-521-85557-0 Hardback
ISBN 978-1-107-40342-0 Paperback

Cambridge University Press has no responsibility for the persistence or
accuracy of URLs for external or third-party internet websites referred to in
this publication, and does not guarantee that any content on such websites is,
or will remain, accurate or appropriate.

Contents

Preface

This book is intended for those people, perhaps final-year undergraduates and research students, who are already familiar with the terminology of stellar astrophysics (spectral types, magnitudes, etc.) and would like to explore the fascinating world of binary stars. I hope it will also be useful to those whose main astrophysical interests are in planets, galaxies or cosmology, but who wish to inform themselves about some of the basic blocks on which much astronomical knowledge is built. I have endeavoured to put into one book a number of concepts and derivations that are to be found scattered widely in the literature; I have also included a chapter on the internal evolution of *single* stars.

In the interest of keeping this volume short, I have been brief, some might say cursory, in surveying the enormous literature on observed binary stars. It is almost a truism that theoretical ideas stand or fall by comparison with observation. My intention is to produce a second volume, with my colleagues Dr Ludmila Kiseleva-Eggleton and Dr Zhanwen Han, in which individual binary and triple stars that rate less than a line in this volume will be discussed in the paragraph or two each, at least, which they deserve. In addition, the synthesis of large theoretical populations of binary stars will be discussed. Some individual binaries can be seen as flying entirely in the face of the theoretical ideas outlined here – see OW Gem, Section 2.3.5. If I took at face value the notion that one well-measured counter-example is all that is needed to demolish a theory, then I would have given up long ago. Rather, I think, it is necessary to persevere: not be paralysed by disagreement with observation, but also not to sweep disagreement under the carpet.

A number of problems that have to be considered may well be capable of being answered only by detailed numerical modelling, constructing three-dimensional models of a whole star, or of a pair of stars in a binary. Massive computer resources will be needed for such investigations; for that reason I moved from Cambridge University to the Lawrence Livermore National Laboratory, California, where such resources are being developed. This Laboratory has started the 'Djehuty Project' – named after the Egyptian god of astronomy – to pursue this long-term goal. We hope that this project will supplement, though it cannot entirely replace, the simple ideas which this book discusses.

I am very grateful to many colleagues who have been generous of their time in discussing the issues of binary-star evolution. Drs Zhanwen Han, Onno Pols, Klaus-Peter Schröder, Chris Tout and Ludmila Kiseleva-Eggleton have kindly supplied some figures, as well as much insight. I particularly wish to thank Prof. Piet Hut for his careful and critical reading of the manuscript, and suggestions for improvement, and Drs Kem Cook and Dave Dearborn for their patience in allowing me to pursue this topic.

This work was performed under the auspices of the US Department of Energy, National Nuclear Security Administration by the University of California, Lawrence Livermore National Laboratory under contract No W-7405-Eng-48; and much use was made of the archive at the Centre de Données astronomiques de Strasbourg.

1

Introduction

1.1 Background

Because gravity is a long-range force, it is difficult to define precisely the concept of an 'isolated star' – and consequently also the concept of a binary or triple star. Nevertheless, many stars are found whose closest neighbouring star is a hundred, a thousand or even a million times closer than the average separation among stars in the general neighbourhood. Such pairings of stars are expected to be very long lived. There also exist occasional local clusterings of perhaps a thousand to a million stars, occupying a volume of space which would much more typically contain only a handful of stars. These clusters can also be expected to be long lived – although not as long lived as an 'isolated' binary, since the combined motion of stars in a large cluster causes a slow evaporation of the less massive members of the cluster, which gain kinetic energy on average from close gravitational encounters with the more massive members. Intermediate between binaries and clusters are to be found small multiple systems containing three to six members, and loose associations containing somewhat larger numbers. Starting from the other end, some clusters may contain sub-clusters, and perhaps sub-sub-clusters, down to the scale of binaries and triples.

Even with the naked eye, a handful of the 5000 stars visible can be seen to be double; and in the northern hemisphere two clusters of stars, the Hyades and the Pleiades, are quite recognisable. But some 2000 naked-eye stars are known to be binary (or triple, quadruple, etc.) by more detailed measurement – astrometric, spectroscopic or photometric. Observation in other wavelength ranges, such as radio, infrared, ultraviolet and X-rays, reveal further and more exotic binary companions, not so many in number, but of unusual interest. The naked-eye stars are only a tiny fraction of all the stars in our Galaxy ($\sim 10^{11}$), but are reasonably representative as far as the incidence of binarity is concerned.

Sometimes the two components are so close together as to be virtually touching; sometimes they are so far apart as to be virtually independent. Measured orbital periods range from hours (or even minutes) up to centuries. Many must have longer periods still, not yet determined but up to millions of years. The evolution of the two components of such pairs has attracted increasing interest over the last fifty years. The presence of a binary companion, if the orbital period is a few years or less, may make the evolution of a star very different from what it would have been if the star were effectively isolated. A number of these differences are now fairly well understood, but although some evolutionary problems which used to trouble astrophysicists, such as the 'Algol paradox', have been largely resolved, several still remain. New observations add new problems considerably faster than they confirm the resolution of older problems. It should be kept in mind that even single stars present many evolutionary problems, and so it is not surprising that many binary stars do.

Questions about binary stars can be divided very loosely into two categories, 'structural' and 'evolutionary'. For a particular type of binary star one can ask what physical processes are currently going on, that give this type of star its particular characteristics. In cataclysmic variables such as novae, for instance, there is little doubt that a fairly normal main sequence star of rather low mass is being slowly torn apart by the gravitational field of a very close white dwarf companion. But one can also ask how such binaries started, and subsequently evolved, so that these processes can currently take place. This evolutionary question can be harder to answer, because most evolutionary processes are very slow. An obvious further evolutionary question is: 'What will the future evolution of such systems be, up to some long-lived final state?' This book attempts to summarise progress in understanding the kind of long-term evolutionary processes involved. In the interest of brevity it will be necessary to quote, and to take for granted rather than to discuss, most of the much more substantial literature on structural problems. However, one aspect of binary stars that might be labelled 'structural', but which is certainly of vital importance for evolutionary discussions, is the determination of such fundamental parameters as masses, radii, etc.

1.2   Determination of binary parameters

If we are interested in determining the masses and radii of stars, then we have to turn almost right away to *binary* stars, since it is only by measuring orbital motion under gravity, and by measuring the shape and depth of eclipses, that we are able to determine these quantities to a good accuracy – one or two per cent in favourable cases; see Hilditch (2001). Analysis of the spectrum of an *isolated* star can determine such useful quantities as the star's surface temperature, gravity and composition. This is done by comparing the observed spectrum, preferably not just in the visible region of wavelengths but also in the ultraviolet (UV) and infrared (IR), with a grid of computed spectra for a range of temperatures, gravities and compositions. However, we do not get a mass from this process, or a radius, only the combination that gives the gravity – except in the special case of white dwarfs, where there is expected to be a tight radius–mass relation (Section 2.3.2) so that both mass and radius are functions only of gravity.

If we have an accurate parallax, as from the Hipparcos satellite, we can get closer to determining the mass of an isolated star, because the distance, the temperature (from spectral analysis), and the apparent brightness give us the radius; and hence the gravity (also from spectral fitting) gives us the mass. However, even if the parallax is good to ∼1%, the gravity is much less accurate, because spectra are usually nothing like so sensitive to gravity as they are to temperature. Perhaps an accuracy of ∼25% is achievable.

The parameters of binary systems are generally obtained from astrometric, or spectroscopic, or photometric observations, and in favourable cases by a combination of two, or even all three, of these methods. Note that terms such as 'astrometric' and 'photometric', coined originally to refer to observations in the visible portion of the electromagnetic spectrum, are now generally used to cover all parts of the spectrum, for instance radio and X-rays. If the two components of a binary are so far apart in the sky as to be resolvable from each other, which means at visual wavelengths more than ∼0.1″ (0.5 μrad) apart, then the system is a 'visual binary' or 'VB', and careful astrometry, sometimes over a century or more, can reveal the orbit. Visual binaries tend to have long periods because short-period orbits are generally not resolvable. Only for systems within ∼5 pc of the Sun (about 50 in number) could a separation of 0.2″ correspond to a period of ≲1 year. The upper limit of well-determined

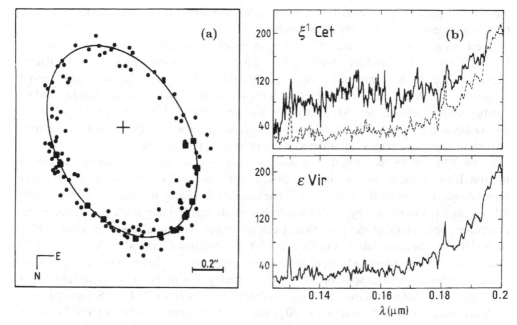

Figure 1.1 (a) The orbit of HR 3579 (F5V+G5V) from visual (dots) and speckle (square) measurements of relative position. The scatter of speckle points about the best-fit curve ($P = 21.8$ yr, $e = 0.15$, $a/D = 0.66''$, $i = 130°$) is much less than for the visual points. From Hartkopf *et al.* (1989). (b) The UV spectrum of the G8III stars ϵ Vir (bottom panel) and ξ^1 Cet (top panel, with ϵ Vir repeated). For 0.18–0.7 μm (not all shown here) the spectra are very similar. The UV excess evident in ξ^1 Cet for 0.13–0.17 μm is attributable to a white dwarf companion. From Böhm-Vitense and Johnson (1985).

visual orbital periods is about 100 years, because good accuracy is only achievable if the VB has consistently been followed for at least two full orbits. There are many orbits in the literature with periods up to 1000 years, or even longer, but these must be considered tentative – extremely tentative if the period is greater than 200 years.

Visual orbits are usually *relative* orbits, the position of one component being measured relative to the other (Fig. 1.1a). Visual orbits have been catalogued by Worley and Douglas (1984), and speckle measurements by McAlister and Hartkopf (1988). These and many other relevant catalogues can be found on the website of the Centre des Données astronomique de Strasbourg (http://cdsweb.u_strasbg.fr). From visual orbits one can determine the period (P), the eccentricity (e), the inclination (i) of the orbit to the line of sight, and the *angular* semimajor axis, i.e. the ratio of the semimajor axis a to the distance D. One can then determine M/D^3, where M is the total mass, from Kepler's law:

$$\frac{GM}{a^3} = \left(\frac{2\pi}{P}\right)^2, \quad \text{so} \quad \frac{GM}{D^3} = \left(\frac{2\pi}{P}\right)^2 \left(\frac{a}{D}\right)^3. \tag{1.1}$$

If the VB is near enough, D may be obtainable from the parallax. For Earth-based measurements, parallaxes of less than $0.1''$ are not reliable, but this has been improved by more than an order of magnitude with space-based measurements from the Hipparcos satellite. If the orbits of both the components of a visual binary can be measured *absolutely*, i.e. each orbit relative to a background of distant and approximately 'fixed' stars, then the mass ratio of the two components can further be determined. We still do not obtain the individual masses, however, unless D is separately determinable.

Even if only one component of a binary is visible at all, an astrometric orbit may in favourable cases be found by observing that the position of a star has a cyclic oscillation superimposed on the combination of its parallactic motion and its linear proper motion relative to the 'fixed' stars, i.e. faint stars most of which do not move measurably and so can be assumed to be distant. Such astrometric binaries can yield P, e and i, but information on masses is convolved with the unknown mass ratio, and also with the parallax which may or may not be measurable even if the astrometric orbit is measurable.

Some VBs can be recognised even when neither component shows measurable orbital motion. If two stars, not necessarily *very* close together on the sky, show the same substantial linear proper motion relative to the 'fixed' background, it is likely that (a) they are physically related, and (b) fairly nearby, with measurable parallaxes. Usually these parallaxes agree, confirming the reality of the pair. Such pairs are called 'common proper motion' (CPM) pairs. The two nearest stars to the Sun, V645 Cen (Proxima Cen) and α Cen, are over $2°$ apart, but have the same rapid proper motion and large parallax. To be pedantic, (a) they are so near the Sun, and so far apart on the sky, that actually their proper motions and parallaxes are measurably different at the 1% level, and (b) α Cen is itself a VB of two Solar-type stars, with semimajor axis $17.5''$ and period 80 years, so that the proper motion of V645 Cen has to be compared with the proper motion of the centre of gravity (CG) of the α Cen pair. The period of the orbit of V645 Cen about the CG of the triple system can be expected to be about 1 megayear.

Common proper motion pairs are usually sufficiently wide that they might appear to be of little relevance to this book, which deals with pairs sufficiently close together that one component can influence the other's evolution. However the presence of a CPM companion can often reveal information on both components that would not be available if they were not paired. Several *close* pairs have a distant CPM companion; and if for example this companion has a character that suggests that it is fairly old, then one can reasonably conclude that the close binary is also fairly old. This may not be evident from the close binary alone, since the components in it may have interacted in ways that disguise the age of the system.

Modern techniques such as speckle interferometry (Labeyrie 1970, McAlister 1985), can resolve components with substantially smaller angular separations than conventional astrometry, and thus determine visual orbits of shorter period. The major limitation on resolving close components astrometrically is atmospheric 'seeing', the blurring effect of turbulence in the Earth's atmosphere. This distorts the image on a timescale of ~0.05 s. In the speckle technique the image is recorded many times a second, and so the time variation of the point-spread function can be followed and allowed for in a Fourier deconvolution. The technique of adaptive optics (Babcock 1953, Beckers 1993) is an alternative way of eliminating seeing, by continuously adapting the shape of the mirror in response to the deformation of the image of a reference point source, either a nearby single star or the back-scattered light of a laser beam pointing along the telescope. Both techniques can give resolution down to the limit of diffraction, $\sim0.01''$ at visual wavelengths on a modern 8 m class telescope. By combining the light from two or more separate telescopes, the technique of 'aperture synthesis', long used in radio astronomy, can nowadays be applied to optical wavelengths (Burns *et al.* 1997), and should be capable of sub-milliarcsecond resolution, so that one might hope to see directly both components of nearby short-period binaries.

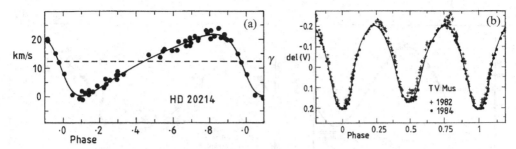

Figure 1.2 (a) The radial velocity curve of the K giant star HD20214. The rms scatter about the mean curve is only ~ 0.2 km/s. Orbital parameters are $P = 407$ days, $e = 0.41$, $f = 0.040\,M_\odot$. From Griffin (1988). (b) The light curve of a contact binary TV Mus ($P = 0.446$ days, $e = 0$, $i = 78.9°$, $R_1/a = 0.59$, $R_2/a = 0.27$, $M_1/M_2 = 7.2$, $T_1/T_2 = 0.98$). A slight variation in brightness over two years, and a small distortion in the secondary eclipse, may be due to starspots. From Hilditch *et al.* (1989).

Systems may be recognisable as spectroscopic binaries (SBs) either because the spectrum is composite (Fig. 1.1b), or because it shows radial velocity variations (Fig. 1.2a), or both. In a composite spectrum, one might see for instance a combination of the relatively broad lines of H and He characteristic of a B dwarf with the narrow lines of Fe and other metals characteristic of a G or K giant. Alternatively, a star whose spectrum at visual wavelengths may seem like a K giant may be found, at UV wavelengths, to have an excess flux that can be attributed to a hot companion, sometimes even a white dwarf (Fig. 1.1b). It is not easy to disentangle composite spectra reliably, since things other than a stellar companion (for example a corona, a circumstellar disc or a dust shell) may contribute to an excess either in the UV or the IR. Even if the spectrum seems definitely a composite of two stellar spectra, we learn only that the star is a binary; we do not obtain information about the orbit unless one spectrum at least shows a variable radial velocity, consistent with Doppler shift due to motion in a Keplerian orbit.

Orbits of 1469 SBs have been catalogued in the important compilation of Batten, Fletcher and McCarthy (1989). The number of orbits is increasing rapidly, perhaps already at a rate of one or two hundred a year, and no doubt with greater rapidity in the future, partly because of cross-correlation techniques and partly because of the much-increased sensitivity of detectors. Commonly SBs are single-lined ('SB1'), but the radial velocity variation of the single spectrum seen (as in Fig. 1.2a) allows P and e to be obtained and also the amplitude K of the radial velocity variation, or equivalently (as is usual for radio pulsars) the projected semi-major axis ($a \sin i \propto K P \sqrt{1 - e^2}$). Information on masses is contained in a single function, the mass function f, convolving both of the masses with the unknown orbital inclination i:

$$f_1 = \frac{M_2^3 \sin^3 i}{(M_1 + M_2)^2} = \frac{K_1^3 P(1 - e^2)^{3/2}}{2\pi G} = 1.0385 \times 10^{-7} K_1^3 P(1 - e^2)^{3/2}$$

$$= 1.0737 \times 10^{-3} \frac{(a_1 \sin i)^3}{P^2}, \qquad (1.2)$$

where $*1$ (pronounced 'star 1') is the observed star and $*2$ the unseen component. Units are: K_1 in km/s, P in days, $a_1 \sin i$ in light-seconds and masses in Solar units. The inclination is not measurable for spectroscopic orbits because we have information on the motion in

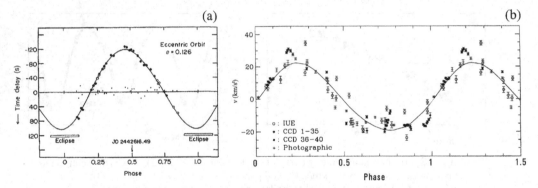

Figure 1.3 Radial velocity curves of both components of the massive X-ray binary Vela X-1 (GP Vel). (a) Doppler shift of the pulses of the X-ray pulsar: note the accurate fit to the Keplerian curve ($P = 8.964$ days, $e = 0.126$, $f_1 = 18.5\,M_\odot$). Small dots near the axis are the residuals multiplied by 2. (b) Doppler shift of absorption lines in the visible spectrum: note the larger scatter, due to irregular pulsations. From these lines $f_2 \sim 0.013$. The ratio f_2/f_1 is the cube of the mass ratio q (~ 0.09). (a) is from Rappaport *et al.* (1976), (b) is from van Kerkwijk *et al.* (1995b).

only one dimension, the line of sight, whereas in visual binaries we have information in two dimensions, both perpendicular to the line of sight. In fact the red giant in ξ^1 Cet (Fig. 1.1b) does show orbital motion ($P = 1642$ days, $e = 0$, $f = 0.035\,M_\odot$, Griffin and Herbig 1981) in addition to being a composite-spectrum binary.

The mass function represents the minimum possible mass for the unseen star, which would be achieved in the somewhat improbable case $M_1 = 0$, $i = 90°$. Slightly more realistically, we might replace $\sin^3 i$ by its average value $3\pi/16 \sim 0.59$ if i is distributed uniformly over solid angle. However the value 0.59 is likely to be an underestimate, because the mere fact that a variation in radial velocity is seen implies that the lowest inclinations can be rejected. For a large ensemble of binaries we might make statistical estimates using a maximum-likelihood procedure. However, for an isolated system, with little else to guide us, we will commonly assume that a reasonable estimate of the reciprocal of $\sin^3 i$ is 1.25. We then take

$$M_1 \sim 1.25q(1+q)^2 f_1, \tag{1.3a}$$

$$M_2 \sim 1.25(1+q)^2 f_1, \tag{1.3b}$$

where $q \equiv M_1/M_2$ is the mass ratio. Sometimes we can estimate M_1 directly from the spectrum of the star, which may be similar to stars whose masses are already known from more favourable binaries (see below); then from Eq. (1.3a) q can be estimated and hence M_2. Alternatively one can often infer that $q > 1$ simply from the probability that the unseen star is less massive than the visible one. In either case both masses could be *considerably* greater than the mass function.

If the system is 'double lined' ('SB2'), and both components have measurable radial velocity variations (Fig. 1.3), we can further obtain the mass ratio, and hence the two quantities $M_1 \sin^3 i$ and $M_2 \sin^3 i$; but we still have no information on i. However, some SBs with $P \gtrsim 1$ year are also VBs, and in favourable cases all four of M_1, M_2, i and D can be separately measured, D in such cases being *independent* of parallax (which may be too small to be measurable).

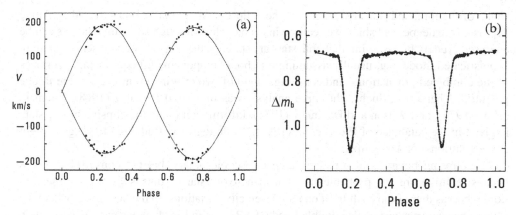

Figure 1.4 (a) The radial velocity curves and (b) the light curve of the eclipsing SB2 system V760 Sco ($P = 1.73$ days). The two components are nearly but not quite identical: in (a), $*2$ has a slightly greater velocity amplitude, and in (b) the second eclipse is slightly shallower than the first. An 'ellipsoidal variation' is seen in the nearly flat portions between eclipses. From Andersen *et al.* (1985).

Among SBs we can include both radio and X-ray pulsars, because the rapid pulsations of these objects, due to rapid rotation of an obliquely-magnetised neutron star, are often very stable and so can reveal a variable Doppler shift due to Keplerian orbital motion. Commonly, pulsar orbits are much more accurate than SB orbits based on spectral lines, so that even companions of terrestrial planetary mass can be detected (Wolszczan and Frail 1992). The much greater accuracy of radio pulsar orbits means that a number of relativistic corrections to Keplerian orbits can be measured (Taylor and Weisberg 1989, Backer and Hellings 1986). Two of these are (a) the rate Z_{GR} of advance of periastron in an eccentric orbit due to general relativity – Appendix C(a):

$$Z_{GR} = \frac{3G(M_1 + M_2)}{c^2 a(1 - e^2)} \frac{2\pi}{P},$$ (1.4)

and (b) a combination γ of gravitational redshift and transverse Doppler shift:

$$\gamma = \frac{G(M_1 + 2M_2)e}{c^2(M_1 + M_2)} \frac{P}{2\pi}.$$ (1.5)

Along with the mass function Eq. (1.2), these two quantities allow one to determine all three of M_1, M_2 and i, even although the orbit is 'single lined'.

X-ray pulsar orbits, though commonly more accurate than radial-velocity orbits from spectral lines (Fig. 1.3), are also commonly less accurate than radio pulsar orbits, because the X-rays come from accretion of gas lost by the companion. The gas flow is normally not steady, and so the neutron star's spin rate is erratically variable by a small amount.

Photometric binaries are stars whose light output varies periodically, and in a manner consistent with orbital motion. Usually they show eclipses, but in some cases where the inclination does not permit an eclipse one may nevertheless recognise 'ellipsoidal variation' or the 'reflection effect' (see below). A light curve (Figs. 1.2b, 1.4b) can yield, in favourable circumstances, P, e and i, the ratios R_1/a, R_2/a of stellar radii to orbital semimajor axis, and the temperature T_2 provided that T_1 is known already, from a spectroscopic analysis of the brighter component. The radius ratios and i come primarily from the duration and shape of the total and partial segments of the eclipse, and the temperature from the relative depths of the deeper and shallower eclipse in each cycle. Although some light curves can be analysed crudely by assuming that both stars are spheres, the majority of eclipsers need more sophisticated modelling, usually assuming that both components fill equipotential surfaces of the combined gravitational and centrifugal field of two orbiting point masses (the Roche potential, Chapter 3). Such light curve analysis was pioneered by Lucy (1968), Rucinski (1969, 1973), and Wilson and Devinney (1971). Information on 3546 eclipsing binary stars is given in the catalogue of Wood *et al.* (1980). A catalogue by Budding (1984) gives light curve solutions for 414 eclipsers.

An eclipsing binary is also usually a spectroscopic binary, but not conversely. This is because eclipses are only probable in systems where one star's radius is $\gtrsim 10\%$ of the separation, whereas there is no such limit on radial-velocity variations. In the best cases, where the system has eclipses and is also double lined ('ESB2', as in Fig. 1.4), we can hope to obtain all of the following fundamental data: $P, e, i, a, M_1, M_2, R_1, R_2, T_1, T_2$ and D (independent of parallax). The last three of these quantities depend not only on good orbital data but also on reliable modelling of stellar atmospheres, so that the effective temperature of at least one component (presumably the brighter) can be determined directly from its spectrum. This is probably reasonable for the majority of stars, but for extremes of effective temperature and luminosity (O and M stars; supergiants and subdwarfs), spectra may be affected by such difficulties as mass loss, instability, convection and metallicity, all of which are not yet well understood. A comprehensive review of data for ESB2 binary stars in the main-sequence band has been given by Andersen (1991); an earlier review by Popper (1980) also gave data for some post-main-sequence binaries. Accuracies of $\lesssim 2\%$ for all quantities are achievable in favourable cases.

Binaries involving evolved stars (giants, supergiants, hot subdwarfs, white dwarfs, etc.) are relatively rare, especially ESB2 systems. Although the photometric and spectroscopic data may be of the same quality, or even better, it is difficult to achieve the same accuracy in the estimation of radii. This is because the two radii are of course very different in giant/dwarf binaries. The information on relative radii, as well as on inclination, is contained in the shape of the ingress/egress portions of eclipses. If one star is so much larger than the other that its occulting edge is virtually a straight line, then the inclination and hence also the ratio of radii are indeterminate. Nevertheless supplementary information from model atmospheres, and from spectrophotometry, the measurement of intensity in several wavebands that may extend from UV to IR, can reduce the indeterminacy. Recent work on such 'ζ Aur' systems (Schröder *et al.* 1997) gives parameters with sufficient accuracy that theoretical models of stellar evolution are seriously tested.

The fact that ESB2 binaries can in principle give a distance measurement that is independent of parallax implies that they could be good yardsticks for measuring distances to external galaxies. Current and developing technology means that at least OB-type binaries may be accessible in fairly nearby galaxies. Of course one does need an estimate of the metallicity in order to relate measured colours to the effective temperature of at least the hotter component.

Because stars in close binaries can be distorted from a spherical shape by the combined gravitational and centrifugal effect of an orbiting close companion, they may show a measurable light variation even when they do not eclipse. This is called 'ellipsoidal variation' – although the stars are only approximately ellipsoidal. Figure 1.4b shows this variation. The system illustrated is in fact at an inclination which also allows eclipses: the ellipsoidal

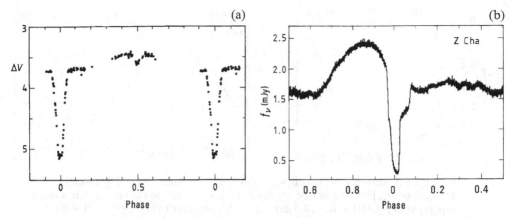

Figure 1.5 (a) The light curve of UU Sge ($P = 0.465$ days), the central star of the planetary nebula Abell 63. The hump centred on the secondary eclipse is due to a 'reflection effect'. The fainter, cooler companion shines partly by reprocessed UV light from the very hot companion; thus it is brightest just before and after it is eclipsed, and is rather faint for half the orbit. From Bond *et al.* (1978). (b) The light curve of Z Cha, an ultra-short-period binary containing a white dwarf and a red dwarf ($P = 0.0745$ days). The hump before the eclipse, the double-stepped nature of the eclipse, and the erratic variation are all due to streams of gas flowing from the red dwarf towards, and round, the white dwarf. From Wood *et al.* (1986).

variation is the slight curvature visible between the eclipses. Such variation even in the absence of eclipses may allow at least P to be determined. Further, if $*1$ (say) is much hotter than $*2$, the hemisphere of $*2$ facing $*1$ may be substantially brighter than the other hemisphere, leading to an orbital variation (Fig. 1.5a) that also does not necessarily involve an eclipse. This is called the 'reflection effect' – although the light (or X-radiation, in some cases) is absorbed, thermalised and reemitted, rather than reflected.

However, not all eclipse light curves, even with high signal-to-noise ratios and with modern light-curve synthesis techniques, lend themselves to accurate measurement of fundamental data (masses, radii, etc.). Neither do all radial velocity curves, even when a non-uniform temperature distribution over the stellar surfaces due for example to the reflection effect is allowed for. This is because stars which are close enough together to have a reasonable probability of eclipse (typically, $R_1 + R_2 \gtrsim 0.2a$) are also quite likely to interact hydrodynamically and hydromagnetically, introducing the complications of gas streams, and of starspots, which are hard to model in any but an ad hoc manner. Figure 1.2b shows a light curve of a contact binary that changed appreciably over time. The changes, and slight asymmetry, can be attributed to transient starspots. Figure 1.5b shows the light curve of a dwarf nova: an eclipse of sorts is clearly recognisable, but the light variation outside the eclipse is due to gas which streams from one component into a ring or disc about the other. Modern methods of analysis such as eclipse mapping (Horne 1985, Wood *et al.* 1986) and Doppler tomography (Marsh and Horne 1988, Richards *et al.* 1995) use image-processing techniques based on maximum-entropy algorithms (Skilling and Bryan 1984). The object of eclipse mapping is to reconstruct the distribution of light intensity over (in the case of Z Cha, Fig. 1.5b) a hypothesised flat, rotating disc of gas around one star that is fed by a stream that comes from the other star. The eclipsing edge of one star as it moves across the disc and stream helps to locate the hotter and

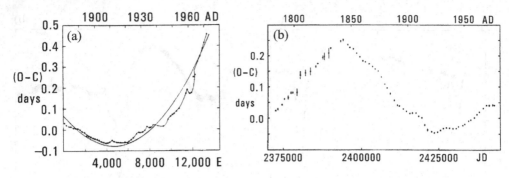

Figure 1.6 Observed times of eclipse, minus computed times obtained by assuming a constant period, plotted against cycle number (epoch) along the bottom and date along the top. (a) U Cep (G8III + B7V; 2.5 days), from Batten (1976). (b) β Per (G8III + B8V; 2.9 days), from Söderhjelm (1980). U Cep shows small erratic variations superimposed on a long term trend of increasing period; β Per also shows erratic fluctuations, but with no clear long-term trend.

cooler parts of the flow. In Doppler tomography, high wavelength resolution across a spectral line, combined with high time resolution, gives a map of intensity on a two-dimensional space of wavelength and orbital phase. This can in principle be Fourier-inverted to map intensity onto a two-dimensional velocity space, and from there one can go via some hypothesised model to a distribution in two-dimensional coordinate space. This might be either a disc-like structure, as in Z Cha, or a distribution of spots over a spherical surface, or even of spots over the joint surface of two stars that are so close as to be in contact (Bradstreet 1985). In this way one can hope to remove the distorting effect of spots and streams from the observational data, and thus be left with accurate fundamental data. But the hypothetical models of spots and streams are not in practice very strongly constrained – for example some systems may contain hot spots as well as cool spots – and so there remains considerable uncertainty in the fundamental data for many, indeed most, interacting systems.

Much information on the statistics of eclipsing binaries (and other types of variable star) comes, as a by-product, from gravitational microlensing experiments (Paczyński 1986). If a relatively nearby star happens to pass very close to the line of sight of a distant star, the apparent brightness of the distant star is temporarily increased by gravitational focusing in the field of the nearby lensing star. Such events are rare, but have been detected by several astronomical groups who monitor photometrically a large number of stars ($\sim 10^6$) in a small area of sky at frequent intervals (e.g. nightly) over several years. The light curve of a lensing event is recognisably different from the light curves of pulsators, eclipsers, novae etc.; but a large number of normal eclipsers shows up as well, and this gives a valuable database from which the statistics of orbital periods can be improved (Udalski *et al.* 1995, Alcock *et al.* 1997, Rucinski 1998). A very few lensing events also exhibit binarity directly: if the lensing object is binary it can produce a marked characteristic distortion on the light curve of a lensing event (Rhie *et al.* 1999).

Some binaries, particularly eclipsing binaries, show a measurable change of period over substantial intervals of time. Period changes are usually demonstrated by 'O − C diagrams' (Fig. 1.6). The difference between the observed time of eclipse, and the computed time based on the assumption of constant period, is plotted as a function of time (or of epoch, i.e. cycle number). One can hope by this method to determine the rates of evolution due to mass transfer or angular momentum loss.

Sometimes the change is periodic. Two possible causes of periodicity (apsidal motion, and a third body) are discussed briefly below. After subtracting such periodic motion if necessary, remaining changes might be an important indication of long-term evolution in the system. But often the long-term behaviour is contaminated by, or even completely obscured by, short-term irregular changes. Figure 1.6a shows the O − C curve for U Cep over the period 1880–1972. If the period were constant we would expect a straight line, and if the period were changing at a constant rate we would expect a parabola as shown. It can be seen that the overall behaviour is roughly parabolic, but with fluctuations of ∼1–2% of the period (∼0.05 days) that are not attributable solely to measuring uncertainty. From the parabolic trend we infer $t_P \equiv P/\dot{P} \sim 1.3$ megayears. The fluctuations are probably due to changes in the distribution of hot luminous gas in this unusually active Algol-like system (Olson 1985). Figure 1.6b shows the same diagram for Algol (β Per) itself over the last 200 years. Unlike U Cep, there is no clear underlying trend: only fluctuations, with possibly the same origin as for U Cep, superimposed on what appears to be a rather sudden period decrease ($\Delta P/P \sim -2 \times 10^{-5}$) around 1845, and a subsequent rather smaller and less sudden period increase around 1920.

O − C curves *ought* to be an important tool for the investigation of the slow changes expected as a result of evolution. One does not have to wait a million years in order to measure a t_P of say 10^8 years quite accurately. If the trend is clearly parabolic, and if individual eclipse timings are accurate to $\pm \delta t$, then we only need observations over a time interval Δt where

$$\Delta t \sim 10 \sqrt{\frac{|t_P|\delta t}{X}}, \tag{1.6}$$

to determine t_P to $\sim X\%$. If the eclipses can be timed to one-minute accuracy, then in a century we can hope to determine an evolutionary timescale of $\sim 10^8$ years reasonably accurately. Unfortunately, rather few binaries show anything like a consistent parabolic trend; we are not helped by the fact that a portion of a parabola can also look like a portion of a periodic third-body effect. If we had observed it only over the last century, β Per might have seemed to show a reasonable parabolic trend. However, the previous century showed quite different behaviour.

Some O − C curves show a clear periodic behaviour that can be attributed to the presence of a distant third body. AS Cam, Fig. 1.7a, is an example, although in this case somewhat marginal. The variable light-travel time due to motion round the third body causes a periodic advance/delay in the eclipse, much as the pulsar orbit in GP Vel (Fig. 1.3a) causes a periodic advance/delay in the arrival time of X-ray pulses. However, orbits of third bodies found by O − C curves are usually very long: an amplitude of 0.1 days translates into an orbital size of about 0.1 light-days or 20 AU, and so a period of ~ 100 years. Such orbits should not be considered reliable unless at least two full orbits have been followed; of course the same qualification applies to any radial-velocity orbit, except for some radio-pulsar orbits where timing can be extraordinarily accurate. In fact Algol itself has a third body in a 1.86 year orbit, but this would not show up in the noise of Fig. 1.6b, even if plotted on a finer scale.

The timing of eclipses is also affected by 'apsidal motion'. The gravitational force between the stars may not be a pure inverse-square law, because (a) general relativity gives a slightly different force and (b) stars can be distorted from the spherical, partly through rotation and partly through the gravitational field of the companion. The line of apses (i.e. major axis)

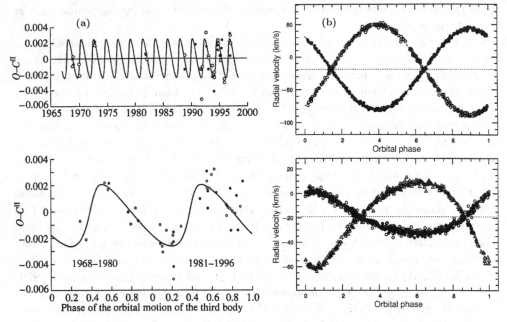

Figure 1.7 (a) O − C curves for the eclipsing binary AS Cam, as a function of date (upper panel) and of phase (lower panel). It shows a roughly periodic variation which may be due to the presence of a third body, in an orbit with $P = 805$ days, $e \sim 0.5$, $f \sim 0.03\,M_\odot$. The inner orbit has $P_1 = 3.43$ days, $e_1 = 0.17$, $(M_{11}, M_{12}) = (3.3, 2.5)\,M_\odot$. After Kozyreva and Khalliulin (1999). (b) The radial velocity curves for the inner and outer orbits of HD109648. Parameters are $P_1 = 5.48$ days, $e_1 = 0.01$, $(M_{11}, M_{12})\sin^3 i_1 = (0.67, 0.60)\,M_\odot$ for the inner orbit (upper panel) and $P = 120.5$ days, $e = 0.24$, $(M_1 \equiv M_{11} + M_{12}, M_2)\sin^3 i = (1.09, 0.54)\,M_\odot$ for the outer (lower panel). From Jha *et al.* (2000).

of a Keplerian orbit is only fixed in space if the force is *exactly* inverse square. Departures make it rotate, and if the orbit is eccentric this means that the eclipses will vary periodically, particularly in the orbital phase of one eclipse relative to the other. The rate of rotation of the line of apses can be measured, and used to check models of internal structure. The rate has also been perceived as a test of GR, but since GR has been verified (see below) to very great accuracy any explanation of discrepancies has to be sought elsewhere.

For example, AS Cam (Fig. 1.7a) shows apsidal motion at a rate inconsistent with GR. Probably this is due to the third body, which affects the apsidal motion as well as introducing a periodic delay (Kozyreva and Khalliulin 1999). Apsidal motion shows up as a slight difference in the period, depending on whether one follows the primary (deeper) eclipse or the secondary. This is because as the major axis rotates slowly the interval between the primary and secondary eclipse changes. Ultimately, the behaviour should be cyclic, with an estimated period (for AS Cam) of ~ 2400 years. The difference in period has however already been allowed for in Fig. 1.7a, where primary eclipses are denoted by heavy dots and secondary eclipses by circles. What remains is not quite constant, but shows (marginally) a periodic fluctuation with an amplitude of ~ 0.002 days and a period of ~ 2 years. This is arguably the 'light-time effect' of a third body, which like a radial-velocity curve (also a Doppler effect) gives a mass function as well as period and eccentricity as listed in the figure caption.

The inconsistency noted above between the measured and theoretically estimated apsidal motion may be due to this third body. Such a body can inject additional apsidal motion (of either sign) into the system, which – somewhat coincidentally – could be of the same order of magnitude (Appendix C).

Figure 1.7b illustrates the radial velocity curves that can be obtained in favourable circumstances from a triple system, HD109648. The spectrum is composed of three separate F stars, two of which show rapid cyclic variations and the third a slower cyclic variation. Not just three but four radial velocity curves can be determined: one is the motion of the centre of gravity of the short-period pair, and mirrors the motion of the third, slowly-moving, spectrum. This gives four mass functions, but unfortunately there are five unknowns: three masses and two inclinations.

Radio pulsars allow enormously greater accuracy to be achieved (Taylor and Weisberg 1989). Some with $P \lesssim 0.4$ days demonstrate the very slow period decrease expected from GR, on a timescale of $\gtrsim 10^8$ years (Section 4.1). For PSR 1913 + 16, the theoretical rate agrees with the observed rate to within one per cent, which is the observational uncertainty. Pulsars near the centre of a globular cluster even show acceleration due to the cluster's gravitational field, and not just a binary companion. What a pity that most stars do not have a pulsar companion!

1.3 Stellar multiplicity

Although only a few thousand stars are well established as binary, with known orbital periods, the incidence of binarity among the most thoroughly observed stars (generally the brightest, but also the nearest) is very high. Conceivably all stars are binary, or of even higher multiplicity. We normally think of the Sun at least as being single, but if there is a continuum of objects from small planets like the Earth ($\sim 3 \times 10^{-6}\ M_\odot$), through massive planets like Jupiter ($0.001\ M_\odot$), to small stars, then perhaps the distinction between single and binary is artificial. Recently detection sensitivity and strategy have improved to the point that three Earth-mass companions to a pulsar (PSR 1257 + 12; Wolszczan and Frail 1992) have been found, and Jupiter-mass companions to about 100 nearby stars, mostly of Solar spectral type (Mayor and Queloz 1995, Marcy and Butler 1998).

A common definition of the term 'star' is that it is an object with mass greater than $\sim 0.08\ M_\odot$, because this is the minimum mass for a self-gravitating hydrostatic spherical gaseous body that can support its radiant energy loss by hydrogen fusion. However, this is a somewhat artificial boundary, because stars in the process of forming will not 'know' that they may come up against this distinction. Low-mass dwarfs are known whose masses are only just above the limit, for example UV Cet (Gl 65AB), a VB where both components are late M dwarfs of $\sim 0.11\ M_\odot$ (Popper 1980). Objects below the critical mass but well above Jupiter's mass are referred to as 'brown dwarfs'. Some are known to exist, but they are hard to detect. An example of a binary containing a star so cool and faint that it is almost certainly below the critical mass is Gl 229AB (Nakajima *et al.* 1995). Some recent low-amplitude orbits of Solar type stars (e.g. HD140913, Mazeh *et al.* 1996) point to companions of $\lesssim 0.05\ M_\odot$, though of course with the ambiguity that the inclination can only be guessed, i.e. assumed not to be improbably small. Observations in the IR (Rebolo *et al.* 1995) have recently been turning up a wealth of probable brown dwarfs in, for instance, the Pleiades cluster.

Recent SB1 detections of companions down to about a Jupiter mass suggest a bimodal distribution, with a fairly rapid drop in numbers to lower mass in the range 0.3–$0.07\ M_\odot$, a low plateau in the brown-dwarf region 0.07–$0.01\ M_\odot$, and then a peak for major planetary masses below $\sim 0.01\ M_\odot$ (Marcy and Butler 1998). This is consistent with the likely hypothesis that the formation mechanism of binary stars is very different from that of planetary systems. The two processes are not exclusive, however. Some systems are known to have both a planetary companion *and* a stellar companion: τ Boo (Butler *et al.* 1997, Hale 1994), 16 Cyg (Cochran *et al.* 1997) and υ And (Lowrance *et al.* 2002). The last has three massive planets and a distant M-dwarf companion.

Most stars are members of binaries. Petrie (1960) showed that 52% of a sample of 1752 stars, independent of spectral type, have variable radial velocities. Since not all short-period binaries can be detected due to finite measuring accuracy, it follows that substantially more than 50% of stars are in relatively short-period binaries. After considering unseen companions, Poveda *et al.* (1982) concluded that nearly 100% of stars are in binaries, including long as well as short periods.

For the sake of terminology, we assume here that there *are* such things as single stars, distinct from binary stars. In other words, we accept the presently-known multiplicity of a particular system, not withstanding the possibility, even probability, that more detailed measurement will mean that small or distant companions will be detected. Thus if a star is not presently known to have a companion, we will speak of it as single. Furthermore, if there is a binary companion but it is too far away ever to have an effect on the evolution of the target star, we shall often use the term 'effectively single'.

Many systems once thought to be binary turn out to contain three or more stars. According to Batten (1973), double-star systems are roughly twice as common as single-star systems, but for $2 < n \leq 6$ the number of systems containing n stars falls off very roughly as 4^{-n}. This means that $\sim 25\%$ of all systems, and $\sim 15\%$ of all stars, are single, while $\sim 20\%$ of all systems ($\sim 30\%$ of stars) are in triples or higher multiples; the average system contains about two stars. Duquennoy and Mayor (1991) found a slightly lower incidence of multiplicity in a sample of 161 F/G-type systems: 92 single, 61 binary, 6 triple and 2 quadruple, but with a proviso that 18 components in this sample showed significant radial velocity variations that might indicate further multiplicity. Tokovinin (1997) has catalogued 612 triple and higher-multiple systems. In this book we will make the assumption, when illustrative numbers are necessary, that $\sim 30\%$ of systems are single to present levels of accuracy, $\sim 60\%$ are binary and $\sim 10\%$ are at least triple.

The incidence of multiplicity is probably not independent of the kind of star being sampled. The 42 nearest stellar systems (within ~ 5 pc; excluding the Sun itself) are mostly M dwarfs, with less than half the mass of the Sun. They contain at least 14 multiples – 10 binary and 4 triple. They also contain at least one massive planet, around an M dwarf star. On the other hand, the 48 brightest systems ($V \leq 2.0$; from Hoffleit and Jaschek 1983 and Batten *et al.* 1989) are mostly B and A stars, typically more than twice the mass of the Sun. They contain at least 22 multiples – 14 binary, 3 triple, 4 quadruple and 1 sextuple. The statistics are not compelling, of course, but seem to imply that more massive systems are more highly multiple. For both these samples, small as they are, the data are far from complete, and the actual multiplicity could well be higher.

It is always difficult to compare distance-limited samples of stars with magnitude-limited samples, because binaries are inherently brighter than single stars, although not by much unless the masses are fairly closely equal. Obviously the first kind of sample is to be preferred where possible, but distances are much harder to measure than magnitudes. In the above two samples the effect is probably quite small.

Multiple ($n > 2$) systems tend to be 'hierarchical' (Evans 1968, 1977), i.e. they consist for example of two close 'binaries' whose centres of gravity rotate around each other in a wide 'binary'. Such a configuration is expected to be stable on a long timescale, provided that the period of the wide 'binary' is several times greater than the period of either close 'binary'. Just how much greater the longer period must be for stability depends strongly on the eccentricities, and also the inclination of the outer orbit to the inner orbit, but for orbits which are nearly circular and coplanar it is typically in the range 3–6, assuming that all the masses are comparable. Observed systems usually have a much greater period ratio than this (10^2–10^4, and even more), and are therefore likely to be extremely stable even allowing for orbital eccentricity and non-coplanarity. Figure 1.7b shows the inner and outer orbits of the triple system HD 109648 (Jha *et al.* 2000). This system of three rather similar F dwarfs has an unusually small period ratio of about 22. One of the two velocity curves in the lower panel of Fig. 1.7b is the velocity of the centre of gravity (CG) of the inner pair.

The well-known sextuple system α Gem is a microcosm which contains within itself two VBs, two SB1s and an ESB2. Its components are organised as follows, using a notation of nested parentheses to emphasise the hierarchical nature:

$$((((A1V + ?; 9.2\,d, e = 0.5) + (A2 : m + ?; 2.9\,d); 500 : yr, e = 0.36; 7'')$$

$$+ (M1Ve + M1Ve; 0.8\,d); 70'').$$

The outermost orbit of the α Gem system is too slow to be measurable, but there is no doubt that the M dwarf pair is related to the other four components, by virtue of the fact that they have a common proper motion – more precisely, the M dwarf pair has almost the same proper motion as the centre of gravity of the pair of A stars. The outermost orbital period can be expected to be of the order of 10^4 years. Each A star is an SB1, one of which has a fairly eccentric orbit. The unseen companion in each SB1 is likely to be a red dwarf, although in principle it could be some other faint object such as a white dwarf. The mass functions are known (0.0013 and 0.01 $M_\odot$ respectively), but do not rule out a substantial range of masses and inclinations. The M dwarf pair is an ESB2 – with a separate variable-star name, YY Gem – and is one of the very few systems from which M dwarf masses and radii can be determined directly.

A few multiple systems are 'non-hierarchical': three or more stars are seen which are all at comparable distances from each other. This *could* be simply a projection effect, but the probability is not large. If it is not due to projection, then such systems cannot be stable in the long run, and indeed the few that are known are groups of young stars, such as the Trapezium cluster in Orion, that have simply not yet had time to break up. If N stars of total mass M are fairly uniformly distributed in a volume of radius R, we expect the system to break up in a time comparable to the 'crossing time' $\sqrt{R^3/GM}$. The final product will typically be a series of ejected single stars, and a remaining close binary; but we might have one or more binaries ejected or a hierarchical triple left over. Usually the stars ejected will be the less massive members, and the remaining binary is likely to contain the two most massive stars.

Although most of this book is concerned with systems of only two components, there are many triple systems and a few quadruple systems known where *all* the components are close enough to interact at some stage. For the most part, when we mention a binary we are thinking of only those binaries at the bottom of the hierarchical pyramid: three binaries in the case of α Gem.

1.4   Nomenclature

In discussions of binary and (necessarily) more highly multiple stars, we should probably be careful to use the word 'system' rather than 'star', since the latter term is often ambiguous – commonly used to mean either the individual components, or alternatively the whole ensemble of components. We should also be cautious about using the words 'primary' and 'secondary': some authors use 'primary' to mean the more luminous star (at least in a particular wavelength range), others to mean the hotter star, and others still to mean the component that has the lowest right ascension.

When discussing the behaviour of a binary, it will often be convenient to refer to the components as $*1$ and $*2$ – to be pronounced 'star 1' and 'star 2'. I will use two somewhat contradictory conventions throughout this book, depending on context. Sometimes, for example when discussing an SB1 binary, $*1$ will mean the star we see and $*2$ the star we do not see, as in Eq. (1.2). At other times, for instance when discussing an SB2 binary, I will take $*1$ to be the component which we infer to have been *initially* the more massive, and $*2$ to have been *initially* the less massive. For example, in Fig. 1.1a $*1$ would be the F5V star and $*2$ would be the G5V star; in Fig. 1.1b $*1$ would be the white dwarf and $*2$ the G8III star. When I am discussing the theoretical behaviour of one component of a binary, in Chapter 3 and later, I will usually call that component $*1$, often considering, for illustrative purposes, that $*2$ is just a point mass with no structure. But when I am discussing the long-term evolution of both components of a binary, I will adhere again to the principle that $*1$ is the initially more massive component. I do this because any discussion of the evolution demands that we identify the same component as $*1$ throughout several substantial changes in mass ratio, luminosity ratio, etc. I do not apologise for possible confusion because I consider that there is no convention which (a) I could adhere to rigorously and (b) would not cause confusion at some point.

It may appear that we are bound to have even less information on *initial* masses than on *current* (measurable) masses. However, this is really not so, given the rather clear theoretical understanding (Chapter 2) that the rate of evolution of a star is principally determined by its initial mass, rather than its current mass, provided only that mass loss or gain did not begin at a *very* early stage in the life of the star. If we see a combination of white dwarf and red giant we can be reasonably sure that the white dwarf is $*1$ in the second sense defined above, even though we may have little or no information on the *current* masses. Only a modest fraction of binaries poses any real challenge to this assumption.

A further convention that I will impose throughout most of this book is that the mass ratio q is M_1/M_2, rather than its reciprocal. Thus, in the *second* convention above, $q \geq 1$ at zero age – but at a later stage of evolution q may drop below unity because of evolutionary changes of mass in one or other (or both) components. However, in this chapter alone, I will use Q defined as $1/q$. In this preliminary discussion I concentrate mainly on unevolved binaries, where the brighter and hotter component can be reasonably assumed to be $*1$. In such systems the mass ratio M_2/M_1 is usually what is discussed in the literature, and this is Q, not q. In Chapter 3 and later, however, Q is a dimensionless quadrupole moment of a distorted star.

In a hierarchical triple system logic demands that I refer to the outermost pair as $*1$, $*2$, and the inner pair (if it is $*1$ that is binary) as $*11$, $*12$. The periods would be P for the outer pair and P_1 for the inner pair. By extension, in α Gem above the unseen close companion to the A2:m star is $*122$, for example, and $P_{12} = 2.9\,\mathrm{d}$. However, logic does not quite dictate which of the two A-type SB1s is $*11$ and which is $*12$:

(a) the more massive *pair*?
(b) the pair that has the most massive component?
(c) the more highly multiple (supposing that the multiplicities were different)?

I shall duck this issue in its fullest generality. For triples, I shall

(a) attempt to identify the *originally* most massive of the three components, as argued above for binaries, and
(b) name the subsystem that contains it as ∗1, which might be either single or binary.

I have used this convention in the caption of Fig. 1.7, regarding the two triple systems AS Cam and HD109648. The data given there for the second system show that the two orbits within it are not parallel to each other: $\sin i_1 = 1.06 \sin i$. Unfortunately even if $\sin i_1 \sim \sin i$ we cannot assume that the orbits are parallel. In the system β Per referred to previously both the inner (eclipsing) orbit and the outer (visual) orbit are inclined at nearly 90° to the line of sight. But radio interferometry with a very long baseline (VLBI) shows that the orbits are actually inclined to each other at ∼100° (Lestrade *et al.* 1993). Nature appears to be playing a rather cruel joke, since the probability that the two orbital axes and the line-of-sight axis are all (nearly) mutually orthogonal must be rather small.

Although Nature is probably logical most of the time, the human perception of it is often influenced by customs that are historical and cultural rather than logical. Consequently, we shall in practice refer quite often to '∗3' and even '∗4', as distant companions to some binary of interest, or as recently-discovered close companions to components of a wide binary. It is unfortunate but unavoidable that whenever a new component is discovered the names of some at least of the previous components will have to change.

1.5   Statistics of binary parameters

The statistical distributions of masses, orbital periods, mass ratios and eccentricities are not well known: see for example discussions by Heintz (1969), Griffin (1983, 1985), Zinnecker (1984), Trimble (1987), Halbwachs (1983, 1986), Hogeveen (1990), Duquennoy and Mayor (1991) and Halbwachs *et al.* (2003). That many systems (perhaps ≳10%) are at least triple makes it even harder to arrive at a firmly-based distribution of these parameters. I concentrate in this book primarily on systems whose periods are short enough to allow for some kind of binary interaction.

1.5.1   Binary interaction

Although most stars are as small as the Sun, or smaller, they are capable of growing in radius by a factor of ∼1000 during their evolution (Chapter 2; Table 3.2). The Sun may well fill the orbit of Mars or even Jupiter before collapsing to a white dwarf. A substantially more massive star than the Sun could grow to an even larger radius, before exploding as a supernova. Only if the period is longer than ∼10^4 days (∼30 year) is there a reasonable probability that the two components go through their entire evolution almost independently (Plavec 1968, Paczyński 1971). However, some interaction (in addition, of course, to the basic gravitational one) can take place in even wider systems. The prototype Mira variable o Cet has a white-dwarf component (VZ Cet, ∗1) in a roughly 400 year orbit about the M supergiant pulsating variable (∗2). The white dwarf flickers rapidly, unlike normal single white dwarfs, and this is probably because it is interacting with the copious wind that is being ejected by the Mira. It may be accreting only a small fraction of this wind, but that could be enough to affect the white dwarf's future evolution.

Although stars in systems with orbital periods substantially in excess of 100 years are not likely to undergo very much mutual interaction, this is not to say that their orbits remain uninterestingly constant for all time. A supernova explosion, or the ejection of large amounts of gas by a blue or red supergiant, may change or even disrupt the orbit. So also can the random perturbations imposed by interaction with nearby systems. This last effect imposes a loose upper bound on the orbital separation of individual systems. The orbital separation can hardly be larger than the mean distance between independent systems, ~ 1 pc in the Solar neighbourhood, and this translates by Kepler's law into an upper limit on orbital period of $\sim 10^{10}$ days for a system of mass $\sim 1\ M_\odot$. In practice the upper limit is likely to be at least an order of magnitude less, since many near collisions of such a system with adjacent systems can be expected in the course of the Galaxy's lifetime.

Within a dense cluster of stars, such as a globular cluster, and also near the Galactic centre, it is possible for binaries of shorter period to be disrupted by near collisions. It is even possible in such an environment for binary stars to be *formed*, for example by tidal capture. This can happen if, in a close approach of two single stars, large tides are raised on at least one. Such tides can dissipate energy, and so allow the stars to move from a hyperbolic to an elliptical orbit (Fabian *et al.* 1975). Thus some interesting binaries in globular clusters need not be the products of long-term evolution of a primordial binary. In a dense stellar environment binaries can also, and in fact more easily, be modified by 'exchange' interactions, where one star in a near collision with a pre-existing binary may eject one component and replace it, perhaps in a much closer orbit. In the bulk of our Galaxy, however, such interactions are not likely because of the low stellar density, and so it is reasonable to suppose that a star which is presently binary has always been binary.

Several mechanisms are identified in Chapters 3–6 that can result in 'mergers', two components of a binary becoming merged into one. Thus the mere fact that an observed star presently appears to be single does not exclude the possibility that formerly it was binary. By extension, a system which is now a binary (but presumably a fairly wide one) may be a former triple. Mergers can be the result either of slow evolution or of some rapid dynamical event.

Returning briefly to the issue of nomenclature, it is unfortunate that the terms 'close binary' and 'wide binary' can have very different meanings depending on context. To someone, say, using speckle techniques to resolve binaries in a star-forming region at a distance of 500 pc, a binary with a separation of $0.05''$ is close, if not very close. But the linear separation is ~ 25 AU, which for present purposes makes it a rather wide, if not very wide, binary – probably too wide to interact. In this book we will generally use 'close' to mean a period of a few days, and 'wide' to mean a few years; 'very wide' will mean too wide to interact seriously, i.e. a period in excess of ~ 30–300 years.

Recently it has become clear that the evolution of a binary can be seriously modified by the presence of a third body, even if that body is in a *wide* orbit – perhaps 10^4 years – and even if the third body is of quite low mass so that it is hard or impossible (at present) to observe. The main requirement for important interaction ('Kozai cycles'; Kozai (1962); Section 4.8) is only that the outer orbit be substantially inclined to the inner ($\geq 39°$). In a Kozai cycle the inner orbit's eccentricity fluctuates cyclically between a small and a large value, while the period remains roughly constant. The cycle time is $\sim P_{\text{outer}}^2/P_{\text{inner}}$, multiplied by a factor

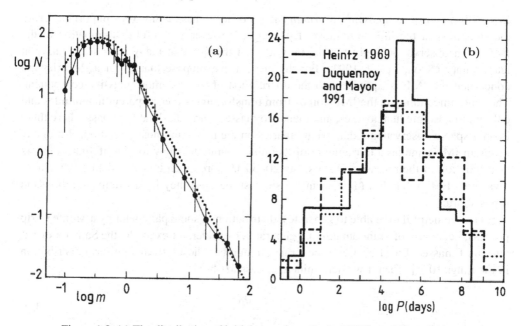

Figure 1.8 (a) The distribution of initial mass from Scalo (1986), and (dotted) the approximation of Eq. (1.10). The latter is displaced slightly upwards for clarity.
(b) Histograms of the period distribution, from Heintz (1969) (solid line), Duquennoy and Mayor (1991) (dashed line) and the approximation of Eq. (1.14) (dotted line) all normalised to the same total area.

(total mass)/(third-body mass). If the periods were 10^2 and 10^4 years, and the masses all comparable, the Kozai time would be $\sim 10^6$ years; and if the mutual inclination were $70°$ then the inner eccentricity would peak at 0.9, if it were zero to start with. Thus the *periastron* separation at this peak would be equivalent to a circular binary with a period of only ~ 3 years. If the inner and outer orbits have a random inclination, a reasonable but by no means certain hypothesis for fairly wide orbits, the average inclination would be $60°$, and $70°$ would be by no means unusual (cf. β Per, $100°$, in Section 1.4). This increases substantially the scope for 'binary' interaction.

1.5.2 Masses

Stellar masses show a distribution which (per unit volume of the Galaxy) favours low masses; although, because massive stars are very much brighter than low-mass stars, the distribution down to a given *apparent* brightness favours higher masses. The Salpeter IMF (i.e. initial mass function) is the following approximation to the distribution $N(M)$ of zero-age masses as a function of mass (Salpeter 1955):

$$N \, \mathrm{d}M \propto M^{-2.35} \, \mathrm{d}M \qquad (M \geq M_0 \sim 0.1 \, M_\odot),$$

$$= 0 \qquad (M < M_0). \tag{1.7}$$

This distribution has to be truncated at a low mass (say $\sim 0.1 \, M_\odot$), to keep the number finite. More recent IMFs (Miller and Scalo 1979, Scalo 1986, Basu and Rana 1992) show a turnover at a mass of about $0.3 \, M_\odot$, as shown in Fig. 1.8a. Whether, for binary stars, a Salpeter-like

IMF is thought of as applying to *primary* mass or *total* mass is not very important, given the steepness of the IMF over most of its range. However, a careful determination of the IMF from observation ought to take into account the fact that many stars are actually at least binary (Kroupa *et al.* 1991). For the present, we suppose that so far as binaries are concerned the IMF is equivalent to the distribution of M_1, the more massive component. The main uncertainty in the IMF comes from transforming stellar apparent magnitudes and colours to absolute luminosities, and thence to masses, particularly at low mass where there is only sparse observational data from binaries on the mass–luminosity relation. But in any event an IMF contains a fair amount of theoretical input, to allow for the lifetimes of stars as a function of their masses. O stars, say 20–50 $M_\odot$, are much less abundant relative to G dwarfs ($\sim 1\ M_\odot$) than Eq. (1.7) seems to suggest, because they have lifetimes a thousand times shorter.

It is often helpful to be able to generate a distribution of some parameter by a Monte Carlo process, i.e. by use of a random number generator. Consider, for example, the Salpeter distribution of masses, Eq. (1.7). Let X be a random number chosen from a uniform distribution in the range $[0, 1]$. Then if we determine the mass M_1 by

$$M_1 = \frac{M_0}{(1 - X)^{0.75}}, \tag{1.8}$$

we generate the Salpeter distribution. We require $M_0 = 0.1$ if the distribution is to be truncated at $0.1\ M_\odot$, as suggested for distribution (1.7).

The physical significance of the inverse function $X(M_1)$ is that it is the fraction of all stars that have mass less than M_1, i.e. it is the *cumulative* distribution function. It may be helpful to spell out this relationship. Let us integrate and normalise the distribution (1.7):

$$X(M_1) \equiv \frac{\int_0^{M_1} N(M)\mathrm{d}M}{\int_0^\infty N(M)\mathrm{d}M} = 1 - \left(\frac{M_0}{M_1}\right)^{1.35}, \quad M_1 > M_0. \tag{1.9}$$

This $X(M_1)$ relation is just the inverse of the $M_1(X)$ relation (1.8), to the extent that 0.75 is approximately the reciprocal of 1.35. The mass spectrum $N(M_1)$ (normalised) is therefore just $N(M_1) = \mathrm{d}X/\mathrm{d}M_1 = 1/(\mathrm{d}M_1/\mathrm{d}X)$. A small but important point, often overlooked, is that it is better, in order to approximate an observed distribution, to start by approximating the cumulative distribution $X(M_1)$, or equivalently the inverse function $M_1(X)$, than by approximating the *differential* distribution $N(M_1)$. Coincidentally, it is also much more convenient numerically: it is usually easier to differentiate a function than to integrate it.

Believing that the Scalo (1986) distribution of Fig. 1.8a is a more accurate distribution than Salpeter's (1955), we attempt to approximate it with

$$M_1 = 0.3 \left(\frac{X}{1 - X}\right)^{0.55}. \tag{1.10}$$

This mass distribution is Salpeter-like at $M_1 \gg 0.3\ M_\odot$, but with exponent 2.82 rather than 2.35. The distribution of masses generated by this formula is shown in Fig. 1.8a. It is somewhat coincidental that the slope of the mass distribution below the peak is much the same, but with opposite sign, as the slope above the peak. This allows us to use a single exponent (0.55) in the distribution (1.10). If the two slopes were markedly different one might choose

$$M_1 \propto \frac{X^\alpha}{(1-X)^\beta},$$ (1.11)

but for the observational distribution illustrated this refinement seems unnecessary.

Recent work on very low-mass stars (Jameson *et al.* 2002), including spectral types L and T beyond M, suggests, although not yet with complete conviction, that the IMF continues to rise, but more slowly, below the peak of Eq. (1.10) at $M_1 = 0.3 \, M_\odot$. Values of $\alpha \sim 1.5$, $\beta \sim 0.55$ in Eq. (1.12) might be somewhat better. This value of α implies that $N(M_1) \sim M_1^{-0.33}$ at low M_1, i.e. at low X.

1.5.3 Orbital periods

For spectroscopic binaries, mainly of spectral type G or K, giant or dwarf, Griffin (1985) found an increasing distribution of number N versus log P; so that, very crudely,

$$N dP \propto P^{-0.7} dP \propto P^{0.3} d\log P, \quad (P \lesssim 30 \, \text{years})$$ (1.12)

over a range of periods P from days to decades, the only upper limit to period being set by the patience of spectroscopic observers (i.e. about 30 years; but one hopes this will increase). Heintz (1969) and Duquennoy and Mayor (1991) found something similar, and also found that for still longer periods, in visual rather than spectroscopic binaries, the number per decade of log P falls off again (Fig. 1.8b); the peak in the distribution occurs at roughly 200 years. For systems whose orbital periods are too long to have been measured directly, an order-of-magnitude estimate of the period can be obtained from the observed angular separation α, the distance D based on either a directly measured parallax or on spectral type and apparent magnitude (a 'spectroscopic parallax'), and Kepler's law, Eq. (1.1). Assuming that the sum of the masses is roughly Solar (because the observed masses range over only about two orders of magnitude while periods range over about ten), we can translate crudely but directly from separation and distance to period. The falling-off in number at longer P found by Heintz (1969) and Duquennoy and Mayor (1991) can be represented roughly by

$$N dP \propto P^{-1.3} dP \propto P^{-0.3} d\log P. \quad (P \gtrsim 300 \, \text{years})$$ (1.13)

Figure 1.8b also shows the period distribution found by Duquennoy and Mayor (1991) for 79 binary periods from 161 systems that are within 22 pc of the Sun, that have an F4–G9 IV-V primary, and are north of $-15°$; the median period is at about 180 years.

Both the Heintz and the Duquennoy–Mayor distributions in Fig. 1.8b are fitted well, in the same spirit as Eq. (1.10), by

$$P(\text{days}) = 5.10^4 \left(\frac{X}{1-X}\right)^{3.3}, \quad (\text{F/G dwarfs})$$ (1.14)

where X is a second, independent, random variable distributed uniformly over the range $[0, 1]$. This distribution is also shown in Fig. 1.8b. As with the mass distribution in the previous subsection, a single exponent (3.3) seems in practice to be adequate, since the slopes at short and long periods appear to be much the same but of opposite sign.

We should not assume, however, that the same distribution would be found for binaries with say OB, or M dwarf, primaries as for those with F/G dwarf primaries. In fact, the balance of short to long period systems depends markedly on primary mass. Among the 42 *nearest* systems, mostly M dwarfs, two have $P < 100$ days (excluding a massive planet in a 61 day orbit). On the other hand, among the 48 *brightest* systems (by apparent magnitude), mostly A/B dwarfs or G/K giants substantially more massive than the Sun, 10 have $P < 100$ days – although three of these are within the same sextuple system α Gem. Somewhat greater bias still towards shorter periods is shown by the 227 O-type stars with $V \leq 8.0$ (Mason *et al.* 1998). These contain 52 SB orbits (23%) with $P \leq 0.1$ years. This is hardly consistent with the Solar-dwarf sample of Duquennoy and Mayor (1991), where only 13 out of 161, or 8%, of systems have periods <100 days.

However, the O-star sample of Mason *et al.* (1998) shows a strongly bimodal distribution with a second, even larger, accumulation (80, or 36%) of visual binaries at estimated $P \sim 10^4$–10^6 years. This is also many more than in the G-dwarf sample. Correspondingly, there is a marked shortage of systems (37, or 16%) in the considerable intermediate range 0.1–10^4 years. It can be argued that this is the most difficult range for detection of binarity: firstly, O stars are much more distant than G stars, so they have to be further apart to be recognisable as VBs; and secondly, O stars tend to show erratically variable radial velocities at the level of $\sim$20 km/s, which can obscure the lower radial-velocity amplitudes in the longer-period spectroscopic orbits. Recent advances in interferometry have already increased the numbers in the 'gap', to the percentage quoted above. About 100 systems would have to be found if the gap is to be levelling off, and about 250 if it is to be turned into a modest peak as in the G-dwarf sample. These numbers are not quite as ridiculous as they sound because the high multiplicity typical of massive systems may well mean that '200% of stars are binaries'.

Nevertheless, I adopt here a relatively cautious position. Spectroscopic orbits with periods of 0.1–1 year should not be *much* harder to detect than those in the range 0.01–0.1 year. Their velocity amplitudes will be down by a factor of about 2.2, but still well above the noise level: yet only 5 are known against 33 in the shorter-period bin. The apparent shortfall might also be related to the distribution of mass *ratios*, which I discuss shortly. Perhaps low-mass companions are relatively more likely at longer than shorter periods, but a rather drastic change in the distribution of mass ratios at about 0.1 year would be required. Let us content ourselves with a distribution that peaks at about 15 days, and drops off fairly rapidly on both sides; say

$$P(\text{days}) = 15 \left(\frac{X}{1 - X} \right)^{1.3} \qquad \text{(O stars)}. \qquad (1.15)$$

I do not suggest that this distribution can be used to include the visual binaries at very long periods, but it roughly represents the presently-known binarity among potentially interactive binaries, i.e. those with periods up to 10^2 years, giving some allowance for possible new discoveries in the period range 0.1–10^2 years. Specifically, it predicts that 50% of systems have periods over 15 days, whereas at present only 30% (of SBs) do. Our prescription also predicts that 10% of O star binaries would have periods less than 1 day, which is hardly possible given the estimated sizes of O stars; but several O stars are found with periods in the range 1.4–2.5 days. Somewhat simplistically, we will treat binaries of improbably short period from such distributions as 'merged binaries', or in other words single stars.

At the other end of the mass spectrum, for $\sim$200 G9–M3 dwarfs Tokovinin (1992) found an even smaller proportion (3%) of short periods than among the G/K dwarfs, let alone the O stars. It appears to be reasonable to suppose that the median period of the distribution shifts fairly continuously from short to long periods as mass decreases, and that the distribution also becomes wider and shallower. In my tentative model of Section 1.6, I suggest a distribution like Eqs. (1.14) or (1.15) but with a coefficient, and also an exponent, that is a function of mass.

The distribution of pairs of periods within triple systems is, of course, much more uncertain. I have already suggested that as a first approximation we assume that ~10% of systems are at least triple. In most of these, the outer orbit will be much too wide for serious interaction as described in Section 1.5.1; but I estimate even more provisionally that ~20% of triples, and thus ~2% of systems, may have *both* periods shorter than ~30 yr, and thus be potentially capable of two distinct interactions.

1.5.4   Mass ratios

The distribution of mass ratios is less well known than either of the distributions over period or mass. This is because substantially more orbital data are required for a mass ratio than for a mass or a period (Section 1.2). We need an SB2, rather than an SB1, for a mass ratio, and in many systems ∗2 is too faint relative to ∗1 to be measured reliably.

A common ad hoc model is made by assuming that both of the component masses are given by the same distribution, for example distribution (1.11). This is equivalent to saying that the two components have uncorrelated masses. Duquennoy and Mayor (1991) found this to be an adequate approximation for their sample of binaries whose primaries were all F/G dwarfs like the Sun. However, it cannot be an adequate approximation for massive stars, since these are intrinsically rare and yet are frequently paired with a comparably massive star. Furthermore, Lucy and Ricco (1979) found that among short-period binaries ($P \lesssim 25$ days) there is a much higher proportion of systems with nearly equal masses than can be accounted for by selection effects, strong as these are. Almost certainly the degree of correlation of the two masses is a function of both orbital period and mass, and it appears that there is more correlation at short periods or high masses.

Specifically, Lucy and Ricco (1979) found that among F2–M1 dwarfs with $P < 7.5$ days the number in the range $Q = 0.94$–1 was 50% of the number in the range 0.6–1. The Monte Carlo distribution generated by

$$Q = 1 - X^{\gamma}, \quad \gamma \sim 3, \tag{1.16}$$

with X yet another random number uniformly distributed in [0, 1], has approximately this property. On the other hand, they found for OB stars the smaller fraction 21%, which corresponds to $\gamma \sim 1.2$ instead of 3.

Mazeh *et al.* (1992) analysed more closely the Q-distribution for the 23 short-period ($P < 3000$ days) SB2 and SB1 members of the Duquennoy/Mayor sample, using a maximum-likelihood algorithm. They found a mild concentration to equal masses, with about 60% of systems having $0.5 < Q < 1$, which corresponds to $\gamma \sim 1.4$ in the distribution (1.16). They also suggested that this is significantly different from the distribution in wider orbits still, which favours more extreme mass ratios.

Tokovinin (1992) looked at spectroscopic orbital data for ~200 G9–M3 dwarfs, out of which 13 were SB2 and 9 SB1 with $P \leq 3000$ days. Using a maximum-likelihood method, he found a bimodal distribution of secondary masses: 10% of his systems have secondaries in the range 0.32–0.64 $M_{\odot}$, 3% in the range 0.08–0.16 $M_{\odot}$, and the remainder have no (spectroscopic) secondaries. The first peak more-or-less corresponds to $Q \sim 0.5$–1, suggesting that

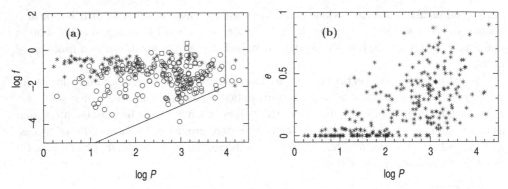

Figure 1.9 The distribution of (a) mass-function and (b) eccentricity against period for 268 G/K giants. In (a), SB1s are shown as circles and SB2s as asterisks. Five triples, the companion being itself a close binary, are shown as squares. Unrecognised triple companions may cause some of the other large values. The sloping line corresponds to a radial velocity amplitude of $K_1 \sim 2$ km/s, roughly the limit of observational detection; the vertical line at $P = 10^4$ days represents an empirical upper limit to period, since longer periods are hard to determine. In (b), all systems are represented by asterisks. The sloping upper boundary and the modest concentration at $e = 0$ probably reflect evolutionary effects rather than primordial properties.

there is still a preference for near-equal masses at moderately short periods. However, the numbers are too few to say if there is a significant departure from distribution (1.16).

The O-star sample of Mason *et al.* (1998) gave a rather different picture from the (much smaller) OB sample of Lucy and Ricco (1979), with about as many systems in the range $Q = 0.4$–0.6 as in the range 0.6–1. In the distribution (1.16) this corresponds to $\gamma \sim 0.8$.

We should emphasise that distributions like (1.16), with just one free parameter (the exponent), can easily be made to fit any statement of the character that a certain percentage of SB2s has $Q > Q_1$, and the remainder have $Q < Q_1$. However it then implies an extrapolation to the smallest mass ratios that may not be warranted, but is very hard to check.

A study of B-type binaries by van Rensbergen (2001), using the catalogue by Batten *et al.* (1989), found distributions slightly biased towards low Q, as in the more complete sample of O stars from Mason *et al.* (1998). His distributions are roughly equivalent to $\gamma \sim 0.8$ and 0.7 for late B and early B systems respectively.

The distinction between distance-limited and magnitude-limited samples is particularly important for mass ratios, since two equal-mass stars can be expected to be twice as bright, and therefore visible over 2.83 times the volume, as systems with a small mass ratio. But for O stars, distances are so great as to be very uncertain. We probably do best to establish the distribution iteratively, using a magnitude-limited sample and then making due allowance for the over-representation of equal-masses down to a given magnitude limit.

Even with data only (or mainly) on single-lined binaries, i.e. a determination only of the mass function and not of the mass ratio, some information can be gleaned about the distribution of mass ratios. Figure 1.9a shows the distribution of measured mass functions – Eq. (1.1) – from a compilation of published data for 268 red giant SBs (G/K, II–IV). SB1s are shown by circles and SB2s by asterisks, but in SB2s only the mass function of the giant is plotted (or the more massive giant, in a few cases where both components are giants). By SB2 in this context we mean stars in which two spectra are seen, even though in many

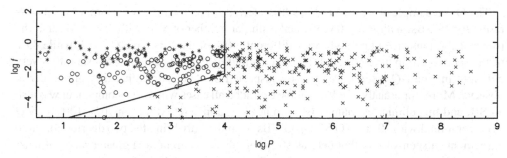

Figure 1.10 A sample of 500 hypothetical binaries from the distributions (1.17)–(1.20), for $M_1 = 1.2 - 4\,M_\odot$. Log mass function ($M_\odot$) is plotted against log period (days). SB2s are indicated by asterisk, SB1s by a circle. Rough observational limits are indicated by straight lines, as in Fig. 1.9a. Those of too long a period or too small a velocity amplitude to be readily measurable are indicated by a cross. The stars might be late B/A/F dwarfs, or, if evolved, G/K giants as in Fig. 1.9a. Unlike in Fig. 1.9a, no triples were included.

cases only the G/K giant has had its radial velocity curve measured. The fact that *2 is seen at all suggests that it is only moderately less massive than *1. It is probably reasonable to assume that most G/K giants of luminosity class II–IV are in the fairly limited mass range 1.2–4 $M_\odot$.

We believe that a slight trend can be seen in Fig. 1.9a: most systems with $P \lesssim 10^{2.5}$ are above $f \sim 0.05\,M_\odot$, and most with $P \gtrsim 10^{2.5}$ are below it. In spite of the fact that f depends on both M_1 and M_2, as well as the inclination, a reasonable interpretation of the trend is that at longer periods there is a greater spread of secondary masses than at shorter periods. The relative shortage of small f at short P is not entirely due to the difficulty of measuring small velocity amplitudes, as is shown by the expected cutoff of observations near the sloping line $K_1 = 2$ km/s. This line slopes in the opposite direction to the observed (but rather marginal) trend. The 'missing' systems of low f and short P should be quite measurable, although no doubt somewhat under-represented.

Several G/K giants with known or suspected white dwarf companions, in particular barium-rich stars (Section 6.3), have been excluded, because the mass of at least one component, and perhaps also the period, will have been modified by evolution. Other binaries, such as Algols, have also been excluded on the basis that their masses are likely to have been much modified by evolutionary interaction.

Figure 1.9a contains five systems (squares) which are known to be triple, the companion to the giant being itself a close pair of stars. Such systems can be expected to have large mass functions, and they do indeed give four of the eleven largest values in the figure. It is quite likely that more SBs, particularly at the upper margin of f, are in fact triple, but have not been recognised as such.

Figure 1.10 is a similar plot to Fig. 1.9a, but based on a *theoretical* model – Section 1.6 – sampled in a way somewhat similar to the way in which real stars are sampled. Random inclinations were also included. We believe that the best way to assess the reliability of a model of the distribution of Q, which can be expected to depend on P and M_1, is to compare a theoretical figure like Fig. 1.10 with an observational figure like Fig. 1.9a, for various ranges of P and M_1. But we still have the problem of what to do about systems where the period is too long or the velocity too small to be measured.

The O star sample of Mason *et al.* (1998), referred to previously, gave information on mass ratios for 49 SBs, with $P \lesssim 10$ years, and a similar number of VBs with $P \gtrsim 10^4$ years. The two distributions are at first sight very different: 90% of the SBs have $Q \gtrsim 0.4$, while 85% of the second group have $Q \lesssim 0.4$. However, a lot hinges on whether the SB sample is very deficient in small-Q systems, as may be suspected on the grounds that they are harder to observe. More importantly, the VB sample excluded all cases where one component was also an SB, and this eliminates many VBs with near-equal masses. But even 50 or 100 small-Q SBs, combined with 20 or 30 O SBs + O VBs, would hardly bring the two distributions into agreement. It seems likely that here as elsewhere there is a trend to a greater range of mass ratios at long periods than at short. In a distribution such as (1.16) the exponent γ may have to range from about 2–3 at short periods and low masses to about 0.7–1 at long periods or high masses.

A recent analysis (Halbwachs *et al.* 2003) of a substantially larger sample of F7–K7 dwarfs than that of Duquennoy and Mayor (1991) suggests modest changes to the discussion here of the distributions in period and mass ratio. Halbwachs *et al.* (2003) analysed 456 stars, 61 of which have orbits with $P < 10$ years. If these are binned as in Fig. 1.8b, they show little change in the upward slope; but in detail, they suggest a slight deficit at $P \sim 200$ days and a slight excess at $P \sim 500$ days. There is substantially more information regarding mass ratios, both from SB2 orbits and from SB1 combined with astrometric (VB) orbits. A total of 45 mass ratios were obtained. These display a rather flat distribution in Q over the range $0.2 \leq Q \leq 0.7$, but with an unexpected deficit at 0.7–0.8. This is followed by an excess at 0.8–1.0, which is due mainly to short-period systems ($P \leq 50$ days). Both the deficit and the excess are by about a factor of 2.

1.5.5   *Eccentricities*

The distribution of orbital eccentricities appears to be roughly uniform in the range from zero to unity, at least for those binaries whose components are sufficiently well separated that they can have had little chance yet to interact with each other, even at periastron. Figure 1.9b shows a plot of e against P for the same collection of spectroscopic binaries as Fig. 1.9a. The absence of eccentric orbits ($e \gtrsim 0.05$) at periods shorter than ~ 10 days probably reflects the fact that the components in these systems are close enough to interact by tidal friction (Section 4.2), a process which should tend to make orbits more circular. Binary interaction may also account for (a) the sloping upper envelope, since eccentric binaries of fairly short period will have small separations at periastron, and so should suffer particularly from tidal friction, and (b) a concentration of systems at $e = 0$, some or all of which may well be the result of considerable binary interaction. Known barium-rich red giants fall into this category, and were excluded, but the systems in Fig. 1.9a with $P \gtrsim 10^{2.5}$ days and $e = 0$ may, like barium stars, contain a white dwarf, although unlike them they show little or no barium enrichment.

If binary orbits were randomly distributed in phase space we would expect that e^2 rather than e would have uniform probability. Possibly this is the case at periods substantially greater than 30 years, since very few of these have measured eccentricities. But although the sample in Fig. 1.9b is not large, and is far from being a complete distance-limited sample, it does not show the marked concentration to $e \sim 1$ that this argument would imply. It may be that some aspect of the formation process for binaries tends to weigh against high eccentricities.

1.6 A Monte Carlo model

Our preferred model, for the time being, of a population of unevolved binary systems has a mass distribution (for $*1$)

$$M_1 = 0.3 \left(\frac{X_1}{1 - X_1} \right)^{0.55}, \qquad (1.17)$$

a period distribution

$$P = \frac{5 \times 10^4}{M_1^2} \left(\frac{X_2}{1 - X_2} \right)^\alpha, \qquad \alpha = \frac{3.5 + 1.3\alpha'}{1 + \alpha'}, \qquad \alpha' = 0.1 M_1^{1.5}, \qquad (1.18)$$

a mass-ratio distribution

$$\frac{1}{q} \equiv Q = 1 - X_3^\beta, \qquad \beta = \frac{2.5 + 0.7\beta'}{1 + \beta'}, \qquad \beta' = 0.1 P^{0.5}(M_1 + 0.5), \qquad (1.19)$$

and an eccentricity distribution

$$e = X_4, \qquad (1.20)$$

where $X_1 \ldots X_4$ are independent random variables uniformly distributed in $[0, 1]$. The mass dependence in the period distribution, and the mass/period dependence in the mass-ratio distribution, are crude attempts to quantify the discussion of the previous sub-sections. I noted in Section 1.5.3 that the distribution of periods among O stars is strongly bimodal, and I emphasise that the peak at long periods ($\sim 10^4$–10^6 yr) has been ignored.

Table 1.1 shows a distribution obtained by generating 10^6 binaries using Eqs. (1.17)–(1.19); eccentricity was ignored. I have not truncated at low mass, so that quite a high proportion of primaries, and an even higher proportion of secondaries, are presumably brown dwarfs. Binaries with $P > 10^9$ days are treated as two single stars, and binaries of such short period that the stars would overlap are treated as one single star with the combined mass; and all 'singles' are listed as if $q > 9.99$. No doubt coincidentally, the proportion of singles to binaries that I generate is not in practice very different from what observation suggests.

Figure 1.10 is, like Fig. 1.9a, a plot of mass function against period, containing 500 theoretical systems with $M_1 = 1.2$–$4\,M_\odot$, a range probably comparable (when evolved) to the observed G/K-giant sample of Fig. 1.9. The distribution was convolved with a random distribution of inclinations, to simulate the observational selection criterion for SBs that the velocity amplitude should be above some threshold ($K_1 \gtrsim 2$ km/s, for G/K giants, which have relatively narrow lines and correspondingly high accuracy). Systems below that threshold, or whose period is longer than the loose practical limit of ~ 30 years (10^4 days), are shown with a cross; asterisks are 'theoretical SB2s', defined by $0.7 < Q < 1$, and circles are 'theoretical SB1s', with $Q < 0.7$. Theoretical SB2s seem to be under-represented, particularly at shorter periods, but this may well be a selection effect in the observed sample, since systems that are seen in the first instance to be double-lined are more likely to be investigated further.

The density of stars in the Solar neighbourhood, projected on to the plane of the Galaxy, is $\sim 50\,\mathrm{pc}^{-2}$, and the effective area of the Galaxy is $\sim 10^9\,\mathrm{pc}^2$, so that the number of stars in the Galaxy is $\sim 5 \times 10^{10}$. The lifetime of the Galaxy is about 10 gigayears, and so we should generate stars from the distributions (1.17)–(1.20) at a rate of about 5 pc^{-2} gigayear^{-1}. We might also generate some single stars, from Eq. (1.17) alone, except that the 'singles' in Table 1.1 (arising from either merged very close binaries or disrupted very wide binaries) may be

Table 1.1. *An estimate of the frequency of binary parameters among* 10^6 *systems*

M_1	q	$P<1$	1–3.2	3.2–10	10–10^2	10^2–10^3	10^3–10^4	10^4–10^9	'single'
<0.5	1.00–1.41	1 441	2 999	4 817	15 450	21 884	25 402	1 06 627	0
	1.41–3.00	624	1 403	2 215	8 554	17 722	31 600	2 19 339	0
	3.00–9.99	348	627	1 097	4 147	9 860	19 971	1 63 776	0
	>9.99	134	252	422	1 634	3 962	8 382	75 797	2 74 648
0.5–1	1.00–1.41	63	386	688	2 038	2 659	2 694	6 903	0
	1.41–3.00	47	182	349	1 437	2 820	4 158	14 487	0
	3.00–9.99	23	104	184	795	1 742	2 826	11 039	0
	>9.99	17	42	74	316	713	1 320	5 097	9 914
1–2	1.00–1.41	9	96	246	749	895	923	1 792	0
	1.41–3.00	5	53	127	606	1 227	1 591	3 761	0
	3.00–9.99	6	35	58	340	745	1 126	3 030	0
	>9.99	1	11	29	150	346	558	1 301	3 451
2–4	1.00–1.41	0	20	68	226	315	295	358	0
	1.41–3.00	0	19	53	293	515	630	762	0
	3.00–9.99	0	8	34	180	394	451	611	0
	>9.99	0	5	12	62	158	189	284	1 228
4–8	1.00–1.41	0	5	25	89	116	64	52	0
	1.41–3.00	0	5	24	103	193	173	97	0
	3.00–9.99	0	2	11	67	135	123	74	0
	>9.99	0	0	4	43	79	64	46	355
8–16	1.00–1.41	0	0	4	39	35	16	7	0
	1.41–3.00	0	1	7	43	59	29	7	0
	3.00–9.99	0	1	8	33	64	27	12	0
	>9.99	0	1	1	19	16	9	3	176
16–32	1.00–1.41	0	0	2	7	3	0	0	0
	1.41–3.00	0	0	1	16	15	6	2	0
	3.00–9.99	0	0	3	18	15	4	1	0
	>9.99	0	0	1	4	7	2	0	84
>32	1.00–1.41	0	0	0	1	0	0	0	0
	1.41–3.00	0	0	0	5	1	1	1	0
	3.00–9.99	0	0	2	5	0	1	0	0
	>9.99	0	0	1	0	0	0	0	74

Period is in days, primary mass in Solar units

sufficient in number already. In fact we should also add a population of triples and quadruples, at least, at about a tenth of the rate or more; but the statistical distribution of triples over three masses and two periods is completely uncertain. We think of the distribution of binaries in Table 1.1 as representing only those binaries which are furthest down the hierarchical pyramid. Several of the shorter-period systems, say with $P \lesssim 10^4$ days, will be resident in wider triples.

The surface density of massive stars is, of course, to be modified according to their short lifetimes (Chapter 2). Stars above a mass $\sim 5\,M_\odot$ only live for $\lesssim 100$ megayears, so that their abundance is reduced to $\sim 1\%$ of what is predicted above. The remaining 99% will now be white dwarfs, neutron stars or black holes, but they may nevertheless have companions of initially lower mass that are still normal stars.

We should not suppose that all parts of the Galaxy, including its collection of satellite globular clusters, would conform to a single distribution. In particular, the distribution in globular clusters is likely to be truncated at periods of $\sim 10^4$ days. A simple criterion is that 'soft' binaries, whose orbital velocities are less than the velocity dispersion in the cluster, are likely to be destroyed.

Although the mean rate of star formation suggested above may be a reasonable average over a long time, evidence mainly from other galaxies suggests that there are short periods of rapid star formation ('starbursts') presumably separated by long periods of relative quiescence. For present purposes however it is probably good enough to assume a fairly steady rate.

1.7 Conclusion

A substantial proportion of stellar systems have periods short enough ($P \lesssim 10^4$ days) for future interaction. Low-mass systems ($M_1 \lesssim M_\odot$) might be supposed incapable of serious interaction, because their evolutionary lifetimes are longer than the age of the Universe. But this is illusory since, at least at periods of a few days, they can interact by magnetic braking and tidal friction (Chapter 4) to produce contact binaries and (probably) merged single stars. Given that low-mass stars are much more common than high-mass stars, there is arguably more interaction of this sort than of any other.

In an ideal world, we would have at least half-a-dozen distance-limited samples each containing about 2000 unevolved systems – including single stars as 'systems' – covering perhaps the following ranges of spectral type: O, early B, late B, A, F/G, K/M. They should be unbiased towards binarity, i.e. they should not be selected on the basis of known radial-velocity variation or other potential binarity indicators. In addition, some evolved samples of similar size, e.g. G/K giants, F/G/K supergiants and M giants would be added, but we should be wary that some members of these will have already undergone evolutionary interactions and so have parameters differing from their values at age zero. In fact we must be wary even for the 'unevolved' sample, since a proportion of unevolved stars will be coupled with white dwarfs, neutron stars and black holes that are not always easy to identify. These samples would have to be examined spectroscopically, photometrically and astrometrically for at least 30 years, and preferably 60 years, to determine the appropriate distributions. Many will turn out to have multiplicity higher than two, but we might hope with samples of this size to get some significant statistics on multiplicity as well as on binary orbits.

We would attempt to model these samples in an iterative manner. Having an a-priori estimate from these samples of the distributions discussed above, we would construct an ensemble of at least 10^6 (and preferably 10^{11}) theoretical systems. We might need somewhat more sophisticated distributions than the ones used here with only one or two parameters, but the size of the samples would still not justify more than three parameters. We would attempt to model the evolution of M_1, P, q in individual systems according to our present understanding, as outlined in Chapters 3–6. We can 'observe' the theoretical samples at a variety of ages, allowing theoretically for such selection effects as we think we can quantify.

In the course of evolving such a theoretical ensemble we should find ourselves populating a zoo of exotic binaries – and indeed of possibly exotic single stars that come from binaries that are either disrupted or merged according to some of the processes modelled. We need further observational sets of these exotic stars – cataclysmic binaries, X-ray binaries, contact binaries, symbiotic stars, barium stars, double-white-dwarf binaries, B-sub-dwarf stars, neutron stars, double-black-hole binaries, etc. – for comparison with the theory. We would hope that after one or two iterations we might get some degree of convergence. Even 10^6 theoretical systems would not be enough to produce samples of some of these species of a size adequate for statistical analysis.

In the real world, we will certainly not get complete convergence. Already some well-determined binaries, of by no means exotic character, exist which cannot possibly be explained on the basis of our present understanding of binary stars – see Section 2.3.5. This might be seen to mean that we should throw the present understanding out of the window. But there also exist some, in fact many, that seem to accord very well with present understanding; what is one to make of that? Some difficult binaries may be explicable as triple or formerly triple systems; others may require exotic processes not yet determined. Do we need a major paradigm shift, as from Ptolemaic to Copernican orbits, or can we cope by tinkering about the edges? Do we have to invoke magnetic fields, as the last refuge of the charlatan? At any rate further iterations will be required.

But an important preliminary is to attempt to model evolution in systems on a one-by-one basis, since this will test at least some of the evolutionary processes that we expect. After discussing evolution of effectively single stars in Chapter 2, I describe what I believe to be the main binary-interactive processes in Chapter 3 and later.

2

Evolution of single stars

2.1 Background

The evolution of single stars, and of those stars which are in binaries sufficiently wide that the effect of a companion can be ignored, has been much studied, especially with the aid of increasingly powerful computers over the last 50 years. This is not to say, however, that every problem has been solved; in the final section of this chapter I emphasise some of the outstanding problems.

Figure 2.1 shows a comparison between recently computed models, and data obtained by observation. They are shown in a Hertzsprung–Russell diagram (HRD) where luminosity, i.e. the total energy output of the star, is plotted against surface temperature; the latter is plotted backwards, for traditional reasons. Our theoretical understanding of the internal structure and evolution of single stars is based on the concepts of hydrostatic equilibrium, thermodynamic equilibrium and the consumption of nuclear fuel, mainly hydrogen. In hydrostatic equilibrium, the inward force of gravity is balanced by the outward push of a pressure gradient. In thermodynamic equilibrium, the heating or cooling of a spherical layer of material is determined by the balance of heat production in nuclear reactions, at temperatures of about 10 MK (megakelvin) and upwards in the deep interior, against heat loss as heat flows down the considerable temperature gradient until it can be radiated into space from the photosphere at temperatures observed to be about 2–100 kK. The heat flux is carried either wholly by radiation, or by a combination of convection and radiation, depending on whether the temperature gradient that would be required to carry the heat entirely by radiation is less than or greater than the critical (i.e. adiabatic) temperature gradient at which convective instability sets in. Most stars contain some region or regions that are predominantly convective and some that are wholly radiative.

The nuclear reactions that provide the heat also change the nuclear composition of the star on a slow time scale, megayears at least; although in stars substantially less massive than the Sun the time scale can be longer than the Hubble time ($\sim$15 gigayears), which is presumed to be about the age of the Universe, so that little nuclear evolution is to be expected in such stars. The nuclear reactions principally burn hydrogen to helium, with subsidiary reactions that modify the abundances of carbon, nitrogen and oxygen, and also of deuterium, ^{3}He, lithium and beryllium, for instance. In later stages much of the helium is itself burnt to form a mixture of carbon and oxygen. In very late stages a large number of nuclear reactions can take place, involving all the elements from carbon upwards. Provided the products of these reactions can be returned to the interstellar medium via stellar winds, for instance, or via the outbursts of novae and supernovae, stars appear to be able in principle to produce all known nuclear species, apart from hydrogen and most helium, which is thought to be

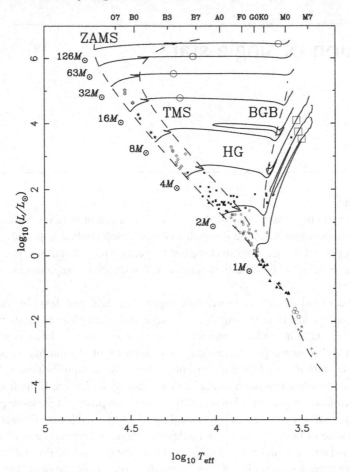

Figure 2.1 Observed stars (symbols) and theoretical models (lines) in the HRD. Bolometric luminosity is plotted against surface temperature; equivalent spectral types for the main sequence (Popper 1980) are shown along the top. Different symbols represent stars with observed masses in the ranges 0.125–0.25, 0.25–0.5,..., 16–32 $M_\odot$; different tracks are 1, 2, ..., 128 $M_\odot$ (more precisely, log M = 0.0, 0.3, ..., 2.1). The observational data are from ESB2 systems (Andersen 1991, Pols *et al.* 1997, Schröder *et al.* 1997) and low-mass VBs (Popper 1980). Theoretical models, with indicated masses, are by Drs O. R. Pols, K.-P. Schröder, C. A. Tout, and the author. The zero-age main sequence, terminal main sequence, Hertzsprung gap and beginning of the giant branch (see text) are indicated as ZAMS, TMS, HG and BGB. For the higher masses, large circles indicate where He ignited. For the lower-mass stars, large open squares indicate where the evolution is likely to terminate due to mass loss, which was not included in these calculations. Kinks in the tracks near these squares are due to the coarseness of the opacity grid at low temperature.

'primordial'. It is by no means yet clear, however, what are the detailed stellar evolutionary mechanisms by which nuclear species are produced in the abundances observed. Certainly it is necessary to have an equally detailed understanding of Galactic evolution before the problems of nucleosynthesis can be considered solved.

Important as Galactic evolution is, it is outside the scope of this book. For present purposes it will, we hope, be sufficient to suppose that all stars have an initial 'zero-age' composition, which is described by just two parameters X and Z. X is the fraction by mass of hydrogen, and Z the fraction by mass of all elements together ('metals') apart from hydrogen and helium: the abundance of helium is then $Y = 1 - X - Z$. X varies little from star to star and is about 0.7–0.75. Z ranges from about 0.0001 in old stars ('Population II', found in globular clusters, in the Galactic halo, and in nearby high-velocity stars) to about 0.02 in stars like the Sun and younger ('Population I', found in the thin Galactic disc). Within the metallicity parameter Z it is possible for the balance of, say, heavy metals to light metals or of oxygen to carbon to vary with age, but for the time being we ignore this possibility. Probably X correlates with Z to some extent, older stars having somewhat more hydrogen and substantially less metals: say, $X \sim 0.76 - 3Z$. It is commonly hypothesised that the earliest-formed stars ('Population III') should have had virtually zero metals, but there is no observational evidence at present for such a population.

Before a star can settle into the state of hydrostatic and thermal equilibrium described above, it first has to condense out of a pre-stellar gas cloud. The star-formation process is much less well understood than the later evolution – partly because of the *absence* of equilibrium in condensing protostars (the more absent the earlier the phase), and partly also because formation takes place, hardly surprisingly, in relatively dense gas clouds whose very density obscures our direct view of what is going on. One of the most dramatic advances of instrumental astronomy in the last quarter of the 20th century has been in the infrared region of the spectrum, about 0.7–100 µm, where radiation is much better able than visible light to penetrate through these gas clouds. This has already enormously improved the quality and quantity of information on star formation. But it has still not given us an understanding of the formation process as compelling as our understanding of later processes, even though these later processes are themselves not definitively understood. In Section 2.2.7, we return briefly to star formation.

A set of partial differential equations, outlined in Appendix A, can be written down to model the physical processes that occur within stars that have settled into near-equilibrium. These equations are not as definitive as one might reasonably suppose; for example, in circumstances where the heat content of material is changing simultaneously with the composition (as a result either of nuclear reactions or of convective mixing) one is liable to find different formulations in different computer codes, and even in different published accounts. The least definitive part of the stellar evolution equations is probably the treatment of convection (Section 2.2.3), and the next least definitive is mass loss by stellar wind (Sections 2.3, 2.4). The problem of turbulent compressible convection is both extremely important and extremely difficult; it includes the problems of semiconvection, and of convective overshooting. We would not expect it to be solved on a timescale of less than decades. But even if one sets aside this difficulty, by using a very simple treatment of convection such as the mixing-length theory of Böhm-Vitense (1958), and of semiconvection with a simplistic diffusion approach (Eggleton 1972, 1983b), solving the equations computationally remains a substantial problem.

Figure 2.2 illustrates, in the temperature/density plane, regions where different physical processes dominate the equation of state (EoS) of stellar material. Although a considerable variety of processes have to be included in a definitive equation of state, it is fortunate that in most stars, at most stages of evolution, the main contributions to pressure come just from radiation pressure, the perfect gas law and degenerate electron pressure. On most of the main sequence (Section 2.2) even the third of these can usually be ignored. A very efficient equation

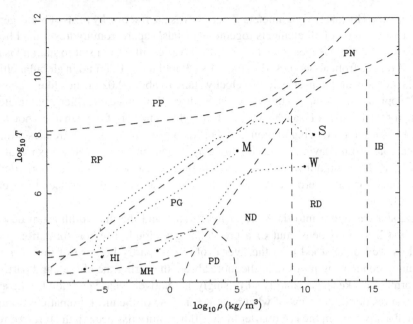

Figure 2.2 Plot of log T (K) versus log ρ (kg/m^3, showing dominant effects in the equation of state: RP – radiation pressure; PG – perfect gas; HI – hydrogen and helium partial ionisation; MH – molecular hydrogen; PD – pressure dissociation; ND – non-relativistic electron degeneracy; RD – relativistic electron degeneracy; IB – inverse β-decay (neutronisation); PN – photo-dissociation of nuclei; PP – pair production. Dotted lines are a 2 $M_\odot$ star on the main sequence (M), a 4 $M_\odot$ star in a highly evolved late supergiant state (S), and a 0.6 $M_\odot$ white dwarf (W).

of state was used here, based on the model of Eggleton, Faulkner and Flannery (1973) but updated (Pols *et al.* 1995) to include electron screening and an improved treatment of pressure dissociation. A brief description is given in Appendix A.

For the computed models used here (Fig. 2.1; see also Table 3.2), the opacity coefficient was taken from the work of Rogers and Iglesias (1992) and (for low temperatures at which molecules contribute importantly) from Alexander and Ferguson (1994). Nuclear reaction rates were taken from Caughlan *et al.* (1985) and Caughlan and Fowler (1988), and neutrino energy-loss rates come from Itoh *et al.* (1989, 1992). A degree of convective overshooting (Section 2.2.4) was used as suggested by the analysis of Schröder *et al.* (1997) and Pols *et al.* (1997).

An 'implicit, adaptive' distribution of meshpoints (Eggleton 1971) was used in the numerical scheme, since this gives great numerical stability and allows the models to be computed with only 200 meshpoints between the centre and the photosphere. This procedure for distributing meshpoints works as follows. Instead of deciding in advance where the meshpoints should go in the next timestep, which is difficult to do since the regions that need the most meshpoints will move with time at a rate that is not easy to predict, we require that the meshpoints in the *next* timestep satisfy the condition that the root-mean-square change in the variables is the same from one space-like interval to the next. This means that an extra implicit second-order equation has to be solved, along with the four first-order equations of structure.

In fact the number of equations becomes larger still, because it is important to treat the composition variables in the same implicit manner as the structure and mesh variables; the optimal position of meshpoints depends as much or more on the composition distribution as on the distribution of structure variables (density, temperature, luminosity and radius). Fortunately only a handful of composition variables are really important in determining the structure (hydrogen, helium, carbon, oxygen and neon). The equations for the abundances include convective mixing, which is treated as a diffusion process (Eggleton 1983b) with a rather simple estimate of the convective diffusion rate that is based on the mixing-length approximation to convection; consequently the composition equations are all second order. Thus we end up with four first-order and six second-order equations to be solved implicitly and simultaneously. The code used has a powerful and general difference-equation solution package, which can handle a general mixture of first and second-order equations. All the difference equations are differentiated numerically, in order to set up a Newton–Raphson iteration.

The resulting code can evolve a star from the zero-age main sequence to the onset of carbon burning in less than a thousand timesteps, although modern computation is so fast that we can comfortably allow three or four thousand timesteps, which probably gives slightly greater accuracy. However the accuracy of evolved stellar models is much more affected by approximations for convective heat transport, convective and semiconvective mixing and mass loss than by discretisation.

A further advantage of an implicit adaptive mesh is that it becomes easy, even trivial, to include both mass loss by stellar wind in single stars (Section 2.4) and mass transfer between components in a close binary (Chapter 3). Unfortunately our knowledge of what these rates (especially of stellar wind) should be is not commensurate with the ease of including them. In the models of Fig. 2.1 no winds were included, but they are included in some later discussions and in Fig. 2.17, which is otherwise similar.

2.2   Main sequence evolution

Theory, that is to say computed models, and observation (e.g. Popper 1980, Andersen 1991) agree that many stars should be in a hydrogen-burning main-sequence (MS) band that crosses the HRD from top left to bottom right (Fig. 2.1). Within the band, such bulk parameters as the radius and the luminosity are determined mainly by the mass of the star, but also partly by its age and initial composition. As the star depletes its hydrogen fuel in and near the centre, it crosses the MS band, from the zero-age main sequence (ZAMS, Fig. 2.1) to the terminal main sequence (TMS), increasing its radius and luminosity by about a factor of about 2.5–3 on a slow nuclear-burning timescale. Once the fuel runs out at the centre, and also in a central core containing about 10% of the star's mass, evolution usually becomes rapid until the star has increased its radius by a larger factor. We can define the main-sequence band loosely as terminating when this rapid phase of evolution begins. However, for stars of initial mass about 1–2 $M_\odot$ the evolution slows again significantly before accelerating again, and as a consequence the TMS becomes hard to define except in a somewhat arbitrary way. We return to this point in Section 2.2.8.

2.2.1   *Approximate formulae for main-sequence (MS) stars*

There is reasonable agreement with observed masses (M), radii (R), surface temperatures (T) and luminosities (L) for MS stars, as seen in the HRD of Fig. 2.1. Approximate formulae relating these quantities, for stars on the theoretical ZAMS with a near-Solar composition of 70% hydrogen by mass, 28% helium and 2% 'metals' (everything else) are (Tout *et al.* 1996)

$$L = \frac{0.397M^{5.5} + 8.53M^{11}}{2.55 \times 10^{-4} + M^3 + 5.43M^5 + 5.56M^7 + 0.789M^8 + 5.87 \times 10^{-3}M^{9.5}}, \quad (2.1)$$

$$R = \frac{1.715M^{2.5} + 6.60M^{6.5} + 10.09M^{11} + 1.0125M^{19} + 0.0749M^{19.5}}{0.01077 + 3.082M^2 + 17.85M^{8.5} + M^{18.5} + 2.26 \times 10^{-4}M^{19.5}}, \quad (2.2)$$

and

$$T = 5.77L^{0.25}R^{-0.5}. \quad (2.3)$$

The range of validity of the first two equations is $0.1 \lesssim M \lesssim 100\,M_\odot$. Here, and throughout, L, R and M are in Solar units, with $L_\odot = 3.84 \times 10^{26}$ W, $R_\odot = 6.96 \times 10^8$ m, and $M_\odot = 1.99 \times 10^{30}$ kg, to sufficient accuracy for present purposes. The temperature T is in units of kilokelvins or kK (i.e. 10^3 K). Equation (2.3) is simply the Stefan–Boltzmann law, which can be used as a definition of the 'surface temperature' of a star. Stellar surfaces radiate, to a fair degree of accuracy, like black bodies, so that although the photons that we see were emitted by several different layers of stellar material with a modest range of local temperatures, the surface temperature defined as above represents the best mean of this range. Tout *et al.* (1996) give generalisations of Eqs (2.1) and (2.2) for a range of metallicities from $Z = 0.0001$ to 0.03.

The luminosity L is dictated mainly by the opacity in the outer layers, and itself dictates the rate of consumption of nuclear fuel: see Section 2.2.2. The available reservoir of fuel is roughly proportional to mass M. Hence the lifetime (in megayears) of a star within the main sequence band is roughly proportional to M/L, and can be approximated in the same spirit as Eqs (2.1), (2.2) by

$$t_{MS} = \frac{1532 + 2740M^4 + 146M^{5.5} + M^7}{0.0397M^2 + 0.3432M^7}, \quad 0.25 \lesssim M \lesssim 50. \quad (2.4)$$

Formulae (2.1) to (2.3) fit the results of computed models to better than 1.2% over the mass range 0.1 to 100 $M_\odot$; Eq. (2.4) fits to better than 3.3% over the mass range 0.25 to 50 $M_\odot$ (J. Hurley, private communication). The computations assume that the composition of a ZAMS star is uniform, and is approximately the same as the present day Solar surface, except that certain light elements (D, Li, Be, C) that burn rather easily at the prevailing core temperatures are assumed to have already burnt to exhaustion or to equilibrium in the core. Equations (2.1) and (2.2) do not give $L = 1$, $R = 1$ for $M = 1$, because the Sun is not a ZAMS star: it is about halfway through its main sequence life.

Equation (2.1) gives the bolometric luminosity of the star, i.e. the energy output integrated over all wavelengths. Since, even today, most investigations are carried out in the visual waveband ($\sim$0.5–0.6 μm), it is often desirable to correct for the fraction of energy released in the visual waveband. This fraction depends mainly on surface temperature T, although there is also a weak dependence on gravity and on composition, because stellar surfaces are not perfect black bodies. But Eq. (2.2) shows that gravity ($\propto M/R^2$) varies little on the main sequence anyway. Consequently we can start by using an approximation depending only on T, viz.

$$\frac{L_V}{L} = \frac{1 + 5 \times 10^{-8} T^{10}}{4 \times 10^4 T^{-7} + 4 \times 10^{-4} T^5 + 3 \times 10^{-10} T^{12.3}}. \tag{2.5}$$

Where T is in kK as before. Equation (2.5) reproduces, to about 10%, the tabulation of Popper (1980), over the temperature range 3–40 kK. A similar formula for the more limited range 4–40 kK corrects for fractional luminosity in the 'blue' waveband ($\sim$0.4–0.5 μm):

$$\frac{L_B}{L} = \frac{1 + 3 \times 10^{-5} T^5}{6 \times 10^7 T^{-12} + 125 T^{-2.5} + 0.8 + 3 \times 10^{-7} T^{7.1}}. \tag{2.6}$$

Beyond the main-sequence band these formulae are less reliable, and equivalent formulae for atmospheres at lower gravity are not yet well determined. These formulae also assume a metallicity Z comparable to the Sun, which is commonly but not universally true.

Observations are generally plotted as $\log L_V$ against $\log(L_B/L_V)$, except that a factor of 2.5 multiplies each of these, for traditional reasons. The latter is a function of temperature only, in the above approximation, but in reality depends (moderately) on gravity and metallicity. It is also affected by distance, since the interstellar medium imposes differential absorption and scattering, which is stronger at shorter than longer wavelengths. The former ($\log L_V$) is of course strongly affected by distance, which is often not well known, but a great deal of information comes from compact clusters of stars, where it is generally reasonable to assume that all members are at the same distance. Two observational cluster HRDs are shown in Fig. 2.10 (Section 2.3.1). For stars in the Solar neighbourhood distances are now relatively well known thanks to the Hipparcos astrometric satellite. Their HRD is shown in Fig. 2.11 (Section 2.3.1).

Computed models of MS stars show that stars with $M \gtrsim 1.25 \, M_\odot$ have convective cores and predominantly radiative envelopes, while stars with $1.1 \, M_\odot \gtrsim M \gtrsim 0.3 \, M_\odot$ have radiative cores and convective envelopes. Less massive stars still are wholly convective. In the narrow range of masses about 1.1–1.25 $M_\odot$, the core starts in radiative equilibrium and becomes convective as the star evolves across the main-sequence band. Convection is discussed in Section 2.2.3 and convective mixing in Section 2.2.4. The presence or absence of central convection, and the fraction of the star's mass that is convective when the hydrogen fuel is exhausted at the centre, affects the way in which one can sensibly define the TMS: see Section 2.2.8.

2.2.2 *Polytropic approximations*

Although the stellar structure equations of Appendix A do not look amenable to elementary solutions, stars in the main sequence band are in fact remarkably well approximated by polytropes, i.e. by gas spheres in which the pressure is proportional to a power of density. This relation can be written parametrically as $\rho = \rho_c \theta^n$, $p = p_c \theta^{n+1}$, or equivalently $p \propto \rho^{1+1/n}$, with n constant, and ρ_c, p_c being central values. Combining this with hydrostatic equilibrium and self-gravity gives the Lane–Emden equation

$$\frac{1}{r^2} \frac{\mathrm{d}}{\mathrm{d}r} r^2 \frac{\mathrm{d}\theta}{\mathrm{d}r} = -\theta^n, \tag{2.7}$$

provided that we scale the radius r by a factor $\{(n + 1)p_c/4\pi G\rho_c^2\}^{1/2}$. Equation (2.7) has 'Emden solutions' which start from $\theta = 1$ at the centre and reach $\theta = 0$ at a dimensionless radius and mass which can readily be computed, provided $n \le 5$; for larger n the radius and mass are unbounded. For a general discussion of polytropes see Chandrasekhar (1939).

The polytropic index n is not, in practice, dictated by the equation of state of the gas (although, exceptionally, it is in the case of a white dwarf; see Section 2.3.1), but rather by the temperature distribution and hence by the heat transport process – in particular, by the way in which the radiative opacity of the material depends on density and temperature. Radiative equilibrium dictates that a value of $n \sim 3$ is not far wrong, in practice, over the whole of the main sequence above $M \sim 0.5 \, M_\odot$.

This is illustrated by the fact that Eq. (2.1) agrees quite closely with a result that can be obtained analytically, subject to certain approximations that may seem drastic but that must clearly be reasonable in practice. Let us assume the following:

(a) Pressure is a combination of perfect gas and radiation pressure only, i.e.

$$p = \frac{\Re \rho T}{\mu} + \frac{1}{3} a T^4 \equiv \frac{\Re \rho T}{\mu}(1 + \zeta), \quad \text{say.} \tag{2.8}$$

Eddington (1926) took the gas pressure to be a fraction β, and radiation pressure the remaining fraction $1 - \beta$, of the total pressure, but we prefer to use the closely related ratio ζ of radiation pressure to gas pressure: $\beta \equiv 1/(1 + \zeta)$.

(b) The opacity $\kappa(\rho, T, \text{composition})$ has the form

$$\kappa = \kappa_{\text{Th}} + \kappa_{\text{Kr}} \frac{3\Re \rho}{a T^3} = \kappa_{\text{Th}} + \frac{\mu \kappa_{\text{Kr}}}{\zeta}. \tag{2.9}$$

Here κ_{Th} is the Thomson-scattering opacity, a constant. The second term (with κ_{Kr} constant) is a rough replacement for Kramers' opacity law $\kappa \propto \rho/T^{3.5}$, which crudely approximates the absorption of photons by bound–free electronic transitions in highly ionised gas. For material with composition similar to the Sun's, $\kappa_{\text{Th}} \simeq 0.034 \, \text{m}^2/\text{kg}$ and $\kappa_{\text{Kr}} \simeq 0.015 \kappa_{\text{Th}}$.

(c) The nuclear energy generation rate is *uniform* throughout the entire stellar interior, rather than concentrated towards the centre as in a realistic model.

Under these assumptions, the equations of hydrostatic and of radiative thermal equilibrium lead to the following consequences:

(d) The star is *exactly* an $n = 3$ polytrope, with $p \propto T^4$, $\rho \propto T^3$ throughout.

(e) Eddington's quartic equation applies, which in terms of ζ above is

$$\zeta (1 + \zeta)^3 = \frac{\mu^4 M^2}{M_{\text{Edd}}^2}, \tag{2.10}$$

where ζ by (d) and (a) is a constant throughout the star, and where μ is the mean molecular weight ($\mu \sim 0.62$ for Solar material). The Eddington mass M_{Edd} is a mass whose value is determined solely by fundamental constants, including the dimensionless mass (2.01824) of the $n = 3$ polytrope:

$$M_{\text{Edd}} \equiv 2.01824 \, \frac{\Re^2}{G} \left(\frac{48}{\pi a G} \right)^{1/2}$$

$$= 2.01824 \, \frac{12\sqrt{5}}{\pi^{3/2}} m_{\text{U}} \left(\frac{G m_{\text{U}}^2}{\hbar c} \right)^{-3/2} \simeq 18.3 \, M_\odot, \tag{2.11}$$

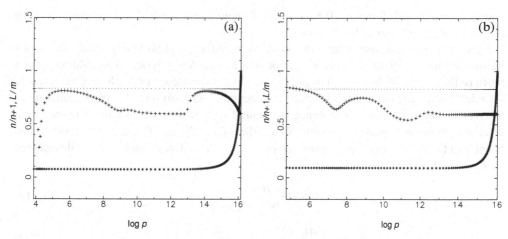

Figure 2.3 The variation of the local polytropic index n (in the form of the softness index $s \equiv n/n + 1$, pluses), and of $L(r)/m(r)$ (asterisks) in arbitrary units, as functions of pressure (N/m^2), in ZAMS stars of 1 $M_\odot$ (a) and 0.25 $M_\odot$ (b). The left-hand star is convective from the surface down to log $p \sim 13$; the right-hand star is fully convective. Departures from $s \sim 0.6$ in the convective regions are due to ionisation, molecular dissociation, or inefficient convection very near the surface. The dotted line in each panel is $n = 5$; cf. Fig. 2.8.

where m_U is the atomic mass unit (for most purposes the mass of a baryon). The Eddington mass is closely related to the Chandrasekhar mass, Section 2.3.2: they differ by a factor of $32\sqrt{15}/\pi^2$.

(f) The star's luminosity is given by

$$L = \frac{4\pi acGMT^4}{3\kappa p} = \frac{4\pi cGM\zeta^2}{(\zeta\kappa_{Th} + \mu\kappa_{Kr})(1 + \zeta)}. \tag{2.12}$$

Equations (2.10–2.12) imply a unique $L(M)$ relation, which can easily be compared numerically with the 'empirical' relation (2.1), and will be found to be within a factor of two for all masses in the range 0.5 to 400 $M_\odot$; over this range log L ranges from -1.5 to 7. The maximum error in this range is at about 10 $M_\odot$, where Eq. (2.12) is too large by a factor of 2.

For the highest masses, with $M \gg M_{Edd}/\mu^2$, Eq. (2.10) gives $\zeta \gtrsim 1$ and then Eq. (2.12) gives $L \propto M$. For lower masses the relation becomes steeper: $L \propto M^3$ provided $1 \gtrsim \zeta \gtrsim \mu\kappa_{Kr}/\kappa_{Th}$. It is steeper still, $L \propto M^5$, at even lower masses where $\zeta \lesssim \mu\kappa_{Kr}/\kappa_{Th}$. Only for $M \lesssim 0.5 M_\odot$ does Eq. (2.1), giving $L \propto M^{2.5}$, begin to deviate significantly from the analytic result, giving $L \propto M^5$. This departure is due to the fact that convective rather than radiative equilibrium is dominant at low masses, and the analytic approach based on the dominance of radiative transport breaks down. We discuss the reasons for convection in Section 2.2.3.

Figure 2.3 shows how the effective *local* polytropic index n, defined in terms of the slope of the curve followed by the star in the log p, log ρ plane, varies in zero-age stars of 1 $M_\odot$ and 0.25 $M_\odot$. This slope, which we define as the 'softness index' s, relates to n by $n/n + 1 \equiv s$. In the 1 $M_\odot$ star, although n is hardly constant it varies in a fairly limited range with a minimum of about 1.5 (apart from a very narrow region near the surface) and a maximum of about 4; equivalently $0.6 \lesssim s \lesssim 0.8$. A value of about $n \sim 3$, $s \sim 0.75$ is a surprisingly good approximation to the overall structure.

Note that in the regime where the radiative postulate approximately works the mass-luminosity law (2.10) to (2.12) is independent of any *nuclear* physical input, although for a more realistic opacity formula than Eq. (2.9) a slight dependence on nuclear physics would intrude. It is mainly the opacity which determines the luminosity, and that in turn determines the rate of nuclear burning, and hence the central temperature and density. The following three equations determine the radius R and the values T_c, ρ_c of central temperature and density in the $n = 3$ polytropic model above, for a given M from which ζ, L are determined by Eqs (2.10) to (2.12):

$$RT_c = 0.8543 \frac{\mu GM}{\Re(1 + \zeta)},$$ (2.13)

$$\rho_c = \frac{a\mu}{3\Re} \frac{T_c^3}{\zeta},$$ (2.14)

and

$$L = \frac{M\epsilon_c}{(2 + \eta/3)^{3/2}} = \frac{AM\rho_c T_c^\eta}{(2 + \eta/3)^{3/2}}.$$ (2.15)

The first two equations are exact (for $n = 3$ polytropes). In the third, there are two approximations. Firstly, the factor $A\rho_c T_c^\eta$ is an estimate, of a type commonly used in back-of-the-envelope analyses of stellar structure, for the rate ϵ of generation of energy by nuclear reactions at the stellar centre. For a given nuclear reaction, η is taken to be constant, although more realistically the effective η is a function of temperature that diminishes slowly as the temperature increases. The physics of nuclear reactions in the context of astrophysics is developed in detail by Clayton (1968); tables of reaction rates can be found in Caughlan and Fowler (1988). On the upper MS, i.e. $M \gtrsim M_\odot$, hydrogen burning by the CNO cycle dominates, and we can take for illustration $\eta \sim 16.5$, $A \sim 3.6 \times 10^{-19} XZ$ (T_c in MK); for the lower main sequence the p–p chain dominates, and we can take $\eta \sim 3.85$, $A \sim 9.6 \times 10^{-5} X^2$. Secondly, the factor $(2 + \eta/3)^{3/2}$ in Eq. (2.15) is an approximation to allow for the fact that the energy generation rate is *not* uniform through the star, as was assumed in (c) above, but is peaked more sharply at the centre the larger η is. To justify the η-dependent factor in Eq. (2.15), let us model the temperature distribution, and consequential ρ and ϵ distribution, by a Gaussian:

$$T \propto e^{-r^2/a^2}, \quad \rho \propto e^{-3r^2/a^2}, \quad \epsilon \propto e^{-(3+\eta)r^2/a^2}.$$ (2.16)

Then by integrating ρ and $\rho\epsilon$ over the star, we find that the ratio of the central value of L/m to its surface value is just $(2 + \eta/3)^{3/2}$. We would get a similar result if we assumed alternatively that $T \propto (1 + r^2/a^2)^{-1}$. Equations (2.12) and (2.15) require that the centre of an MS star should lie approximately on the following curve in the (ρ, T) plane:

$$\frac{3\rho\kappa\epsilon}{T^4} = 4\pi acG(2 + \eta/3)^{3/2}.$$ (2.17)

Figure 2.4 is a plot of contours of constant $\rho\kappa/T^4$ in the (ρ, T) plane. Opacities κ have not been determined for a substantial portion of this plane in the lower right, where the physics of cool dense gas is very difficult. At the upper left both κ and p/T^4 are constant, the

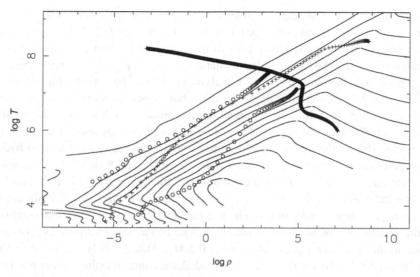

Figure 2.4 Contours of constant $p\kappa/T^4$ in the temperature (K)/density (kg/m^3) plane. Contours are in decades, with high values towards the lower middle. Radiative portions of a star follow these curves fairly closely; convective portions follow adiabats which are usually steeper, but may be shallower in ionisation zones near the surface. ZAMS stars of 1 and 6 $M_\odot$ are plotted as circles (lower and upper, respectively), and a highly-evolved 4 $M_\odot$ model as pluses. The last two stars have much the same value of L/m outside their main energy-producing regions. The solid line is Eq. (2.17), an approximate locus on which the centres of ZAMS stars should lie.

former because of Thomson scattering and the latter because of radiation pressure. Our very simplistic model above would make these contours straight lines, and would put a (main-sequence) stellar interior on one of these contours. In fact for the 6 $M_\odot$ model illustrated in Fig. 2.4 this is quite closely the case: the contour is not quite straight, but the stellar model keeps quite close to it all the same, and the centre of the model falls almost exactly on the plotted curve (2.17).

We show in the next section that we should have $p\kappa/T^4$ constant in any part of the star that is in radiative equilibrium. In Fig. 2.4 each of the three stars plotted does follow such a contour in at least a part of the star; where it departs from such a contour is either in a surface convection zone, a central convective core, or below the main energy-producing region, as we discuss shortly.

Mathematically, a polytrope has a surface at $p = T = \rho = 0$, but physically the surface is at finite p, T and ρ, determined by the condition that the mean free path λ of a photon should be comparable to the pressure scale height H_p at the surface:

$$\lambda \equiv \frac{1}{\kappa\rho} \sim H_p, \quad \text{where } H_p \equiv -\frac{dr}{d\log p} = \frac{p}{g\rho}, \quad \text{hence } p\kappa \sim g. \quad (2.18)$$

A slightly better estimate, based on a Milne–Eddington radiative atmosphere with constant opacity, is $p\kappa = \frac{2}{3}g$. Along with the boundary condition $L = \pi a c R^2 T^4$ – effectively Eq. (2.3) – this gives surface p and T typically smaller than central values by at least about 10^{12} and about 10^3 respectively, so that a complete polytrope (out to $T = 0$) is in fact a very good approximation.

Fully convective stars, although not covered by the above analysis, can nevertheless also be represented approximately as polytropes, this time of index $n \sim 3/2$. This value of n comes from thermodynamics of gas, rather than from radiative heat transport, since convective stellar interiors tend to be completely adiabatic ($p \propto T^{5/2}$, $\rho \propto T^{3/2}$) except perhaps for a small region near the photosphere. An analytic model can be set up which is similar in plausibility to the above radiative approximation, but somewhat more messy because here it is the *surface* opacity which is important. It determines which adiabat the star lies on. The opacity coefficient near the surface is less convincingly expressible as a power-law approximation than in the interior; it is dominated by the degree of ionisation of hydrogen, which drops dramatically with temperature below $\sim 10\,$kK, and which itself determines the (small) abundance of the negative hydrogen ion (H^-; binding energy $0.75\,$eV). This loosely bound ion is the major contributor to opacity in the temperature range 5–9 kK, as the number of free electrons, and their contribution by Thomson scattering, drops off. A further complication is that the equilibrium between molecular hydrogen and atomic hydrogen is progressively more important as the mass goes below about $0.5\,M_\odot$. This has only a minor effect on the mean polytropic index, but a major effect on the adiabatic constant in the relation $p = \text{constant}$ $T^{5/2}$ that prevails throughout most of the interior. The right-hand panel of Fig. 2.3 shows how n varies in practice in a fully convective star ($0.25\,M_\odot$). There are two major bumps that are due to molecular dissociation near the surface, and partial ionisation further in. A dip and bump at $\log p \sim 11$–13 is due to *pressure* dissociation, a highly non-perfect-gas effect at high density and low temperature. The physics in this region is particularly hard to model accurately.

A useful model of lower main sequence stars can be made by constructing an $n = 3/2$ polytropic envelope around an $n = 3$ core (Rappaport *et al.* 1983). Such models can be scaled to agree with results of detailed computations. The latter show that below $M \sim M_\odot$ the convective $n \sim 3/2$ envelope grows strongly at the expense of the $n \sim 3$ radiative core, until the whole star becomes convective below about $0.3\,M_\odot$. This critical value may be rather dependent on the opacity coefficient, both at the surface and near the centre. It is not, however, very accurate to think of lower MS stars as $n = 3/2$ polytropes, since n is quite considerably increased, over quite a substantial region, by ionisation and by the dissociation of molecular hydrogen (Fig. 2.3(b)).

The main sequence as usually defined terminates at about $0.08\,M_\odot$. In fact there is a bifurcation, with two sequences at higher masses; the second sequence being thermally unstable and having degenerate cores along with nuclear burning. Although there are no solutions below $\sim 0.08\,M_\odot$ that are in thermal equilibrium, objects that form below this mass can be in hydrostatic equilibrium while cooling indefinitely, as 'brown dwarfs', towards a third equilibrium series at zero temperature which is essentially the white dwarf sequence continued to low mass (Section 2.2.6).

2.2.3   Convection

There are very few stellar models which do not contain a convective region somewhere. In fact this is something of a coincidence: upper MS stars have convective cores, and lower MS stars have convective envelopes, and although the physics dictating these two facts is quite different, the latter begins at almost exactly the same point on the MS where the former ends. The result is a very small region (about 1.1–$1.25\,M_\odot$) where there is significant surface convection as well as core convection.

Classically, three important temperature gradients are definable at any layer within a star:

$$\nabla \equiv \frac{d \log T}{d \log p} \equiv \frac{\partial \log T / \partial r}{\partial \log p / \partial r}, \tag{2.19}$$

which is the actual gradient of $\log T$ with respect to $\log p$ as one travels through the star;

$$\nabla_r \equiv \frac{3 \kappa p L}{16 \pi a c G m T^4}, \tag{2.20}$$

which is the gradient which would be necessary to carry all the local heat flux by radiative transport alone; and

$$\nabla_a \equiv \left(\frac{\partial \log T}{\partial \log p} \right)_{S, \text{composition}}, \tag{2.21}$$

the adiabatic gradient. If $\nabla_r < \nabla_a$ in a certain layer, then that layer is stable against convection, and so $\nabla = \nabla_r$; all the heat is carried by radiation. If $\nabla_r > \nabla_a$ in another layer, then that layer is unstable to convection, and to a reasonable level of approximation $\nabla = \nabla_a$, because convective heat transport is so efficient (see below) that a very small excess of temperature gradient over adiabatic will allow convection to carry many times as much heat as the radiative flux. In the lowest level of approximation we can therefore say that

$$\nabla \approx \min (\nabla_r, \nabla_a). \tag{2.22}$$

In a radiative region, since $\nabla = \nabla_r$ we can differentiate Eq. (2.20) logarithmically with respect to $\log p$, to obtain

$$\frac{d \log \nabla}{d \log p} = (4 - \kappa_T)(\nabla_0 - \nabla) + \nabla_{L/m}, \tag{2.23}$$

with the definitions

$$\nabla_0 \equiv \frac{1 + \kappa_p}{4 - \kappa_T}, \quad \nabla_{L/m} \equiv \frac{d \log(L/m)}{d \log p} \tag{2.24}$$

and

$$\kappa_p \equiv \left(\frac{\partial \log \kappa}{\partial \log p} \right)_T, \quad \kappa_T \equiv \left(\frac{\partial \log \kappa}{\partial \log T} \right)_p. \tag{2.25}$$

The opacity derivatives κ_p, κ_T have, for example, the values $1, -4.5$, if the opacity is Kramers' ($\kappa \propto \rho / T^{3.5}$) and the pressure is perfect gas ($p \propto \rho T$); and they are both zero for Thomson scattering ($\kappa = \text{constant}$). To a reasonable approximation therefore, all of κ_p, κ_T and ∇_0 can be thought of as given constants. The quantity $\nabla_{L/m}$ is gratifyingly small throughout main-sequence stars. Whereas pressure decreases by a factor of at least 10^{12} from its central value to its surface value, L/m decreases by a factor of only about 5–100: this factor is approximated as $(2 + \eta/3)^{3/2}$ in Eq. (2.15). Only in the innermost two decades of pressure (in a main sequence star) does L/m begin to depart at all from its surface value: Fig. 2.3. Thus $\nabla_{L/m}$ can be neglected everywhere except within four or five pressure scale heights of the centre, where it is positive and of order unity.

If $\nabla_{L/m}$ is neglected in Eq. (2.23), if κ_T, ∇_0 are taken as constant, and if $\kappa_T < 4$ (which it usually is by a wide margin), then the equation is readily solved to show that $\nabla \to \nabla_0$ as p increases. The value that ∇ starts from at the surface is almost irrelevant, since convergence is quite rapid. But the fact that the photosphere has to be radiative means that ∇ must start with a small value: a simple plane-parallel radiative atmosphere with negligible radiation pressure has $\nabla = 0$ at $p = 0$. Even if there is convection in a surface envelope (for reasons indicated below), the radiative zone below this envelope will tend rapidly to $\nabla = \nabla_0$. We have $\nabla_0 = 0.235$ or 0.25 in the above two cases of Kramers' opacity and Thomson scattering. For a perfect-gas equation of state, these imply polytropic indices of 3.25 and 3 respectively. Then returning to Eq. (2.20), we see that in a radiative portion of a star $\kappa p/T^4 = $ constant, except within four or five pressure scale heights of the centre where L/m can increase. It is this increase in L/m (going inwards) which can push ∇ above ∇_a and so drive central convection, in stars where η is moderately large.

We can estimate analytically a minimum value for η that is necessary (but not sufficient) for the star to have an infinitesimal convective core (Tayler 1952, Naur and Osterbrock 1952). In such a star Eq. (2.23) must hold just outside the core (but effectively at the centre, since the core is infinitesimal). Furthermore the right-hand side of the equation must be positive there, since $\nabla = \nabla_r < \nabla_a$ just outside, and must increase to $\nabla = \nabla_a$ just inside. But $\nabla_{L/m}$ can be calculated at the centre by expanding quantities there (T, ρ, L, m) to first order in r^2, and using the fact that (for a perfect gas) $\partial \log \rho / \partial r = (\nabla^{-1} - 1) \partial \log T / \partial r$ then:

$$\nabla_{L/m} = \frac{3}{5} (\nabla + \eta - \eta \nabla) = \frac{3}{25} (2 + 3\eta), \quad \text{if} \quad \nabla = \nabla_a = \frac{2}{5}. \tag{2.26}$$

Then the right-hand side of Eq. (2.23) is positive if

$$\eta > 1 - \frac{5}{9} (5\kappa_p + 2\kappa_T) = \frac{29}{9} \quad \text{(Kramers' opacity).} \tag{2.27}$$

This, as we said, is necessary but not sufficient; numerical modelling shows that $\eta \gtrsim 5$ is sufficient.

Figure 2.4 shows 'radiative' contours of constant $\kappa p / T^4$ in the (ρ, T) plane. These are intersected by adiabats (not shown) which are usually steeper ($\rho \propto T^{1.5}$), although in the surface layers with $T \sim 10^4$–$10^{4.7}$ the ionisation of hydrogen and helium make them less steep. Travelling inwards from the surface, the interior must follow whichever of the two curves is shallower in slope (negative slope counting as steep). A major feature of Fig. 2.4 is a 'knee' in the radiative curves, at $T \sim 10^4$ K; it is caused by the fact that κ_T is strongly positive in the region where hydrogen is partially ionised. If the photosphere is below this knee, then the structure curve must start off along an adiabat, and will only reach the radiative curve again (the *same* curve that passes through the photospheric point) some distance above the knee. But if the photosphere is itself above the knee, then the entire envelope can follow the radiative curve, at least until near the centre where convection tends to be driven by the $\nabla_{L/m}$ term, if $\eta \gtrsim 5$ (but depending slightly on the opacity law). A small departure from this simple picture is caused by the fact that in the very outermost layers convection is not fully efficient at carrying heat, and so ∇ may exceed ∇_a by a modest amount within one or two pressure scale heights of the photosphere.

The $6 M_\odot$ star in Fig. 2.4, starting from a photosphere which is above the knee, follows a contour of $p\kappa / T^4$ almost exactly, until very near the centre where the increase in L/m drives the gradient up to the convective value. On the other hand the $1 M_\odot$ model starts below the knee, and so cannot follow the contour. Instead it follows an adiabat, but finally when the adiabat recrosses the *same* contour the interior becomes radiative again and follows the contour until L/m increases close to the centre. The centre does not, in fact, reach the convective slope of another adiabat, but almost does.

In a convective region we can estimate the excess of ∇ over ∇_a a little more seriously, using the 'mixing-length theory' (MLT; Böhm-Vitense 1958), which I describe here in a rather simplified version. This estimates the mean velocity of convective eddies, w, from the difference between the total luminosity (obtained by integrating the nuclear energy generation over the interior) and the luminosity that can be carried by radiation within the convective layer, supposing that convection is so efficient that the temperature gradient ∇ can be approximated (Appendix A) as the adiabatic gradient ∇_a:

$$L = L_{\text{rad}} + L_{\text{con}}, \quad L_{\text{rad}} \approx \frac{16\pi acGmT^4\nabla_a}{3\kappa p}, \quad L_{\text{con}} \approx 4\pi r^2 \frac{\rho w^3}{\alpha}, \tag{2.28}$$

and so, using Eq. (2.20)

$$w^3 \approx \frac{4acT^4 g\alpha}{3\kappa p\rho}(\nabla_r - \nabla_a). \tag{2.29}$$

The constant α is the 'mixing-length ratio', a fudge parameter which carries all the uncertainty in the physics, and is assumed to be of order unity. It is normally taken to be the ratio of the supposed typical mean free path l of an eddy to the local pressure scale height $H_p = p/g\rho$, Eq. (2.18). However, since this definition of H_P is infinite at the centre, which is inconvenient and probably also unphysical, we use here a definition

$$l = \alpha \, \min\left(\frac{p}{g\rho}, \sqrt{\frac{p}{G\rho^2}} \right). \tag{2.30}$$

The convective heat flux comes from the proposition that the heat energy carried in an eddy is comparable to its kinetic energy. This in turn comes from the proposition that the kinetic energy acquired by the eddy is equal to the work done by buoyancy over the mean free path. Assuming that the velocity of the eddy is strongly subsonic, it should be in pressure equilibrium with the ambient material, so that an upward-rising eddy is both hotter and less dense than its surroundings. It carries thermal energy because it is hotter, and gains mechanical energy because it is less dense. Thermodynamics ensures that the relative temperature excess and density deficit are comparable.

The excess of actual over adiabatic gradient, which relates to the entropy gradient, is estimated from

$$\frac{p}{g\rho \, C_p} \frac{\partial S}{\partial r} = \nabla - \nabla_a \approx \frac{w^2}{v_{\text{sound}}^2 \alpha^2}, \tag{2.31}$$

where $v_{\text{sound}}^2 \sim p/\rho$ is roughly the sound speed squared and C_p is the specific heat at constant pressure. This assumes that buoyancy accelerates the eddy until, after a mean free path, it dissolves back into the ambient fluid. Normally the Mach number w/v_{sound} is indeed much less than unity, as it ought to be for the validity of the model. However it can approach unity in the photospheric layers.

By convention, standard mixing-length theory has certain specific coefficients of order unity within it, leaving α as the only free parameter. Appendix A includes the standard prescription, which leads to a slightly more elaborate cubic equation for w and hence $\nabla - \nabla_a$. Stars earlier than F are not much affected by α, because they have little convection in their envelopes, and in their cores the Mach number is barely different from zero. For a whole range of cooler stars with convective envelopes, from G/K/M dwarfs through the Sun to G/K/M giants, a value for α of 2.0, used in the models of Fig. 2.1, gives adequately good agreement with observation. This does not mean that mixing-length theory is right, but it does help to make it at least very useful. It is much the simplest recipe available, being entirely 'local': the actual temperature gradient at a point is determined only by values of quantities at that point, such as L, r, m, T and ρ.

The quantity l/w defines a local convective timescale, and leads to an estimate of the convective envelope turnover time t_{ET}. This timescale is important in investigations of tidal friction (Section 4.2) and magnetic dynamo activity (Section 4.4). In the latter, the Rossby number, or ratio of rotation period to convective turnover time, is considered important. Several different estimates of t_{ET} can be made, depending on how one averages l or w. A fairly unbiased estimate is $\int dr/w$, taking the integral over the whole convective zone. This is the time it would take fluid to rise from bottom to top; it is $\sim$35 days for the Sun. Although the integrand diverges at both ends, the integral is finite since, from Eq. (2.29), $w \sim |r - r_0|^{1/3}$ near a boundary at r_0. However, the integral is probably an overestimate, since it is likely to be the larger eddies near the base which contribute most to dynamo action. A more conventional estimate is the value of l/w at a height one half of a mixing length ($l/2$) above the base of the convection zone. In the case of the Sun, this gives a value of $\sim$15 days (Rucinski and Vandenberg 1986), i.e. 3/7 of the integrated value. We take this ratio as 'canonical', and so define

$$t_{ET} \sim 0.43 \int_{CZ} \frac{dr}{w}. \tag{2.32}$$

At a cruder level, it is convenient to define a global turbulent velocity w_G and global convective time scale t_G thus:

$$L = 4\pi R^2 \overline{\rho} \, w_G^3, \quad M = \frac{4\pi}{3} R^3 \overline{\rho}, \quad t_G = \frac{R}{w_G} = \left(\frac{3MR^2}{L}\right)^{1/3}, \tag{2.33}$$

L, R and M being surface values in SI, not Solar, units. This gives $w_G \sim 36$ m/s and $t_G \sim 250$ days for the Sun. The envelope turnover time t_{ET} is less than t_G for the Sun because w is somewhat larger at the lower-than-mean density in the envelope and the scale height is substantially less than the overall radius. For fully convective stars, at the bottom of the main sequence, we expect $t_{ET} \sim t_G$. For less-than-fully convective stars, an empirical estimate can be made in terms of the actual radius R and the 'Hayashi Track' radius R_{HT}, discussed in more detail below (Section 2.3.1). R_{HT} is the largest radius that a star of given luminosity and mass can have (in hydrostatic equilibrium, but not necessarily in thermal equilibrium), and is reached if the star is fully, or at least very largely, convective. R/R_{HT} is also a function – Eq. (2.49) – of only global quantities L, R and M. $R < R_{HT}$ if the star is partly radiative: $R \sim 0.55 \, R_{HT}$ for the Sun. A rough empirical fit to both low-mass ZAMS stars and to red giants, as well as to the Sun, is

$$t_{ET} \sim 0.33 \, t_G \left(\frac{R}{R_{HT}}\right)^{2.7}. \tag{2.34}$$

2.2.4 *Convective mixing, entrainment, semiconvection and overshooting*

The convective motion not only transports heat but also mixes the composition of the star. In a first approximation, convective mixing is assumed to be confined to those regions of a star where the temperature gradient is steep enough to cause convective instability. But it seems likely that turbulent motion starting within an unstable region may continue under its momentum into at least the edges of an adjacent stable region, so that convective mixing of composition takes place over a larger region than we at first expect. Before discussing this 'convective overshooting', however, it is helpful to consider briefly the classical model of convective mixing, which necessarily involves also the concepts of 'convective entrainment' and of 'semiconvection'.

Classical convective mixing, along with entrainment and semiconvective mixing, can be modelled by a diffusive transport equation. We imagine convective eddies which move with typical speed w for a typical mean free path l (which we take for illustration to be the pressure scale height, i.e. $\alpha \sim 1$) before losing their identity by merging with other eddies. Then the appropriate diffusion coefficient is approximately wl, which can be estimated from Eqs (2.29 and 2.30) as

$$wl = \sigma \left(\nabla_r - \nabla_a\right)^{1/3}, \quad \sigma = \frac{p}{g\rho}\left(\frac{4acT^4g}{3\kappa p\rho}\right)^{1/3}. \tag{2.35}$$

We can now write the composition equation, for a nuclear species with abundance X, as

$$\frac{1}{\rho r^2}\frac{\partial}{\partial r}\left[r^2\rho\sigma\left(\nabla_r - \nabla_a\right)^{1/3}\frac{\partial X}{\partial r}\right] = \frac{DX}{Dt} + R, \tag{2.36}$$

where D/Dt represents a Lagrangian time derivative and $R(X, \rho, T)$ is the rate of nuclear burning. Of course, in regions stable to convection $(\nabla_r < \nabla_a)$, we must put $\sigma = w = 0$. It is important to note that ∇_r is itself a function of the abundance X, via κ in Eq. (2.20). Thus Eq. (2.36) for X is much more non-linear than it appears to be. The nuclear evolution of a star is actually governed by a whole set of equations like Eq. (2.36), one for each nuclear species; but in practice there are only a few species sufficiently important that their abundances can themselves modify the structure to a significant extent.

Normally σ/l^2, an estimate of the convective mixing rate – comparable to t_G^{-1}, Eq. (2.33) – is so large in comparison with R, the nuclear burning rate (or, more significantly, R averaged over the stellar interior), that we can take $\partial X/\partial r \simeq 0$ as a good approximation to the solution of Eq. (2.36). This is the usual assumption of convective-mixing algorithms, that the composition is uniform within an unstable region. Then the rate of change in time of this uniform composition is obtained by integrating Eq. (2.36) over an entire unstable region. However, making the composition gradient nearly vanish is not the only way in which Eq. (2.36) can be balanced for very large σ/l^2: an alternative possibility is that $\nabla_r - \nabla_a$ may be very small, but still positive and not zero, in which case $\partial X/\partial r$ need not be small. This latter kind of solution is 'semiconvective'. There is not actually an ambiguity in Eq. (2.36) as to which kind of solution is achieved. A typical evolutionary calculation in which the set of composition Eqs (2.36) is solved *simultaneously* with the structure and mesh-spacing equations that determine such other variables as p, T, r, L, m will automatically ensure (a) that $\partial X/\partial r \simeq 0$ in some part of an unstable region, (b) that $\nabla_r - \nabla_a \simeq 0$ in the remaining part of the unstable region and (c) that $\sigma = 0$, and hence $DX/Dt + R = 0$, in stable regions. Figures 2.5 and 2.6

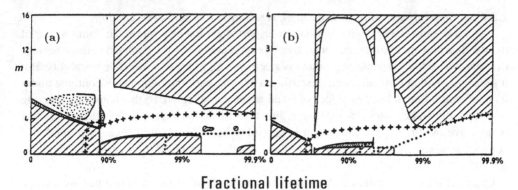

Fractional lifetime

Figure 2.5 The movement of boundaries of radiative, convective (shaded; $\nabla_r - \nabla_a > 0.01$) and semiconvective or weakly convective (dotted; $0 \le \nabla_r - \nabla_a \le 0.01$) regions, with mass coordinate plotted vertically against age as a fraction of the total life of the star. Also shown are lines where hydrogen (pluses) and helium (heavy dots) have been depleted by burning to 10% by mass. (a) $16\,M_\odot$ star; (b) $4\,M_\odot$ star. In (a), a large semiconvection zone develops during the main-sequence evolution: its apparently ragged edge mainly reflects the coarseness of the mesh used. In addition, hidden in the spike at somewhat less than 90% of the lifetime, there is a convection/semiconvection zone, which is shown in more detail in Fig. 2.6. In (b), there is a substantial semiconvection zone outside the convective helium-burning core, which is also shown in more detail in Fig. 2.6. During the last 1% of the star's lifetime, the two burning shells and the base of the convective zone are indistinguishable on this scale, although they are separated by several pressure scale heights.

illustrate the way in which convection or semiconvection zones can grow or disappear in the course of evolution of a $16\,M_\odot$ and a $4\,M_\odot$ star.

At *any* boundary between a convective region and a stable region there will be at least a thin layer where the turbulent velocity is very small. If the composition is different between the stable and unstable regions there will necessarily be a transition layer where there is a composition gradient. Whether this layer is very thin, or on the other hand quite substantial, depends on how the opacity varies with the composition. Only if the layer turns out to be substantial would we bother to call it 'semiconvective', although in principle thin as well as thick layers are semiconvective. This is illustrated in the upper panels of Fig. 2.6, discussed shortly. For purposes of illustration, *any* zone which has $0 \le \nabla_r - \nabla_a \le 0.01$ is treated as semiconvective, so that a normal convective zone appears to be bounded by a 'semiconvective' region.

When solving the structure along with the composition equations, it is naturally found that the boundary between an unstable and a stable region moves in time with respect to the mass coordinate. If the movement is such that the boundary encroaches on the stable region, we have the phenomenon which we refer to here as 'convective entrainment'. Whether the convective region shrinks, or grows by entrainment, in the course of evolution, and whether some part of the unstable region is semiconvective or not, is not easy to predict a priori. Experience shows that most main-sequence models have convective cores which shrink, and which avoid semiconvection. An exception is the range of models of about 1–1.5 $M_\odot$, where the convective core grows during the first half of the main-sequence evolution because the dominant hydrogen-burning reaction gradually switches from the p-p chain to the CNO cycle. In the lower part of this mass range the core is radiative to start with, developing convection

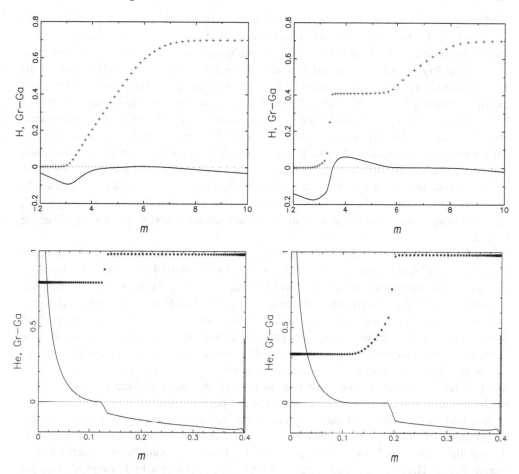

Figure 2.6 The growth of a semiconvective region at the edge of a convective region, in (upper panels) a TMS star of 16 $M_\odot$, and (lower panels) a helium-burning core of 0.4 $M_\odot$. In each case both the composition (pluses) and the convective parameter $\nabla_r - \nabla_a$ (continuous curve) are plotted against mass coordinate, at an early stage (left) and a late stage (right) in the development of the semiconvective region. All of the action in the upper panels takes place inside the vertical spike in Fig. 2.5a located at just before the 90% fractional lifetime, i.e. in the rapid transition between core hydrogen burning and core helium burning.

as the CNO cycle grows in importance. In massive stars, $\gtrsim 15\ M_\odot$, a semiconvection zone may appear somewhat beyond the outer edge of the convective core, at an intermediate stage in the star's evolution across the main-sequence band, as illustrated in Fig. 2.5a.

Figure 2.6 illustrates two of the more common situations in which semiconvection is found to occur. The first is the situation referred to above in massive stars towards the end of main sequence evolution. The second is the situation found in a helium-burning core. Whereas hydrogen-burning convective cores on the main sequence tend to shrink as the star evolves, mainly because the opacity decreases as the helium abundance rises, helium-burning convective cores tend to grow, mainly because the opacity increases as the carbon abundance rises. The increase may become so great at the outer boundary of the convective core that a semiconvection zone is forced to develop beyond it.

The top panels of Fig. 2.6 show two profiles of hydrogen abundance X against mass coordinate m, for two models during evolution of a 16 $M_\odot$ star. They occur during the rapid (thermal timescale) evolution at the end of the main sequence – both models are hidden in the spike in Fig. 2.5a located at just before 90% fractional age. In the earlier model (left) X rises steadily from zero in a central core to 0.7 in the surface layers. The corresponding run of the parameter $\nabla_r - \nabla_a$ is also shown. It is negative, implying no convective mixing, everywhere except (by a very slight margin) at four points near 6 $M_\odot$ which have just become unstable. They have barely begun to affect the composition profile. In the later model (right), the convection has spread over a much greater region, from about 3.4 to 8.5 $M_\odot$. The inner part of this region is fully convective, with $\nabla_r - \nabla_a$ well above zero and the composition profile almost flat, except for one point at the inner boundary which is still moving inwards. Entrainment is going on at this lower boundary. The outer part has $\nabla_r - \nabla_a$ very small, so that mixing is slow and the composition profile rises smoothly; this is a semiconvective region.

The lower panels of Fig. 2.6 similarly show the growth, by a combination of convective entrainment and semiconvection, of the helium-burning core of a 0.4 $M_\odot$ helium star. Between the models the outer boundary of the convective carbon-enriched core has moved outwards by roughly 0.09 $M_\odot$; this growth is driven mainly by the fact that carbon opacity is higher than helium opacity at the same pressure and temperature, so that a temperature gradient that would be stable if helium alone were present becomes unstable once carbon is mixed in. However, in this case the growth of the core in effect *dilutes* the carbon as the boundary moves out, so that part of the entrained region becomes semiconvective rather than fully convective. Most of the extra core is semiconvective, as seen by the fact that there is a considerable gradient of composition along with a very small but positive value of $\nabla_r - \nabla_a$.

Standard linear analyses of stability against convection are done on the assumption that there is no composition gradient. The situation where there *is* a composition gradient requires a more detailed analysis, but its results are quite straightforward (Kato 1966, Eggleton 1983b, Spruit 1992). They tell us, firstly, that if $\nabla_r < \nabla_a$, the 'Schwarzschild criterion', then the situation is stable; secondly, if ∇_r lies in an intermediate regime

$$\nabla_a < \nabla_r < \nabla_a + \left(\frac{\partial \log T}{\partial \mu}\right)_{p,\rho} \frac{d\mu}{d \log p}, \tag{2.37}$$

then the situation is 'overstable', by which is meant that the motion is oscillatory on a short (dynamical) timescale, but with an amplitude that grows on a longer (thermal) timescale. However, the linear overstability in this regime is not in fact the predominant instability, since in the same regime a non-linear mode, involving the successive overturning of a large number of thin layers, is more favoured energetically. The consequence of the instability is that mixing (possibly very slow mixing) will take place as long as ∇_r exceeds ∇_a. This mixing need not proceed to completion, i.e. until the composition gradient has been reduced to zero, but only until ∇_r has been reduced to fractionally above ∇_a, since the mixing then becomes slow. Thirdly, if ∇_r is larger still, outside the range indicated on the right in expression (2.37), the situation is dynamically unstable to much the same degree that it would be without the composition gradient. This is the 'Ledoux criterion' for dynamical instability. However, one does not expect this criterion to be interesting in practice, because before the gradient has grown large enough for dynamical instability it would have already been large enough for the non-linear mode of successive overturning to have taken place, and so would be prevented from rising much further. The upshot is that we should use equations like (2.35) and (2.36) whether there is or is not a gradient of composition. An exact value for the coefficient σ in wl is not necessary, provided that wl contains a factor with the property that it goes to zero with $\nabla_r - \nabla_a$; but see Eq. (2.38) below.

All the above behaviour takes place within the classical model of convective mixing, where it is an axiom that mixing does not take place in regions where $\nabla_r < \nabla_a$. But it is unlikely that the motion of convective eddies will actually fall exactly to zero at the boundary $\nabla_r = \nabla_a$. Eddies which move on the unstable side of the boundary will be accelerated by the buoyancy force right up to the boundary. Of course, they will also be subject to the decelerative force of turbulent viscosity. But although the forces may (somewhat naively) vanish at the boundary, the velocity presumably does not, and so the eddies may be expected to overshoot somewhat into the stable region.

In addition to such a process, called 'convective overshooting', there are other possible reasons why uniformly-mixed cores might be larger than the ones determined by current models with the 'standard' assumptions. For example, stellar rotation can cause circulation currents (Section 3.2.1), and these can be expected to be particularly strong at a boundary between a convective and a radiative region. They may therefore cause mixing across the classical boundary, and thus contribute to overshooting. Waves perturbed in the stable region by motions in the unstable region can lead to mixing. Dynamo generation of magnetic field in the convective core might lead to buoyant toroidal flux loops which float up through the stable envelope and cause mixing. 'Convective overshooting' will be used here in a very general sense, to mean any process that produces mixing *beyond* the classical boundaries of convective and semiconvective mixing. Possibly the term 'enhanced mixing' would be better, because less specific, but convective overshooting is the process most commonly discussed. Unfortunately, in some literature the term 'convective overshooting' is also used, confusingly, to describe the process by which a *classical* convective region may grow in the course of stellar evolution; this is the process which in this book is referred to as 'convective entrainment'.

We would not, of course, be discussing possible mechanisms for convective overshooting if there were not rather clear reasons, coming from the comparison of observed with computed stars (Andersen 1991), for believing that stars have larger mixed cores than the standard models produce. The observational evidence suggests that the main-sequence band is broader than is indicated by models without overshooting, at least for masses above $\sim 1.8\ M_\odot$; and also that core-helium-burning giants are more luminous at a given mass than the standard models. Figure 2.1 shows a small group of stars that are quite far above the ZAMS, at about spectral type A. Standard models would evolve very rapidly through this area, so that such stars should be rare. But the assumption of convective overshooting broadens the main-sequence band, and allows the observed area to be more heavily populated.

The effect of overshooting is to prolong main-sequence evolution (because a greater amount of nuclear fuel is accessible to the central nuclear furnace), to broaden the MS band, and to make later evolutionary stages more luminous because their He cores are more massive. Let's model convective overshooting here by using a modified formula (cf. Eq. 2.35) for the diffusion coefficient of turbulent mixing:

$$wl = C\sigma(\nabla_r - \nabla_a + \nabla_{os})^2, \quad \nabla_{os} \equiv \frac{\delta_{os}}{2.5 + 20\zeta + 16\zeta^2}, \tag{2.38}$$

where ζ, as in Eq. (2.8), is the ratio of radiation pressure to gas pressure. We find (Schröder *et al.* 1997, Pols *et al.* 1997) that a value $\delta_{os} = 0.12 \pm 0.04$ for the overshooting parameter is necessary, by comparison of computed models with certain highly evolved binaries which have well-determined masses, luminosities, temperatures and radii.

The ζ-dependence of (2.38) is an ad hoc modification to ensure that very massive stars, where $\nabla_r \sim 0.25$ because Thomson scattering dominates, and where also $\nabla_a \sim 0.25$ because radiation pressure dominates, do not become wholly convective. The ζ-dependent factor is equivalent to saying that mixing takes place in a band where the entropy drops by a fixed amount from its constant value in the adiabatic core.

Equation (2.38) differs in two other ways from Eq. (2.35). Both changes are for numerical convenience, and have no physical basis. The exponent 2 replaces $1/3$, because differentiation – necessary in the iterative solution of the equations – would give a singularity at the boundary $\nabla_r = \nabla_a - \nabla_{os}$. The constant C, which should be unity, is weakened to 10^{-2} or 10^{-4} because even in double precision the composition gradient is so slight that it is poorly defined numerically. We would hope to be able to report that the changes make little difference in practice, since it matters little whether the composition changes by one part in 10^6 or in 10^{10} within a convective region. However, in the $8\,M_\odot$ star of Fig. 2.1, the 'blue loop' (see later), where the star after having reached the giant branch retreats temporarily into the Hertzsprung gap, is quite significantly shortened in length as one goes from $C = 10^{-2}$ to $C = 10^{-4}$. Apparently minor changes in the helium composition profile can change blue loops rather significantly.

It appears to be necessary for overshooting to disappear rather abruptly when the convective core is itself small below about $1.8\,M_\odot$, since some old Galactic star clusters such as M67 appear to contain a turn-off region which is better modelled without overshooting (Morgan and Eggleton 1979, Pols *et al.* 1997, 1998). By contrast, younger clusters such as IC 4651 are in better agreement with models containing overshooting. It seems possible that a model of overshooting adequate to describe both older and younger clusters will need two fitting parameters in it, which is an unfortunate complexity. An ad hoc prescription that does this is

$$\nabla_{os} \equiv \frac{\delta_{os}}{2.5 + 20\zeta + 16\zeta^2} \frac{U}{U_0 + U}, \quad U = \frac{Gm^2}{4\pi r^4 p}, \tag{2.39}$$

U being a homology invariant (i.e. a dimensionless quantity) that vanishes at the centre but increases strongly outwards in the core. Along with $\delta_{os} = 0.12$, the constant U_0 is taken as 0.1, which crudely ensures that the overshoot disappears when the core is suitably small.

Theoretical estimates of the amount of overshooting are very uncertain, for the present. We may have to wait until three-dimensional hydrodynamical codes have enough spatial resolution, as well as numerical reliability, to solve this problem convincingly – but even then we should probably include rotation and magnetic fields, as a minimum. It is also possible that yet more detailed calculations of the opacity coefficient will change the size of the 'standard' convective core, and hence the size of the overshoot region needed to give agreement with observation. It is not so much the magnitude of the opacity as the rate of change of opacity with temperature and pressure – the quantities κ_T, κ_p of Eq. (2.25) – which determines the size of a core. These could change significantly even if the mean opacity level is about right.

Probably the most stringent test at present of models of stellar evolution, but so far only for the Sun, comes from helioseismology, the study of the very rich spectrum of low-amplitude oscillations at $\sim$1–5 mHz that are observed on the Sun (Claverie *et al.* 1979, Isaak 1986,

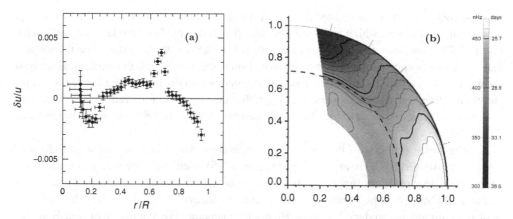

Figure 2.7 (a) The difference between measured and computed sound speed (squared), as a function of radius, for a Solar model. Errors are largest (~0.4%) near the base of the convection zone at ~0.7 $R_\odot$. From Christensen-Dalsgaard and Däppen (1996). (b) Contour plot of the rate of rotation as a function of latitude and depth in the Sun. The base of the convection zone is shown as a dashed line. Rotation is rapid near the equator, and slower towards the poles. It is indeterminate in the blank region to the left. From Schou *et al.* (1999).

Schou *et al.* 1999). These oscillations are driven by the random perturbing motion of convective eddies near the surface, but like earthquakes the waves can penetrate quite deeply (more deeply, the longer the period), be refracted back to the surface, and thus yield information about the deeper layers. Figure 2.7a shows the close but not perfect agreement of the observed spectrum with a theoretical model (model S of Christensen-Dalsgaard and Däppen, 1996). Agreement is best when it is assumed that there is a slight diffusion (Proffitt and Michaud 1991) of helium towards the centre within the radiative core (~70% of the Sun by radius).

Selective diffusion (Chapman and Cowling 1958) takes place in stars, in those regions which are not mixed to uniformity by convection or other processes, because the usual structure equations (Appendix A) do not include the fact that different nuclear species (and even different atomic states of the same species) are acted on by different forces. Hydrostatic equilibrium (Eq. A1) is an accurate *average* over all species, but a particular species of particle experiences a particular extra force relative to the mean force. This extra force is (loosely) a linear combination of the pressure gradients of all species (including photons), with different linear combinations for each species. Some species experience a net outward force, some a net inward force. The selective force translates to a selective velocity, via atomic collisions – much as in conduction by electrons in a wire. The fact that photons diffuse outwards relative to the mean fluid can be seen as an example of selective diffusion, and is due to their extremely low (i.e. zero) mass relative to the other species, mainly H. He atoms are heavier than H, and so diffuse inwards; except that the problem is actually much more complicated, as *all* the partial pressures, including that of radiation, are involved, for all species. If a massive species has an unusually high cross-section to photons, it may be dragged outwards by the photon flux even though otherwise it would tend to settle inwards.

In a non-rotating star there would be considerable degeneracy among the normal modes of oscillation, but this is lifted by rotation. The observed spectrum of the Sun is so rich that the internal rotation can be determined as a function of position over most of the interior except for a small central core which is not significantly penetrated by those modes measured so far. Figure 2.7b shows the angular velocity determined. There is a strong variation on the surface, that was already known from the rotation rates of sunspots; but the helioseismological analysis shows that the surface variation persists with depth down to the base of the convection zone, and then rather abruptly disappears to give a rather uniformly rotating radiative core, at least to the depth which is measurable. The peak angular velocity is $\sim$8% greater than the mean interior value.

Another test of models of the Sun is from measurements of its neutrino output (Bahcall 1964, Kuzmin 1966, Davis *et al.* 1968, Haxton 1995). Neutrinos are side-products of the chain of nuclear reactions that converts hydrogen to helium. Some come from the basic p-p reaction (or rather its pep variant), others from β-decays of ^{7}Be and ^{8}B. Although two neutrinos are always produced for each He nucleus generated by the fusion of four H nuclei, the energy spectrum of the neutrinos is a strong function of temperature as different parts of the nuclear-reaction network contribute at different temperatures. There is a marked disagreement here with theory. The shortfall is of about 40% in detectors based on ^{71}Ga and about 70% for the detector based on ^{37}Cl (Haxton 1995; Bahcall 2000); the shortfall depends on the parameters chosen for the theoretical model as well on the nature of the detector. Before we throw out stellar models, however, we have to ask whether the theory of the neutrino is more complex than at first supposed: 'flavour oscillations', where the neutrinos (supposing their rest mass is not exactly zero) oscillate between e, μ and τ states as they travel through matter, or magnetic field, may possibly be the cause of the discrepancy (Wolfenstein 1978, Mikheyev and Smirnov 1985, Hata and Langacker 1995). We shall adopt the point of view that helioseismology largely vindicates stellar modelling, and that therefore the neutrino problem requires better neutrino physics.

2.2.5   *Anomalous main sequence stars*

If the standard equations of stellar structure, involving strict hydrostatic equilibrium of spherical objects with Solar composition at age zero, were literally true, there would be no main-sequence stars showing (a) emission lines, (b) peculiar (i.e. strongly non-Solar) composition, (c) short-term variability, e.g. pulsation, flares, rotational asymmetry, (d) excess radiation, relative to the visual, in XR, UV, IR, radio or other wavebands or (e) magnetic fields. In fact many MS stars show one or more of these properties. Some of these anomalies may well be related to binarity. But it will be convenient to summarise here, very briefly, some of the main types of anomaly that require explanation. In approximate order of decreasing luminosity down the main sequence, we find the following:

(a)   Of stars. These are O stars which show emission lines, both in the visual and the UV. The lines have a 'P Cygni' shape (red-shifted emission bordering a broad blue-shifted absorption line), which is indicative of a roughly spherical wind flowing outward with high terminal velocity and substantial mass-flux (Section 2.4).

(b)   WR (Wolf–Rayet) stars. These show very strong emission lines, especially of ionised He, N, C; as for Of stars the lines have a P Cygni shape, but are so much stronger that they actually dominate the visual spectrum. WR atmospheres are so affected by wind that it is difficult to locate the stars with certainty in the HR diagram, but they appear to be in the top left corner, with $L \gtrsim 10^5 \, L_\odot$. Their masses, determined with some reliability from spectroscopic orbits, are about 30–70% of what is expected from their luminosities ($\gtrsim$30–50 $M_\odot$, Fig. 2.1). The enhancement of nitrogen seen in the WN subset of WR stars is roughly consistent with mass loss, if the star has been stripped down to a core where nuclear burning formerly took place (Gamow 1943). The WC subset (carbon-enriched) presumably represent a later phase where products of helium burning are revealed, and it may be that WC stars, and perhaps the hotter ('earlier') WNs, are actually post-MS objects. Possibly all stars of high enough luminosity are subject to such winds, perhaps as a result of some major instability triggered by high radiation pressure at the Humphreys–Davidson limit (HDL; Section 2.3.4), but the physics of the mass-loss process is not yet well understood.

(c) OBN stars. Some late O and early B stars show anomalously strong nitrogen. This could be due to mass loss, as in WN stars, but the winds observed are by no means as strong. It could more probably be due to unusually strong mixing of the outer layers with the nuclearly-processed interior, perhaps driven by rotation (Paczyński 1973). There is evidence that binarity is involved (Walborn 1976).

(d) β Cep, 53 Per and SPB stars. These show pulsation in radial or non-radial modes. Most or all stars in a limited range of spectral type (B0.5–B2 II–IV) are radial β Cep pulsators. The pulsations appear to be driven by a mechanism which derives from an enhancement of opacity in a limited region of the density–temperature plane near the surface, due to partially ionised Fe and other heavy elements (Cox *et al.* 1992, Dziembowski and Pamyatnykh 1993, Dziembowski *et al.* 1993). They relate to the bump seen in Fig. 2.4 at about 10^5 K and low density. The 53 Per stars (Smith 1977) occupy a larger region of the HR diagram roughly centred on the β Cep region, extending to mid-B, and perhaps are a less extreme form of the same mechanism. They exhibit periodic changes in line shape, attributable to non-radial pulsation. Slowly pulsating B (SPB) stars (Waelkens 1991), in the range B3–B8, have low-amplitude multi-periodic pulsations of period $\sim$1.5–4 days, which may be high-order non-radial gravity modes, driven by some destabilising mechanism in the energy-producing core. The β Cep pulsators are not to be confused with δ Cep pulsators; the latter are not main-sequence stars and so are not discussed here, except tangentially in (h) below.

(e) Be (B-emission), λ Eri, ζ Oph stars. Although Be stars range from late O to late B they are most common (about 20%) at early B. The emission, unlike in WR stars, is usually double peaked and roughly centred at rest. It appears to be due to a rotating equatorial ring or disc of gas. The emission is often episodic, on a timescale of about a decade, as if occasionally shells of gas are thrown off at or near the equator. Infall is sometimes seen, as well as outflow, but circulatory motion dominates. Be stars are all in very rapid rotation, so that equatorial gravity is strongly reduced though not to zero. An excitation in the outer layers near the equator, perhaps due to magnetic activity or pulsational instability (or both), may be intermittently overcoming the rather weak gravity there. Some Be stars are λ Eri variables (Balona 1990), whose variability appears to be due to the rotation of a starspotted surface, as in the cooler α CVn and much cooler BY Dra stars (below). ζ Oph stars appear to show non-radial pulsation (as do 53 Per stars) as well as Be characteristics (Kambe *et al.* 1993).

(f) Bp (B-peculiar), α CVn, roAp stars. These chemically peculiar stars are quite common ($\sim$10%) at late B, but range from early B to A and even early F. Different elements show marked overabundances (by about 10^3–10^{10}), roughly correlating with mean surface temperature. Helium, and sometimes specifically ^{3}He, can be anomalous at early B, and a range of metals (Sr, Eu, Cr, Ho) at later types. The overabundances are usually concentrated at strong magnetic poles, and since the magnetic field is normally oblique to the rotation axis the peculiar abundances as well as the magnetic field appear to vary periodically as the star rotates. The poles may also be so large and cool that the light output is rotationally modulated, in α CVn variables. The overabundances are also presumably concentrated very much to the outermost layers; they are not thought to represent an *overall* overabundance of the element. Selective diffusion of elements in the local temperature and pressure gradients of the photosphere (Section 2.2.4), particularly as modified by very large magnetic fields in starspots, is thought to be the cause (Michaud 1970). Bp stars are often called Ap stars, because the excess of metals gives a first impression that the spectrum is later, and the surface cooler, than is actually the case – but there is a minority of such stars to be found genuinely at spectral type A. Some of the cooler Bp/Ap stars show rapid oscillations ('roAp stars' – Kurtz 1990), with periods of about 10–20 min. They may be 'oblique pulsators', with the pulsation and magnetic axes aligned. The roAp stars are in the δ Sct instability region (see below) and so the pulsations are presumably driven by helium ionisation, but in a surface whose behaviour is dominated by strong starspots.

(g) Hg/Mn, Am (A-metallic) stars. From late B to early F there is a high proportion of stars ($\sim$20%) showing overabundances (by about $10–10^2$) of elements like Y, Ga, Hg, Mn, and also underabundances of He, Ca. Unlike the Bp stars, there is no evidence of magnetic fields, and the abundance anomalies appear to be distributed isotropically, not patchily. Selective diffusion is probably also the cause here, but without the extra effect of starspots. Selective diffusion might be suppressed by rotationally driven mixing, but presumably in these stars this mixing is ineffective because these stars have slower-than-average rotation – see (h). Am stars are usually in binaries with 2.5 days $\lesssim P \lesssim 100$ days (Abt 1983), which suggests that tidal friction (Section 4.2) is involved.

(h) δ Sct, AI Vel, δ Del, γ Dor stars. Most A stars that are not Am stars are δ Sct pulsators. The pulsations are akin to the δ Cep pulsations of evolved stars, and are driven by the zone some way below the photosphere where helium is undergoing second ionisation ($T \sim 50$ kK). The pulsation is presumably absent in Am stars because selective diffusion has largely drained the helium from this zone. The δ Sct stars probably counter this selective diffusion with rotationally-driven mixing, being faster-than-average rotators. AI Vel variables appear to be δ Sct stars that are at the upper edge of the MS band. A few stars (δ Del stars) show both Am and δ Sct characteristics. Balona *et al.* (1994) have identified a γ Dor class of variable at or just beyond the cool edge of the δ Sct instability strip. They are early F stars apparently pulsating non-radially, but probably related to δ Sct stars.

(i) β Pic stars. These show a considerable excess IR flux, which can be interpreted as a cool disc some hundreds of AU across (Aumann 1985). Such discs are presumably left over from the formation process, and indicate a fairly young star. They may be sites for the formation of planetary systems.

(j) Blue stragglers. These stars are not especially anomalous in themselves, but are anomalous in relation to those star clusters in which they are found. In principle they can be any spectral type earlier than about Solar, but most are of types late B to F. They lie on a cluster main sequence above the turn-off point, where normal stars should have evolved into giants or white dwarfs (Fig. 2.10b). Apart from binary mechanisms, two suggested explanations are (a) recently-born stars, younger than the bulk of stars in the cluster, and (b) anomalously mixed stars, whose MS lives have been extended through large-scale mixing of the envelope with the core. Neither single-star mechanism is especially satisfactory.

(k) λ Boo stars. These are mildly anomalous stars of spectral type A, which are fairly normal in C, N, O and S, but depleted in many metals. They may be related to β Pic stars (above) in having cool discs around them, and the abundance anomalies may be due to depletion of metals in the disc by selective condensation into grains. These grains are driven out of the system by radiation pressure, and accretion of the remaining disc material on to the photosphere gives the anomaly. They may represent ≲1% of main-sequence A stars.

(l) Strong λ4077 stars. These are mildly anomalous F (and some A) stars showing overabundances of Zr, Ba and other elements possibly related to the s-process of nucleosynthesis (Section 2.3.2). They may be main-sequence analogues of the classical red-giant barium stars, for which a binary-star mass-transfer mechanism is strongly indicated (Section 6.4). If the same mechanism applies to λ4077 stars, we would expect them all to be binaries with a white-dwarf companion.

(m) FGKM 'subdwarfs'. These are Population II main-sequence stars, and 'anomalous' in our definition only because, being old, they formed with substantially lower metallicity than the Sun. They are hotter and smaller than 'normal', i.e. Solar-composition, main-sequence dwarfs, and so on a Hertzsprung–Russell diagram appear below the normal main sequence, but by no means as far below as white dwarfs.

(n) BY Dra, AB Dor stars. Some K dwarfs, but also F–M, show a quasi-periodic variation due to the rotation of one or two large starspots or starspot clusters into and out of the line of sight. Individual spots persist for months. Such stars tend to rotate rapidly (∼0.5–5 days), and also to show flaring activity. Rapid rotation is probably the cause of dynamo activity, in turn causing the flares and spots. Binarity is sometimes the cause of the rapid rotation, as for evolved RS CVn red subgiants (Section 4.6), but rapid rotation may also be simply an indication of youth. We use BY Dra as a prototype of such behaviour as produced by binarity, and AB Dor as a prototype of such behaviour in single stars (or at least stars with no *close* companion).

(o) Flare stars. A proportion, increasing to about 50% at latest types, of K/M dwarfs have 3–5 magnitude outbursts of a few seconds duration, at intervals of weeks or months. These outbursts are typically more energetic than Solar flares, and seen against the background of a star that may be about 1000 times fainter. The flares are due to magnetic activity, which is probably related to more rapid rotation than average; but not all flare stars show the rotational modulation characteristic of BY Dra stars (above). Mass loss due to flaring, combined with the magnetic field, will lead to 'magnetic braking' of the rotation; so the rotation and all the activity consequent on it should diminish with age. This suggests that flare stars are simply younger-than-average M dwarfs.

Almost all of the above types of anomalous behaviour have been attributed at some time to the influence of a binary companion, but the case has not always been sustainable. The best cases for the importance of binarity can probably be made for classes (c), (g), (h), (j), (l) and (n), and will be discussed later; while for classes (d), (f), (m) and (o) there is little or no case to be made.

2.2.6 *Brown and black dwarfs*

Towards the lower end of the MS, where stellar interior conditions involve substantially lower temperatures and higher densities than in the Sun, the equation of state (EoS) becomes rather complicated. It is high density rather than high temperature that causes material near the centre to be ionised ('pressure ionisation'), and the electron gas may be substantially degenerate. The effect of degeneracy can be to allow the star to support itself in hydrostatic equilibrium without the help, necessary in more massive stars, of a temperature gradient due to nuclear reactions. Thus models are possible in which the heat flux derives only from cooling of the interior. Such stars are called 'brown dwarfs'; they are expected at $M \lesssim 0.08 \, M_\odot$, the exact value depending on what approximation is used for the EoS. In principle such a star can continue cooling to zero temperature, since either degenerate electron pressure, or else the pressure of the liquid or solid state (as in planets), can support the star against gravity at arbitrarily low temperature. If the temperature is so low as to contribute negligibly to pressure support the object is sometimes called a 'black dwarf'. The radius of a black dwarf will be determined purely by its mass and chemical composition; a brown dwarf can be somewhat larger, since the internal temperature also contributes to the pressure support.

Zapolsky and Salpeter (1969) constructed black-dwarf models using a relatively simple EoS. They obtained a radius–mass relation which can be approximated, for $M \gtrsim 10^{-5} \, M_\odot$, thus:

$$R = R_{ch} \left[\left(\frac{M_{ch}}{M} \right)^{2/3} - \left(\frac{M}{M_{ch}} \right)^{2/3} \right]^{1/2} \left[1 + 3.5 \left(\frac{M_{pl}}{M} \right)^{2/3} + \frac{M_{pl}}{M} \right]^{-2/3}. \quad (2.40)$$

The characteristic radius R_{ch} and masses M_{ch} and M_{pl} are given in terms of the composition by

$$R_{ch} = 0.0228 \, \langle Z/A \rangle \, R_\odot, \quad M_{ch} = 5.83 \, \langle Z/A \rangle^2 \, M_\odot,$$

$$M_{pl} = 0.0016 \, \langle Z/A \rangle^{3/2} \, \langle Z^2/A \rangle^{3/4} \, M_\odot, \quad (2.41)$$

where Z, A are the atomic number and atomic weight of the chemical constituents, and angular brackets mean that the quantities are averaged over the different constituents, by weight. Thus material consisting of 70% hydrogen and 30% helium by weight has $\langle Z/A \rangle = 0.85$, $\langle Z^2/A \rangle = 1.0$. Equation (2.40) gives $R \propto M^{1/3}$ for low (planetary) masses ($M \lesssim M_{pl}$), where the EoS gives virtually an incompressible liquid. For higher masses, approaching stellar, where electron degeneracy pressure dominates, Eq. (2.40) gives $R \propto M^{-1/3}$; except that the radius goes to zero as $M \to M_{ch}$, the Chandrasekhar limit (see Section 2.3.2). A maximum radius of about $0.1 \, R_\odot$ is attained at about the mass of Jupiter ($\sim 0.001 \, M_\odot$). Brown dwarfs, being hot enough for the internal temperature to increase the pressure, will have somewhat larger radii than is given by Eq. (2.40) for black dwarfs.

Although brown dwarfs have been postulated for a considerable time, it is only relatively recently that, thanks to better detectors in the IR, they have been found in substantial numbers. An early clear candidate was Gl 229B (Nakajima et al. 1995), but recent advances in IR detection show them to be common (Rebolo et al. 1995, Jameson et al. 2002). Objects of brown-dwarf mass are sometimes to be found as companions of white dwarfs in some cataclysmic or related binaries of very short period (Tables 5.1 and 6.3). Some of these may be remnants of initially more massive main-sequence stars, which have lost substantial mass by binary interaction, but others may be primordial.

Objects of major planetary mass (~ 1–$50 \, M_{pl}$), or equivalently of low brown-dwarf mass, have been detected around a number of nearby Solar-type stars (Mayor and Queloz 1995, Butler et al. 1997) by very accurate measurement of radial velocities; and objects of terrestrial planetary mass by the even more accurate process of pulsar timing (Wolszczan and Frail 1992). In practice, Eq. (2.40) extrapolates reasonably well to terrestrial planetary masses.

2.2.7 Star formation

One blessing of stellar astrophysics is that, although it is very difficult to arrive at a clear understanding of how stars form, remarkably little of the later evolution of stars appears to depend on the details of their formation process. Throughout most of this book we shall assume that stars 'start' on the ZAMS. Although purists might feel that we should start with a star-forming gas cloud, they should acknowledge that such clouds show many signs of having been processed by earlier generations of stars, so that we have a chicken/egg situation. We choose to start with the chicken, but in this sub-section briefly refer to the egg.

The formation of stars (Lada 1985, Shu *et al.* 1987, Pringle 1989, Matthieu 1994) is probably less well understood than any other portion of stellar evolution. This is partly because the expected lifetimes of pre-main-sequence stars are short – $\lesssim 0.001 \, t_{MS}$, Eq. (2.4) – so that such stars should be, and are, relatively uncommon; partly because many pre-main-sequence stars tend to show complicated spectra, and erratic variability, e.g. Herbig Ae/Be stars and T Tau stars (Herbig 1960), which makes them harder to quantify than main sequence stars; and also partly because star formation takes place predominantly, even almost exclusively, in gaseous and dusty 'star-forming regions' (SFRs) where the protostellar gas cloud itself blocks most of the radiation from the stars, at least at visual and shorter wavelengths. The 1980s and 1990s have, however, produced a vast improvement in the quality and quantity of information on star formation, mainly thanks to observational work in the infrared and millimetre wavelength ranges.

Many SFRs are very massive collections of gas and stars called giant molecular clouds (GMCs), of which the Orion SFR, at a distance of $\sim$500 pc, is an example. This SFR fills much of the constellation of Orion, but with a concentration to the central part of Orion's sword, where the visible star is in fact a collection of about a dozen massive newly-formed stars. On a more modest scale is the SFR in Taurus-Auriga, at $\sim$140 pc. This consists largely of low-mass T Tau stars.

Perhaps the most basic problem for theorists of star formation is the absence of any sensible 'initial conditions'. There is no particular moment in the life of a star-forming region at which one can, for example, assume that it is spherical with a given density/temperature distribution, and thus investigate its stability to fragmentation on various lengthscales. In contrast, there are reasonably good 'final conditions', i.e. stars on the ZAMS, so that theories of star formation will have to be tested primarily by their ability to produce the right distributions of masses, and of binary (and triple, etc.) parameters. Unfortunately this is not likely to be a powerful restriction. Different SFRs, or different regions in the same SFR, may produce different IMFs and different period distributions, which merge ultimately to give some more global properties as in the Solar neighbourhood.

Within an SFR there are many 'cores', which appear to be the actual sites of star formation. If there were rough spherical symmetry near a core, the centre of the core should collapse first to a low-mass protostar, with infall of material further out subsequently increasing the mass, often by a substantial factor. But typically there will be considerable rotation, which requires stars to solve the 'angular momentum problem' (Bodenheimer 1978, 1991; Boss 1991, 1993). A cloud of gas should spin up, on contraction, because of conservation of angular momentum, and the ratio of centrifugal to gravitational force at the equator (loosely defined) should increase roughly as $1/r$, where r is the equatorial radius of the cloud. The apparently trivial amount of rotation in an initial cloud that would be due to galactic rotation should make it impossible for the cloud to contract to stellar radii by the necessary factor of 10^7–10^8. Even contraction to fairly close binaries, those with periods of $\sim$1–10^4 days, would be difficult, and yet these constitute $\sim$10% of low-mass stars, and $\gtrsim$30% of massive stars. We appear to need a dissipative process, i.e. viscosity, to allow some regions to transfer their angular momentum outwards, and thus contract further inwards. The viscosity presumably has a magnetohydrodynamic (MHD) or turbulent basis, since molecular viscosity is too weak to have much effect.

The transition from a gas cloud of say 10^6 $M_\odot$, a few parsecs across, to a group of newly-formed stars and systems, and from there to a broad distribution of stars and systems as seen in the Solar neighbourhood, along with some remaining older clusters, probably involves a considerable number of processes, some acting at an early stage and some at later stages. We identify something like six steps, overlapping in time:

(a) contraction of the gas cloud, and fragmentation into several (say $\sim$100) smaller, denser accumulations – proto-sub-clusters
(b) viscous (primarily MHD-driven) interaction within the fragments, allowing substantial contraction within proto-sub-clusters which further fragment into proto-sub-sub-clusters, and so on iteratively to individual multiple systems with $\sim$2–10 protostellar components
(c) slowing down of contraction locally when central material becomes opaque enough to establish a temperature gradient sufficient for hydrostatic equilibrium
(d) gravitational interaction, leading to the concentration of the more massive fragments towards the centre, to mergers (collisions) of some close pairs of proto-sub-clusters, etc. and to ejection of some of the lighter members, at each hierarchical level
(e) ejection of the remaining ambient gas
(f) dissipative-tidal and evolutionary interaction of close pairs and triples.

Process (a) might divide the initial gas into a non-hierarchical assembly of fragments which would not yet be on a stellar scale but rather on a sub-cluster scale of say 10^3–10^4 $M_\odot$ each. Presumably the regions which are particularly poor in angular momentum would be prone to contract farthest. They would not be able to contract very far, in the first instance, before reaching the centrifugal barrier.

However, process (b) allows further angular momentum to be extracted from some already denser regions, permitting further contraction, of a more hierarchical character, down to the scale of sub-sub-clusters of perhaps 10–10^2 $M_\odot$, and by iteration on a smaller scale still to fairly hierarchical multiple systems of perhaps 2–10 protostars. It is necessary that we have some kind of unstable process that depletes angular momentum in regions that are already somewhat low in it, and transfers it to ambient regions. This allows regions which are capable of more substantial contraction to continue contracting farther. It is also necessary that this takes place on a considerably smaller length-scale than the entire cloud, so that what we get is 10^6 condensations rather than one large condensation.

We believe that MHD 'viscosity', i.e. process (b), is likely to be the most effective means of the necessary redistribution of angular momentum. Gravitation has the property that low angular momentum goes along with high angular velocity: a close binary has less angular momentum but more angular velocity than a wide binary. Gravitational contraction and spin-up therefore encourages differential rotation, which will rapidly amplify locally any pre-existing magnetic field. Then MHD interaction can lead to an effective viscosity (Shakura and Sunyaev 1973; see Appendix F) and a further loss of angular momentum from regions already short of it, and so lead to further contraction. At a fairly late stage in the formation process the angular momentum loss may take place primarily in disc-like structures (see below), but in the earlier stages the process is likely to be more unstructured, yet general enough to cause local instabilities where angular momentum is siphoned out of regions that are already somewhat low in it, and so where greater contraction is possible.

Provided that angular momentum can be extracted fairly efficiently from local condensations, the process of hierarchical fragmentation will stop, or at any rate slow down, only as protostars reach hydrostatic equilibrium – process (c). At early stages this does not happen because the gas is fairly transparent to the low-temperature radiation released by gravitational contraction (although it is highly opaque to visible radiation). Isothermal spheres cannot reach hydrostatic equilibrium; but as the opacity goes up, more heat is retained and a temperature gradient grows which can allow hydrostatic equilibrium to develop.

Although the angular momentum problem means that it is difficult for a local blob of gas to contract radially by a large factor, the blob is not inhibited in principle from contracting by a large factor parallel to the local angular momentum vector, to form a disc. We then need viscosity within the disc to allow material to spiral inwards, on to the hydrostatically supported central protostar, while the angular momentum is transported outwards. As discussed in Section 6.2, the main agent for producing viscosity in a disc is likely to be MHD. However, we are not thinking of MHD viscosity as taking place exclusively in thin discs around protostars that are already in hydrostatic equilibrium: it will probably already be contributing to the solution of the angular momentum problem as soon as there is any marked local differential rotation, and it contributes by unstably taking angular momentum out of those regions which are already low in angular momentum, but high in angular velocity and *differential* rotation.

Infrared observations have shown that some stars are surrounded by cool equatorial discs of $\sim 100\,\mathrm{AU}$ in radius. The disc absorbs part of the radiation of the star and reradiates it at longer wavelengths. Such discs are particularly in evidence around pre-main-sequence stars, although they can persist to the main-sequence stage (e.g. β Pic, Aumann 1985). A typical 'core' in a star-forming region probably consists of (a) several protostellar nuclei (perhaps $\sim 10^2$–$10^4\,\mathrm{AU}$ apart) that are already of roughly stellar density, and supported in hydrostatic equilibrium, (b) accretion discs around each nucleus (~ 10–$10^2\,\mathrm{AU}$ in radius), and (c) a roughly spherical envelope of $\gtrsim 10^4\,\mathrm{AU}$ around the whole system. Material is accreted from the envelope on to the disc or discs, and viscosity within these discs allows material to spiral in and be accreted by the nuclei. At a fairly late stage in this process, it is possible that the remaining disc is capable of condensing into planets.

We anticipate that something analogous to common-envelope evolution in evolved binaries (Section 5.2) may be at work, to produce the closest young binaries. Common-envelope evolution is invoked to explain how in a highly evolved binary, containing an AGB star in an orbit of a few years, there can be interaction with a companion to produce the very close white dwarf / M dwarf pairs with period less than a day that are quite often found in planetary nebulae and elsewhere. The interaction can be perceived as two small objects spiralling in towards each other within an envelope that is $\sim 10\,\mathrm{AU}$ across and contains $\sim 1\,M_\odot$ of gas. Dynamical friction may be sufficient, but there must be some dissipative agent, which could also involve MHD-driven turbulence, to achieve the transport of angular momentum from the stellar pair to the envelope. A somewhat similar situation may arise if two protostars have a near-collision (at $\sim$1–10 AU), while their surrounding discs interact to provide the common envelope. This might similarly lead to spiralling-in, and the formation of a *close* pair with a period of a day or so. The analogy cannot be *very* close, since common-envelope evolution in evolved stars appears always to lead to a circular orbit whereas close pairs of young stars are often eccentric. But the situation with protostars may be more chaotic than with evolved stars, and inhomogeneities may allow eccentricity to survive, or even be amplified.

Purely gravitational interactions – process (d) – between clumps of protostars (and their surrounding gas), at the sub-cluster and deeper levels, is probably not effective at encouraging highly hierarchical contraction, but may nevertheless cause substantial evolutionary progress. More massive concentrations will tend to settle towards the centre, and less massive ones to be ejected. This will happen at all levels of hierarchy, such as sub-sub-clusters. Of course there is not in practice a sharp distinction between sub-clusters and sub-sub-clusters; rather there will simply be a range of scales, perhaps of a fractal character, which we divide artificially into quantised sizes for the sake of exposition. At the deepest levels, actual collisions of stars, or at least protostars, might occur: not so much direct collisions of previously independent stars, but rather situations where a binary has an interaction with another star or binary, and the binary orbit is perturbed to a high eccentricity, that could lead to a collision at periastron (Section 5.4). In places where the protostellar density is unusually high, we might even have runaway mergers (Portegies Zwart and McMillan 2002). Stars that have just collided will probably puff up temporarily to rather large radii, and may therefore be more likely to have a further collision.

At some stage the remaining ambient gas is likely to be driven out as a result of stellar activity and evolution – process (e). Massive stars create strong winds, and also lead to supernovae within 3–4 megayears of formation; less massive stars may contribute to gas ejection through such energetic phenomena as bipolar jets from Herbig–Haro objects (Schwartz 1983). The overall efficiency of star formation – whether say 10% or 90% of the gas condenses into stars before the remaining gas is expelled – is not clear, and may vary from region to region. But on the whole we expect that more than half of the original gas is expelled, since this will help to unbind the overall cluster, and to ensure that the population of stars is mostly spread through the Galactic disc rather than left in the clusters where they originated.

Process (e) will obviously terminate process (b) but not process (d). Strongly hierarchical systems will be harder to form once the gas is expelled – but see the discussion of process (f) below. However, those formed already will interact gravitationally with others, the more massive concentrations on any scale tending to drift to the centre and the less massive ones tending to be ejected into the general Galactic field. In some clusters and sub-clusters this process may mean that the entire cluster has dissolved in a few megayears, but evidently some clusters survive for a few gigayears, and globular clusters for many gigayears. This may be mainly a function of the initial mass, but may also depend on the orbit of the cluster relative to the Galaxy. Some clusters are tidally stripped of their outer layers (by what can be seen as a version of Roche-lobe overflow – Chapter 3 – on a Galactic scale), and others will lose some further members by gravitational interaction with the stars of the Galactic plane on occasional passages through it.

There do however remain processes, under (f), which can increase the hierarchical depth of some multiples. We note in Section 4.8 that within triple systems the combination of Kozai cycles and tidal friction can cause the inner binary to shrink, transferring much of its orbital angular momentum to the outer orbit. This requires, as we now expect, a dissipative agency, but it is no longer necessary at this level that MHD play a major role. Ordinary tidal friction – Section 4.2 – may be enough. For Kozai cycles to work, we require only that the inner binary be quite highly inclined to the outer binary.

Since SFRs are themselves gravitationally bound clouds of gas, it may seem at first glance that they would only produce gravitationally bound clusters of stars such as Galactic or globular clusters, but not field stars as in the Solar neighbourhood. However, the disintegration of a cluster as it condenses out of a star-forming region can be understood as the result of much of the gas being expelled from the cloud by processes involving newly-formed stars: for example massive stars with strong winds, or supernova explosions. Low-mass stars can also have strong winds at an early stage in their lives, producing such energetic phenomena as Herbig–Haro objects. Provided such processes occur while still only a fraction (say 10–20%) of the SFR's mass has been condensed into stars, they may be able to put in enough energy to eject a considerable fraction of the remaining gas in the SFR. If a cluster (gas plus stars) is in roughly virial equilibrium, i.e. its gravitational energy is twice its kinetic energy (with opposite sign, so that the total energy is negative), and if subsequently the cluster has half its mass, presumably gas rather than stars, expelled on a short timescale, with the energy for this coming from internal stellar processes rather than from the kinetic or gravitational energy of the cluster, then the cluster will become gravitationally unbound. This is because gravitational energy goes roughly as the square of the mass, while kinetic energy goes linearly with the mass. This is much the same as the reason (Section 5.3) why a binary is disrupted if more than half of the mass is expelled in a supernova explosion. Expanding OB associations such as III Cen, I Ori and Sco-Cen (Blaauw 1964) probably represent intermediate stages in the unbinding process.

Another process which might help to unbind small clusters ($N \lesssim 100$ stars) is two-body gravitational relaxation, i.e. the gravitational scattering of less massive members into escape orbits by simple two-body encounters with more massive members. However, the persistence of some *large* clusters, such as globular clusters, shows that this process is less effective for large clusters.

2.2.8 The terminal main sequence (TMS)

The point at which the main sequence life of a star terminates is not as easy to define as might be supposed. For most stars with $M \gtrsim 1.1\,M_\odot$ evolution abruptly becomes rapid once hydrogen is exhausted more-or-less instantaneously throughout the convective core. However, for $M \lesssim 2\,M_\odot$ the evolution slows again temporarily while hydrogen burns in a thick shell around the helium core. During this phase the star is only moderately larger and more luminous than in the first slow phase. But when the core reaches about 11% of the star's mass (see Eq. (2.42) below and Table 3.2) the evolution accelerates again. Unfortunately this acceleration is not very abrupt, and thus it is difficult to assign its position on the HRD unambiguously. This *second* acceleration continues, as discussed in more detail in the next section, until either (a) the envelope becomes substantially convective, at a point which is fairly clearly identified by a local minimum of luminosity on the cool side of the HRD (Fig. 2.1), or (b) helium ignites at the centre as its temperature reaches about 120 MK. Option (b) may happen either before or after (a), but either way it is also relatively easy to locate unambiguously.

In this book we adopt, for the sake of argument, the following definition of the TMS. Let t be the time, measured from the ZAMS, at which either (a) or (b) above, whichever is first, occurs. Then the TMS is taken to be at age $t_{MS} \equiv 0.99t$ (Eq. 2.4). Although there is of course an element of arbitrariness in this, the definition can be applied uniformly to all masses down to about $1\, M_\odot$, despite the fact that the degree to which central convection is important or not changes markedly, particularly in the range 1–$2\, M_\odot$. Unfortunately, when the influence of a binary companion is included (Chapter 3) it is once again difficult to formulate a sensible definition of the TMS.

2.2.9 *Rotation and magnetic fields*

Rotation in stars is inferred both from the broadening of lines by Doppler shift, and (in a subset of stars) by periodic variation of light output, or spectral line shape, that can be attributed to some anisotropy that rotates into and out of the line of sight. Such anisotropy can in fact be due to magnetic fields, which may also show up as Zeeman splitting of certain lines.

For a star in *uniform* rotation – just an assumption, for the moment – there is an upper limit to rotation, such that centrifugal acceleration at the equator balances gravity (Fig. 3.1d). Roughly, for zero-age main-sequence stars, the shortest possible period of rotation varies over ~ 0.1–2 days from the lowest to the highest masses. By the end of the main sequence the range is roughly 0.6–30 days also depending on mass. Few, if any, stars are known to be rotating at very close to break-up, but some, particularly Be stars, are rotating at up to 70% of break-up. However, many stars, such as the Sun, are rotating much more slowly, at only $\sim 1\%$.

Surface magnetic fields on main sequence stars are found as large as ~ 1–3 T, which is probably an upper limit dictated by the balance of gas pressure near the surface with magnetic pressure. The distribution of magnetic field within the interior is hard to guess at: some stars may have a roughly dipolar magnetic field, but many have higher-order fields. On the Sun, there is both a weak dipolar field, roughly parallel to the rotation axis and reversing every ~ 11 years, but also many small transient spots coming and going on a timescale of days to weeks, where the field may reach its pressure-equilibrium limit, more or less.

Rotation and magnetic fields are almost certainly related, but the relationship is very complicated. We discuss some aspects of this in a little more detail in Chapter 4. The reason for deferring this discussion is that probably, though not certainly, these processes are more significant for the long-term evolution of binaries than for single stars. It is not yet clear how important these processes are for single stars, but a great deal of stellar modelling ignores both processes, and seems to reach reasonable agreement with observation. For example, Fig. 2.1 ignores them, and yet seems to get reasonable agreement between observed and computed masses, temperatures and luminosities.

It is probably futile to attempt to deal with either process in isolation from the other: they must couple very strongly, particularly since the hot ionised gas of a star is a very good conductor, and therefore the magnetic field can be expected to be 'frozen in' to the gas it threads. But what makes the joint problem particularly difficult is that both processes appear to be much influenced by turbulent convection, which is itself very difficult to model.

Perhaps the best reasons for ignoring them in the first instance are the following:

(a) The magnetic field tends to be buoyant and thus to float to the surface. It might reach pressure equilibrium at the surface, but it is difficult to imagine that flux tubes could stay deep in the interior with the much stronger fields that would be necessary for equilibrium at the much higher pressures there. A strong flux tube in the interior would expand or contract on a dynamical timescale towards pressure equilibrium, and since heat conduction will tend (more slowly) to keep the internal temperature near to the ambient temperature the density, according to the perfect gas law, would be less and so the tube would rise.

(b) Stability analysis of differentially rotating stars suggests that there is stability on a thermal timescale only if the rotation is constant on cylinders *and* the angular momentum per unit mass increases outwards.

(c) Internal magnetic fields, even of trivial strength initially, would be rapidly amplified by differential rotation, to the extent where magnetic torque would inhibit the differential rotation.

None of these reasons is entirely convincing on its own, but they all seem to argue against the nightmare scenario of a star which seems normal on the surface but whose interior is a seething mass of huge magnetic and velocity fields for which the simple non-magnetic non-rotating models are completely invalid.

Spruit (1998) has argued that point (c) above may be the reason why the Sun (Fig. 2.7b) shows little sign of differential rotation in the core, despite the fact that evolution should have caused the inner part of the core to contract and the envelope to expand. He argues cogently for what might seem a rather extreme position, that magnetic torque may enforce uniform rotation even in the much later evolutionary stages (discussed later in this chapter) when the core may have contracted in radius by two or more orders of magnitude, and the envelope expanded by a similar amount.

But although magnetic stress may suppress differential rotation within the radiative core of the Sun, it patently does not suppress it within the convective envelope (Fig. 2.7b). One of the odder features of convection is that it apparently *generates* differential rotation (at the level of $\sim 10\%$ in the Sun) – whereas one might expect that turbulence, acting like viscosity, would *suppress* it.

Since we discuss this, and develop a very tentative model of convective dynamo activity, in Section 4.5, we will not pursue it further here. We accept provisionally the view of Spruit (1998) that stars will evolve in a state of near-uniform rotation for much of their lives, and as a result rotation is likely to have little effect on their overall evolution. Uniform rotation can in fact, somewhat surprisingly, be incorporated into a code for spherical models – Appendix B and Section 3.2.1 – and can be shown to be rather unimportant. Further, from point (a) above, it is likely that magnetic field plays an important role only in the surface layers, which are themselves not very important for the nuclear evolution going on deep in the interior. We therefore continue our discussion on the assumption that the evolution of stars can be reasonably well understood in the non-rotating, non-magnetic approximation.

2.2.10 *Examples from observed binaries*

Table 2.1 gives a small selection of binary and multiple systems which are on or still approaching the main sequence. T Tau, the prototypical pre-main-sequence star, is itself in a multiple system, which appears recently to have been in the process of breaking up. There is a cool infrared companion (T Tau S), probably dominated by an accretion disc rather than

Table 2.1. *Some binaries before and on the main sequence*

Name	Spectra	P	e	M_1	M_2	R_1	R_2	Reference
T Tau	K0Ve + (IR + …)	0.72″						Ghez *et al.* 1991
T Tau S	(IR + IR?) + M1	20:years[a]	0.8:					Loinard *et al.* 2003
T Tau Sa	IR + IR?	2:years						Loinard *et al.* 2003
TY CrA[b]	B9Ve + K0IV:	2.89		3.16	1.64	1.80	2.08	Casey *et al.* 1998
TY CrA[b]	(B + K) + ?	270:	0.5:	4.8	1.3:			Beust *et al.* 1997
BM Ori	B3V + A7IV	6.47		5.9	1.8	2.9	4.7[c]	Popper and Plavec 1976
EK Cep	A1.5V + G:	4.43	0.11	2.02	1.12	1.58	1.32	Popper 1987
XY UMa	G0V + K5Ve	0.48	0	1.0	0.6	1.15	0.65	Hilditch and Bell 1994
V624 Her	A7m + A8m	3.90	0	2.28	1.88	3.0	2.2	Andersen 1991
RR Lyn	A7m + F0	9.95	0.08	2.00	1.55	2.50	1.93	Popper 1980
δ Cap	F2m + G/K:	1.02	0.01	0.038[d]				Lloyd and Wonnacott 1994
EN Lac	B2IV + F6-7	12.1	0.04	10	1.3	5.3	1.3	Garrido *et al.* 1983
DI Her	B4 + B5	10.6	0.49	5.18	4.53	2.7	2.5	Guinan *et al.* 1994
SZ Cen	A7 + A7	4.11	0	2.32	2.28	4.55	3.62	Grønbech *et al.* 1977
GG Lup	B7V + B9V	1.85	0.15	4.12	2.51	2.38	1.73	Andersen *et al.* 1993
SS Lac[b]	A2V + A2V	14.4	0.14	2.93	2.85	3.4	3.2	Torres and Stefanik 2000
SS Lac[b]	(A + A) + ?	679	0.16	5.78	0.80			Torres and Stefanik 2000
η Ori[b]	B1 + B2e	7.99	0	11	10.6	6.3	5.2	De May *et al.* 1996
η Ori[b]	(B1 + B2e) + B	3500:	0.4:	1.4:[a]				Waelkens and Lampens 1988

A period may be replaced by a separation (″) in visual systems
[a] recently destroyed; see text
[b] close triple system
[c] polar, equatorial radii
[d] mass function

a star, 0.7″ to the south of the main K0V component. T Tau S is at least two components, one of which (T Tau Sb) appears to have been in an eccentric 'visual' (actually, radio VLBI) orbit of period ∼20 years around the other, but to have been ejected around 1996 into a hyperbolic orbit with velocity ∼20 km/s towards the east. This suggests that T Tau Sa is itself a closer binary, with estimated period ∼2 years, and that an interaction at the periastron of the eccentric outer orbit between Sb and the binary Sa led to an ejection, as is commonly seen in N-body gravitational simulations of non-hierarchical systems (Anosova 1986). It is likely that the entire Solar neighbourhood has been populated by stars or systems ejected in a somewhat similar manner from star-forming regions. We are very fortunate to see, before our very eyes, an instance of dynamical breakup during star formation.

TY CrA is an eclipsing double-lined system, and so gives rather precise fundamental data. Qualitatively, there is agreement with evolution of pre-main-sequence stars, but there is a problem: ∗1 is significantly cooler and less luminous (by ∼10% and ∼40% respectively) than expected for its mass, whether on the ZAMS or still approaching it (Casey *et al.* 1998). This may perhaps be broadly accounted for with a higher metallicity than Solar; but it is difficult to match both stars with the same isochrone, and we may need to entertain the possibility that while ∗2 is as young as 3 megayears, ∗1 is as old as 10 megayears or even older. This need not be surprising if the current binary was created by a dynamical encounter among more primordial binaries. The system is triple, with a low-mass companion in an orbit (but a very tentative orbit, so far) of less than a year. Although the outer orbit is tentative – Beust *et al.* (1997) suggest four possible periods, ranging from 126 to 270 days – it is double-lined in the sense that the CG of the close pair, and Li lines in the third body, give complementary orbits.

BM Ori is a somewhat enigmatic system in which the fainter, less massive, but larger component can be interpreted (Popper and Plavec 1976) as a flattened differential rotator which has not yet reached the main sequence. Its apparent ratio of equatorial to polar radius, as determined from its eclipse light curve, seems more extreme than is permissible for a star in uniform rotation, and may imply that the core is rotating substantially more rapidly than the envelope. This ought to be unstable on a thermal timescale (Section 3.2.1). On this picture, the age of the system should be $\lesssim 1$ megayear, and so it is not surprising that thermal instability or tidal friction have had little effect on the less massive component. What does seem surprising is that the more massive component is not rotating about equally rapidly.

BM Ori (θ^1 Ori B) is a member of the Trapezium Cluster (M42), the central concentration of young stars in the Orion Nebula cluster, at $\sim$470 pc distance. The Trapezium can be viewed as a multiple system, but of a non-hierarchical character, i.e. several stars are at comparable distances from each other. Such systems are gravitationally unstable. Almost all components will be ejected and will ultimately escape, leaving one system behind that is likely to be a *hierarchical* multiple; but several of the escapers may themselves be binaries or hierarchical multiples. Such breakup is a process of which T Tau above seems to be an actual example. The timescale for the cluster to break up is roughly the dynamical crossing time of the system, i.e. $\sim 2\pi \sqrt{a^3/GM}$ where a is the linear size of the cluster and M the total mass. The timescale is typically $\lesssim 1$ megayear, even for quite wide multiples.

The Trapezium Cluster is about a dozen OB stars, and several hundred lesser members. Specifically (Preibisch *et al.* 2001), there are 13 OB stars, which between them have at least 14 companions closer than $\sim 1''$. Four of them have spectroscopic orbits, with periods in the range 6.5–65 days. The principal members are $\sim$10–100$''$ apart. BM Ori is in fact in a non-hierarchical *sub-multiple* of the Trapezium: five components are grouped in two binaries and a single, and these three sub-sub-systems are all $\sim$0.5–1$''$ apart. Probably most of the star formation in the Trapezium has taken place in the last $\sim$1 megayear (Herbig and Terndrup 1986), but I shall argue (Section 5.4) that some distant, high-velocity stars escaped from it 2.5 megayears ago, and of these one is about 10 megayears old.

EK Cep is a rather less surprising system than BM Ori. Here, the main evidence that $*2$ at least is pre-main-sequence comes from the fact that it is significantly larger than a ZAMS star of its mass.

The remaining systems in the table appear to be on the main sequence and in some cases significantly evolved across the MS band. XY UMa is a low-mass, very close, system in which the stars are very active presumably because of their rapid rotation – Section 2.2.5 (n). V624 Her and RR Lyn contain Am stars – Section 2.2.5 (g) – and δ Cap is an interesting combination of activity in the cooler component and metallic lines in the hotter component. EN Lac is a combination of a β Cep pulsator – Section 2.2.5 (d) – and a *much* less massive companion. Among close binaries with components that have not yet interacted, EN Lac appears to have the largest mass ratio known so far.

In the range $1 - 4\,M_\odot$ the TMS, or equivalently the blue edge of the Hertzsprung gap, is not clearly defined. The evolutionary tracks make a 'hook' at the TMS in the HR diagram (Fig. 2.1), during which evolution is rather rapid as central hydrogen is exhausted; but stars can spend a significant fraction of their main sequence life in a shell-burning phase which is somewhat beyond the 'hook', while not far advanced into the Hertzsprung gap. Whether such stars are to be considered as late main-sequence stars or early Hertzsprung-gap stars is a semantic question. I prefer to include them as main sequence stars, particularly since in the range $\sim$2–1 $M_\odot$ the hook progressively disappears, and the distinction between stars before it and stars after it becomes progressively less relevant. However, the issue is further complicated by the fact that in the same mass range the Hertzsprung gap itself progressively disappears. By $\sim$1 $M_\odot$, a star evolves on something like a nuclear timescale from a dwarf to a sub-giant to a giant, with only a mild temporary acceleration in the sub-giant region. As indicated in Section 2.2.7, I will take the term 'main sequence' to include any star in a long-lived hydrogen-burning state, but exclude those with deep convective envelopes, i.e. stars on the first giant branch (GB). The triangular region between the main sequence thus defined and the giant branch is then to be defined as the HG. But it is still inevitable that the MS/HG/GB boundaries are rather indeterminate, especially around $\sim$1–1.5 $M_\odot$ (and still more especially if *1 has lost a considerable amount of mass through Roche-lobe overflow, Section 3.3).

Andersen (1991) has drawn attention to a handful of eclipsing binaries at spectral type $\sim$A, which appear to be well above the main sequence (Fig. 2.1). SZ Cen is an example, where *1 is $\sim$3 times as large as a ZAMS star; *2, of almost the same mass, is $\sim$2.5 times as large. The mere fact that there are $\sim$4 such systems, out of 45 well-determined systems tabulated by Andersen (1991), suggests that they are in a relatively slow stage of evolution, rather than in the Hertzsprung gap. If classical stellar models are to be believed, then it is difficult to account for these stars even as post-hook, but still slowly evolving, objects, let alone as pre-hook stars. Andersen has therefore suggested that they support the existence of convective overshooting (Maeder 1975, 1976; Section 2.2.4). The theoretical models shown in Fig. 2.1 contain a degree of convective overshooting.

Pols *et al.* (1997) compared theoretical models with 49 ESB2 systems having relatively well-determined data, and found better agreement with models incorporating overshooting than with 'classical' models. Twelve systems out of 49 were in rather poor ($\chi^2 > 5$) or very poor ($\chi^2 > 8$) agreement with either kind of model; V624 Her is an example of very poor agreement, with *2 too large by about 8%. On the other hand 30 systems gave $\chi^2 < 2$, for the overshooting models. The six systems of lowest mass ($M_1 \leq 1.24\,M_\odot$) were all problematic in some regard, usually because *2 was larger than the models allowed. Possibly this is because they have not fully contracted to the main sequence; several are active BY Dra-like objects such as XY UMa. Three examples of good agreement, DI Her, SZ Cen and GG Lup are in Table 2.1. The last two are relatively stringent tests of models, because either the components are quite evolved (SZ Cen) or the mass ratio is quite severe (GG Lup, $q \sim 1.6$). For the many systems in which both components are quite closely equal in radius and temperature as well as mass, the constraint on modelling is not severe: one can usually adjust the two unknowns, metallicity and age, to give the two data, radius and temperature, at a given mass. But if the masses are very different one has to fit four data values with the same two unknowns. A high degree of evolution can mean that even a small difference in mass can make a substantial difference in radius and temperature, and thus we again have to fit four data with two unknowns.

It is not quite clear what conclusion to draw from the balance of agreement and disagreement; perhaps other parameters than age, mass and metallicity can influence the structure of some stars, but not all, at the $\sim$5–10% level. It is encouraging for a theorist to note that over the last 30 years agreement has become substantially better, without the *theoretical* models changing by as much as the observational data.

SS Lac is an interesting system that used to eclipse until about 1950, but then stopped. This is well explained by the recent discovery of a third body, which must be in an inclined orbit and thus causes the close pair's orbital axis to precess (Section 4.2).

η Ori is an interesting multiple system with at least five and possibly six components:

η Ori: $((((B1 + B2e; 8\,days) + B; 0.05'', 9.5\,years, e = 0.4:) + B; 1.65'') + faint; 115'')$

The innermost pair ($*111$ in the notation of Section 1.4) is an ESB2 system with rather good data leading to the masses and radii in the table; $*11$ has been resolved by speckle interferometry, and is also an SB1 by virtue of the motion of the CG of $*111$. The very similar luminosities for all three components of $*11$ suggest similar masses. The fourth B star ($*12$), similar to the previous three, can be expected to orbit in $\sim$1000 years. One of the four B stars, but *not* one of the eclipsing pair (Waelkens and Lampens 1988), varies photometrically with an amplitude of 0.3 magnitudes and a period of 0.43 or 0.86 days: this may mean that $*12$ or $*112$ is a close binary, probably with an inclination a good deal less than 90°; but several other interpretations (pulsation, rotational modulation) are possible.

Star 1112 (B2e) is a rapid rotator, from its rotationally broadened lines, with period $\sim$2 days; hence it is supersynchronous by a factor of four. It is also a non-radial pulsator – Section 2.2.5 (d) – with line-profile variations on a period of 0.133 days. By contrast, $*1111$ is *sub-synchronous* by a factor of two, and shows no detectable pulsation. It is puzzling that two such similar stars have such different rotational and pulsational properties. This may may be linked to the existence of the relatively close third body ($*112$). A somewhat similar phenomenon – very different rotation rates within the same close pair – is in fact also seen in TY CrA.

2.3   Beyond the main sequence

Once a star of mass $M \gtrsim 0.8\,M_\odot$ reaches the terminal main sequence (Fig. 2.1, Table 3.2), the mass M_c of the hydrogen-exhausted helium core is roughly

$$M_c \simeq \frac{0.11M^{1.2} + 0.0022M^2 + 9.6 \times 10^{-5}M^4}{1 + 0.0018M^2 + 1.75 \times 10^{-4}M^3} \quad (M \lesssim 100\,M_\odot), \tag{2.42}$$

in models with convective overshooting as given by Eq. (2.39). Some details of evolved stars, including M_c at various evolutionary stages, are given in the next chapter (Table 3.2), in the context of the influence of a binary companion.

2.3.1   The Hertzsprung gap and Hayashi track

Beyond the terminal main sequence, computed models show a progressive contraction of the helium core, along with an expansion of the envelope. The contraction of the core once it has burnt up all its hydrogen is fairly easy to explain qualitatively. Nuclear burning in the deep interior demands a substantial temperature gradient over the whole star, so that the heat released can flow outwards, and this temperature gradient contributes strongly to the pressure gradient which supports the star against its own gravity. Removing the energy source therefore weakens the pressure gradient and allows gravitational contraction to gain the upper hand – though only by a narrow margin, since the contraction of the core will itself release (gravitational) energy, which largely but temporarily compensates for the loss of nuclear energy. The contraction is liable to be rapid at first, taking place on the thermal or Kelvin–Helmholtz timescale of the core, which is related to the main-sequence lifetime t_{MS} of Eq. (2.4) by

$$t_{KH} \sim \frac{GM_c^2/R_c}{L} \sim 10^{-3} \frac{E_{nuc}M_c}{L} \sim 10^{-3}t_{MS}. \qquad (2.43)$$

The factor 10^{-3} is roughly the ratio of thermal or gravitational energy to nuclear energy. Thermal and gravitational energy per unit mass are usually comparable, thanks to hydrostatic equilibrium, and have values $\Re T_c/\mu \sim GM_c/R_c \sim 2 \times 10^{11}$ J/kg. The nuclear energy E_{nuc} from hydrogen as it burns to helium is $E_{nuc} \sim 0.007Xc^2 \sim 4.5 \times 10^{14}$ J/kg.

The rapid contraction of the core can be either slowed down, if the core density increases sufficiently for electron degeneracy to become important (low-mass stars, Section 2.3.2), or even reversed temporarily, though only by a small amount, if the core temperature rises sufficiently to ignite helium *before* the core density has increased to degeneracy (intermediate and high mass stars, Sections 2.3.3, 2.3.4). Figure 2.9a illustrates the way in which density and temperature at the centre vary with evolution, for a range of stellar masses.

The considerable expansion of the outer envelope, which takes place at the same time as the contraction of the core, is much less easy to account for in back-of-the-envelope terms, despite the fact that it is a near-universal feature of computed stellar evolution. It is easier, in fact, to say what it is *not* due to (Eggleton and Faulkner 1981): for example, it is not due to:

(a)   the onset of degeneracy in the core
(b)   the onset of convection in the envelope
(c)   the release of gravitational energy by the contracting core
(d)   a thermal instability of the envelope
(e)   the fact that the polytropic index n tends to have the value $n \sim 3$ in the radiative part of the envelope.

Nor is large expansion an *inevitable* concomitant of core contraction: in helium 'MS' stars (Section 2.5) of $\lesssim 0.7\,M_\odot$ the core also contracts by a substantial factor, but the envelope only expands by a factor of two or three.

In essence, there are two main reasons for the envelope's considerable expansion. Firstly, a gradient in mean molecular weight μ develops as the hydrogen burns: μ rises by about a factor of two in a zone between the wholly unburnt outer envelope and the burnt-out core. Secondly the nuclear burning zone, previously at the centre of the star, shifts outwards to the base of the envelope where the fuel is not exhausted, allowing a nearly isothermal and (in some circumstances) non-degenerate layer to develop at the outer edge of the core. Neither reason is an *obvious* reason, however, but some insight can be gained by the following argument.

What mainly distinguishes an evolved giant structure from a dwarf-like main-sequence structure is the degree of 'central condensation' C, defined as

$$C \equiv \frac{\rho_c}{\langle\rho\rangle} \equiv \frac{4\pi R^3 \rho_c}{3M}, \qquad (2.44)$$

where ρ_c is the central density, and $\langle\rho\rangle$ is the mean density derived from the surface radius and mass. Typically $C \sim 50$ for ZAMS dwarfs, whereas $C \sim 10^8$–10^{15} for red giants and super-giants. For simple polytropes C is listed along with other polytropic constants in Table 3.4. Eggleton and Cannon (1991) have proved the following result (see Appendix A):

(a) Define $n(r)$, the local polytropic index, and $s(r)$ the local 'softness' index, as before (Section 2.2.2) by

$$s \equiv \frac{n}{n+1} \equiv \frac{\partial \log \rho / \partial r}{\partial \log p / \partial r}. \tag{2.45}$$

Although the polytropic index n is more familiar, the softness index s turns out to be more significant. Further define $C_{poly}(s)$ as the readily calculable central condensation of a *complete* polytrope of softness index s: a complete polytrope is an entity where n and s are constant throughout the star. A reasonable approximation to numerical values is

$$C_{poly}(s) \sim \frac{0.025}{(5/6 - s)^3} + \frac{0.86}{5/6 - s}, \quad \text{for } 0 \leq s < 5/6. \tag{2.46}$$

Only the first term, which dominates as $s \to 5/6$, is important for the present discussion. The singularity at $s = 5/6 \, (n = 5)$ is due to the fact that the $n = 5$ polytrope or 'Plummer sphere' has a finite mass but infinite radius.

(b) Then if s_{max} is the largest value of s within a star, and if $s_{max} < 5/6$, we can *prove* that the star must be less centrally condensed than a polytrope of constant softness s_{max}:

$$C[s(r)] \leq C_{poly}(s_{max}), \quad \text{if } 0 < s(r) \leq s_{max} < 5/6 \text{ for all } r. \tag{2.47}$$

The proof is given in Appendix A.

For ZAMS stars, with $n \sim 3 \, (s \sim 0.75)$, this is consistent with $C \sim 50$. But for a structure with (say) $C \sim 7 \times 10^{14}$, we *must* have $s \gtrsim 0.83333 \, (n \gtrsim 4.99988)$ somewhere in the interior. Such a high value can usually only be obtained, and then only in a fairly narrow region, as a result of either (a) a μ-gradient, or (b) a nearly isothermal, non-degenerate ($n \sim \infty$, $s \sim 1$) zone immediately below the nuclear-burning shell (Fig. 2.8), as indicated above. Only processes that contribute to a 'softening' of the effective equation of state, i.e. to increasing n or s beyond the typical $n \sim 3$, $s \sim 0.75$ of a ZAMS star, push a star towards a very centrally condensed structure, and then only if they push n very close to or preferably beyond the value $n = 5$, over a sufficiently significant part of the star.

The fact that a μ-gradient 'softens' the star is not obvious at first, but comes from the specific definition (2.45) of softness. 'Softness' does *not* have the intuitive meaning that material is 'soft' if pressure is only weakly dependent on density *at constant temperature*, or *at constant entropy*. It means that as the density increases going inwards, the pressure increases slowly rather than rapidly. To illustrate this, suppose that the molecular weight increases inwards at exactly the same rate (logarithmically) as the temperature, over a region where the temperature increases by say a factor of two. Then μ / T is independent of r locally, and so, in a perfect gas where $P \propto \rho T / \mu$, we have just $P \propto \rho$. From this s is unity (and n infinite). Softness in our context is not an inherent property of material, but is a quantity that is only known a posteriori in a model that has been constructed in hydrostatic equilibrium. Nevertheless we can make some predictions about its behaviour without actually solving a model: the statement that a molecular weight gradient in some region softens that region is such a prediction: see Eq. (2.48) below.

I prefer to talk in terms of s rather than n because it is quite possible to have a region where $s > 1$, a very soft region, but this actually implies a *negative* n. A discontinuity in molecular weight (decreasing outwards) counts as 'infinitely soft', but corresponds to $n = -1$. Although loosely the statements $n > 5$ and $s > 5/6$ appear to be equivalent, actually they are not.

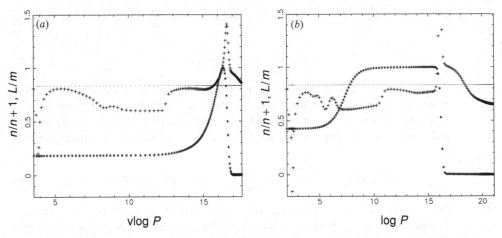

Figure 2.8 The behaviour of $L(r)/m(r)$ in arbitrary units (asterisks) and the softness index $s \equiv n/n + 1$ (pluses) in a star of 1 $M_\odot$, at the beginning of the giant branch (a) where the degree of central condensation $C \sim 10^5$, and the top of the giant branch, i.e. at degenerate helium ignition (b) where $C \sim 10^{13}$. The fact that both models have condensed cores and extended envelopes is partly explained by the fact that $s \gtrsim 5/6$ in regions near the burning shell, where L/m drops and μ rises abruptly. The value $s = 5/6$ ($n = 5$) is shown by the dotted line. Compare with Fig. 2.3 for two ZAMS stars.

Figures 2.3 and 2.8 show the very different behaviour of $s = n/n + 1$ in a main sequence star and in red giant. In the former, s never gets substantially above the value 5/6, shown by a dotted line; but in the latter there is a region well above this value at the hydrogen/helium boundary ($\log p \sim 16 - 17$). The sharpest part of this spike is due to the molecular weight gradient, but there is a substantial shoulder to the right which is in the isothermal but non-degenerate outer layers of the helium core. This shoulder is due to the fact that L/m has become small, so that the region is nearly isothermal while also non-degenerate ($s \sim 1$). But deeper still the core, though still nearly isothermal, becomes degenerate ($s = 3/5$).

In Section 2.2.3 we saw how Eq. (2.23) can be used to infer that (a) going inwards from the surface, ∇ tends rapidly to a constant, ∇_0, that depends only on the temperature and pressure dependence of the opacity, and (b) as we approach the central burning core ∇ rises further, possibly to the convective value ∇_a, because of an increase in L/m there. Approaching a burning *shell*, on the other hand, (a) remains true, but at the shell L/m *decreases*, very sharply and indeed more or less like a step-function. Thus $\nabla_{L/m}$ is like a negative delta function, and the effect on ∇ is to decrease it, rather than increase it.

For a perfect gas, but allowing the molecular weight to be a function of position,

$$s \equiv \frac{n}{n+1} = 1 - \nabla + \nabla_\mu, \qquad (2.48)$$

where ∇_μ is the logarithmic gradient of molecular weight relative to pressure. In the shell, ∇ drops and at the same time ∇_μ rises, so that both effects increase the 'softness', quite possibly to well above unity (Fig. 2.8).

Generally, to understand why a particular model is very centrally condensed, we need only locate the region or regions where n (or more precisely s) is largest. This largest value of s *must* be close to or above 5, or the model could not in fact be very centrally condensed; and then whatever causes that large s can reasonably be said to be the 'cause' of the giant-like structure.

A piece of pedantry that we reluctantly bring forward is that we must not assume that all red giants are very centrally condensed. A pre-main-sequence Hayashi-track star can be a red giant, but although its mean density is low so is its central density. It is in fact mostly an $n = 3/2$ polytrope. We often use the terms 'giant-like' and 'centrally condensed' as if they are equivalent, but actually they are not.

Although we can prove that if n is never greater than, say, four the star cannot be more centrally condensed than a polytrope of index $n = 4$; unfortunately we cannot make a converse claim that, for instance, if $n > 5$ somewhere the star *must* be very centrally condensed. A *small* region in which $n > 5$ by a *small* amount will not necessarily produce a large degree of central condensation. For a star with $C \sim 7 \times 10^{14}$ the condition that s exceeds 0.83333 somewhere is *necessary*, but not *sufficient*. It is not difficult to construct stellar models which have $s \gtrsim 1$ somewhere but are not especially centrally condensed (though they are usually more centrally condensed than main-sequence stars). These do not violate the theorem, but they limit its usefulness. For example, the two cases in Fig. 2.8, though both strongly centrally condensed compared with main-sequence stars, have widely different degrees of central condensation despite apparently similar distributions of s at least so far as the regions with $s \geq 5/6$ are concerned. Empirically, it seems as though a peak has less effect when near either the centre or the surface than when somewhere near the middle (in terms of mass): perhaps there is a further theorem to be developed here.

As the star crosses the Hertzsprung gap (Fig. 2.1), the surface temperature drops. Once it drops significantly below about 10 kK, the corresponding drop in opacity (Fig. 2.4) as the free electrons recombine causes the surface to become convective, just as for main-sequence stars. According to the analysis of Hayashi *et al.* (1962), the convective envelope deepens until the atmospheric structure converges to a nearly unique radius for a given luminosity and mass (whereas if the envelope remained radiative there might be a wide range of possible radii). For different luminosities and a given mass, the star must lie on a locus in the HRD, the Hayashi track, and this is found to agree very well with the observation that cool giants lie on a well-defined track, the giant branch (GB, Figs. 2.1, 2.10). Although the theoretical location of the GB is uncertain because it depends on the convection theory, and on poorly-known low-temperature opacities and bolometric corrections, the existence, observationally, of a well-defined giant branch confirms the general concept of deep convective envelopes.

For massive stars the expansion from the terminal main sequence to the Hayashi track is by a factor of $\gtrsim 100$ in the radius, but for low-mass stars it may only be a factor of less than two. The Hayashi track is the locus for stars that are cool enough to have deep convective envelopes; properly, they should be fully convective to the centre, but in practice the track is little different provided the envelope is deep. The radius on the Hayashi track can be approximated, in the same spirit as Eqs (2.1)–(2.5), by

$$R_{HT} = (1.65L^{0.47} + 0.17L^{0.8})M^{-0.31}, \qquad (2.49)$$

with R_{HT}, L and M still in Solar units. This formula is a reasonably good empirical fit to computed giants, but agrees reasonably well also with low-mass dwarfs – Eqs (2.1) and (2.2) – for $M \lesssim 0.5\,M_\odot$; these are also largely or wholly convective. The luminosity L of a star when it first reaches the Hayashi limit is only perhaps a factor of about 1–3 above what it was on the ZAMS (Eq. 2.1), although in subsequent evolution it may increase very considerably.

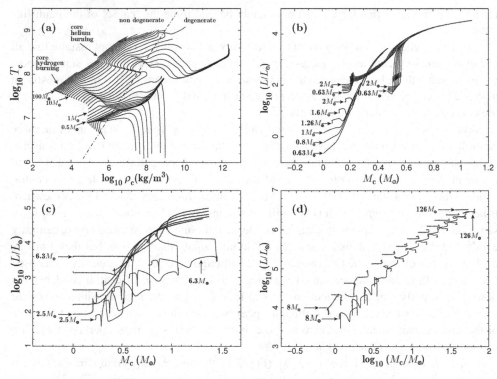

Figure 2.9 (a) The density–temperature plane, showing the evolution of central conditions in stars of various masses, with steps of 0.1 in log M. Stars of mass 0.5–2 $M_\odot$ converge to a locus for He white-dwarf cores: evolution was terminated at the He flash. In the mass range 2.5–6.3 $M_\odot$ they converge to a C/O white-dwarf locus: mass loss was ignored, and evolution was terminated at degenerate C ignition. More massive stars ultimately ignite C in (nearly) non-degenerate conditions. (b)–(d) Luminosity as a function of mass M_c at the boundary of the helium core (continuous line) and of the C/O core (dotted line), for (b) low-mass, (c) intermediate-mass and (d) high-mass stars. In (b), the steeply sloping asymptote to which the cores converge for $M_c \lesssim 0.45\,M_\odot$, if joined smoothly to the shallower asymptote for $M_c \gtrsim 0.7\,M_\odot$, is well approximated by Eq. (2.50). In (c) and (d), for some masses the He core may shrink temporarily, because surface convection eats into it. No mass loss was included in any of these models.

The surface temperature may still be estimated from Eq. (2.3). In Fig. 2.1 the locus where stars of various initial masses first develop substantial convective envelopes is marked BGB (beginning of giant branch). The position of this locus depends on the detailed approximation to turbulent convection that is used, but mixing-length theory with $\alpha \sim 2$ seems adequate at present.

2.3.2 Low-mass stars ($M \lesssim 2\,M_\odot$)

For stars with $M \lesssim 2\,M_\odot$ electron degeneracy becomes important in the helium core at an early stage, soon after the hydrogen is exhausted (Fig. 2.9a). It stabilises the core, preventing any rapid further contraction. This happens, somewhat fortuitously, at about the same time that the Hayashi limit starts to prevent substantial further expansion of the envelope. But the star still evolves by *slow* core contraction, and envelope expansion, on a nuclear rather than a thermal timescale. Since the Hayashi limit approaches close to the main sequence at about 1 $M_\odot$, the 'Hertzsprung gap' between the terminal main sequence and the Hayashi limit is much less marked than at high masses. It is almost non-existent at and below about 1.2 $M_\odot$.

Once degeneracy sets in, the star's core is virtually a white dwarf, with an 'initial' mass at the base of the giant branch of 0.12–0.2 $M_\odot$, Eq. (2.42). It is fed by hydrogen burning which continues at the core's surface. As the core mass M_c grows, the burning shell gets thinner but also hotter, allowing the nuclear luminosity to increase quite strongly, though on a slow (nuclear) timescale. Conditions in the burning shell are dictated almost entirely by M_c. The total mass M is barely relevant to the shell, although it does of course affect the outer radius of the star: see Eq. (2.49). The luminosity L of the burning shell, and the time scale t_{RG} of red giant evolution, defined by $t_{RG} \equiv M_c / \dot{M}_c$ are roughly (Fig. 2.9b)

$$\frac{dM_c}{dt} = \frac{M_c}{t_{RG}} \simeq 10^{-5} L \simeq \frac{2.2 M_c^6}{1 + 6.7 M_c^5}, \tag{2.50}$$

with L, M in Solar units and t in megayears as usual. This formula is crudely justified below.

The core radius R_c is not very different from the radius of a cold white dwarf (Chandrasekhar 1931), which is well approximated (Nauenberg 1972) by

$$R_c = R_{ch} \left[\left(\frac{M_{ch}}{M_c} \right)^{2/3} - \left(\frac{M_c}{M_{ch}} \right)^{2/3} \right]^{1/2},$$

$$R_{ch} = 0.0114 \left(\frac{2}{\mu_e} \right) R_\odot,$$

$$M_{ch} = 1.457 \left(\frac{2}{\mu_e} \right)^2 M_\odot, \tag{2.51}$$

also in Solar units; $\mu_e^{-1} \equiv \langle Z/A \rangle$ is the mean molecular weight per free electron. Equation (2.51) is the same as Eqs (2.40) and (2.41) for black dwarfs, with $M \gg M_{pl}$. Clearly as $M \to M_{ch}$ (the Chandrasekhar limit) the radius goes to zero; hence the core's mass cannot grow without limit. The Chandrasekhar mass, like the Eddington mass (Eq. 2.11), depends only on fundamental constants. Because both masses are determined by special-relativistic effects, related to photons and to electrons respectively, both involve $n = 3$ polytropes, and so their ratio is just $32\sqrt{15}/\pi^2$. That this factor is not unity reflects the fact that photons are bosons, while electrons are fermions. Most white dwarfs are expected to be made of He, C, O, Ne, Mg or Si, for which $\mu_e = 2$, but cores of ^{56}Fe, which can be expected to develop in a late evolutionary stage in massive stars (see below), will have μ_e larger and so a significantly smaller limiting mass (1.256 $M_\odot$). In practice the Chandrasekhar limit is about 4% smaller than the value above because of a number of physical effects such as inverse β-decay and general relativity. The physics of white dwarfs has been thoroughly reviewed by Koester and Chanmugan (1990).

The relation (2.51) is a good empirical approximation to detailed models; but it is also obtainable by a surprisingly elementary argument. The Fermi–Dirac integrals which give the relation between pressure and density for a degenerate gas (Chandrasekhar 1939) can be approximated as

$$\rho = \rho_0 x^3, \quad p = p_0 \int_0^x \frac{y^4}{\sqrt{1+y^2}} \, dy \simeq p_0 \frac{x^5}{(25 + 16x^2)^{1/2}}, \tag{2.52}$$

with

$$\rho_0 \equiv \frac{8\pi}{3} \frac{\mu_e m_U}{\lambda_c^3}, \quad p_0 \equiv \frac{8\pi}{3} \frac{m_e c^2}{\lambda_c^3}, \quad \lambda_c \equiv \frac{m_e c}{h}. \tag{2.53}$$

The constants p_0, ρ_0 are determined by atomic physics: ρ_0 is the density at which the exclusion principle for electrons demands relativistic momenta (about 3×10^9 kg/m^3). The quantity λ_c is the Compton wavelength. Let us approximate hydrostatic equilibrium crudely by

$$\frac{GM\rho}{R^2} \sim \frac{Gm\rho}{r^2} = -\frac{dp}{dr} \sim \frac{p}{R}, \tag{2.54}$$

and estimate the mass by

$$M \simeq \frac{4\pi}{3} R^3 \rho. \tag{2.55}$$

Eliminating p, ρ and x between Eqs (2.52)–(2.55), we get an $R(M)$ relation of exactly the same functional form as Eq. (2.51), although the numerical constants are not quite correct.

We can set up a simple model of the evolution of a red giant interior comparable in plausibility to the main-sequence model of Section 2.2.2. As for the MS model, the justification is not that it can be demonstrated from first principles to be correct, but rather that it can be shown to agree reasonably well with computed models. The latter are found to contain five different zones: (1) a degenerate, possibly partly relativistic, nearly isothermal helium core; (2) a non-degenerate nearly isothermal helium shell; (3) a hydrogen-burning shell; (4) a radiative zone covering several pressure scale heights and (5) a convective zone reaching to the surface, and containing most of the mass that is not in the degenerate core. Zones (2) to (4) all contain rather little mass. Apart from the burning zone (3), the zones are all closely polytropic, with $n \sim \frac{3}{2}, \infty, 3$, and $\frac{3}{2}$, respectively. This analytical model is obtained by fitting an exactly $n = 3$ radiative zone (with pressure and opacity given by Eqs (2.8) and (2.9), and therefore with ζ, κ constant as for the MS model) directly on to a degenerate core, Eq. (2.51). All the nuclear energy is approximated as coming from the base of the radiative region, within about one pressure scale height of the core boundary. Further out, the luminosity and mass are almost constant for several pressure scale heights. The luminosity in this $n = 3$ zone is therefore given – cf. Eq. (2.12) for main-sequence stars – by

$$L = \frac{4\pi a c G M_c T^4}{3\kappa p} = \frac{4\pi c G M_c \zeta^2}{(\zeta \kappa_{Th} + \mu \kappa_{Kr})(1 + \zeta)}. \tag{2.56}$$

Hydrostatic equilibrium in this zone, where $m \sim M_c = $ constant and $p \propto T^4$, $\rho \propto T^3$, gives, to a similar level of approximation,

$$T = \frac{\mu G M_c}{4\Re r (1 + \zeta)} = \frac{T_{sh} R_c}{r}, \quad \rho = \frac{a\mu T^3}{3\Re \zeta} = \frac{\rho_{sh} R_c^3}{r^3}, \tag{2.57}$$

throughout the radiative zone, ρ_{sh} and T_{sh} being values at the shell, i.e. where $r = R_c$. We can now crudely integrate the nuclear energy generation rate $\epsilon = A\rho T^\eta$ over the radiative zone to get

$$L = \frac{4\pi\rho_{sh}R_c^3 \cdot A\rho_{sh}T_{sh}^{\eta}}{\eta + 3}. \tag{2.58}$$

Equations (2.51), (2.56) and (2.58), along with Eqs (2.57) evaluated at $r = R_c$, $T = T_{sh}$ for T_{sh} and ρ_{sh}, are now five equations from which the five unknowns R_c, T_{sh}, ρ_{sh}, ζ and L can be determined as functions of the independent variable M_c. The $L(M_c)$ relation in particular can be compared with the 'empirical' relation (2.50).

Realistic cores of red giants have somewhat larger radii than Eq. (2.51) gives, because of zone (2) above. This zone grows significantly as the degenerate core shrinks, keeping R_c nearly constant at

$$R_c \sim 0.03\, R_\odot. \tag{2.59}$$

Using this value instead of Eq. (2.51) improves the agreement between the analytical and numerical values of $L(M_c)$. Adopting $\eta \sim 13$, a somewhat lower value than is appropriate for the main sequence because shell-burning tends to take place at higher temperatures, we find that for successively lower regimes of M_c (below the Chandrasekhar limit) we get $L \propto M_c$, $L \propto M_c^6$ and $L \propto M_c^{8.5}$. In practice M_c is never small enough for the last approximation to be valid. The 'empirical' Eq. (2.50) agrees with the first two of these three power laws.

Equations (2.56) to (2.59) tell us that a burning shell should be located approximately on the following curve in the (ρ, T) plane:

$$\frac{\kappa\rho^2\epsilon}{T^4} = \frac{4ac}{3R_c^2}(\eta + 3) \simeq \text{constant.} \tag{2.60}$$

So the fact that R_c empirically is nearly constant means that burning shells should lie on a unique locus – cf. Eq. (2.17) for ZAMS stars where burning occurs at the *centre*.

Equation (2.50) shows that the star's luminosity should increase strongly with M_c, and hence with time. Since the envelope remains close to the Hayashi limit, the star ascends the 'giant branch' (GB) of the HR diagram (Figs. 2.1, 2.9b). Early in the star's climb up the GB, the convective envelope deepens to the point where it begins to entrain material that was partly burnt during MS evolution (Fig. 2.5a). This is the 'first dredge-up' phase, the first opportunity for material processed by nuclear reactions near the centre to be observed at the surface. In particular, ^{13}C and ^{14}N are enhanced, the latter at the expense of ^{12}C and ^{16}O. Once the burning shell moves out close to the base of the convective envelope, however, the deepening of the envelope is reversed. There is always a radiative buffer between the burning shell and the convective envelope, which may contain very little mass but covers several pressure scale heights. Consequently no further dredge-up should occur. There is, however, observational evidence to suggest that ^{14}N is progressively enhanced beyond this point (Sneden *et al.* 1991). This appears to mean that some slow or perhaps intermittent mixing can occur in the convectively stable zone, as is also suggested by the existence of OBN stars (Section 2.2.5). Possibly such mixing is driven by the interaction of rotation, especially differential rotation, and magnetic field. We would expect the core to spin up as it shrinks, and the envelope to spin down as it expands, but on the other hand we also expect even a rather weak internal magnetic field to preserve uniform rotation. The resulting redistribution of angular momentum perhaps can mean a certain degree of mixing across the stable radiative mantle separating the burning shell from the convective envelope. An alternative might be that thin H-burning shells could be subject to a mild form (Bolton and Eggleton 1973) of the shell flashes which occur in thin He-burning shells at a later stage of evolution – see below.

Although rotationally driven mixing seems an attractive possibility, we should note the argument of Spruit (1998) that it would take very little magnetic field to enforce uniform rotation. Evolution in the Sun, with the core contracting and the envelope expanding, has long been supposed to produce non-uniform rotation in the Solar core, and yet helioseismology (Fig. 2.7b), while showing non-uniform rotation in the convective envelope, shows remarkably little in the radiative core. This is plausibly due to magnetic field. Differential rotation in the core at the level of one part in a million would roughly double the magnetic field every million rotations, or in $\sim 10^5$ years. This is very short compared with the age of the Sun. Taking this concept to its logical conclusion, even red giants would fail to generate differential rotation by core contraction and envelope expansion.

Well before the helium white dwarf core can reach the Chandrasekhar limit, however, it reaches temperatures of about 100 MK (when $M_c \sim 0.47\,M_\odot$ and $L \sim 2500\,L_\odot$) at which helium can ignite. This ignition, in contrast to the more massive stars ($M \gtrsim 2\,M_\odot$, see below), is explosive (Mestel 1952), because the degeneracy of the electrons means that the pressure in the core is insensitive to temperature and therefore the temperature can run away. However, once the temperature has risen in the explosion by a factor of about three, the degeneracy begins to lift and the core 'flash' is brought under control; the core begins to expand because the pressure increases, and the nuclear energy release is channelled into expansion against gravity rather than further heating. The core then settles down in a steady helium-burning configuration, which is non-degenerate by a narrow margin. This configuration is rather independent of the star's initial mass (in the range $M \lesssim 2\,M_\odot$) because of the convergence, illustrated by Eq. (2.50) and Fig. 2.9a,b, of degenerate cores to a unique evolutionary track. During core helium burning, the star's luminosity is about $50\,L_\odot$, with about 20% coming from helium burning in the core and the rest from hydrogen burning in a shell surrounding the core. Helium burning in its initial phases produces mainly carbon, through the triple-α reaction, but at a late stage, when the abundance of helium is reduced to $\lesssim 15\%$ by mass, the reaction $^{12}C(\alpha, \gamma)^{16}O$ comes to dominate, and the abundance of ^{16}O can eventually exceed ^{12}C.

Provided that the star has not lost substantial amounts of mass during its first journey up the giant branch, the star remains close to the Hayashi limit after the He flash, with radius given by Eq. (2.49), or perhaps about 10–20% smaller. It will be little different in outward appearance from a star ascending the giant branch in the previous evolutionary phase, although there may be some subtle changes in surface composition. Some HRDs of Galactic clusters and the Solar neighbourhood show a 'clump' of stars at about the right level on the giant branch (Fig. 2.11), which can reasonably be identified with the core He-burning phase. The He fuel in the core lasts about 10^2 megayears, after which the core lapses back into degeneracy as a C/O white dwarf of about 0.3 $M_\odot$. He burning continues in a shell whose luminosity rises rapidly while the shell moves out, but then decreases again when it has nearly caught up with the H-burning shell; by this time both have moved out to about 0.55 $M_\odot$.

Many globular clusters contain a horizontal branch (HB) at a luminosity of about $50\,L_\odot$ (Fig. 2.10b). This must be the region occupied by core He-burning stars, but whereas theoretical models ought to be close to the GB – even for the lower-than-Solar metallicities that are typical of globular clusters – observation shows that the HB can extend a long way to the blue side of the GB. The best explanation of this is mass loss, as a result of stellar wind,

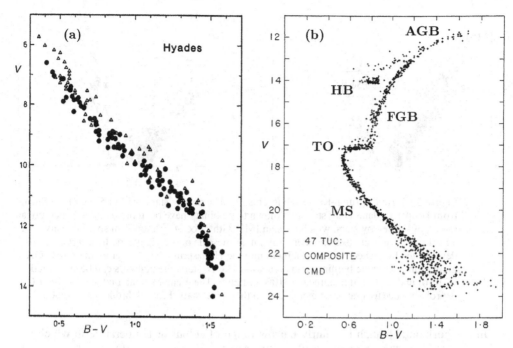

Figure 2.10 Hertzsprung–Russell diagrams. (a) Main sequence of the Hyades, a Galactic cluster: open triangles are known spectroscopic binaries, which can be up to twice as luminous at the same colour as single stars; filled circles are single stars, or at least SBs below the detection threshold, from Griffin *et al.* (1988). (b) The globular cluster 47 Tuc; from Hesser *et al.* (1987). The main sequence (MS), turn-off (TO) from the MS, first giant branch (FGB), horizontal branch (HB) and asymptotic giant branch (AGB) are indicated.

most probably on the GB prior to the helium flash (Faulkner 1966). The mass of envelope above the shell at this point, expected to be about 0.35–0.5 $M_\odot$ (depending mainly on initial metallicity) if there were no mass loss, has to be reduced to about 0.05–0.2 $M_\odot$, to explain the blue extent of some observed HBs. This is not unreasonable with empirical estimates of mass loss rates from red giants, as given in Section 2.4. Those HB stars which achieve the lowest masses of envelope and so populate the extreme horizontal branch (EHB stars) are probably little different from helium stars (Section 2.5), and will evolve fairly directly to white dwarfs without a further transition back to red giants. It is even possible for a star to lose its entire envelope on the first giant branch and yet manage to ignite He as an EHB star, provided that it is quite near to the He flash when all the envelope is lost (d'Cruz *et al.* 1996). But probably the great majority of HB stars, and certainly their 'clump giant' analogues in Galactic clusters and in the Solar neighbourhood (Fig. 2.11), should return to and re-ascend the GB once central He has been exhausted.

There are several sub-dwarf B (SDB) stars in the Solar neighbourhood rather than in clusters. These are small hot stars that are probably much the same as EHB stars. It has been shown recently (Maxted *et al.* 2001) that they are commonly, and arguably always, in binaries: and so it may be that the action of a binary companion, rather than just greater-than-average stellar wind, is responsible for removing almost the entire envelope.

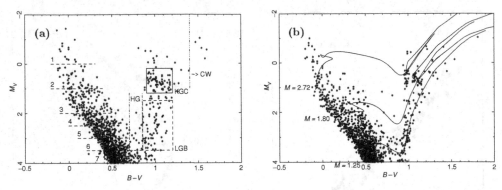

Figure 2.11 Hertzsprung–Russell diagrams. (a) The nearby stars (within 50 pc of the Sun), from Hipparcos data. Only stars not known to be binary have been plotted. (b) A theoretical model of the nearby stars, which uses an IMF with $N \propto M^{-2.8}$, uniform space density, and a birth-rate that decreases exponentially on a timescale of 6.3 gigayears. In several regions demarcated by dashed lines in (a), including the Hertzsprung gap, lower giant branch, GK giant clump (core He burning) and cool wind (AGB) stars, the numbers can be compared with (b) and are found to agree to ~10% or better. The greater spread on the observed LGB is probably due to a range of metallicity in the oldest stars. From Schröder *et al.* (2000).

On the horizontal branch is a fairly narrow range of colour or temperature in which the atmosphere is unstable to radial pulsations. The RR Lyr pulsating variables are found there, with pulsation periods of ~0.25–0.75 days. The pulsation is a relaxation cycle driven primarily by the second ionisation of He. The zone in the interior where this occurs is at about 10^5 K. In hot stars, this zone is too near the surface to contain sufficient mass to drive the pulsation. In cool stars, the atmosphere may already be convective down to this depth, and the convective instability appears to overwhelm the radial pulsational instability. Thus the instability strip has rather well-defined blue and red edges. Stars may cross this strip once (blue to red), or twice (red to blue, and later blue to red) during their evolution on the HB. One might hope to observe this evolution by determining rates of period change; as noted in Section 1.2 one does not need to wait for a substantial change before the rate of change is measurable. However, rates of change are found to be one or two orders of magnitude larger than expected, and of either sign, and may be due to some chaotic influence on the relaxation cycle rather than to underlying nuclear evolution.

At a late stage during the He depletion of a convective He-burning core in stars below about $4\,M_\odot$, an instability sometimes occurs in which the convective core abruptly grows larger, mixing in fresh He from further out. The convective core may almost double its mass briefly, before shrinking again, and may (in a 1 or 2 $M_\odot$ star) repeat this process (called 'breathing') two or three times. This is in marked contrast to the depletion of H in an H-burning core, which appears to proceed very steadily. However, whether 'breathing' occurs or not appears to depend on the details of the opacity and nuclear-reaction data used; the models used in this book did not usually display such behaviour, in contrast to models with the same code but using previous data. However some 'breathing' behaviour can be seen in the low-mass HeMS models of Fig. 2.15 (Section 2.4).

The usual outcome of core H burning, central H exhaustion, core He burning and central He exhaustion in low-mass stars is a degenerate C/O core with a mass of about 0.55 $M_\odot$, surrounded by a shell of much less mass containing mainly He and no H, and then an envelope of giant proportions (about 30–50 $R_\odot$) consisting of material that has been largely though not entirely unprocessed by nuclear reactions. The star contains two burning shells which, after some transient adjustment, begin to march outwards together through the star, increasing the core mass until it can, in principle, approach the Chandrasekhar mass (Fig. 2.8b). The evolution of the luminosity as a function of core mass returns fairly closely to Eq. (2.50), and the surface and core radii are again given by Eqs (2.49) and (2.51). The star re-ascends the Hayashi track, to greater luminosities than it reached on the first ascent (the FGB). This continuation is called the 'asymptotic giant branch' or AGB.

When the helium-burning shell has almost caught up with the hydrogen-burning shell, at $M_c \sim 0.6\,M_\odot$ (or perhaps earlier, see below), the inner of the two shells becomes thermally unstable (Schwarzschild and Härm 1965), and drives a series of 'He shell flashes', not to be confused with the He core flash above. A complex mixture of minor nuclear reactions can be set in train by the episodic mixing of material between the comparatively cool ($\sim 70\,$MK) base of the hydrogen envelope and the hot ($\sim 250\,$MK) outer edge of the C/O core, which can produce neutrons and so convert some of the normally inert Fe nuclei into traces of Zr, Tc, Ba and other heavy metals (the s-process). The main reactions are probably $^{12}C(p, \gamma\beta^+)^{13}C$ in the H burning region, and $^{13}C(\alpha, n)^{16}O$ in the He burning region. The details are not yet well understood, but it is difficult to avoid the conclusion that some such process must take place.

The He shell flash instability is driven mainly by the steep dependence of the He-burning reaction on temperature ($\epsilon \sim \rho^2 T^{50}$), which allows the temperature, and hence the nuclear luminosity, to run away temporarily. This initiates convection, which mixes outwards material with a high concentration of ^{12}C, the primary product of He burning. It is only at a late stage of He burning, or equivalently only at the bottom of the He-burning shell, when the He is already reduced to $\lesssim 15\%$ by mass, that the production of ^{16}O becomes dominant. To produce an enhancement of carbon and s-processed material as is observed (see below) the ^{12}C-rich layers have to mix with the base of the H-burning shell, where the H-abundance is low. This combination of high ^{12}C and low ^1H allows burning, which ensures that the major product is ^{13}C. Less ^{12}C and more ^1H would favour the production of ^{14}N, which not only reduces the amount of ^{13}C available for producing neutrons, but also introduces a 'neutron poison' since ^{14}N is particularly good at absorbing neutrons and so preventing them from interacting with Fe, etc. Thus rather delicate circumstances have to prevail, but apparently they do. Once the the ^{12}C material has been enriched by the H reaction to be ^{13}C rich, the convection due to the flash must carry the material back down for the further $^{13}C(\alpha, n)$ reaction, liberating one neutron for each H atom absorbed further up. Although Fe is not the most abundant species of potential neutron absorber, it has the highest value of cross-section times abundance, and so it, and the successively heavier elements it produces, tend to absorb most of the neutron flux. Provided there are more than about 80 neutrons produced per Fe atom, the Fe can be transformed to Ba and beyond, but with abundance peaks at Zr and Ba because these elements have 'magic' numbers of neutrons and hence local minima of neutron capture cross-section.

The convection in the outer envelope extends intermittently down into the region which is occupied by C-rich and s-processed material from a slightly earlier episode of shell flash and deeper mixing, and so can bring the heavy elements, and accompanying carbon, to the surface. Such elements are observed to be enhanced in a proportion of red supergiants (main sequence stars, S stars, C stars). Although C stars as a fraction of all giants amount to only about 1% in our own Galaxy, they are common in the Large Magellanic Cloud ($\sim 60\%$) and very common in the Small Magellanic Cloud (96%, Blanco and McCarthy 1981). Since the

LMC is deficient in metals, relative to our own Galaxy, by about a factor of two, and the SMC by about a factor of four, it is clear that the late stages of evolution of stars are very sensitive to the metallicity.

The difference between the various spectral classes M, MS, S and C is primarily due to the ratio of C to O (which is about $1/3$ in normal, i.e. Solar surface, material). This is because the CO molecule is especially tightly bound. If $C/O < 1$, *all* the C is locked up in CO and only O is left to form molecules, such as TiO which is the characteristic molecule producing bands in the spectrum of M stars. However, if $C/O > 1$, it is the O which is all locked up, and C is left to show such molecules as C_2, SiC, which are characteristic of C stars. MS and S stars are intermediate, but presumably still have $C/O \leq 1$.

Enhancement of C and of s-process elements (notably Ba) is also seen in a class of G/K giants, the 'Ba stars'. However Ba stars are usually of too low luminosity to be AGB stars, and hence cannot have produced their s-process enrichment themselves; they appear to be the product of binary interaction (Section 6.4), having accreted s-processed material from a companion that was once an AGB star and is now a nearly invisible WD.

If a star on the AGB evolves without mass loss, then either (a) when the core mass approaches the Chandrasekhar mass, we would expect the core to ignite C in a degenerate thermonuclear explosion, similar to, but much more dramatic than, the He core flash – a thermonuclear supernova explosion (SNEX) – or (b) for lower mass, the star would end up as a C/O WD once the last of its envelope was burnt. However, it is very likely that in fact the star loses substantial mass, which means that possibility (a) is excluded (for initial masses $\lesssim 2\,M_\odot$, as considered in this section).

At some point on the AGB, stars become unstable to radial pulsations – they become Mira variables. These pulsations appear to be driven mainly by hydrogen ionisation; unlike in RR Lyr pulsations, the fact that the envelope is already unstable to convection is apparently not a barrier to pulsational instability. Such stars are capable of driving copious low-velocity winds, which can manifest themselves as cool dust shells in the infrared. At a late stage the star may be obscured by the shell to such an extent that it is only visible in the IR. In this stage the object may be conspicuous as an $OH/H_2O/SiO$ maser source. In the next major stage the star is stripped down to an extremely hot UV-bright core, illuminating and heating the remnant shell so that the shell appears as a planetary nebula (PN). Such planetary nebulae often have a bipolar morphology, which suggests that the material ejected in the slow but copious AGB wind is concentrated to an equatorial ring or disc (perhaps as a result of having a binary companion), and then a very fast but relatively meagre wind from the very hot post-AGB remnant punches out more readily in the polar directions. In addition, many planetary nebulae show a series of rings, suggestive of episodic AGB mass loss. Finally, the core cools down to become a white dwarf, while the envelope dissipates itself.

In at least two stages during the final cooling-down process the star may be pulsationally unstable again, this time at short periods (~ 10 min) that allow for the possibility of astero-seismology. Early in the cooling ($T \sim 100\,$kK, $L \sim 10\,L_\odot$) there are the GW Vir variables. For example, BB Psc (Vauclair *et al.* 1995, O'Brien *et al.* 1998) shows nine modes ($l = 1$ g-modes) with $P \sim 336$–612 s. These are consistent with a fairly uniform spacing of $\Delta P \sim 21$ s (not all such modes are seen), leading to an estimate $M \sim 0.7\,M_\odot$, $L \sim 5\,L_\odot$. The driving mechanism of these modes is not yet understood. Much later in the cooling ($T \sim 13\,$kK, $L \sim 10^{-3}\,L_\odot$), WDs cross the high-gravity extension of the δ Cep/δ Sct instability strip and become ZZ Cet pulsators. ZZ Psc (G29-38; Kleinman *et al.* 1998) shows 20 $l = 1$ g-modes

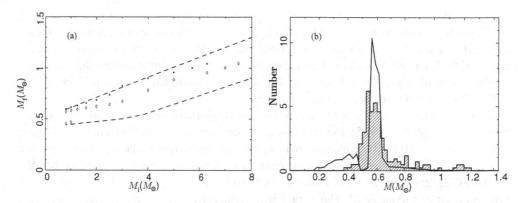

Figure 2.12 (a) Relations between initial (ZAMS) and final (WD) mass, from Eq. (2.62): diamonds – Pop I; filled squares – Pop II; dashed lines – upper and lower limits to the observational distribution of Weidemann and Koester (1983). (b) Convolving the results in Fig. 2.12a with the IMF of Eq. (1.17) gives a distribution of masses for planetary-nebula nuclei (thick curve) similar to the observed distribution of Zhang and Kwok (1993).

with $P \sim 110$–1240 s. These are consistent with a spacing of $\Delta P \sim 47$ s, which may imply $M \sim 0.5$–$0.7 \, M_\odot$. In both types of pulsators the modes depend fairly sensitively on the structure of the outer shell or shells of hydrogen or helium, but one can hope to determine this structure by sufficient astroseismological data. Long-term monitoring of both types may also determine rates of period change that should strongly constrain models of cooling in WDs.

The loss of mass that terminates a star's AGB evolution is probably a compound of two processes: (a) a fairly steady stellar wind, increasing in strength as the star grows in luminosity and radius (Section 2.4), both on the first GB and on the AGB; and (b) a more drastic 'superwind', in which the remaining envelope is driven away very rapidly, possibly quite early on the AGB. Weidemann and Koester (1983) plotted a semi-empirical relation between initial (ZAMS) mass M_i and final (WD) mass M_f, based largely on observations of WDs in a number of galactic clusters of reasonably well-known age. Weidemann and Koester's graphical relation (Fig. 2.12a) can be rendered approximately as

$$M_f \sim 0.4 + 0.05 M_i + 0.00015 M_i^4, \qquad 1 \lesssim M_i \lesssim 8. \tag{2.61}$$

The masses are in Solar units. There is considerable scatter in the semi-empirical data however, and an uncertainty of $\pm 0.2 \, M_\odot$ in M_f at given M_i would be fairly conservative. It is not impossible that some scatter is quite real: perhaps there is something chaotic about the mass-loss process, which makes initially similar stars lose their envelopes at significantly different points on the AGB.

One white dwarf (40 Eri B) appears to have well-determined parameters of $0.50 \pm 0.011 \, M_\odot$ and $0.0136 \pm 0.00024 \, R_\odot$ (Shipman *et al.* 1997), using the Hipparcos parallax and an orbit by Heintz (1974). These values fit well with Eq. (2.51), which applies equally for He and for C/O WDs. They also agree reasonably well with the gravitational redshift. Nevertheless, the mass is uncomfortable: evolutionary models suggest that there should be a gap in white-dwarf masses between about ~ 0.46 and about $\sim 0.55 \, M_\odot$, because stars leave the first GB at the former core mass, and return to much the same point in the HR diagram only when the core mass has increased to the latter value. However, the uncertainties quoted above for the parameters of 40 Eri B seem optimistic in view of the fact that the orbit of $\sim$250 years has only been observed for slightly more than half of one orbit. A more cautious view is that any orbit, spectroscopic or astrometric, should be seen round twice before accuracies of $\lesssim$1% can be claimed (with an exception only for pulsar orbits, where the accuracies can be enormously greater). The mass of 40 Eri B has crept up from 0.43 ± 0.02 in 1974 to 0.50 ± 0.01 in 1997, and may not yet have settled down.

White-dwarf masses of $\lesssim 0.4\,M_\odot$ are known, and are demonstrated to be products of binary evolution (Marsh *et al.* 1995). However, binary interaction fails to explain a mass in the gap, and the mass of 40 Eri B remains something of a problem. If we take it at face value we would say that the progenitor must have reduced its mass by stellar wind to about $0.5\,M_\odot$ on the FGB, ignited helium and settled as an EHB star, and then evolved to a white dwarf without returning to the AGB. This is probably at the extreme of single-star mass-loss rates (Section 2.4), but might happen occasionally.

Paczyński and Ziółkowski (1968) suggested that a star's progress up the AGB might terminate at the point where the binding energy E_B of the envelope (i.e. the integral of the sum of gravitational and thermal energy, the latter including ionisation) becomes negative. This happens because as the envelope expands the gravitational contribution becomes smaller, but the ionisation energy, with the opposite sign, remains substantial. We shall see (next section) that intermediate mass stars converge to much the same evolution from this point on, and so for the remainder of this section we consider stars in the wider mass range 1–8 $M_\odot$. The prescription of Paczyński and Ziółkowski gives an M_f, M_i relation in an implicit form as

$$E_B \equiv \int_{M_f}^{M_i} \left(\frac{Gm}{r} - U \right) dm = 0, \tag{2.62}$$

where we identify M_f with M_c, the mass of the degenerate core. This algorithm gives (Han *et al.* 1994) a slightly more linear relationship than Eq. (2.61), which can be approximated as follows:

$$M_f \sim \max(0.51 + 0.049M_i, \min(0.35 + 0.11M_i, 0.60 + 0.06M_i)),$$

$$0.8 \lesssim M_i \lesssim 7.5\,M_\odot. \tag{2.63}$$

Equation (2.62) also gives, for $M_i \lesssim 1\,M_\odot$, a solution on the first GB, with $M_f \sim 0.47\,M_\odot$. When convolved with a reasonable IMF (e.g. Eq. 1.17), Eq. (2.63) gives a distribution of PNN masses that is strongly peaked at about $0.60\,M_\odot$, roughly in accordance with the observational data of Zhang and Kwok (1993) – see Fig. 2.12b. Note that the models used by Han *et al.* (1994) did not include convective overshooting, and so differ from the models used here; but the tendency of giants to converge to a unique evolutionary locus – Fig. 2.9a,b – as the core evolves means that the difference should not be large.

There is no guarantee that mass loss is sufficiently deterministic that any narrow M_f/M_i relationship is to be expected. Especially if magnetic fields are involved for G/K/M giants, as they certainly are for the Sun, the process may well be chaotic, and one can imagine that chaotically different mass-loss rates might lead to both He and C/O WDs, and to a substantial range of masses, from initially very similar stars.

A substantial difficulty in using the recipe (2.62) is deciding where in a supergiant star is the boundary between 'core' and 'envelope', because the integral, viewed as a function of M_f, is quite sensitive to M_f once M_f approaches close to the degenerate core mass. Because the

Table 2.2. *Final mass and maximum luminosity and radius, in Solar units*

M_i	1	2	3	4	5	6	7
M_f	0.56	0.61	0.68	0.79	0.90	0.96	1.02
L_{max}	4915	7139	11020	17500	23600	26650	29500
R_{max}	290	365	476	628	747	799	843

issue of envelope binding energy is perceived as being particularly important in the evolution of binaries containing red supergiants, I discuss this in some detail in Section 5.2; I also return to mass loss in Section 2.4. Provisionally, let us define the base of the envelope, i.e. M_f in Eq. (2.62), as the place where the hydrogen abundance is 0.15.

From the point of view of binary-star evolution, it is particularly important to be able to estimate the maximum radius R_{max} that a star can achieve, which is the radius on the AGB just before the envelope is lost. For a given initial mass M_i, Eq. (2.63) gives M_f, then Eq. (2.50) gives the corresponding L_{max} (using $M_c = M_f$), and finally Eq. (2.49) gives the radius. We ought to make allowance for the fact that the maximum radius occurs shortly before the stellar mass is reduced to M_f, but in practice using Eq. (2.49) with $M = M_f$ appears to be good enough. We obtain the results in Table 2.2.

It is however something of a problem that, in stars with initial mass 1–2.5 $M_\odot$, the He-shell instability referred to earlier is only just beginning when the core mass reaches the value of about 0.6 $M_\odot$ at which envelope ejection apparently occurs. This means that it is difficult to see how the nuclear enrichment brought about by shell flashing is as common as it is. This may be an indication that shell flashing in reality starts at a somewhat earlier stage than current numerical modelling implies, so that a star can already have quite an enriched envelope when the C/O core mass exceeds, say, 0.5 $M_\odot$.

I have already referred (Section 2.2.9) to Spruit's (1998) suggestion that interior magnetic field, even if quite weak, is liable to keep stars in nearly uniform rotation, even red supergiants where the core has contracted by a factor of nearly 100 in radius while the envelope has expanded by an even larger factor. This has the apparent difficulty that it should lead exclusively to very slowly rotating white dwarfs; yet many isolated white dwarfs are rotating with periods of ∼1 day. This can be explained (Spruit 1998) as a result of the fact that the ejection of the envelope is not a perfectly spherically symmetrical process. The tiny moment of inertia of a white dwarf means that small random anisotropies in the superwind phase can generate a net rotation rate of the right order.

2.3.3   *Intermediate-mass stars* ($M \sim 2$–$8\,M_\odot$)

The expansion of a star as it crosses the Hertzsprung gap between the main sequence and the Hayashi limit is confined to the outer layers. The core contracts, in response to the partial loss of pressure gradient that was formerly provided, in the form of a temperature gradient, by central H burning. Contraction stops when either degeneracy removes the need for a temperature gradient, as in the previous subsection ($M \lesssim 2\,M_\odot$), or else ($M \gtrsim 2\,M_\odot$) when the hydrogen-exhausted helium core contracts and heats up from hydrogen-burning temperatures of 15–50 MK (depending on stellar mass) to temperatures of about 150 MK at which helium can itself burn. This initiates a period of steady core helium burning, which is not unlike the core helium-burning phase of the less massive stars except that the core can be *smaller* in mass, at least in the range 2–$3\ M_\odot$, since the core did not have to grow slowly on the FGB to attain a critical mass for the helium flash. Usually the envelope has reached the Hayashi track shortly before the core ignites helium, and so the star starts its core helium-burning as a red giant; but at some higher masses the helium may ignite while the star is still in the Hertzsprung gap.

For stars with masses in the range 5–$15\ M_\odot$ helium burning may temporarily drive the star back to substantially smaller radii than at the Hayashi limit, though still much larger than main sequence radii. This excursion or 'blue loop' towards the blue side of the Hertzsprung–Russell diagram, seen particularly in the $8\ M_\odot$ star of Fig. 2.1, takes the star on a fairly slow (i.e. nuclear) timescale through a narrow range of temperatures, the Cepheid strip (roughly 6.5–$7.5\ \text{kK}$), in which the surface layers are unstable to pulsation of the δ Cep type. The basic destabilising mechanism here, as for δ Sct stars (Section 2.2.4) and RR Lyr stars (Section 2.3.2), is the second ionisation of helium, which can lower the adiabatic exponent considerably in the outer envelope. The strip is terminated on the hot side by the fact that the second ionisation of helium occurs too near the surface to affect a sufficiently substantial mass of envelope material; and the cooler side is presumably bounded by the fact that turbulent convection becomes much more important than radiative heat transfer in the surface layers.

Once a significant amount of central helium is burnt the star returns from its blue loop in the Hertzsprung gap to the Hayashi limit. The duration of the blue-loop phase might be expected to be about 10% of t_MS (Eq. 2.4), because the nuclear binding energy of helium is about 10% of hydrogen, while the star maintains about the same luminosity as on the MS and burns most of the helium that was produced there. However, in a range of masses (~2–$6\ M_\odot$) the fraction can be as large as 25–40%, because in these stars the He luminosity is actually quite small compared with the luminosity from the H-burning shell, and remains so for most of the core-He-burning phase. For ~2–$5\ M_\odot$, there is hardly a blue loop, but central He burning is presumably responsible for the more luminous members of the K-giant clump (Fig. 2.11), the less luminous members having arrived there after the He flash (previous section).

The core contracts slowly during, and rapidly after, core helium burning. If the initial mass of the star is $\lesssim7\ M_\odot$ the C/O core has less than the Chandrasekhar mass, and will reach densities at which electrons become degenerate (Fig. 2.9a). Further evolution, with a growing degenerate core surrounded by two burning shells, and mass loss from the envelope, should be similar to the AGB stars of the previous section, leading also to a shrouded red supergiant, a planetary nebula, and then a white dwarf. If the C/O core does not become degenerate before it ignites (i.e. if the core has more than the Chandrasekhar mass at the time that it forms) then we consider the star to be 'massive': see next section.

Figure 2.9c shows the relations between luminosity and both the C/O and the He core masses. The former is much the same as for the low-mass regime, assuming as in Figure 2.9b that the low-mass stars do not lose mass. The latter is rather more complicated, because the He core mass can *decrease* temporarily, as the surface convection zone eats into it ('second dredge-up').

The most likely agent for terminating the double-shell-source evolution of intermediate-mass stars is thought to be mass loss, as for low-mass stars, so that the outcome is a C/O white dwarf rather than a thermonuclear supernova explosion. The analysis of Weidemann and Koester (1983), on which the approximation of Eq. (2.61) is based, extended to young clusters (such as the Pleiades, and NGC 2516) with turn-off masses $\gtrsim 5\ M_\odot$, in which white dwarfs have been found with masses approaching $1\ M_\odot$. The uncertainties are very large, however. It is commonly assumed that stars near the upper limit of 'intermediate mass', i.e. about $7\ M_\odot$, will produce white dwarfs near the Chandrasekhar limit, although this is not well supported either by observation or theory.

If mass loss does *not* terminate the evolution of intermediate-mass stars, by reducing the stellar mass below the Chandrasekhar mass, then the degenerate C/O core can in principle undergo a 'carbon flash' similar at least at its inception to the helium core flash of stars with initial mass below about $2\ M_\odot$, once the temperature or density gets high enough ($T_c \sim 1\ \mathrm{GK}$, or $\rho_c \sim 10^{13}\ \mathrm{kg/m}^3$). This requires that the C/O core mass approach fairly close to the Chandrasekhar limit. Possibly some stars less massive than about $7\ M_\odot$ undergo degenerate carbon ignition because they are unable to lose enough mass to prevent it. This ignition will certainly be much more violent than the helium core flash, since the core is much more degenerate: if it takes place, it is expected to be a supernova explosion (Höflich *et al.* 1998).

Supernovae within our own Galaxy are rare (6 in the last 1000 years, Hill 1993), but several tens per year are observed altogether in external galaxies. Broadly, they fall into two spectroscopic classes: Type I, which do not show hydrogen in their spectra, and Type II, which do. Both types are further sub-divided, mainly on the shapes of their light curves. Type Ia supernovae are considered good candidates for explosions driven by carbon detonation. However there are strong reasons for thinking that Type Ia SNe are products of binary evolution rather than single stars, and might for example be due to accretion of mass by a C/O white dwarf from a companion in a novalike binary, or to the merger at a late stage of binary evolution of two C/O white dwarfs of about $0.7\ M_\odot$ each. The reasons for rejecting a single-star origin are that (a) Type Ia supernovae contain no hydrogen in their spectra, by definition of Type I, but a single star of intermediate mass which retained its envelope long enough for the core to reach the Chandrasekhar limit would presumably still have some of its envelope left; and (b) Type Ia supernovae are found in elliptical galaxies and other environments where the stellar population is seen to consist predominantly of old low-mass stars rather than young stars of upper intermediate mass. Some binary scenarios can, in principle, lead to carbon detonation at great age, but not single-star scenarios.

The detonation of a C/O white dwarf, whatever its origin, should lead to the production of large amounts of Ni/Fe – perhaps $\gtrsim 50\%$ of the original mass. Relatively little energy would be released as neutrinos, and the whole white dwarf would be dispersed, leaving no compact (neutron star or black hole) remnant. These features are all in contrast to supernovae driven by core collapse (next section). Although early theoretical work on supernova explosions assumed spherical symmetry for simplicity, much recent work emphasises the role of instabilities, in particular the Rayleigh–Taylor instability due to an inverted molecular weight gradient, which can lead to complicated two-dimensional or three-dimensional behaviour (Falk and Arnett 1973, Nagataki *et al.* 1998).

2.3.4 High-mass stars ($M \gtrsim 8\ M_\odot$)

For stars more massive than about $8\ M_\odot$ an explosive fate is more certain, but is delayed by the fact that the C/O core does not become degenerate before carbon ignition. Thus the carbon can ignite reasonably quietly and then burn hydrostatically at a temperature of about 1 GK. Most of the energy from nuclear reactions at such high temperatures gets converted almost directly into neutrinos, via the weak interactions

$$\gamma + \gamma \leftrightarrow e^+ + e^- \to \nu + \bar{\nu} \quad \text{(pair production)} \tag{2.64}$$

and

$$p + e^- \to n + \nu, \quad n \to p + e^- + \bar{\nu} \quad \text{(Urca process)}. \tag{2.65}$$

The protons and neutrons participating in reaction (2.65) are in practice embedded in heavy nuclei, rather than free particles. Neutrino losses accelerate the evolution in a vicious spiral, so that in a few hundred years at most the C/O mixture burns to a mixture of O, Ne and Mg. In a narrow range of initial masses, perhaps about 6–8 $M_\odot$, it is possible that the core's evolution will terminate here, provided that stellar wind or a binary companion removes the remaining envelope in this short time. Such a core can in principle cool down and become an O/Ne/Mg white dwarf with a mass quite close to but below the Chandrasekhar limit (Nomoto 1984). There is observational evidence that some white dwarfs in classical nova eruptions are of such a character.

For greater initial masses it is difficult for the core to avoid going on to a further stage of nuclear burning (the α-process) in which (γ, α) and (α, γ) reactions come nearly into equilibrium, turning the lighter α-nuclei (O, Ne, Mg) into heavier and more tightly-bound α-nuclei (Si, S), and these in turn to 'iron peak' elements, principally Fe, Ni. These last elements, being more tightly bound than either lighter or heavier nuclei, cannot continue the chain of energy production. Core collapse must happen at this stage. However, it is not clear whether, and in what circumstances, we should expect a black hole or a neutron star remnant and an ejected supernova envelope. We do expect some compact remnant since, unlike in the case of degenerate carbon ignition (Section 2.3.3), there is not enough available nuclear energy in the core material to blow the core out of its much deeper gravitational potential well. The approach to a supernova explosion (SNEX) is reviewed, from theoretical and observational directions, by Mazurek and Wheeler (1980), and Woosley and Weaver (1995).

For some purposes later it will be convenient to restrict the term 'high mass' to those stars with $8 \lesssim M \lesssim 35 \, M_\odot$, and introduce a further category of 'very high mass' for still higher masses. Partly this is intended to distinguish between stars that produce neutron-star remnants and those that produce black-hole remnants; and partly to distinguish between those where mass loss by stellar wind is not very important until a late stage of evolution, and those where mass loss may be important earlier, even on the main sequence. Obviously these two boundaries need not coincide, but since both are very uncertain, and yet both probably seem to be of the same order, we will ignore this.

Figure 2.9d shows that there is a somewhat tighter correlation between L and M_c than for the intermediate-mass stars. However, this is mainly because neither the core mass nor the luminosity changes very much during evolution, and both are largely determined by the Eddington limit.

A major problem with our theoretical understanding of supernova explosions is to determine the mechanism whereby the outer envelope is ejected, while the core collapses. It is not difficult to see why the core should collapse once it has exhausted all its available nuclear energy. But it has been difficult, indeed impossible so far, to determine clearly the mechanism which will prevent much or all of the outer envelope from following the core into collapse.

A large amount of energy is available; the gravitational energy released by the collapsing core. But this energy comes out almost entirely in neutrinos, which interact rather weakly with matter farther out. It is not clear whether they can deposit enough energy sufficiently rapidly to turn the inflow round into an outflow. Nevertheless, observation makes clear that considerable amounts of matter are ejected at very high speed ($\sim 0.1c$) in supernova remnants. It may be that instabilities and asymmetries in the explosion are the key, and that fully 3D modelling of the process is necessary.

Timmes *et al.* (1996) estimated the remnant masses to be expected from ZAMS stars in the mass range 8–40 $M_\odot$. They evolved stars up to the point of core collapse, and then, by imposing a piston-like outward impulse to material just outside the iron core, the envelopes were exploded outwards. Not all of the envelope escaped, however: a proportion was slowed hydrodynamically, failed to reach escape velocity, and fell back into the core. With a ZAMS mass of 35 $M_\odot$, the amount falling back on to the $\sim 2\,M_\odot$ iron core ranged from almost zero to over 5 $M_\odot$ as the piston energy ranged over 1.2–2.2×10^{44} J. But for models at ZAMS masses 11–28 $M_\odot$, the lowest piston energy in this range was enough to eject most of the material except for part of the silicon shell immediately outside the iron core. Baryonic masses of the remnants ranged over 1.3–2 $M_\odot$. In the small but important range 8–11 $M_\odot$, cores, although they ignite carbon non-degenerately, are degenerate for later burning stages and tend to produce remnants fairly close to the Chandrasekhar limit, at baryonic masses of $\sim 1.39\,M_\odot$.

A baryonic mass has to be translated into a final gravitational mass, i.e. the mass that would be determined by observation of a body in orbit around it. The gravitational mass is smaller because the mass equivalent of the (negative) gravitational energy of the collapsed core has to be added. This gives approximately a quadratic relationship

$$M_g \sim M_b - \frac{GM_g^2}{2c^2 R} \sim M_b - 0.075 M_g^2, \tag{2.66}$$

in Solar units (Lattimer and Yahil 1989), assuming a reasonable average for R. The remnant baryonic mass was a very non-monotonic function of ZAMS mass because convective zones in the carbon-burning core, and later in the carbon-burning shell, would appear and disappear somewhat chaotically, and influence the remnant mass significantly. ZAMS masses above 19 $M_\odot$ were little influenced by carbon burning and tended to be higher.

By convolving the ZAMS masses with a Salpeter IMF, Eq. (1.8), Timmes *et al.* (1996) estimated that the distribution of gravitational masses would be bimodal, with peaks at 1.27 and 1.76 $M_\odot$. The latter peak comes from ZAMS masses $\gtrsim 19\,M_\odot$. Arguably the remnants in the higher peak may be black holes rather than neutron stars; certainly they would be if the equation of state of neutron-rich material sets an upper limit to neutron-star masses at, say, 1.7 $M_\odot$.

It is not yet clear whether it is simply the initial mass of the progenitor star, or some more complicated criterion, that determines whether the remnant of a supernova explosion is a neutron star or a black hole. Nor should one assume that the process is monotonic, with all black-hole progenitors initially more massive than all neutron-star progenitors. Both kinds of remnant are observed. Many neutron stars are known: about 100 of them are members of close binaries with normal stars and show up as X-ray pulsars, but many more (~ 1000) show up as radio pulsars, of which only a small proportion are in binaries. A few radio pulsars are in binaries where the companion is another neutron star, though presumably too old to be still pulsing, and have extremely well-determined orbits from which *both* masses can be determined (Section 1.2). All such yield masses in the range $1.39 \pm 0.06\,M_\odot$ (Brown *et al.* 1996). This is somewhat above the lower peak of Timmes *et al.* (1996).

The best way to distinguish observationally between a neutron star and a black hole is to determine its mass, as can sometimes be done from a binary orbit; although a mass function gives only a lower limit, several lower limits are already clearly in excess of plausible neutron-star masses. No radio pulsar has yet been found with a black-hole-mass companion, but several faint and presumably compact objects with low-mass stellar companions have very large orbital velocities that at least give a mass function. Bailyn *et al.* (1998) show that six out of seven black-hole candidates in low-mass X-ray binaries have mass functions which, with plausible inclinations and mass ratios, are consistent with a narrow range about $7\,M_\odot$; the seventh requires $\sim 11\,M_\odot$. Isolated black holes must surely exist, but will be very hard to detect.

Evidence coming from the consideration of X-ray binaries can be expected to cast light on the propositions that (a) neutron stars can be obtained from massive stars, say $\gtrsim 35\,M_\odot$, and (b) black holes can be obtained from lower-mass stars, say $\lesssim 20\,M_\odot$ (Ergma and van den Heuvel 1998). If this is true, it appears to imply that some property other than total mass is important: for example, magnetic field or rotation. However, we feel that the evidence is unclear, and for the present we will stick to the simple view that there is a unique critical mass, probably in the range 35–$40\,M_\odot$, below which neutron stars are formed and above which black holes are formed.

Not only is the nature of the ultimate type of remnant uncertain, but so is the prior evolutionary track taken in the HRD. This is because, although it is clear that mass loss by stellar wind (next section) is an important process, it is not by any means clear how this affects the evolutionary track. Empirically, there is evidence that stars of $\gtrsim 30\,M_\odot$ initially do not evolve into red supergiants, as would be expected if they evolved without mass loss. Evolutionary tracks of the most massive stars tend to be almost horizontal in the HRD (Fig. 2.13a). Helium burning may begin not long after the TMS, but (for Pop I theoretical models, if there is no mass loss) most of the core helium-burning phase is spent as a red supergiant. Observationally, however, there is an almost complete absence of red supergiants with bolometric luminosities $\gtrsim 3 \times 10^5\,L_\odot$ (Humphreys and Davidson 1979), for which appropriate masses are $\gtrsim 30\,M_\odot$ – see Fig. 2.13a. There is no corresponding shortage of blue supergiants at luminosities up to about $3 \times 10^6\,L_\odot$, whose masses must range up to about $100\,M_\odot$. The sloping line in Fig. 2.13a, above which there are no stars, is called the Humphreys–Davidson limit (HDL).

Mass loss may be capable of explaining the HDL, since there is a tendency for stars that lose a good deal of mass on or shortly after the main sequence to remain relatively blue during core helium burning, rather than to make a complete excursion to the red supergiant domain. Indeed, a star that contrives to blow off all of its hydrogen-rich outer layers at the end of its MS life will, in effect evolve, as a helium star (Section 2.5), which is always hot and small if its mass is $\gtrsim 2.7\,M_\odot$. Some O stars, the Of sub-type, show strong winds. However, the empirical mass-loss rate for O stars in the next section, Fig. 2.15 and Eq. (2.71), is not large enough to achieve a stripping down to the core, except perhaps at $\gtrsim 100\,M_\odot$. What is needed is something like the much higher rate attributed to P Cyg stars, and to the related class of luminous blue variables (LBVs), which are indeed found near the Humphreys–Davidson limit.

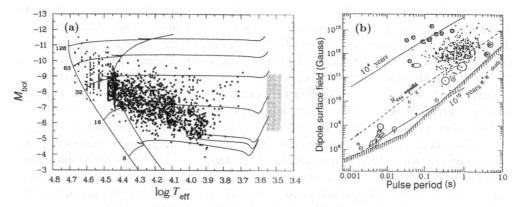

Figure 2.13 (a) HRD of the brightest supergiants of the LMC. The Humphreys–Davidson Limit (HDL) is the upper envelope of the observed stars (dots); the lower envelope is simply an observational cut-off. Some evolutionary tracks from Fig. 2.1 are reproduced; these tracks do not include mass loss. Several hundred M supergiants are subsumed in the shaded region. The theoretical tracks need a larger mixing-length ratio ($\alpha \sim 2.5$) to get there. The gap between $\log T_{\text{eff}} \sim 3.7$ and 3.55 appears to be real. From Fitzpatrick and Garmany (1990). (b) Magnetic field, estimated by Eq. (2.70), plotted against pulsar rotation period, for ~ 550 radio pulsars in the Galaxy and the Magellanic Clouds. Single pulsars are dots, binary pulsars are circles or ellipses. Heavily-circled pulsars at the top centre are in supernova remnants. From Phinney and Kulkarni (1994).

We can hypothesise that as a very massive star evolves horizontally across the Hertzsprung gap its envelope becomes unstable when it reaches the HDL. This may be because the luminosity in the interior becomes very close to the Eddington limit – the limit given by Eq. (2.12) as $\zeta \to \infty$, i.e.

$$L_{\text{Edd}} = \frac{4\pi c G M}{\kappa_{\text{Th}}}, \tag{2.67}$$

which is the maximum luminosity that a star can transmit while still in hydrostatic and radiative equilibrium. This may trigger mass loss at a rate in excess of $10^2 \, M_\odot$/megayear, which continues (erratically on a timescale of decades to centuries, as in η Car, S Dor in the LMC and P Cyg itself) until the mass is so reduced that the instability is largely removed. This apparently happens when the star is stripped down to its helium core. Some Wolf–Rayet (WR) stars appear to be such stripped-down stars. Many, though not all, appear to have little or no hydrogen in their spectra. Several are in binaries from which masses can be determined, and their masses are low for a main-sequence star of their luminosity, but not for a stripped-down remnant of a star formerly two or three times more massive.

Wolf–Rayet stars are normally divided into two main classes, the WNs, which show nitrogen apparently enhanced relative to carbon and oxygen, and the WCs, which show enhanced carbon (and there is also a very small class of WOs, showing enhanced oxygen). The WCs normally show no hydrogen, which at least is consistent with the view that mass loss has stripped them down right to the carbon-enriched helium burning core. The WN subclass can be further subdivided, somewhat loosely, into 'late' (WNL) and 'early' (WNE), i.e. cooler and hotter, with the former showing some hydrogen in the spectrum and not the latter. Naively, therefore, they may represent two earlier steps on the road to WC stars.

The evolution of the interior of a WR star is probably not much different qualitatively from what it would be if it remained as the core of a more massive star. Helium burning, lasting about 10^5 years, will be followed by C-burning and later nuclear stages on a much shorter timescale ($\sim 10^3$ years), the evolution being accelerated because neutrino losses take away 99% of the nuclear energy. Mass loss does not cease, but arguably slows down by about a factor of 10, which may mean that in the limited time available they only lose a modest fraction of their remaining mass. A supernova explosion still seems inevitable.

Wolf–Rayet stars are famous for their strong stellar winds, which completely dominate the visible spectrum and make it very difficult to determine stellar surface parameters: the photosphere is somewhere in the wind itself, perhaps at several times the radius of the underlying star. But the above scenario requires that mass loss be substantially stronger during the preceding P Cyg/LBV phase, while the star is located fairly centrally in the Hertzsprung gap, than in the later WR phase when the underlying star is a more compact, hotter object masked by an expanding envelope. We can hypothesise that the most massive stars follow a route that can be abbreviated as

$$\text{Of} \rightarrow \text{WNL} \rightarrow \text{PCyg, LBV} \rightarrow \text{WNL} \rightarrow \text{WNE} \rightarrow \text{WC} \rightarrow \text{SNEX}. \tag{2.68}$$

The star gets to its furthest right-hand position in the HRD during the LBV/P Cyg phase, on the HDL. However, for masses of $\lesssim 30\,M_\odot$ the mass loss is not so important at any stage, and the star evolves from the TMS across the Hertzsprung gap to the Hayashi track. The supernova explosion is expected when the star is a red supergiant.

Single stars with initial masses of about 7–$30\,M_\odot$ are expected to be red supergiants when their cores collapse, and to have substantial hydrogen-rich envelopes even if they have lost some mass by stellar wind. They should therefore produce Type II supernovae. But a star might lose all of its H-rich envelope if either it was very massive initially and passed through a substantial P Cyg stage as above, or if it lost its envelope to a binary companion. In this case it can be Type I. However Type Ia is found in elliptical galaxies, and so may represent a specific kind of binary interaction peculiar to low-mass stars (such as the merger of two white dwarfs in a very close binary). By contrast, Types Ib and Ic are mainly found in spiral arms, as are Type II, where there are young massive stars, and so may be the result of envelope stripping either by P Cyg wind or by binary interaction. However, Hill (1993) warns that five of the six Galactic supernovae, for whose remnants we have more detail than most extragalactic supernovae, do not fit particularly comfortably into the usual classification scheme. In addition, SN 1987A in the LMC, although spectacularly confirming the importance of neutrinos (Bratton *et al.* 1988, Hirata *et al.* 1988, Arnett *et al.* 1989), was a very atypical supernova in most respects. At spectral type B3II before the supernova explosion, it was also at an unexpected place in the HRD; neither a red supergiant, LBV, nor a WR-like object. This could be the result of binary interaction.

Once again, note Spruit's (1998) hypothesis that internal magnetic field enforces slow uniform rotation even within stars whose cores have contracted by large factors while their envelopes expanded by comparably large factors. It may, therefore, seem difficult to explain why neutron stars at birth are rotating rapidly (Fig. 2.13b). However, there is evidence, principally from the rapid space motions of pulsars, that supernova explosions are asymmetric, and result in a 'kick' of typically 300 km/s in a random direction. Such asymmetry will presumably give an impulsive couple in addition to an impulsive force, and the tiny moment of inertia of a neutron star means that it could easily acquire its angular velocity in this way.

Of course, once the core starts to implode on a hydrodynamic timescale it will no longer be prevented by magnetic coupling from spinning up.

Pfahl *et al.* (2002) have suggested a modification of the above picture. Possibly when the core is contracting on a *thermal* timescale immediately prior to the supernova it is able to spin up substantially, and the explosion may be more symmetrical as a result. There is evidence from binaries that some neutron stars receive a substantial kick and others do not. Possibly the difference is due to the amount of rotation in the core at the onset of the supernova explosion.

A qualitative picture of the evolution of single massive stars as a function of their initial mass, but without a definitive basis either in observation or theory, may run something like this:

(a) Stars with initial mass $\gtrsim 50\, M_\odot$ may lose substantial mass (say 10–30%) while crossing the main sequence band, which will be broader as a result, and then lose considerably more mass (perhaps a further 40–60%) much more rapidly as a P Cyg star at the HDL, so that at some point in the left-hand Hertzsprung gap the evolutionary direction reverses towards a smaller, hotter WR configuration. After core helium burning there is rapid evolution towards a supernova explosion (Type I); at no stage is the star a red, or even a yellow, supergiant. The remnant may be a black hole rather than a neutron star.

(b) Stars with initial mass in a range about 30–50 $M_\odot$ lose mass more slowly (relative to their nuclear time scale, which anyway is slower), so that the star is able to evolve some way across the HRD, perhaps to types AI–KI, before rapid mass loss as a P Cyg star pushes it back to the blue. The star becomes a WR object, and then experiences a supernova explosion, perhaps also Type I as in (a) although more probably Type II, but having lost a rather smaller proportion of initial mass. The remnant may be a neutron star or a black hole.

(c) Stars with initial mass in a range about 15–30 $M_\odot$ are able, with relatively slower mass loss still, to evolve to the red supergiant region and spend significant time there. Probably the mass loss does not move the star back to the blue before the supernova explosion (Type II); the total amount of mass lost might only be about 10–20%. A neutron-star remnant is expected.

(d) Stars with initial mass in a range of about 7–15 $M_\odot$ may perform a 'blue loop' during core helium burning, bringing the star back from spectral type about MI at He ignition to BI/AI, before returning to type MI at supernova explosion. Alternatively, they may either ignite helium, and burn it, entirely as red supergiants; or at some masses they may ignite helium and burn some of it while still blue, and then complete the burning while red. The behaviour can be quite sensitive to input physics, and perhaps also to the computational procedure. This may be due to the importance of radiation pressure, which means that much of the star is rather close to convective or semiconvective instability. In any event, the amount of mass lost might be no more than about 10% altogether (apart from the final supernova explosion, Type II), and may not have a significant effect on the star's location in the HRD. The remnant should be a neutron star.

Although in many contexts a neutron star can be seen as a stationary (i.e. non-evolving) end-point of evolution, neutron stars that are observed as pulsars do in fact evolve at least to the extent that their rotation rates, and arguably magnetic fields, evolve. Direct observation reveals spin-down timescales that are typically ~ 1 megayears. These rates can be used to estimate the magnetic field, assuming that the loss of rotational energy is due principally to the radiation rate of a rotating magnetic dipole in vacuo:

$$\frac{d}{dt}\frac{1}{2}I\Omega^2 = -\frac{\mu_0}{6\pi c^3}|\ddot{\mathbf{m}}|^2, \quad |\ddot{\mathbf{m}}| = \Omega^2 m, \quad m = \frac{4\pi}{\mu_0}BR^3, \tag{2.69}$$

where R, I, Ω are the radius, moment of inertia and rotation rate of the neutron star, $\mathbf{m}$ is the magnetic dipole moment, and B is the magnetic field on the surface at the magnetic equator. It is assumed here that the dipole axis is through the centre of the star and perpendicular to the rotation axis. Using reference values $R = 10$ km, $I = 10^{38}$ kg m^2, and putting $\Omega = 2\pi/P$, this gives

$$B^2 \approx 10^{31} P\dot{P}, \tag{2.70}$$

with B in tesla (10^4 gauss) and time in seconds.

Figure 2.13b shows the 'pulsar HR diagram' of B plotted against P, for radio pulsars. The great majority of pulsars are in the top right quarter, and are almost all isolated, i.e. not in binaries. About 5% lie in the bottom left-hand corner and are almost all in binaries. Both groups lie to the left of a sloping 'death line', where the combination of period and magnetic field is too weak to support the radio emission that makes them detectable. This radiation comes from electron–positron pairs generated by the rotating magnetic field above but fairly near the pulsar surface. Although this is the radiation that is detected, it is presumably a small fraction of the energy flux emitted directly by the rotating dipole – Eq. (2.69).

Two things are reasonably clear: (a) since a high proportion of massive stars are in binaries, and since few radio pulsars are, binaries must typically be disrupted by a supernova explosion (Section 5.3); and (b) many of those that do remain in binaries are spun up to short periods, presumably by accretion from the companion.

Two important aspects are not so clear: (c) the equation of state (EoS), and (d) the evolution (if any) of the magnetic field. Although the equation of state is much simplified by the fact that temperature is almost irrelevant, and hence $p = p(\rho)$, the strong interaction is not yet sufficiently well-known to determine an $R(M)$ relation analogous to the relatively simple one for white dwarfs (Eq. 2.51). Equations of state range from 'hard', where p depends strongly on ρ, to 'soft', where the dependence is relatively weak. The latter will, at a given mass, produce a neutron star that is smaller and more centrally condensed than the former, but both will have an upper limit to the possible mass, analogous to the Chandrasekhar limit. The softest hypothetical equations of state can be ruled out on the basis that the upper mass limit is less than some well-determined NS masses (1.33–1.45 $M_\odot$). Too hard an equation of state would give relatively large radii at such masses, and could be rotationally unstable at the shortest rotational period observed (1.6 ms); this constraint is weaker, however, because the fastest pulsars do not (yet) have well-determined masses, and might in principle be 2–3 $M_\odot$.

Whether the magnetic field evolves or not is a matter of considerable debate. One might suppose, a priori, that the field could change (a) its strength, (b) its orientation, relative to the rotation axis and (c) its topology, e.g. a dipole component decaying slower or faster than a quadrupole component. Any or all of these might happen in isolated pulsars, and they might also happen for different physical reasons during accretion in a binary.

Several neutron stars and black holes are detectable as X-ray sources rather than as radio pulsars, their X-radiation coming from accretion of gas donated by a close companion. Some (neutron stars, but not black holes) show pulsed radiation, the accretion being funnelled by the magnetic field on to the magnetic poles. Rates of rotation are often much slower than in

Table 2.3. *Some binaries with red giant components*

Name	Spectra	P	e	M_1	M_2	R_1	R_2	q	Y^a	Reference
RZ Eri	K2III + F5m	39.3	0.35	1.62	1.68	7.0	2.8	0.96	1.9	Popper 1988b
HR2030	K0IIb + B8IV	66.5	0.02	4.0	4.0	41	5.9	1.00	2.5	Griffin and Griffin 2000
RS CVn	K0IV + F4IV–V	4.80		1.44	1.41	4.0	2.0	1.02	1.45	Popper 1988a
TZ For	G8III + F7III	75.7		2.05	1.95	8.3	4.0	1.05	2.5	Andersen *et al.* 1991
α Aur	G8III + G0III	104		2.61	2.49	11.4	8.8	1.05	4.9	Barlow *et al.* 1993
η And	G8II-III + do.	116	0.006	2.39	2.26	10.5	8.5	1.06	5.0	Hummel *et al.* 1993, Schröder *et al.* 1997
93 Leo	G7III + A7IV	71.7		2.2	2.0	8.7	2.7	1.09	1.68	Griffin and Griffin 2004
α Equ	G7III + A4Vm	98.8		2.3	2.0	9.2	2.6	1.15	1.67	Griffin and Griffin 2002
δ Sge	M2Ib-II + B9V	3720	0.40	3.4	2.7	157	3.3	1.26	1.77	Schröder *et al.* 1997
ζ Aur	K4Iab + B6.5IV–V	972	0.41	6.6	5.2	151	5.1	1.27	1.89	Schröder *et al.* 1997
V2291 Oph	G9II + B8-9V	385	0.31	3.86	2.95	32.9	3.0	1.31	1.53	Schröder *et al.* 1997
V695 Cyg	K4Ib + B4V	3784	0.22	7.2	5.5	170	4.0	1.31	1.44	Schröder *et al.* 1997
γ Per	G8III + A3V	5350	0.79	2.5	1.86	21:	4:	1.34	2.6	Pourbaix 1999
τ Per	G8IIIa + A2V	1516	0.73	2.8	2.0	15.8	2.2	1.40	1.37	Griffin *et al.* 1992
OW Gem	F2Ib-II + G8IIb	1259	0.52	5.8	3.9	30	32	1.49	14	Terrell *et al.* 2003
V415 Car	G6II + A1V	195		3.1	2.0	31	1.9	1.55	1.16	Brown *et al.* 2001
QS Vul	G5Ib-II + B8V	249		5.4	3.4	77	3.3	1.59	1.56	Griffin *et al.* 1993
V1488 Cyg	K5Iab + B7V	1145	0.30	7.2	4.1	170	3.1	1.76	1.32	Schröder *et al.* 1997

^{a} Ratio of R_2 to ZAMS radius for same mass M_2.

isolated pulsars, at any rate in the wider binaries, and there is indirect evidence to suggest that rotation is much *faster* in short-period low-mass binaries. It can be seen in Fig. 2.13b that the small proportion of pulsar binaries with very short ($\sim$millisecond) periods are almost all in binaries. These are also the pulsars with the weakest fields, and so it is plausible that the same mechanism that speeds them up encourages their fields to decay.

Pulsar physics is beyond the competence of the author, and therefore outside the scope of this book. But some relatively simple aspects important to an understanding of the evolution of binaries are mentioned briefly later.

2.3.5 *Some observed binaries with evolved components*

Table 2.3 lists 18 binaries containing substantially evolved stars (red giants and supergiants), with rather well-determined parameters; they are ordered by increasing mass ratio. They potentially provide a quite stringent test of stellar evolution models, and it cannot be said that the results are satisfactory. R. E. M. Griffin (private communication 2002) has noted that in many of these systems the secondary is significantly oversized.

The lifetime of a star as a red giant is substantially shorter than its lifetime as a main sequence star; less than, and at high mass much less than, $\sim$40%. Since evolutionary lifetime is a strong function of mass, in a binary where a red giant has, say, 1.5 times the mass of its main sequence companion the latter should be very little evolved from the ZAMS. In fact for a given mass ratio q the ratio, Y say, of R_2 to the ZAMS radius corresponding to M_2 should be constrained between two values: the lower value applies if $*1$ is just beginning its red-giant evolution, and the higher value if it is at the end of its evolution. In Fig. 2.14a, two solid lines give the two limits if $M_1 = 2.8$, and two broken lines apply to $M_1 = 6.3$. The lower solid and broken lines almost coincide, but the upper lines differ fairly substantially.

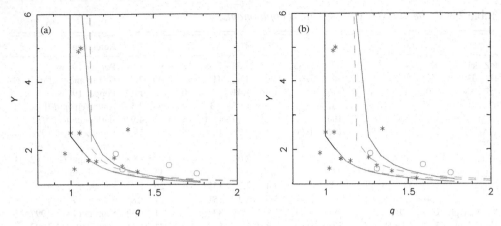

Figure 2.14 (a) Binaries with at least one red giant component, plotted with radius R_2 (relative to ZAMS radius) upwards and mass ratio horizontally. Circles – $M_1 > 4\,M_\odot$; asterisks – $M_1 < 4\,M_\odot$. Also plotted are the theoretical minimum and maximum Y. The minimum is when $*1$ is at the beginning of its giant life, and the maximum when it is at the end. These curves are plotted for two values of M_1: 2.8 and 6.3 $M_\odot$. The two minimum curves are almost the same, but the maximum curve is usually lower for the more massive $*1$. (b) The lifetime of the post-main-sequence phase has been artificially increased by a factor of 2.5 relative to the main sequence life, raising the maximum Y but leaving the minimum unchanged.

The systems in Table 2.3 are all plotted in the (q, Y) plane in Fig. 2.14a, except for OW Gem, which is far off the scale. Asterisks correspond to $M_1 \sim 2$–$4\,M_\odot$, and should lie roughly in the area between the two solid lines and the vertical at $q = 1$; circles correspond to $M_1 \sim 4$–$8\,M_\odot$ and should lie roughly between the two broken lines and the vertical. It can be seen that only ten of the eighteen systems lie in the expected regions; two lie below and six above. Four of the last six (including OW Gem) lie *far* further than can be plausibly attributed to measuring uncertainty, which might possibly be as large as 10%.

At least the two systems *below* the expected region can be reasonably accounted for. The two giants have probably lost mass as a result of stellar wind, this wind being enhanced substantially over that expected for single giants by the fact that the binary giants are being forced to rotate perhaps ten times more rapidly than would be expected in single giants. I will discuss this 'enhanced wind' in Section 4.6. This must surely be the case in RZ Eri, where the giant is slightly the *less* massive component; and it is quite reasonable for RS CVn, even though the giant is fractionally the more massive. The other binaries (except HR 2030) are generally considerably wider. If there *were* significant enhanced wind in these systems, it would make the disagreement worse.

HR 2030 is interesting in that it is probably the only giant that is (a) still not at helium ignition, and (b) evolving redwards on a *thermal* timescale. A giant of this mass would not ignite central helium until substantially larger than the binary system allows. Although it may also be suffering from 'enhanced wind', on something like the nuclear timescale as in RZ Eri and RS CVn, its rapid evolution may mean that its integrated mass loss is not yet significant.

In Section 4.6 I argue that OW Gem is a former triple. The F2 component, I suggest, is the merged remnant of a former close binary. While direct evidence for this is not strong, there do exist many known triples with something like the right parameters. In Section 3.5 I show that many close binaries, in particular those with an initial mass ratio $q_0 \gtrsim 2$, are likely to merge as a result of evolutionary processes. If OW Gem started with parameters $((4 + 1.8 \, M_\odot; 2 \, \text{days}) + 3.9 \, M_\odot; 1250 \, \text{days})$, it might reasonably end up as presently observed.

A possibility to be considered for the remaining five discrepant systems is that the lifetime of a giant in the core helium-burning phase is substantially larger than present modelling suggests. Perhaps the core's helium-burning luminosity, relative to the hydrogen-burning shell, should be less, and this might in turn be due to greater opacity in the core. Alternatively, perhaps core overshooting is much more substantial in helium-burning cores than in hydrogen-burning cores. This is crudely explored in Fig. 2.14b: the lifetime of the giant stage has been artificially assumed to be 2.5 times larger than the models implied. But even this very substantial change only brings two of the defaulters, δ Sge and ζ Aur, within the compass of the theory; γ Per, QS Vul, V1488 Cyg and OW Gem remain well outside.

Another possibility that we might consider is that the components are non-coeval. Within a dense star-forming region, or a somewhat less dense expanding OB association, dynamical encounters can take place (Section 5.4) in which, for example, an older single star might eject and replace one component of a younger binary. But such encounters are probably limited to perhaps the first 10–20 megayears of a star's life, when the stellar density is still high, and so should result in no larger age discrepancies than this. Only ζ Aur looks like a reasonable candidate: the ages of the two components considered separately may be ~ 65 and 80 megayears.

We do not have a persuasive answer to this problem. We emphasise that such a problem is only recognisable because of (a) the high quality of the observational data for these systems, and (b) the likelihood that the components have evolved without interaction so far. Most binaries (apart from ESB2 binaries that have well-detached main sequence components) have data of substantially lower quality, or else have undergone a major interaction that has altered the masses or period, and so problems such as 'oversized secondaries' can be overlooked.

2.4 Stellar winds and mass loss

Many stars show some evidence of loss of mass from the surface by way of a wind. For most stars, such as the Sun, this wind is rather meagre in evolutionary terms. The Sun is losing mass at a rate of about $10^{-7.6} \, M_\odot/\text{megayear}$, and so is expected to lose $\lesssim 10^{-4}$ of its mass in the remainder of its main sequence life (about 5 gigayears). But some stars show evidence of much more copious winds, especially stars of high luminosity. Winds from cool supergiants can affect evolution strongly, as described in Section 2.3, by allowing white dwarfs to be remnants of stars whose initial masses may have been up to five times the Chandrasekhar limit. Winds can also be important in hot, blue, massive stars. Most stars with surface temperatures above about 25 kK or luminosities above $10^5 \, L_\odot$ have spectroscopic indications (P Cyg line profiles) of a roughly radial outflowing wind, of sufficient density and speed in some cases to be significant on evolutionary time scales.

Mass loss rates for OB stars, which are probably best derived from radio or infrared measurements of the expanding gas cloud, have been estimated by many workers, for instance Olson and Castor (1981), Garmany *et al.* (1981), Abbott *et al.* (1981), Lamers (1981) and

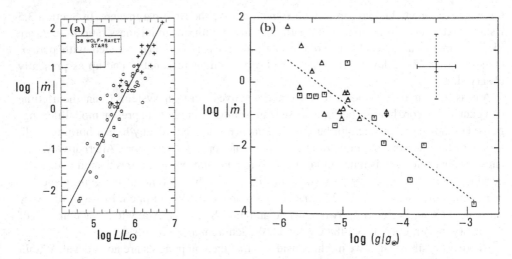

Figure 2.15 Rates of mass loss ($M_\odot$/megayear) by stellar wind, from (a) OB and Wolf–Rayet stars (after Conti 1982), and (b) red giants and supergiants (after Judge and Stencel 1991). The mean lines are given by Eqs (2.71) and (2.75).

Chlebowski and Garmany (1991). They are roughly consistent (Fig. 2.15a) with

$$\dot{M} \sim -2 \times 10^{-11} L^{1.9} \quad (4 \times 10^4 \lesssim L \lesssim 5 \times 10^6), \tag{2.71}$$

with time in megayears and M, L in Solar units. There appears to be a real spread about this rate, on top of any systematic and measuring errors: stars with quite similar photospheres may differ in $\dot{M}$ by more than an order of magnitude. Comparing Eq. (2.71) with Eqs (2.1) and (2.4), and allowing for the considerable spread seen in Fig. 2.15a, it is possible that the most mass-lossy stars of over about 60 $M_\odot$ can have their masses halved in their main-sequence life-times. Unfortunately theoretical evolutionary tracks of massive MS and post-MS stars depend quite sensitively on the assumed relation between $\dot{M}$ and variables such as L in Eq. (2.71), or R, as well as on other mass loss rates that have been proposed. A convincing theoretical model of winds, capable of predicting $\dot{M}$ as a function of surface quantities (which probably should include rotation rate and magnetic field, for instance), does not yet exist. But great strides have been made in modelling hot atmospheres with spherically-symmetrical steady winds (Lucy and Solomon 1970, Pauldrach *et al.* 2001). These have to include the driving effect of radiation pressure on a multitude of spectral lines, and the fact that abundances of ionised species are not necessarily in local thermodynamic equilibrium. Such models can predict the rate of mass loss (and also the terminal velocity of the wind), and agree to within an order of magnitude with Eq. (2.71).

An interpolation formula, based on observed mass-loss rates from the literature, was given by de Jager *et al.* (1988). They found that in the upper part of the HRD ($\log L \geq 2.5$) $\log |\dot{M}|$ is mainly a function of L and T only, given in terms of Chebyshev polynomials

$$T_i(x) \equiv \cos(i \cos^{-1} x), \quad -1 \leq x \leq 1, \tag{2.72}$$

by

$$-\log |\dot{M}| \approx \sum_{i=0}^{5} \sum_{j=0}^{5-i} a_{ij} T_i \left(\frac{\log T - 4.05}{0.75} \right) T_j \left(\frac{\log L - 4.6}{2.1} \right), \tag{2.73}$$

Table 2.4. *Coefficients for the mass-loss rate of de Jager et al. (1988)*

	$j = 0$	1	2	3	4	5
$i = 0$	6.349 16	−5.042 40	−0.834 26	−1.139 25	−0.122 02	0.0
1	3.416 78	0.156 29	2.962 44	0.336 59	0.575 76	
2	−1.086 83	0.419 52	−1.372 72	−1.074 93		
3	0.130 95	−0.098 25	0.130 25			
4	0.224 27	0.465 91				
5	0.119 68					

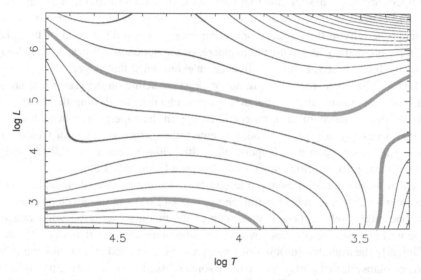

Figure 2.16 Contours of constant mass-loss rate, according to de Jager *et al.* (1988). Contours are in decades, with heavy lines for 10^{-6} $M_\odot$/megayear (lower left) and 1 $M_\odot$/megayear (centre and right).

with L in Solar units, T in kelvins and $\log |\dot{M}|$ in Solar masses per megayear. The coefficients a_{ij} are given in Table 2.4; de Jager *et al.* (1988) do not list a value for a_{05}, but the value zero appears to be adequate. The scatter between observed and computed values is about ±0.5. Much of this scatter is, no doubt, real. A contour plot of this mass-loss rate is given in Fig. 2.16.

For luminous stars that are cool, i.e. red giants and supergiants ranging in luminosity 10^2–10^4 $L_\odot$, and radii 10–100 $R_\odot$, mass loss rates were estimated by Reimers (1975) and are roughly consistent with

$$\dot{M} \sim -10^{-6.4} \frac{\eta L R}{M}, \tag{2.74}$$

in megayears and Solar units as before. The parameter η is a fudge factor, which we choose in order to reach reasonable final core masses. Judge and Stencel (1991) give a slightly different formula, also based on observational data (Fig. 2.15b):

$$\dot{M} \sim -10^{-7.6} \left(\frac{R^2}{M} \right)^{1.43}. \tag{2.75}$$

In practice, this differs little from the previous formula, since L relates to R, M on the RG branch via Eq. (2.49). Curiously, Eq. (2.75) extrapolates successfully to the Sun. Just as for massive hot stars, there is considerable spread, some of which is probably intrinsic.

In Section 4.5 I develop a model for mass loss as a result of dynamo activity in cool rotating stars. I obtain a result – Eq. (4.84) – that is very like Eq. (2.74), but contains two extra factors. One is a factor $(R/R_{HT})^2$, which is unity on the Hayashi Track – Eq. (2.49) – and decreases rather rapidly going into the Hertzsprung gap. The other depends on the Rossby number, the ratio of the rotational period to the convective envelope turnover time – Eq. (2.34).

There is something of a dichotomy in cool stars between those that have hot chromospheres, coronae not unlike the Sun, and hot, fast (300–500 km/s), low-density winds, and those where high-temperature gas is absent, and that have cool, slow (20–30 km/s), high-density winds (Linsky and Haisch 1979). At least for stars with atmospheres like the Sun's, it is probable that dynamo activity in the surface convection zone, and the dissipation of this magnetic energy in flaring activity above the photosphere, is a major cause of mass loss; although the mass loss itself is a relatively minor influence on evolution at that stage.

For the more luminous red giants, radiation pressure acting on the grains that are able to form at low temperatures may be the dominant mechanism for driving the wind, although some deposition of mechanical or magnetic energy in the superphotospheric layers would seem to be necessary to start the wind. Extreme red supergiants, many of which are Mira variables, may enter a phase of 'superwind', with winds of order 10–$100\,M_\odot$/megayear, which rapidly strip the envelope down to the hot core (Section 2.3.2).

I argued in Section 2.3.2 that asymptotic giant branch stars terminate their evolution roughly at the point where the integrated binding energy – Eq. (2.62) – of the envelope changes sign. Naively one might well suppose that as soon as the binding energy becomes negative the envelope will be lost. The physics however must be more complex than that; it is not clear how efficiently the available (un)binding energy can be converted into outflowing motion. It seems more plausible that it may start to drive some relaxation cycle, during part of which the energy is converted to heat and radiated away. This may be at least a contribution to the onset of Mira oscillations. But the oscillations become strong enough to drive an increasing cool wind, with grains forming and with radiation pressure on grains contributing to the strength of the wind.

The form of Reimers' law – Eq. (2.74) – suggests that a constant fraction of the stellar luminosity is used to provide the gravitational energy necessary for escape ($\sim GM/R$, per unit mass). This fraction is 1.3×10^{-5}, if we convert Eq. (2.74) into SI units. Let us generalise by using the binding energy rather than just the gravitational energy; then we might try

$$-\dot{M} = 1.3 \times 10^{-5}\, \frac{\eta M L}{\int_{M_c}^{M} \left(\frac{Gm}{r} - U\right)\, dm}, \qquad (2.76)$$

in SI units. We find that $\eta \sim 0.2$–0.5 gives reasonable remnant core masses. Equation (2.76) will clearly lead to rapid mass loss as the denominator approaches zero on the AGB, but the denominator will never actually reach zero (at least until M is reduced to M_c) because the binding energy per unit mass increases again as the envelope is stripped down to a small hot core. Equation (2.76) can be seen as combining the concept of slowly increasing wind on the giant branch with rapidly increasing 'superwind' at a late stage on the AGB.

An issue left unclear, however, is what value of M_c to use as the lower limit in the integral. The integral in Equation (2.76) only represents a physically meaningful binding energy if the region interior to M_c is completely unaffected, in its distribution of $r(m)$ and $U(m)$, by the progressive loss of the layers above M_c. There is no value apart from $M_c = 0$ for which this is literally true. However, we suggest that it is reasonable to take as M_c the mass coordinate where, say, the hydrogen abundance has been reduced to 15% by mass, since the core inside this is compact and largely degenerate, at least in the highly evolved red giants considered here. Some crude experimentation suggests that the fraction 15% is not very critical. I will return to this point, in some detail, in Section 5.2.

Similar rates of mass loss are seen in WR stars, but the underlying physical regime is quite different, since temperatures are at least ten times higher. Perhaps more significantly, the velocity with which the material is ejected is ~ 100 times greater. This WR mass loss presents a considerable challenge to theorists. With a mass flux of $\gtrsim 30 \, M_\odot$/megayear (Willis 1982), and ejection velocity ~ 2000km/s, the kinetic energy flux in the wind may be as much as 5% of the total energy flux of the star. The momentum flux is relatively even larger, at several times the momentum flux of the radiation field, which makes it difficult to understand how radiation pressure alone drives the wind. It is not yet clear how even binary interaction, let alone a single-star process, can cause such winds. Pulsational instability or strong turbulent motion in the outer layers, perhaps combined with dissipation of magnetic energy produced by dynamo activity, might help, but these would have to tap the nuclear energy of the star rather than just, say, the rotational energy, since the latter is too feeble. De Jager *et al.* (1988) note that WR stars have mass-loss rates enhanced over their formula – Eq. (2.73) – by an average of $10^{2.2}$.

The momentum problem referred to above may, in fact, be solved by 'multiple scattering': photons trying to escape may be scattered several times before actually escaping, and so contribute more to the momentum flux. Lucy and Abbott (1993) show that such multiple scattering can take place if there is a sufficient stratification of different degrees of ionisation; de Koter *et al.* (1997) argue that mass-loss rates of some very massive stars in the LMC can be modelled in this way. They conclude that mass-loss rates normally thought of as characteristic of WR stars can be maintained in the earliest main sequence stars, whose spectra may be characterised as O3f/WN.

Here we adhere to the view (Conti *et al.* 1983) that WRs are evolved stars, most of which have lost a great deal of mass. Underhill (1983, 1984) has argued that (a) the underabundance of hydrogen, and overabundances of nitrogen and carbon (assumed to be evidence of late hydrogen burning, or helium burning) are not real, but are artefacts of the very crude abundance analyses; (b) the very high mass-loss rates inferred, especially from radio and infrared observations, are also not real; and (c) WRs are analogous to the Herbig Be/Ae objects (Herbig 1960), which appear to be young or pre-main-sequence stars. However, the crucial evidence is that WRs are much too luminous, relative to their masses, and much too common, relative to O stars, to be either fairly normal (though young) main sequence stars, or pre-main sequence stars. In many binaries the WR luminosity is comparable to, or greater than, that of the O-star companion, while the mass is often a half to a quarter of the companion's mass. And since pre-main-sequence contraction must be ~ 1000 times more rapid than main sequence evolution (at the same luminosity), the proportion of WRs to O stars ($\sim 10\%$, Conti *et al.* 1983) is too high for a pre-main-sequence picture to be sustained. However, we must acknowledge that analyses of WR atmospheres to date may well be misleadingly oversimplified. They tend to assume a steady, spherically symmetrical gas outflow, driven primarily by radiation pressure, where the reality may well be a fairly turbulent,

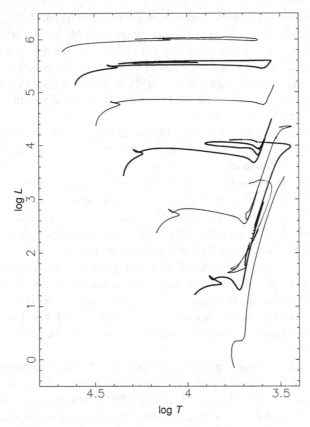

Figure 2.17 A theoretical Hertzsprung–Russell diagram for stars including mass loss according to the larger of Eqs (2.73) and (2.76). The masses are, from the bottom, 1, 2, 4, 8, 16, 32 and 64 $M_\odot$. Alternate masses are thin or thick lines. The tracks were terminated when the timestep dropped below 0.1 year. Although the most massive star evolved well to the cool side of the Humphreys–Davidson limit – Fig. 2.13a – its visit there was very rapid and short-lived.

non-steady flow, driven at least in part by hydromagnetic stresses and energy release. If dynamo activity does play a part, producing a magnetic field that perhaps influences the flow out to several stellar radii, and that may initiate the flow even if radiation pressure is important in accelerating it, it would not be surprising if binarity is also significant, as it is for example in RS CVn stars.

Figure 2.17 shows an HRD that includes mass loss according to the combination of Eq. (2.73) for luminous stars and Eq. (2.76) for cool stars; for stars that are both luminous and cool we used the larger of these two rates. For the three lowest masses the cores at the tip of the AGB were 0.58, 0.67 and 1.01 $M_\odot$. For the two highest masses the remnants at the end of the plotted tracks (where the timestep became uncomfortably short) were 14 and 29 $M_\odot$. The two intermediate masses (8, 16 $M_\odot$) were changed by only a modest amount. Some details at various stages are given in Table 3.2. The 64 $M_\odot$ star was found to evolve briefly into the red supergiant region, but spent only 1% of its life redder than 10 kK. The 32 $M_\odot$ star oscillated several times, rapidly, across the top of the diagram.

We find that the evolution of massive stars is very sensitive to the assumed mass loss rate. It is probably also very sensitive to (a) metallicity and (b) the numerical procedure, particularly for convective and semiconvective mixing. The latter may be important because in massive stars both ∇_r and ∇_a are very close to 0.25, so that convective and semiconvective zones can appear, move or disappear in an apparently capricious way. In my models this is probably exacerbated by the fact that I discretise the star with only 200 meshpoints, at all stages of evolution. We cannot take the detailed behaviour of the most massive models very seriously, but it seems clear that the mass loss rate (2.73) can account reasonably well for the observed shortage of stars in the uppermost right-hand corner of the HRD.

Over the whole of the HR diagram we can recognise something like nine types of mass loss, where we distinguish in particular whether the wind is 'copious' or 'meagre':

(a) a fast, hot, meagre, Solar-like wind in cool (GKM) dwarfs and in some GK giants
(b) a slow, cool, meagre wind in M giants and some GK giants
(c) a slow, cool, copious wind (superwind) in late M giant (AGB) stars
(d) a very fast, hot, meagre wind in PN nuclei (post-AGB, pre-WD)
(e) an episodic, meagre, rotating wind in Be stars
(f) a fast, meagre wind in Of stars
(g) a fast, copious, episodic wind in LBVs (P Cygs)
(h) a very fast, somewhat less copious wind in WRs
(i) an almost instantaneous, copious wind in a supernova explosion.

Probably (c), (g) and (i) are the most important for overall evolution; they can change the mass significantly in less, even much less, than the nuclear lifetime of the star. Process (b) is probably very important in old clusters where it may determine the distribution of stars on the HB subsequent to the helium flash. Processes (f) and (h) may be marginal, affecting the evolution to some extent, but perhaps not crucially, except for very massive Of stars. Processes (a), (d) and (e) probably have little effect on the evolution, though they can be conspicuous observationally. Processes (d) and (h) may, in practice, be much the same in physical origin, but WR stars are typically several times more massive and several times more luminous than PN nuclei. I would like to emphasise, however, as in Section 4.5 and subsequently, that the presence of a close binary companion may enhance some of these winds, and make them copious where they would be meagre in a single star.

A comprehensive theory of mass loss would have to link together such properties of a star as its stability, rotation, magnetic field generation and dissipation, differential rotation (which is especially effective at assisting dynamo activity), turbulent convection and radiative driving. These processes cannot properly be modelled in isolation from each other, since each influences the others. Furthermore, the wind produced will itself influence the other processes; wind interacting with magnetic field will carry off angular momentum ('magnetic braking'), which will influence both the rotation and the differential rotation of the star. Binarity is almost certainly an extra factor, affecting (by way of 'tidal friction') both rotation and differential rotation, and so presumably activity and mass loss. Consequently a solution to the overall problem appears to be still a long way off, and so we have to content ourselves for the time being with very empirical and approximate formulae like (2.71) to (2.76), possibly modified – presumably enhanced – by a binary companion.

The situation may be rather better, paradoxically, for the more drastic mass loss episodes: (c), (g) and (i). It should only be necessary to know (a) the stage in evolution where the episode occurs and (b) the amount of mass to be lost; the details of the mechanism by which it is lost might to some extent be secondary. For example, with AGB stars one might postulate that the envelope becomes unstable when the core reaches a certain mass that depends only on the initial mass, perhaps using an empirical relation, or else a theoretical relation like Eq. (2.62). The amount of mass lost will be (almost) all the difference between the core mass and the initial mass. With LBVs, one might postulate that the envelope becomes unstable when the star crosses or attempts to cross a line (the HD limit) in the HRD; very rapid mass loss continues until it retreats back across that line. For a supernova explosion, one might postulate an initial/final mass relationship, as for AGB stars, and so proceed in a similar manner.

2.5   Helium stars

There appear to be stars whose surface layers show a complete, or almost complete, absence of hydrogen. Such stars would be hard to understand in terms of 'normal' stellar evolution, but they can be understood, at least qualitatively, in terms of mass loss, whether by means of stellar winds as in the previous section, or of mass transfer between components of a binary as in the remaining chapters. Wolf–Rayet stars, at least of the WNE and WC types, show little or no hydrogen in their spectra.

There is a rather different kind of star, the 'hydrogen-deficient carbon star' or HdC star, which also appears to consist principally of helium, but often with an excess of C (relative to N, O) as well. An important group of these are R CrB variables, stars which at intervals of a few years show an abrupt *decrease* in luminosity, followed by a more gradual return to normal luminosity. Such stars are typically yellow supergiants when they are in quiescence. The variability appears to be due to erratic episodes of mass ejection, during which the ejected mass, rich in carbon, expands and cools until carbon-based dust forms, which temporarily obscures the star, at least at visual wavelengths. The HdC stars are not associated with young, massive stars, unlike WRs; they appear, rather, to be associated with the older, low-to-intermediate mass, population of the Galaxy. This makes them harder to understand, since significant mass loss in such stars is thought to take place only at such a late (AGB) stage that the remaining core should evolve directly to the white dwarf region, with little time spent in the region where HdC stars are actually found. For this reason, a binary-star mechanism for forming them – the merger of a C/O white dwarf and a He white dwarf – is attractive.

A kind of He star is expected, and indeed found, in some binary stars where one star has had its hydrogen-rich envelope stripped off by its companion. This does not appear to be a general explanation of either WR or HdC stars, however, since some of the former and most of the latter are not known to have binary companions. Such companions ought to be fairly easily visible.

Ideally, we might construct models of helium stars by starting with normal hydrogen-rich stars, and following their evolution subject to mass loss. Since, however, we do not have an a priori understanding of the mass-loss history of such stars, it is helpful to consider instead the evolution of stars starting from a hypothetical zero-age *helium* main sequence. The simple approximations (Eqs 2.16–17) of Section 2.2.2 apply equally to such stars, the only differences being that (a) μ is larger, and κ smaller, by about a factor of two from their values in Solar-mixture stars, and (b) the nuclear reaction rate formula in Eq. (2.15) involves a factor A which is a great deal smaller for helium burning than for hydrogen burning, and an η which is substantially larger ($\eta \sim 50$). Equations (2.10) and (2.12) say that we will have the same ζ, and therefore about half the luminosity, for a He star one quarter of the mass of a hydrogen MS star. Since luminosity depends steeply on mass, ZAHeMS stars are much more luminous than ZAMS stars of the same mass. The fact that the nuclear constant A in Eq. (2.15) is very much smaller, while L/M is larger, makes the central temperature hotter by about 3–5, and the radius smaller by a comparable factor from Eq. (2.13).

Although the central temperature is higher, the central density is higher still (from Eq. 2.14): cores are nearer to electron degeneracy, because this depends on the ratio $\rho/T^{3/2}$. The helium main sequence, like the hydrogen MS, terminates at low masses because electron degeneracy becomes important. Consequently the helium main sequence terminates at about $0.3\,M_\odot$, instead of about $0.08\,M_\odot$ for Solar composition.

Empirical fits, accurate to a few per cent, for the luminosity, radius and 'main sequence' lifetime of helium stars (cf. Eqs 2.1–2.4) over the range $0.32 \lesssim M \lesssim 20$ are

$$L = \frac{M^{10}}{1.2 \times 10^{-6} + 1.08 \times 10^{-3}M^5 + 2.63 \times 10^{-3}M^7 + 1.42 \times 10^{-4}M^{8.5}}, \quad (2.77)$$

$$R = \frac{M^2 + 0.1M^3}{0.36 + 3.24M + 1.75M^2}, \quad (2.78)$$

and

$$t_{\text{HeMS}} = \frac{2.985 + 51.88M^6 + 43.95M^{7.5} + M^9}{0.3597M^4 + 6.217M^{9.5}}, \quad (2.79)$$

with L, R, M in Solar units and t in megayears, as usual (Z. Han, private communication 1998).

We can evolve helium stars with the same numerical procedure as hydrogen stars. Their evolution is very similar in principle to the evolution of the helium core of an originally hydrogen-rich star, except that in the latter case the core is likely to increase its mass by 20–50% as a result of the H shell-burning which takes place at the same time as the He core-burning (Fig. 2.5b). There is also similarity between the evolution of He stars and of H stars, at least for an intermediate range of masses of the former, about 0.9–$2.7\,M_\odot$. These stars evolve from the helium main sequence to a red supergiant region which is hotter, but not by much, than the AGB of ordinary H rich stars (Figs. 2.1, 2.17). In the lowest part of this mass range stars can evolve to C/O white dwarfs, even without mass loss. Stars of moderate to high mass ($\gtrsim 2.4\,M_\odot$) ignite carbon non-degenerately, and go on presumably to a supernova explosion; while those in between develop degenerate C/O white dwarf cores which will be forced to C detonation if there is no mass loss, but which may settle down as white dwarfs if there is in fact sufficient mass loss from the cool supergiant envelope.

In an interesting contrast with H-rich stars, however, He stars *outside* the range of about 0.9–$2.7\,M_\odot$ do not expand to red giant dimensions. They remain always small and hot. Low-mass He stars expand by a modest factor up to central He exhaustion, but then as the He-burning shell eats its way outwards the star evolves steadily towards the white dwarf region. The reason for this, so far as the low-mass ($\lesssim 0.9\,M_\odot$) helium stars are concerned, is probably a combination of two things: firstly, the increase in molecular weight from He to C/O is more modest than from Solar mixture to He; and secondly, the C/O core is degenerate all the way to the He-burning shell (which is usually on the borderline of degeneracy in these low-mass stars), so that the degenerate core is not separated from the burning shell by an isothermal non-degenerate zone as in H-rich stars on the first GB. There is therefore no

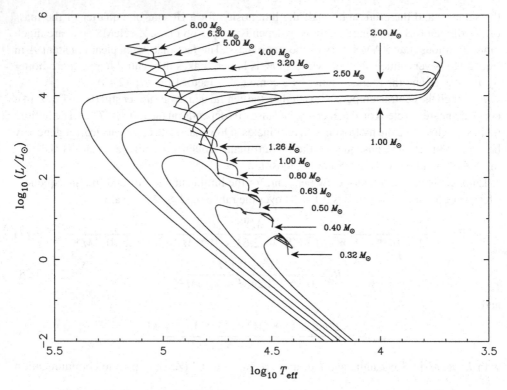

Figure 2.18 The theoretical HR diagram for He stars. Low-mass stars ($M \lesssim 0.63\,M_\odot$) evolve to white dwarfs without going to the red-giant region. Those of intermediate mass (0.8–$2.0\,M_\odot$) evolve to red giants, and those of $M \gtrsim 2.5\,M_\odot$ reach a supernova explosion while still very blue (Courtesy of Z. Han).

substantial 'soft' region in the star, i.e. one with $s \gtrsim 5/6$ (crudely $n \gtrsim 5$); such a soft region is necessary, though not sufficient, for an evolved star to become very centrally condensed, as outlined in Section 2.3.1.

He stars above about $2.2\,M_\odot$ also avoid expanding to the red supergiant region. They evolve from the ZAHeMS to dimensions that are comparable (by coincidence) to the H-rich MS, but not larger (Fig. 2.18). The lack of drastic expansion may be due to the fact that the C-burning core produces a luminosity comparable to the He-burning shell, so that the shell does not dominate the overall structure as it tends to in the intermediate-mass He stars, and in H-rich stars of all masses. A shell which is weak, either because of too slight a change in molecular weight, or too small a contribution to the total luminosity, or both, will not lead to a substantial soft region, such as could cause a star to become giant-like in structure.

2.6   Unsolved problems

An ideal stellar evolution code would (a) implement a believable set of mathematical–physical propositions and (b) give good agreement with observed stars. We are some way from this at present, but apparently not a *very* long way. To accommodate (b), several ad hoc recipes have to be incorporated that may violate (a). They are:

(i) a model of convective heat transport – the mixing-length theory with only one parameter does surprisingly well, but cannot be given much theoretical weight

(ii) a model of convective overshooting – although we would suppose that a reliable model of convective heat transport would automatically include the right amount of overshooting

(iii) a model of semiconvective mixing – although this may not be separate from the first two. If one had a mathematical–physical model of convection with no free parameters, and satisfying (b), a model for semiconvection would probably be implicit within it

(iv) a model for the diffusive separation of elements, starting with the gravitational settling of helium but ultimately including all species; such a model exists (Richer, Michaud and Turcotte 2000), but I have not incorporated it here

(v) a model for mass loss, including the very different regimes of hot and cool, luminous and faint, in the Hertzsprung–Russell diagram

(vi) a model for rotation, and its redistribution within a star in response to evolutionary changes

(vii) a model for dynamo activity

(viii) a model for photo spheres, including wind effects.

Most or all of these processes are interdependent, although I may have, of necessity, to treat them separately for the time being.

Two aspects of stars that I have barely touched are their rotation and their magnetic fields. It is not clear to what extent these may actually influence the long-term evolution. It seems reasonable to suppose in the first instance that their long-term importance is small, at least in the context of (effectively) single stars. One or other or both are implicated in most of the 'meagre' types of mass loss described in Section 2.4. But they may be substantially more important in the context of binaries. I will therefore discuss them in Section 4.4. In single stars, one possible long-term effect of a magnetic field (coupled with rotation, as it usually is) might arise in supernovae. Whether the entire envelope gets ejected, leaving a neutron-star remnant, or whether, alternatively, much of the envelope follows the core's implosion into a black hole, may be dictated by the field strength and the rapidity of the core's rotation as it collapses. Thus it could be the case that two stars of similar initial mass, say $30\,M_\odot$, could leave very different compact remnants, even if their evolution before core collapse was rather similar (Ergma and van den Heuvel 1998). The same reservation applies to AGB stars that become white dwarfs: it is possible that the rotational/magnetic history of the star plays a role in deciding at what core mass the envelope is finally stripped off.

The significance of photospheres – point (vii) – is that although the usual model of a plane-parallel Milne–Eddington atmosphere is probably good for most stars it becomes very unreliable for really distended stars, such as red supergiants. The region above optical depth unity may be more than 20% of the stellar radius in simplistic models, and may contain half the envelope mass. It is not clear that any spherically symmetrical, let alone plane-parallel, model will do: the atmosphere of a red supergiant may more closely resemble the flames from a log fire than the surface of an electric hot-plate. Any conclusions that we draw from stellar models are particularly uncertain in this area.

I would like to draw particular attention to the problem described in Section 2.3.5, which contrasts surprisingly with the results quoted in Section 2.2.10. Although non-interacting binaries *with high-quality data* accord reasonably well with theoretical models when both components are on the main sequence (provided that 'convective overshooting' of a judicious amount is included), the substantially fewer non-interactive binaries with data of almost equally high quality that contain evolved components (red giants and supergiants) give rather poor agreement, at least in five out of eighteen systems. We can invoke a merger in a former triple system to explain a sixth defaulter, but it is hard to believe that this can be the explanation of five more. I must emphasise here the importance of high-quality data. Any theory can cope with the observations if the uncertainties are of order 50%, which is quite common. But when data have an accuracy of order 5% or better it is not so easy to wriggle out of discrepancies.

One line of investigation that needs to be carried out in the future is direct three-dimensional numerical simulation of stellar interiors and envelopes. It should include magnetohydrodynamic effects. This is beginning to be within the range of modern hardware, but the software is a very challenging problem. Even with say 10^{10} meshpoints, which is about the minimum necessary to resolve a relatively simple star like the Sun, it will be necessary to have algorithms that efficiently move meshpoints to the regions where they are necessary. It is not always easy to to say a priori where these are. Of course three-dimensional simulations would be a supplement to, rather than a replacement of, one-dimensional simulations. In principle we might learn from three-dimensional modelling how to approximate a mixing-length parameter, a mass-loss parameter, a magnetic activity parameter or a rotational parameter in a simple one-dimensional formulation.

3

Binary interaction: conservative processes

3.1 The Roche potential

Interactions in close binaries are generally discussed in terms of the 'Roche potential' (Roche 1873, Kopal 1959, Kruszewski 1966). Suppose that two point masses M_1 and M_2, or equivalently two spherically symmetrical masses, orbit their centre of gravity (CG) in circles, their constant separation being the semimajor axis a. Then the angular velocity of the system is ω, and the orbital period is P, where

$$\omega^2 = \frac{GM}{a^3} = \left(\frac{2\pi}{P}\right)^2, \quad M = M_1 + M_2. \tag{3.1}$$

In a frame that rotates with the same angular velocity ω as the binary, a stationary free particle feels an effective force per unit mass (i.e. acceleration) $\mathbf{f}$, which is given by

$$\mathbf{f} = -\nabla\phi_R, \quad -\phi_R = \frac{GM_1}{|\mathbf{s} - \mathbf{d_1}|} + \frac{GM_2}{|\mathbf{s} - \mathbf{d_2}|} + \frac{1}{2}|\boldsymbol{\omega} \times \mathbf{s}|^2. \tag{3.2}$$

Here, $\mathbf{d_1}$ and $\mathbf{d_2}$ are the positions of the centres of the two stars, with

$$\mathbf{d_1}/M_2 = -\mathbf{d_2}/M_1 = \mathbf{d}/M, \quad \boldsymbol{\omega} \cdot \mathbf{d} = 0, \tag{3.3}$$

$\mathbf{d}$ being the vectorial separation of the two stellar centres of gravity ($\mathbf{d} \equiv \mathbf{d_1} - \mathbf{d_2}$), which is constant in the rotating frame. A non-stationary particle in the same frame will, in addition, experience a Coriolis acceleration $-2\boldsymbol{\omega} \times \dot{\mathbf{s}}$.

Note that:

(a) the gradient operator differentiates with respect to $\mathbf{s}$. I am using the symbol $\mathbf{s}$ rather than $\mathbf{r}$ because I will use $\mathbf{r}$ later to represent the position vector within a star, relative to the centre of the *star*. The origin of $\mathbf{s}$ is at the centre of gravity of the *binary*.

(b) $|\mathbf{d}| = a$. I use this notation because in non-circular orbits (see later) d, i.e. $|\mathbf{d}|$, varies with time while a, the semimajor axis, is constant.

In the same vein, but in contrast, I will always use ω to mean $2\pi/P$, even though in eccentric orbits (again, later) $\boldsymbol{\omega}$ and $|\boldsymbol{\omega}|$ are time-varying. The excuse, apart from the shortage of appropriate letters, is that $\mathbf{d} \times \dot{\mathbf{d}}/d^2$ is equal to and more useful than $\boldsymbol{\omega}$, in an eccentric orbit: see for example Eq. (4.7) below.

If the fluid of which either star is made is in hydrostatic equilibrium, then it must fill up a volume that is bounded by a closed equipotential surface of ϕ_R. It is easy to see that the nature

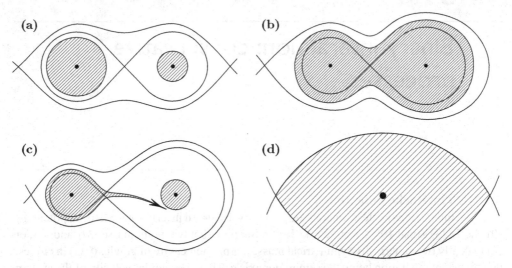

Figure 3.1 Some equipotential surfaces of the Roche potential – Eq. (3.2). For each of four different mass ratios ($q = 1, 2, 4, \infty$, $*1$ being to the right) the critical equipotentials passing through the inner and outer neutral points are indicated; some other equipotentials are also sketched. The case $q = \infty$ (d) has a neutral point all round the equator. A star can be in hydrostatic equilibrium only if it fills a closed equipotential surface, as in (a), (b) and (d). The left-hand star in (c) is unstable, and material flows towards the companion, being deflected by the Coriolis force. In (d), the rotation axis is in the plane of the paper, but in (a)–(c) it is perpendicular to the plane of the paper, in an anticlockwise sense.

of this system of equipotentials depends only on the mass ratio, since by virtue of Eqs (3.1),

$$\phi_R = \frac{GM}{a} f(s/a, q), \quad q = M_1/M_2. \tag{3.4}$$

Some examples are sketched in Fig. 3.1. For all finite $q > 0$, there exists a critical equipotential which is figure-of-eight shaped. The neutral point where this equipotential crosses itself (a saddle point) is called the 'inner Lagrangian point', L1, and the two lobes of the surface it encloses are the 'Roche lobes'. If both stars are sufficiently small, relative to a, that they can fit into closed equipotentials wholly within their respective lobes (Fig. 3.1a) we have a situation that can be expected to be stable. There are, however, two more neutral saddle points (L2, L3) collinear with L1 and the stellar centres, and a family of closed equipotentials exists, which surrounds both centres and also L1. Thus, we can expect a kind of star that is shaped like a peanut, as illustrated in Fig. 3.1b, where L2 is to the left. L3, which is of little physical significance, is beyond the surface through L2, and to the right (not shown).

If one star is too large to fit inside its Roche lobe, but the other is substantially smaller, as would happen in Fig. 3.1a if the left-hand star tried to evolve to substantially larger radius, then the envelope of the oversized star cannot be dynamically stable, and is liable to lose mass. This mass will fall through or near L1 into the potential well of the companion (Fig. 3.1c), being deflected by Coriolis force as it gains velocity.

These three situations are apparently all found among observed binaries. They are called 'detached', 'contact' and 'semidetached' configurations respectively (Kopal 1959), or D, C and S.

Strictly speaking, I might distinguish 'contact' and 'overcontact' systems (Wilson 2001): in the former, the stars exactly fill their lobes, and in the latter they overfill them, up to the same potential surface if they are indeed in hydrostatic equilibrium. However, the former is likely to be very rare, and so I will follow common practice in using the term 'contact' to cover both possibilities.

To start with, let us assume that although material may be transferred from one component to the other in the semidetached and contact cases, there is no net loss of material from the system; and also that no angular momentum is lost either. This is called the 'conservative' model. I will discuss a number of non-conservative processes in Chapters 4–6.

A single star that is rotating uniformly (i.e. all parts having the same angular velocity) can also be described by the Roche potential, using the limit $q = \infty$ (or 0), as in Fig. 3.1d. In this limit, L1, L2 and L3 degenerate into a single equatorial ring. Effectively this means that there is an upper limit to the possible rotation rate of a single, uniformly rotating star of given volume: with a greater rate of rotation the star would begin to shed matter at its equator. This matter would not necessarily flow away, despite the downhill slope of the Roche potential outside the critical surface; for the Roche potential assumes that all the material corotates, and there is no reason why the 'loose' matter shed at the equator should continue to corotate, even if it is pushed slightly further out on to the apparent downward slope.

For completeness, note briefly that there are two other neutral points, L4 and L5, which – somewhat remarkably – for all finite q make equilateral triangles with the stellar centres in the equatorial plane. These neutral points are minima, and so might be expected to be stable. However linear stability analysis in the rotating frame demands a Coriolis term in addition to the potential force, and a detailed analysis shows that L4 and L5 can be stable only if the mass ratio is fairly extreme: $q + 1/q > 25$. This is probably not commonly satisfied in stellar systems, but can be easily satisfied in star–planet systems. In the Solar System, the Trojan asteroids are located near the L4 and L5 points of the Sun–Jupiter binary.

The discussion of Chapter 2 was based on spherical (i.e. single, non-rotating) stars, but it can be applied fairly accurately to the distorted models required by Roche geometry. For each closed equipotential up to and including the Roche lobe, let us define an effective radius which is the radius of a sphere of the same volume as the interior of the equipotential. We then suppose as a first approximation that the structure of the star is the same at a given effective radius as if it were spherical, and that in particular the effective surface radius is the same – see Section 3.2.1 for a second approximation. For the critical equipotential, i.e. the Roche lobe, the effective radius R_L is given (Eggleton 1983a) by

$$\frac{R_L}{a} \equiv x_L(q) \approx \frac{0.49q^{2/3}}{0.6q^{2/3} + \ln(1 + q^{1/3})}, \qquad 0 < q < \infty, \tag{3.5}$$

$$\approx \frac{0.44q^{0.33}}{(1+q)^{0.2}}, \qquad 0.1 \lesssim q \lesssim 10. \tag{3.6}$$

The first approximation is accurate to better than 1% for all q. The second approximation is rather less accurate, but more convenient. It gives the ratio of the two lobe radii as

$$R_{L1}/R_{L2} \approx q^{0.46}, \qquad 0.1 \lesssim q \lesssim 10. \tag{3.7}$$

Table 3.1. *Roche-lobe radii and related functions*

q	R_L/a	$P_{cr}\sqrt{\bar{\rho}}$	R'_L	q	R_L/a	$P_{cr}\sqrt{\bar{\rho}}$	R'_L	a/a_{min}	P/P_{min}
∞	0.8149	0.157	∞	0	0	0.336	-1.67	∞	∞
50.0	0.6857	0.202	102	0.020	0.1259	0.363	-1.64	169	2200
20.0	0.6308	0.226	40.3	0.050	0.1670	0.370	-1.58	30.4	168
10.0	0.5803	0.250	19.5	0.100	0.2054	0.375	-1.48	9.15	29.7
8.00	0.5626	0.259	15.3	0.125	0.2192	0.376	-1.43	6.41	16.2
6.25	0.5423	0.269	11.6	0.160	0.2353	0.377	-1.35	4.42	9.29
5.00	0.5233	0.279	8.99	0.200	0.2506	0.377	-1.27	3.24	5.83
4.00	0.5039	0.290	6.87	0.250	0.2667	0.376	-1.16	2.44	3.81
2.50	0.4621	0.312	3.68	0.400	0.3036	0.372	-0.839	1.50	1.84
2.00	0.4420	0.322	2.61	0.500	0.3207	0.368	-0.623	1.27	1.42
1.60	0.4218	0.332	1.75	0.625	0.3392	0.364	-0.353	1.12	1.18
1.25	0.3997	0.342	0.996	0.800	0.3604	0.357	$+0.025$	1.03	1.04
1.00	0.3799	0.350	0.457	1.000	0.3799	0.350	$+0.457$	1.00	1.00

R_L/a from numerical integration, and critical period P_{cr} in days; mean density $\bar{\rho}$ in Solar units. $P_{cr}\sqrt{\bar{\rho}}$ is from Eq. (3.10); a/a_{min} and P/P_{min} from Eqs (3.13) and (3.14); R'_L, i.e. $d\log R_L/d\log M_1$, from Eqs (3.5), (3.13) and (3.16). The first four columns refer to $q \geq 1$, the next four to $q \leq 1$, and the last two to either.

At the limit $q = \infty$ the volume can be integrated analytically:

$$\frac{x_L^3(\infty)}{3} = \sqrt{3} - \frac{4}{3} - \ln\frac{2 + \sqrt{3}}{3}, \quad x_L(\infty) = 0.814\,885\,7. \tag{3.8}$$

The values of R_L/a given in Table 3.1 were all computed by direct integration; they can readily be compared with the approximations (3.5) and (3.6).

Over the whole range of q, the sum of the two radii, $x_L(q) + x_L(1/q)$, is within 5% of 0.78. This number illustrates the fact that while Roche lobes are not *very* spherical – which would give a value unity – they are also not *very* aspherical.

In several semidetached binaries, it may be the case that q cannot be determined directly from the radial velocity curves of both stars, because *2, the 'gainer' (of mass from the lobe-filling *1), may be surrounded by gas streams that distort or obscure its spectrum. Sometimes an estimate for q can, nevertheless, be made on the basis that the 'loser' (*1) not only fills its Roche lobe but also corotates with the binary. It may be possible to measure the rotational broadening of the relatively uncontaminated lines of *1, i.e. $V_{rot}\sin i$, i being the inclination of the rotation axis to the line of sight. The ratio of $V_{rot}\sin i$ to the orbital velocity amplitude K – which also contains a factor $\sin i$, and which also may be relatively uncontaminated – is a direct function of q:

$$\frac{V_{rot}\sin i}{K} = \frac{\omega R_L}{\omega a_1} = (1 + q)\,x_L(q), \tag{3.9}$$

a_1 being the radius of the orbit of the loser about the centre of gravity. From Table 3.1 it can be seen that the right-hand side is a fairly rapidly varying function of q, ranging from 0.23 at $q = 0.1$ to 0.76 at $q = 1$. Thus q may be estimated in the absence of clearly measurable motion of *2.

Either of formulae (3.5) or (3.6), with the aid of (3.1), gives a useful relation between the mean density $\overline{\rho}$ of a star that just fills its Roche lobe (so that $M_1 = 4\pi R_L^3 \overline{\rho}/3$), and a critical orbital period P_{cr}:

$$P_{cr} = \left(\frac{3\pi}{G\overline{\rho}}\right)^{1/2} \left(\frac{q}{1+q}\right)^{1/2} x_L^{-3/2} \sim 0.35 \sqrt{\frac{R^3}{M_1}} \left(\frac{2}{1+q}\right)^{0.2}. \qquad (3.10)$$

The quantities R_L/a and $P_{cr}\sqrt{\overline{\rho}}$ are tabulated as functions of q in Table 3.1. P_{cr} is in days if $\overline{\rho}$ is in Solar units (i.e. $\overline{\rho} = M_1/R^3$, M_1 and R being in Solar units). P_{cr} is the shortest period possible for a binary of given mass ratio into which a star of given mean density $\overline{\rho}$ can be fitted without overflowing its Roche lobe. In the approximate version of Eq. (3.10) we will usually take $q \sim 1$, since the q-dependence is very weak.

Throughout this chapter I will write the radius and luminosity of *1 for brevity as R, L rather than R_1, L_1, because I shall consider only *1 to have any internal structure; *2 can for most analytical purposes be treated as a point mass. Anything I derive for *1 can of course be generalised to *2 if *2 *does* have structure. However, I still have to distinguish M_1, M_2, and I will use M for the *total* mass.

The fact that P_{cr} (for a particular star of mean density $\overline{\rho}$) varies by less than a factor of 2.4 over the entire range of q, and by no more than a factor of 1.5 over the more restricted but realistic range $0.1 < q < 8$, makes Eq. (3.10) very useful. For example, the Sun (at its present radius) cannot fit into a binary with $P < 0.157$ days without overflowing its Roche lobe, and it cannot overflow its Roche lobe in a binary with $P > 0.377$ days; a value of 0.35 days, appropriate to $q = 1$, is wrong by less than 10% over the range $0 \lesssim q \lesssim 2$.

In Chapter 2 we noted that the radius normally increases, and so the mean density decreases, with age, and rather drastically in the Hertzsprung Gap and red giant stage. Thus we gain a little information on the evolutionary status of a star simply by knowing that it is in a binary of a particular period. For a binary of period P containing a star of radius R, the ratio P/P_{cr} is related directly to R/R_L:

$$\left(\frac{R}{R_L}\right)^3 = \left(\frac{P_{cr}}{P}\right)^2. \qquad (3.11)$$

Table 3.2 gives data regarding (theoretical) stars at up to seven significant points in their evolution. These are the stars whose Hertzsprung–Russell diagram is shown in Fig. 2.17. Core masses, ages (in megayears), radii, luminosities, temperatures and spectral types are indicated, and also the period P_{cr} in days: this is the orbital period such that the star would just fill its Roche lobe at the corresponding stage in evolution. The stages are the beginning (ZAMS) and end (TMS) of the main sequence, the beginning (BGB) of the giant branch, where the atmosphere switches from mainly radiative to mainly convective, and central helium ignition (HeIgn). The stellar radius usually then shrinks to a temporary minimum during core helium burning (CHeB), increases again but passes through another temporary minimum on the early asymptotic giant branch (EAGB), and then increases to a final maximum (RMAX) before the star collapses to a white dwarf (1–$4\,M_\odot$), or explodes as a supernova ($\geq 8\,M_\odot$). These models include an estimate of mass-loss rate, as described in Section 2.4. The stellar mass at the last stage tabulated is shown at the foot of the table; it will decrease further in the post-AGB evolution of masses 1–$4\,M_\odot$, and presumably also in the WR and post-WR stages of masses 32–$128\,M_\odot$, but perhaps not much further for the 8, $16\,M_\odot$ stars.

Table 3.2. *Critical binary periods for stars of various masses and evolutionary stages*

	Mass	0.25	0.50	1.00	2.00	4.00	8.00	16.0	32.0	64.0	128
ZAMS:	$M_{c,conv}$	–	–	–	0.293	0.906	2.360	6.48	18.2	47	109
	$\log R$	−0.571	−0.340	−0.051	0.210	0.376	0.555	0.726	0.894	1.07	1.27
	$\log L$	−2.023	−1.411	−0.148	1.258	2.378	3.440	4.372	5.146	5.77	6.28
	$\log T$	3.542	3.579	3.750	3.971	4.168	4.344	4.492	4.602	4.67	4.70
	spectrum	M3V	M0.5V	G6V	A0V	B5.5V	B1.5V	O9.5V	O5.5V	O3V	O3V
	$\log P_{cr}$	−1.012	−0.815	−0.532	−0.296	−0.192	−0.075	0.030	0.132	0.24	0.40
TMS:	$M_{c,He}$	–	–	0.117	0.254	0.634	1.640	4.70	12.9	31.3	59.5
	age	–	–	11 000	1 128	172.3	35.44	11.27	5.432	3.442	2.727
	$\log R$	–	–	0.231	0.683	0.782	0.962	1.182	1.505	2.335	1.714
	$\log L$	–	–	0.351	1.544	2.732	3.859	4.774	5.467	5.989	6.332
	$\log T$	–	–	3.734	3.806	4.054	4.246	4.364	4.377	4.091	4.488
	spectrum	–	–	G8IV	F5III	B8III	B2.5II	B0.5Iab	B0.5Ia	B6Ia+	B0Ia+
	$\log P_{cr}$	–	–	−0.110	0.418	0.417	0.535	0.720	1.067	2.201	1.22
BGB:	$M_{c,He}$	–	–	0.124	0.260	0.638	1.645	4.06	10.24	33.2	–
	age	–	–	11 110	1 140	174.2	35.79	11.39	5.487	3.474	–
	$\log R$	–	–	0.251	0.736	1.409	2.081	2.720	3.045	3.236	–
	$\log L$	–	–	0.351	1.308	2.541	3.678	4.761	5.448	5.979	–
	$\log T$	–	–	3.724	3.721	3.693	3.641	3.592	3.601	3.639	–
	spectrum	–	–	G9IV	G2III	G2II	K0Ib	K4Iab	K3Ia	K2Ia+	–
	$\log P_{cr}$	–	–	−0.079	0.498	1.355	2.214	3.028	3.378	3.562	–
HeIgn:	$M_{c,He}$	–	–	0.469	0.390	0.639	1.644	4.06	10.3	32.8	–
	age	–	–	11 990	1 177	174.8	35.86	11.38	5.486	3.477	–
	$\log R$	–	–	2.266	1.788	1.845	2.424	2.014	2.956	3.402	–
	$\log L$	–	–	3.425	2.908	3.139	4.126	4.856	5.482	6.059	–
	$\log T$	–	–	3.485	3.595	3.624	3.581	3.969	3.654	3.575	–
	spectrum	–	–	M3.5III	K3III	K2II	K4Ib	A2Ia	G8Ia	K7Ia+	–
	$\log P_{cr}$	–	–	3.003	2.078	2.010	2.729	1.969	3.245	3.818	–
CHeB:	$M_{c,He}$	–	–	0.519	0.411	0.800	2.053	4.79	12.7	16.9	–
	age	–	–	12 080	1 190	190.5	38.05	11.95	5.792	3.512	–
	$\log R$	–	–	1.056	1.037	1.493	1.738	2.853	1.765	2.334	–
	$\log L$	–	–	1.772	1.785	2.685	3.038	4.904	5.559	6.047	–
	$\log T$	–	–	3.677	3.690	3.687	3.902	3.562	4.269	4.104	–
	spectrum	–	–	G5III	G3III	G2II	A8Iab	M0Iab	B0Ia	B6Ia+	–
	$\log P_{cr}$	–	–	1.189	0.953	1.483	1.703	3.237	1.612	2.323	–
EAGB:	$M_{c,He}$	–	–	0.536	0.544	0.885	2.249	–	–	–	–
	$M_{c,CO}$	–	–	0.296	0.311	0.515	1.347	–	–	–	–
	age	–	–	12 140	1 365	209.9	40.40	–	–	–	–
	$\log R$	–	–	1.469	1.447	1.843	2.424	–	–	–	–
	$\log L$	–	–	2.354	2.421	3.145	4.132	–	–	–	–
	$\log T$	–	–	3.616	3.644	3.626	3.583	–	–	–	–
	spectrum	–	–	K2.5III	G9III	K0II	K5Iab	–	–	–	–
	$\log P_{cr}$	–	–	1.811	1.571	2.010	2.736	–	–	–	–
RMAX:	$M_{c,He}$	–	–	0.575	0.669	1.010	2.235	5.58	–	–	–
	$M_{c,CO}$	–	–	0.551	0.660	1.009	1.445	3.91	–	–	–
	age	–	–	12 250	1 372	211.3	40.48	12.75	–	–	–
	$\log R$	–	–	2.317	2.624	2.865	2.674	3.019	–	–	–
	$\log L$	–	–	3.591	3.978	4.488	4.486	5.126	–	–	–
	$\log T$	–	–	3.501	3.444	3.451	3.546	3.534	–	–	–
	spectrum	–	–	M3III	M6II	M6II	M1Iab	M1Ia	–	–	–
	$\log P_{cr}$	–	–	3.119	3.497	3.749	3.111	3.508	–	–	–
	M	–	–	0.629	0.928	1.77	7.74	13.5	13.7a	28.5[a]	61[a]

Masses, radii, luminosities in Solar units; critical orbital period in days; age in megayears; abbreviations defined in the text.

[a] Mass at last evolutionary state tabulated.

Figure 3.1b shows that stars can be larger than their Roche lobes, and still in hydrostatic equilibrium, provided that both components fill the same equipotential. But there is an 'outer critical lobe' which cannot be exceeded even in this case, as shown by the outermost of the three curves in Fig. 3.1b. If this outer surface is divided (somewhat crudely) into two by a plane through the inner Lagrangian point perpendicular to the line of centres, the effective radius of each of its two portions can also be defined and computed, as a function of q. It is, to better than 2%,

$$\frac{R_{OL}}{a} \equiv x_{OL}(q) \approx \frac{0.49q^{2/3} + 0.27q - 0.12q^{4/3}}{0.6q^{2/3} + \ln(1 + q^{1/3})}, \quad q \le 1,$$

$$\approx \frac{0.49q^{2/3} + 0.15}{0.6q^{2/3} + \ln(1 + q^{1/3})}, \quad q \ge 1.$$

(3.12)

The discontinuity of gradient at $q = 1$ is real, not an artefact of the approximation. However, the outer lobe is of much less practical significance than the inner lobe, i.e. the Roche lobe, even for contact binaries (Section 5.4).

Mass transfer from one component to the other by 'Roche-lobe overflow', or RLOF, is going to be an inevitable consequence of the evolutionary expansion of a star, for orbital period $P \lesssim 1000$ days (Table 3.2). For massive stars the limit is substantially larger. When the more massive star, which evolves faster, reaches its Roche lobe it will begin to shed its surface layers. As a first approximation, which we will have to reconsider subsequently, we suppose this mass transfer is slow, steady and 'conservative', i.e. that no mass or angular momentum leaves the system altogether. The mass lost by $*1$ (the loser) is assumed to be accreted by $*2$ (the gainer). Note that in this chapter we take $*1$ to be the star which is nearer to filling, or perhaps already fills, its Roche lobe, rather than to be the initially more massive star, although quite often they are the same. We also assume for the time being that the orbit remains circular; this, in fact, follows from the other assumptions since the eccentricity is an adiabatic invariant. Then using the basic Keplerian Eq. (3.1), the separation a and the period P are given in terms of the constant orbital angular momentum H_0, the constant total mass $M = M_1 + M_2$, and the varying mass ratio q by

$$a = \frac{H_0^2 M}{G M_1^2 M_2^2} = a_{min} \frac{(1+q)^4}{16q^2} = a_{min} \left(\frac{M^2}{4M_1 M_2}\right)^2,$$

(3.13)

$$P = \frac{2\pi}{\omega} = \frac{2\pi H_0^3 M}{G^2 M_1^3 M_2^3} = P_{min} \frac{(1+q)^6}{64q^3} = P_{min} \left(\frac{M^2}{4M_1 M_2}\right)^3.$$

(3.14)

The variation of a and P with q, relative to their minimum values a_{min}, P_{min} at $q = 1$, is also shown in Table 3.1. We normally define q as *greater* than unity initially, so that as q decreases through unity the separation at first decreases and subsequently increases. Thus for conservative RLOF, the separation and period reach their minima as the two masses pass through equality.

For a few semidetached binaries, a rate of period change can be measured. If this is roughly constant over decades or, better, centuries then it may allow us to estimate the rate of mass transfer during RLOF:

$$\frac{\dot{M}_1}{M} = -\frac{\dot{M}_2}{M} = \frac{\dot{P}}{P} \frac{q}{3(q^2 - 1)}.$$

(3.15)

Unfortunately, there are rather few systems in which the period changes at a steady rate for a long time (Fig. 1.6); and even when it does, there may be non-conservative processes at work (stellar winds, magnetic braking, gravitational radiation; Chapter 4) on something like the same timescale.

We should note in addition that the Keplerian relation (3.1), and consequential relations (3.13, 3.14), are *not correct* for stars which are distorted from spherical. A correction is necessary – Eq. (3.51) below – which depends on the quadrupole moment of each star, to lowest order. Thus a small variation of P with time might be due to varying quadrupole moment rather than varying mass. This might result from variation of the internal magnetic field during a Solar-like cycle (Applegate and Patterson 1987). Such cycles can be decades long, and possibly centuries long. A further effect is that the spin of the stars affects the orbital angular momentum. If the moments of inertia fluctuate, perhaps for the same reason, then the period could fluctuate slightly.

An important quantity is the rate R'_L at which the Roche-lobe radius responds to the mass M_1 ($\propto q/1+q$) of the star within the lobe, at constant H_o and M. We define R'_L as the *logarithmic* derivative, a convenient dimensionless expression obtainable from Eqs (3.13) and either (3.5) or (3.6):

$$R'_L \equiv \frac{d \log R_L}{d \log M_1} = (1+q) \cdot \left(\frac{d \log R_L/a}{d \log q} + \frac{d \log a}{d \log q} \right), \tag{3.16}$$

$$\approx 2.13q - 1.67, \quad 0 < q \lesssim 50; \tag{3.17}$$

R'_L is also given in Table 3.1. Note that $R'_L = 0$, i.e. the Roche-lobe radius is a minimum, at $q \approx 0.788$, using the more accurate expression (3.5).

The significance of R'_L is that it can be compared with R', the equivalent (logarithmic) response of a star's radius to its mass as determined by its internal structure. For example, Eqs (2.2) and (2.51) give $R(M)$ for a ZAMS star and for a WD respectively, and when differentiated logarithmically yield corresponding values for R'. As we shall see in Section 3.3, the rate, and the stability, of the mass-transfer process depends importantly on a comparison of R'_L with R'.

We shall see later that the simple picture of Roche geometry outlined above is hardly adequate in some cases, though it may well be adequate in most cases. I will therefore emphasise the assumptions on which it is based:

(a) The stars are treated as spherically symmetrical masses, so far as their gravity in Eq. (3.2) is concerned, despite the fact that they may fill, or even overfill, their Roche lobes. However, this is probably the least worrying assumption. Between 70% (on the lower main sequence) and 90% (on the upper main sequence) of a star's mass is within the inner 50% of its radius, and the equipotentials (3.2) become rapidly nearly spherical as one goes inwards from the Roche lobe. In fact a rather simple correction, based on hydrostatic and thermal equilibrium in a non-spherical potential field, can be applied (Section 3.2.1), and can be shown to be rather small in relation to other uncertainties.

(b) The stars are assumed to rotate uniformly, and with the same angular velocity as the system, i.e. all the material of the system is assumed at rest in the corotating frame. Tidal friction, which I will discuss briefly below (Section 4.2), is likely to enforce this, at least for the outer layers, but only in relatively close binaries. However, tidal friction, especially in fluid as distinct from solid bodies, is by no means well understood yet. Internal magnetic fields might also play an important role in enforcing solid-body rotation, since even a slight amount of differential rotation will lead to rapid amplification of any internal magnetic field.

(c) The orbit is taken to be circular, although it might be a Keplerian ellipse. Tidal friction can also be expected to circularise orbits; for tidal friction is a dissipative process, and two bodies in Keplerian orbits about their centre of gravity have the least energy, for a given angular momentum, if their orbits are circular. Most observed systems where one star is near to filling its Roche lobe are found to have circular orbits within the limits of observational accuracy (which may be rather wide, however).

(d) In Eqs (3.13) and (3.14) the intrinsic angular momentum of the stars is ignored compared with the orbital angular momentum. Once again, the fact that stellar mass is concentrated towards the centre makes this reasonable. We noted in Section 2.2.2 that main sequence stars are well approximated by polytropic gas spheres with $n \sim 3$. The radius of gyration k of $*1$ is

$$\frac{k^2}{R^2} \equiv \frac{I}{M_1 R^2}$$

$$\equiv \frac{2}{3M_1 R^2} \int_0^{M_1} r^2 dm \simeq 0.4 \left(1 - \frac{n}{5}\right)^{1.734} e^{-0.0133n - 0.0182n^2 + 0.0041n^3}$$

$$\sim 0.076 \quad \text{for} \quad n \sim 3, \tag{3.18}$$

if we use a polytropic approximation (requiring $n < 5$); see Table 3.4. Comparing the value for $n = 3$ with the value 0.4 for a uniform sphere ($n = 0$), we see that an $n = 3$ polytrope is quite centrally condensed. It is only when (i) the mass ratio is rather extreme, and (ii) the more massive component is close to its Roche lobe, that spin angular momentum can become comparable to orbital angular momentum. This is further discussed in Sections 4.2 and 5.1.

(e) Eqs (3.13) and (3.14) are true even if H_0 and M are not constant, provided the eccentricity remains zero. Nevertheless the entries of Table 3.1 assume constancy. However, several processes, outlined below (Chapter 4), may work to remove mass or angular momentum (or both) from the binary. Some of these processes are likely to reduce H_0 relatively faster than M, causing the orbit to shrink and speed up; but some may expand the orbit, or even disrupt it entirely.

3.2   Modifications to structure and orbit

3.2.1   *Effect on structure of a non-spherical potential*

There is a convenient semi-analytical model for a star that is distorted from the spherical by (a) supposedly uniform rotation, and (b) a binary companion. The model suffers from a number of disadvantages, outlined below, and cannot be considered as anything more than a recipe that yields a plausible but by no means definitive estimate of the first order consequences of such distorting effects. Nevertheless, the model has the advantage of surprising simplicity that may even outweigh its disadvantages, and so I will present its basis and its conclusions here, with the analysis relegated to Appendix B.

We suppose here that one of the two stellar components, $*2$ say, is a point mass, while only $*1$, whose internal structure is being considered, is an extended body. We work in a frame which rotates with $*1$, noting that this is not necessarily the same as a frame which rotates with the binary (although corotation was assumed in the previous section). We take $\mathbf{\Omega}$, the rotation of ∗1, to be a constant, or at any rate to vary slowly compared with the orbital timescale (unlike the instantaneous orbital angular velocity, if the orbit is eccentric). We also assume, rather less convincingly, that we can in the first instance neglect the velocity of material relative to the rotating frame, so that we take ∗1 to be in hydrostatic equilibrium. If the orbit is in fact eccentric, or if ∗1 is not corotating with the binary, the velocity cannot be strictly zero since there will be a time-dependent tidal distortion; its velocity field is determined in Appendix B, and given below – Eq. (3.36). However, the degree of distortion is really quite small even in the extreme when a star is close to filling its Roche lobe, and so it is not unreasonable to neglect it and assume that the star is always in hydrostatic equilibrium, even with a time-dependent potential due to the change of relative position of ∗2.

To maintain thermal equilibrium in a distorted star that is in hydrostatic equilibrium, it is necessary to introduce a meridional velocity field, $\mathbf{v}$, say – Eq. (3.39) below – in addition to the tidal velocity field. Fortunately $\mathbf{v}$ can be estimated to be very small over the bulk of the star, and so we can reasonably assume that the star is still in hydrostatic equilibrium:

$$\nabla p = -\rho \nabla \phi, \tag{3.19}$$

$$\nabla^2 \phi = 4\pi G \rho - 2\Omega^2. \tag{3.20}$$

Here ϕ is a combined gravitational–centrifugal potential, which includes the potential of the companion star (a point mass outside the object star), the centrifugal potential $\frac{1}{2}|\mathbf{\Omega} \times \mathbf{r}|^2$, as well as the self-gravity of the object star. This ϕ is not the same as ϕ_R in Eq. (3.2): firstly, ∗1 (say) is no longer being treated as a point mass; secondly, the origin is now at the centre of gravity of ∗1 rather than of the binary; thirdly, we are working in a frame which rotates with ∗1, and not necessarily with the orbit.

Equation (3.19) has the rather powerful consequence that both p and ρ must be constant on surfaces of constant ϕ. If, to start with, we think only about stars of uniform composition (e.g. ZAMS stars), then the molecular weight μ is constant and hence T, s are also constant on equipotential surfaces. In a radiative (i.e. non-convective) zone this means that the heat flux $\mathbf{F}$ is given by

$$\mathbf{F} = \chi_{\text{rad}} \nabla \phi, \quad \chi_{\text{rad}}(\phi) \equiv -\frac{4acT^3}{3\kappa\rho} \frac{dT}{d\phi}, \tag{3.21}$$

with χ being constant on equipotentials, as are ρ, T, $dT/d\phi$, and also κ, since we assume uniform composition.

In a convective zone the heat flux vector is not so straightforward. However, by compounding the simple but unjustifiable mixing-length theory (Section 2.2.3) with an equally simple and unjustifiable generalisation to distorted stars (Appendix B), we can approximate the heat flux by

$$\mathbf{F} = (\chi_{\text{rad}}(\phi) + \chi_{\text{con}}(\phi))\nabla\phi \equiv \chi(\phi)\nabla\phi. \tag{3.22}$$

The details of χ do not matter, as shown in Appendix B, but what does matter – from the point of view of being able to get simple results – is that χ, as with other variables, should be constant on equipotentials.

The equation of heat production and transport in a steady state is then

$$\nabla \cdot \chi \nabla T = \rho\epsilon - \rho T \mathbf{v} \cdot \nabla S = \rho\epsilon - \rho T \frac{dS}{d\phi}|\nabla\phi|v_\perp, \tag{3.23}$$

where ϵ is the nuclear reaction rate and $\mathbf{v}$ the meridional velocity field. The left-hand side is *not* constant on equipotentials, whereas $\rho\epsilon$ is, and so Eq. (3.23) gives the component $v_\perp$ of $\mathbf{v}$ normal to the equipotential that is necessary to balance the two sides; the tangential component is then given by continuity:

$$\nabla \cdot \rho\mathbf{v} = 0. \tag{3.24}$$

An explicit expression for $v_\perp$ is given below – Eq. (3.40).

Let us define the 'volume-radius' $r_*(\phi)$ of an equipotential surface by

$$\frac{4\pi}{3}r_*^3 = V(\phi), \tag{3.25}$$

where $V(\phi)$ is the volume contained within the equipotential. Then following the analysis of Appendix B our simple model gives the structure of a distorted star by

$$\frac{dp}{dr_*} \approx -\rho\frac{Gm}{r_*^2}\left(1 - \frac{2\Omega^2 r_*^3}{3Gm}\right), \tag{3.26}$$

$$\frac{dm}{dr_*} = 4\pi r_*^2 \rho, \tag{3.27}$$

$$\frac{dL}{dr_*} = 4\pi r_*^2 \rho\epsilon, \tag{3.28}$$

$$\frac{d\log T}{d\log p} \sim \min(\nabla_a, \nabla_r), \quad \nabla_r = \frac{3\kappa pL}{16\pi acGmT^4}\left(1 - \frac{2\Omega^2 r_*^3}{3Gm}\right)^{-1}. \tag{3.29}$$

L, m are the nuclear luminosity and the mass contained within an equipotential. Equations (3.26)–(3.29) are seen to be virtually independent of the fact that the star is distorted. Only one factor, omitted in Eq. (3.26) but included in Eq. (B8) of Appendix B, depends on a detailed knowledge of $\phi(\mathbf{r})$; and it is clear that the factor differs from unity in *second* order if ϕ differs from spherical in *first* order. Taking this factor to be unity, therefore, the only effect of binarity on internal structure turns out to be a weakening of gravity by way of the rotation of the star: the factor $4\pi Gm$ needs only to be replaced by $4\pi Gm - 2\Omega^2 V$ in the two places where it occurs. This is very easily incorporated into a stellar evolution code making the conventional assumption of spherical symmetry. Note that although Eq. (3.29) is written with only the lowest-order approximation to the temperature gradient $d\log T/d\log p$, the usual more elaborate mixing-length approximation can be used instead (Appendix A), provided ∇_r is modified as above to include rotation. Note also that the distorting effect of the gravitational field of $*2$ does not enter into Eqs (3.26)–(3.29) at all: this is because $\nabla^2\phi$ for that effect is zero within $*1$, whereas for the centrifugal part of the potential $\nabla^2\phi = -2\Omega^2$.

Table 3.3 shows that the effect on the star of including the rotational modification is not large, even if the star exactly fills its Roche lobe. The table lists some ZAMS models – cf. Table 3.2. Equal masses are assumed. The critical period is increased by nearly $\sim 4\%$ for most masses, but for the lowest-mass stars, which are largely or wholly convective, it is *decreased* by $\sim 1.5\%$. The effect would be larger at higher mass ratio, and also for single stars rotating at break-up (equivalent to infinite mass ratio).

The degree to which an internal equipotential of mean radius r – dropping the asterisk – departs from spherical can be represented by a function $\alpha(r)$, such that the radius in direction

Table 3.3. *Critical period on the ZAMS modified by including rotation: equal masses*

M	0.125	0.25	0.5	1	2	4	8	16	32
$\log P_{cr}$	−1.1998	−1.0118	−0.8158	−0.5314	−0.3039	−0.2183	−0.1139	−0.0099	0.0902
$\log P'_{cr}$	−1.2052	−1.0153	−0.8225	−0.5236	−0.2838	−0.1997	−0.0959	0.0078	0.1091

P_{cr} is the critical period for Roche-lobe overflow on the ZAMS as obtained when the effect of rotation on the structure is ignored – Table 3.2; P'_{cr} is the value when rotation is included according to Eqs (3.26) and (3.29). Periods are in days and masses are in Solar units.

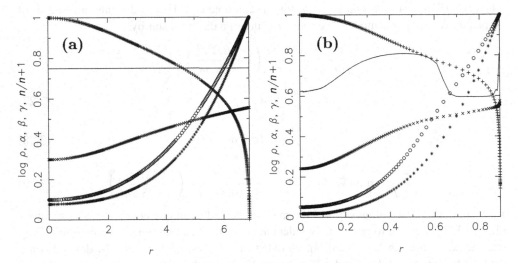

Figure 3.2 The variation of $0.1\log(\rho/\rho_c)+1$ (pluses), and $\alpha(r)/\alpha_1$ (asterisks), which measures the degree of distortion of equipotential surfaces. Also shown are $n/n+1$ (line), where n is the local polytropic index, $\beta(r)$ (circles), which determines the amplitude of the tidal velocity field – Eq. (3.36) – and $0.1\log\gamma(r)+0.5$ (crosses), which determines the tidal dissipation rate – Eq. (4.31). The models are (a) $n=3$ polytrope, (b) $1\,M_\odot$ ZAMS star.

θ from the symmetry axis is a factor $1-\alpha(r)P_2(\cos\theta)$ times the mean radius. As shown in Appendix B, α satisfies Clairault's equation

$$\frac{d^2\alpha}{dr^2} - \frac{6\alpha}{r^2} + \frac{8\pi r^3 \rho}{m}\left(\frac{1}{r}\frac{d\alpha}{dr} + \frac{\alpha}{r^2}\right) = 0. \tag{3.30}$$

Thus α depends only on the zero-order (spherical) distribution of mass, apart from a constant multiplicative factor, which is determined by the strength of the perturbation: see Eq. (3.33) below. For polytropes, $\alpha(r)$ is easily computed, and it can also be computed from tabulated stellar models. Two cases are shown in Fig. 3.2. The first-order aspherical mass distribution can then be allowed for, and generates a quadrupole moment that is best expressed in tensor form, since it consists of two contributions with different axes of symmetry: rotation ($\boldsymbol{\Omega}$), and the companion (**d**). The quadrupole tensor is

$$q_{ij} = q_{ij}^{\text{rot}} + q_{ij}^{\text{comp}}, \qquad q_{ij}^{\text{rot}} = -\frac{A}{6G}\,(3\Omega_i\Omega_j - \Omega^2\delta_{ij}),$$

$$q_{ij}^{\text{comp}} = \frac{M_2 A}{2d^5}\,(3d_i d_j - d^2\delta_{ij}), \qquad A = \frac{R^5 Q}{1-Q}, \tag{3.31}$$

where R is the mean surface radius of $*1$ and Q is a structure constant determined from α:

$$Q = \frac{1}{5M_1 R^2\,\alpha(R)}\int_0^{M_1}\left(5\alpha + r\frac{d\alpha}{dr}\right)r^2 dm. \tag{3.32}$$

This Q is clearly independent of a constant factor in α, so that Eq. (3.30) for α need only be integrated with the initial conditions $\alpha = 1$ (say) and $\alpha' = 0$ at $r = 0$. Then α can be scaled so that its surface value is dictated by the strength of the perturbing effect, either rotation or the companion or both:

$$\alpha^{\text{rot}}(R) = \frac{\Omega^2 R^3}{3GM_1}\frac{1}{1-Q}, \qquad \alpha^{\text{comp}}(R) = -\frac{M_2 R^3}{M_1 d^3}\frac{1}{1-Q}. \tag{3.33}$$

For a polytrope of index n, $0 \le n \le 4.95$,

$$Q \approx \frac{3}{5}\left(1 - \frac{n}{5}\right)^{2.215} e^{0.0245n - 0.096n^2 - 0.0084n^3} \qquad \pm 1.5\% \text{ rms}. \tag{3.34}$$

$Q \sim 0.028$ for an $n = 3$ polytrope, and 0.223 for $n = 3/2$. Q relates to the conventional 'apsidal motion constant' k_{AM} by

$$k_{\text{AM}} = \frac{1}{2}\frac{Q}{1-Q}. \tag{3.35}$$

In much of the literature, k_{AM} is called k_2, but here I will reserve suffix 2 for the other star (even if it has rather little structure).

In the frame that rotates with $*1$, the rotational contribution to the tidal distortion is constant (like the Earth's equatorial bulge) and the companion's contribution is time-varying (like the Earth's lunar tidal bulge). The time-varying part contributes a tidal velocity field **u** say, which, like the meridional field **v** of Eq. (3.23), has to be assumed to be small to justify hydrostatic equilibrium. It is shown in Appendix B to have the quadrupolar form

$$u_i = -\frac{3}{2}\frac{M_2 R^3}{M_1(1-Q)d^3}\,\beta(r)s_{ij}(t)x_j, \tag{3.36}$$

where

$$s_{ij} = \frac{1}{d^3}\frac{\partial d}{\partial t}\,(5d_i d_j - d^2\delta_{ij}) - \frac{1}{d^2}\left(d_i\frac{\partial d_j}{\partial t} + d_j\frac{\partial d_i}{\partial t}\right), \tag{3.37}$$

and

$$\beta = \frac{1}{\rho}\left(\rho(R) + \int_R^r \frac{\alpha}{\alpha(R)}\frac{d\rho}{dr}\,dr\right). \tag{3.38}$$

Figure 3.2 also shows $\beta(r)$ for a polytropic model and a ZAMS model. Contrary to its superficial appearance, Eq. (3.38) satisfies its necessary boundary condition $\beta = 1$ at the surface, even if the surface is approximated as polytropic with $\rho(R) = 0$. Table 3.4 gives a number of constants computed for polytropic models.

Table 3.4. *Various constants for polytropes*

n	0	1	1.5	3	3.5	4
C	1.00	3.29	5.99	53.6	146.9	543
k^2/R^2	0.400	0.261	0.205	0.0759	0.0468	0.0247
Q	0.600	0.342	0.223	0.0285	0.01041	0.00300
β_c	1.00	0.763	0.560	0.1008	0.0402	0.01170
$\overline{\gamma}$	1.00	0.610	0.339	0.0122	0.0020	0.00018
Darwin q_D	2.78	4.01	4.98	11.82	18.05	32.6

C is the central condensation, or central density over mean density – Eq. (2.44); k is the radius of gyration – Eq. (3.18); Q is the dimensionless quadrupole moment – Eq. (3.32); β_c is the tidal velocity coefficient at the centre, relative to unity at the surface – Eqs (3.36, 3.38); $\overline{\gamma}$ is the average dimensionless dissipation amplitude in the inner 25% of mass – Eq. (4.31); q_D is the critical mass ratio for the Darwin instability – Eq. (5.3).

If a star were a perfect (dissipationless) oscillator, its first-order tidal motion would consist of (a) a particular integral, the 'equilibrium tide' of Eq. (3.36), and (b) a collection of normal modes, determined by some initial conditions. Dissipation, i.e. viscosity, even if slight, will tend to remove the normal modes, leaving the equilibrium tide as the dominant motion. But even the equilibrium tide will be subject to dissipation. In most cases (see Section 4.2) this will lead ultimately to uniform corotation of the star with a circular orbit: tidal friction will stop dissipating mechanical energy only if and when the orbit is circular and the star rotates with the orbit. In Section 3.4.2, I will compute the rate of dissipation of the tidal velocity field (3.36).

Returning now to the meridional velocity field $\mathbf{v}$ of Eq. (3.23), the normal component $v_\perp$ is given – Eqs (B21) and (B52) – by

$$\rho T \frac{dS}{d\phi} v_\perp = \left(\frac{1}{|\nabla\phi|} \int |\nabla\phi| \, d\Sigma - |\nabla\phi| \int \frac{d\Sigma}{|\nabla\phi|} \right) \frac{d}{dV} \frac{L}{4\pi Gm - 2\Omega^2 V}, \tag{3.39}$$

$$\approx -\frac{8\pi}{3} \frac{R^3\Omega^2}{GM_1(1-Q)} \frac{r^2}{\alpha(R)} \frac{d(r\alpha)}{dr} P_2(\cos\theta) \frac{d}{dV} \frac{L}{4\pi Gm - 2\Omega^2 V}, \tag{3.40}$$

where $d\Sigma$ is an element of area on the equipotential surface. The distortion due to the companion leads to an additional velocity field of similar mathematical form, but with the *first* Ω^2 in Eq. (3.40) replaced by $-3GM_2/d^3$ and with the polar angle θ measured from the line of centres rather than the rotation axis. The tangential component $v_\parallel$ of the velocity field comes from Eq. (3.24), i.e. continuity.

In the bulk of the star, outside the core, L is constant, and so the V-derivative which is the final term of Eq. (3.40) has a factor $4\pi G\rho - 2\Omega^2$. This will usually vanish somewhere inside the star, unless the rotation is *very* slow, and so will divide the circulation into two distinct parts (Fig. 3.3). The factor $2\pi GM - 2\Omega^2 V$ does not vanish within or on the Roche lobe – Eq. (3.44).

A close look at $v_\perp$ shows some of the weaknesses of the model, however. Firstly, the model predicts that $v_\perp$ tends to infinity at the surface, and not to zero, as $\rho \to 0$. Presumably an infinite 'surface current sheet' must flow there, to close off the streamlines. If we are

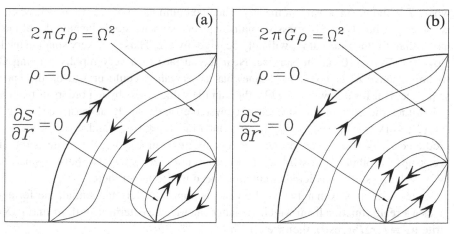

Figure 3.3 Patterns of circulation to be expected in a rotating star, on the basis of Eqs (3.40) and (3.43): (a) massive, with convective core and radiative envelope; (b) low-mass, with radiative core and convective envelope. Heavy lines with double arrows are surface current sheets. The rotation axis is the right-hand edge of each panel.

determined to try to believe the essence of the model, we can argue that (a) the photospheric density is not really zero, but small and given by the surface boundary condition (2.17); and (b) there is a turbulent surface layer of finite speed and thickness (perhaps as thick as the photosphere), the turbulence being driven by the shear that must be relatively large in the surface layers. Secondly, and more importantly, we also see that $v_\perp \to \infty$ at any boundary between a radiative and a convective zone, since $dS/d\phi = 0$ there. Thus, there must also be a surface current sheet at such layers to close the streamlines. Once again, we might imagine that in reality there is a turbulent boundary layer of finite thickness and speed rather than a surface current sheet.

Figure 3.3 gives an artist's impression of the circulation patterns that might be expected for a massive star (convective core, radiative envelope) and a low-mass star (radiative core, convective envelope) in uniform rotation. Note that the factor in parentheses in Eq. (3.39) is positive near the equator and negative near the poles, while the factor to the right of the parentheses is negative near the centre, positive near the surface, and vanishes on the surface $4\pi G\rho = 2\Omega^2$, provided that this is sufficiently far from the centre that $L \sim$ constant there.

From Eq. (3.40) we can estimate the order of magnitude of $v_\perp$. In the central regions we obtain

$$v_\perp \sim \left(\frac{v_{\text{rot}}}{v_{\text{cr}}}\right)^2 \frac{w_{\text{G}}^3}{v_{\text{cr}}^2} \frac{1}{|\nabla - \nabla_{\text{a}}|}, \tag{3.41}$$

where w_{G} (~ 0.035 km/s for the Sun) is the global convective velocity of the star as defined by Eq. (2.28), $v_{\text{rot}} = \Omega R$ (~ 2 km/s for the Sun) is the surface equatorial rotation velocity, and $v_{\text{cr}}^2 = GM_1/R$ (~ 450 km/s for the Sun) is the surface circular velocity. In the outer layers the corresponding estimate is

$$v_\perp \sim \frac{\overline{\rho}}{\rho} \left(\frac{v_{\text{rot}}}{v_{\text{cr}}}\right)^4 \frac{w_{\text{G}}^3}{v_{\text{cr}}^2} \frac{1}{|\nabla - \nabla_{\text{a}}|}, \tag{3.42}$$

where $\overline{\rho}$ is the mean density of the star. The maximum value of v_{rot}/v_{cr} is ~ 1, when the star is rotating at break-up. For massive main-sequence stars w_G could be an order of magnitude larger than for the Sun, but v_{cr} will only be larger by $\lesssim 2$. Thus $v_\perp$ is very small in the interior. However, $\overline{\rho}/\rho \sim 10^7$ at the surface. Near the surface of a very rapidly rotating G dwarf we might have $v_\perp \sim 0.03$ km/s. Somewhat larger values could prevail on the upper main sequence. But for a slow rotator like the Sun the value is less by ten orders of magnitude.

If either the core or the surface is convective (and usually at least one is), then from Eq. (2.28) $|\nabla - \nabla_a| \sim w_G^2/v_{sound}^2$. In the interior, $w_G \ll v_{sound}$ and $v_{sound} \sim v_{cr}$, and so the predicted $v_\perp$ is of the order of the global convective velocity w_G if the star is rotating near break-up. Throughout most of a surface convection zone we also have a small value of $|\nabla - \nabla_a|$, and so the predicted circulation speed is also larger there.

Since $v_\perp$ is approximately radial and $v_\parallel$ is approximately tangential, the former easily gives a stream function from which the latter can then be determined: if, from Eq. (3.40), we write $v_\perp \equiv f(r)P_2(\cos\theta)$, then we obtain

$$v_\parallel \sim -\frac{1}{2\rho r}\frac{d(r^2\rho f)}{dr}\sin\theta\cos\theta, \tag{3.43}$$

so that $v_\parallel \sim v_\perp$ in the interior, but in the surface layers $v_\parallel/v_\perp \sim r/H_P$.

Apart from the somewhat awkward fact that the predicted circulation pattern is singular at the boundaries of convective regions, and at the surface, there are potentially two other problems with the model described here (and in Appendix B). Firstly, the rotation is assumed to be uniform. But the circulation current must itself redistribute angular velocity, even if it was uniform to start with. At least in a binary component, tidal friction can be perceived as an agency that would lead to uniformity from a previous non-uniform state, but it does not necessarily dissipate non-uniform rotation so rapidly that we can always rely on it. Instability on a thermal timescale should occur for any rotation distribution if Ω is not constant on cylinders, or if $\Omega r^2 \sin^2\theta$ increases inwards (Goldreich and Schubert 1967, Fricke 1968), but it is not clear that this instability will tend to redistribute angular velocity towards uniformity. Evolution, leading to contraction of the core and expansion of the envelope, could produce differential rotation; it is often suggested that the evolved cores of massive red-supergiant stars must be rotating much more rapidly than their surfaces, since neutron stars are typically rotating very rapidly at 'birth'. Furthermore, the Sun and the gaseous massive planets are all seen to have non-uniform rotation on their surfaces. However, even if stars are typically rotating non-uniformly, the degree of non-uniformity need not be so drastic that the model is completely irrelevant. Since it is the surface layers that are most distorted by rotation, it is only necessary for the model's approximate validity that in the outer, say, 30% of the star's radius the rotation be uniform to, say, 20%.

Spruit (1998) has suggested that even a very weak magnetic field that permeates the entire star might be sufficient to enforce corotation, including the core. A very small poloidal magnetic field threading the interior would generate a toroidal field quite rapidly if wound up by even a very modest differential rotation. These toroidal loops should become buoyant and float to the surface, where they might rest if the surface is largely radiative or dissipate if it is largely convective. This process could erode the energy of differential rotation on a fairly short timescale. In that case the above problem might be insignificant. Of course, some other mechanism (Spruit 1998) is then needed to explain the rapid rotation of 'young' neutron stars, and some white dwarfs – see Sections 2.3.2 and 2.3.4.

Secondly, the model assumes uniform composition. Obviously this is violated by nuclear evolution. We can argue that, if we assume that μ as well as ρ, p is constant on equipotentials, then so is T from the equation of state, and hence also the nuclear burning rate; therefore the composition change is also constant, thus justifying in a circular manner the assumption about μ. However, this supposes that the circulation currents do nothing to redistribute the composition, which can only be a crude approximation – especially since the circulation is technically singular at the boundary of a convection zone. Perhaps convective overshooting, combined with rotationally driven mixing, does, in practice, keep the composition more or less constant on equipotentials, but this does not seem to be guaranteed. However, as above, we might argue that it is only the outer layers where the distortion is significant, and these outer layers are not normally much affected by mixing; if they are, perhaps they are mixed more or less to uniformity.

Almost certainly for a slow rotator like the Sun the model above of rotationally driven circulation is a complete irrelevance. There does exist an apparently meridional circulation, as shown by the fact that sunspot pairs tend to drift towards the poles (while rotating, and decaying) at a rate of ~ 15 m/s (Wang 1998). This is several orders of magnitude larger than expected from Eq. (3.43). Presumably this circulation has its origin in the hydrodynamics of turbulent convection subject to rotation. There is nothing in the standard model of rotationally driven circulation to explain the marked variation of angular velocity with latitude and depth that is observed in sunspot motions, and by helioseismology (Fig. 2.6b); this variation is presumably also a consequence of the combination of turbulence and rotation, particularly the Coriolis term. This term has been neglected, although it is more important than the included centrifugal term at least in regions that are convective. However, for rapid rotators, which are usually upper main-sequence stars with predominantly radiative envelopes, I believe that the standard model, in particular the two Ω-dependent modifications in Eqs (3.26) and (3.29), may be reasonable for determining the overall structure, and may even give a reasonable estimate of meridional circulation in the outer layers. It is unlikely to give useful insight into the effect of rotation on the central convective core.

The parameter $\alpha^{\rm rot}(R)$ of Eq. (3.33) is a convenient measure of the departure of the simple rotating model from a standard non-rotating single-star configuration. If $\Omega = \omega$ (i.e. corotation), the value of $\alpha_{\rm rot}(R)$ for a Roche-lobe-filling component is

$$\frac{\Omega^2 V}{4\pi GM} = (1 - Q)\,\alpha^{\rm rot}(R) = \frac{1+q}{3q}\,x_{\rm L}^3(q) \sim 0.028(1+q)^{0.4} \quad (q \lesssim 20), \qquad (3.44)$$

from Eqs (3.1) and (3.6). We see that $\alpha^{\rm rot}(R) \lesssim 0.05$, for q between 0 and 2 and $Q \lesssim 0.2$. For larger q it rises, reaching 0.2 at $q = \infty$ (Table 3.1). From Fig. 3.2 we see that $\alpha(r)$ decreases inwards, and so $\alpha(R)$ is the maximum value for $\alpha(r)$. The factor $\alpha(0)/\alpha(R)$ ranges from 0.45 in an $n = 1.5$ polytrope to 0.07 in an $n = 3$ polytrope. It is much smaller still in the cores of highly evolved stars. Thus for all q, $\alpha_{\rm rot}$ is rather small at the centre. Only for quite large q, and then only near the surface, is α not very small. Even then, it is hardly ever as large as ~ 0.1. The effect of $*2$ can similarly be estimated to be never more than moderate: according to Eq. (3.34), the perturbation differs by a factor $3/(1 + q)$ if corotation is assumed. It is therefore more significant, but still not large, if q is small ($M_1 \ll M_2$), and substantially less significant when q is large.

The convenient simplicity of the model with uniform rotation (with or without a binary companion) should not blind us to the possibility that some stars are in non-uniform rotation.

There is indirect evidence that, for example, the cool component of BM Ori (Popper and Plavec 1976), and the more massive but largely invisible component of β Lyr (Wilson 1974) are in non-uniform rotation. I have mentioned above analyses that suggest that such rotation is unstable on a thermal timescale, but some stars, including these two, may be evolving on a thermal timescale.

It may be possible to extend the analysis sketched here to include the case that $\Omega(\mathbf{r})$ is constant on *cylinders*, since in that case the centrifugal term $\Omega \times (\Omega \times \mathbf{r})$ is still derivable from a potential; but to have Ω constant on *spheres*, or on other surfaces, such as ellipsoids, is more difficult. However, it is not clear that this would be a significant advance on the relatively simple case of uniform rotation.

3.2.2   *Perturbations to Keplerian orbits*

A number of processes can modify a binary orbit slightly from the Keplerian model. Examples are:

(a) A distorted star has a quadrupole moment, so that its gravity is not quite that of a point mass. This leads to apsidal motion, i.e. rotation of the semimajor axis about the orbital axis. Distortion can be due both to the mutual gravity of the stars and to their rotation. If the rotation is oblique to the orbital axis, it can also cause precession, both of the spin axes and of the orbital axis.

(b) General relativity modifies the Newtonian gravitation, also leading to apsidal motion – Eq. (1.4).

(c) General relativity also leads to gravitational radiation, which progressively shrinks and circularises the orbit (Section 4.1).

(d) Tidal friction usually leads to circularisation of the orbit (Section 4.2), and also to paral-lelisation and synchronism of the spins with the orbit; but it can, in some circumstances, decircularise the orbit.

(e) Mass loss, in the form of winds presumed to be spherically symmetrical, from either or both stars, and mass transfer between the stars, whether through Roche-lobe overflow or through accretion by one star of part of the wind from the other, can expand or contract the orbit (Sections 4.3 and 4.6) and might change the eccentricity if the rate of mass loss or transfer depends on orbital phase (Section 6.5).

(f) A third body orbiting the binary, even at some considerable distance, can cause preces-sion, apsidal motion, and also periodic or aperiodic fluctuations in the eccentricity and inclination (Section 4.8).

Most of these phenomena can be modelled fairly simply using a procedure outlined in Appendix C. Provided the perturbation is sufficiently weak that it makes only a small change to the orbit in the course of one orbit, its effect can be estimated by averaging the perturbative force over exactly one Keplerian orbit.

A Keplerian orbit can be described compactly by a scalar, the energy $\mathcal{E} \equiv 1/2\,(\dot{\mathbf{d}} \cdot \dot{\mathbf{d}}) - GM/d$, and by two vectors, the angular momentum $\mathbf{h} \equiv \mathbf{d} \times \dot{\mathbf{d}}$, and the Laplace–Runge–Lenz (LRL) vector $\mathbf{e}$ which is given by $GM\mathbf{e} \equiv \dot{\mathbf{d}} \times \mathbf{h} - GM\mathbf{d}/d$. Both $\mathcal{E}$ and $\mathbf{h}$ are per unit reduced mass μ. The LRL vector is a vector parallel to the semimajor axis and of length e, the eccentricity. There is some redundancy: the seven components of $\mathcal{E}$, $\mathbf{h}$, $\mathbf{e}$ satisfy two identities, Eqs (C5). Appendix C shows how to calculate the rates of change of $\mathcal{E}$, $\mathbf{h}$ and $\mathbf{e}$, averaged over an orbit, for a given force $\mathbf{f}(\mathbf{d}, \dot{\mathbf{d}})$ in addition to the usual Newtonian gravitational force. In the case of problem (f) above, it is necessary to average over both the inner and then the outer orbit.

The LRL vector deserves to be better known than it appears to be, at least in the context of binary orbits. Its rate of change due to a perturbative force is quite readily calculated, and leads very directly to the rates of apsidal motion, of precession, and of circularisation (or decircularisation) of the Keplerian orbit. The vector triad $\mathbf{e}, \mathbf{q}, \mathbf{h}$, including a third vector $\mathbf{q} \equiv \mathbf{h} \times \mathbf{e}$, forms a very useful right-handed orthogonal triad fixed in the orbital frame but (possibly) rotating in an inertial frame; we refer to it as the 'orbital frame'. We write the unit vectors of the orbital frame as $\bar{\mathbf{e}}, \bar{\mathbf{q}}, \bar{\mathbf{h}}$.

When we allow for the quadrupole distortion of $*1$ – Eq. (3.31) – due to its rotation and to the tidal effect of $*2$, the force $\mathbf{F}$ on $*1$ is derivable – Appendix B(x) – from a new potential $\Phi(\mathbf{d})$:

$$\mathbf{F} = -\nabla_{\mathbf{d}} \Phi(\mathbf{d}), \quad \Phi(\mathbf{d}) = -\frac{GM_1 M_2}{d} - \frac{GM_2 d_i d_j \left(q_{ij}^{\text{rot}}(\mathbf{\Omega}) + \frac{1}{2} q_{ij}^{\text{comp}}(\mathbf{d})\right)}{d^5}, \quad (3.45)$$

where q_{ij} is the quadrupole tensor of Eq. (3.31). Note that Φ is different from both ϕ_R of Eq. (3.2) and ϕ of Eq. (3.20), partly because Φ is in an inertial frame whereas the others are in a frame that rotates with the binary, or with $*1$, and partly because ϕ_R only includes the monopole gravitational term for $*1$.

The perturbation $\mathbf{F}$ leads to a couple which has the effect – Appendix C(b) – of making the $\mathbf{e}, \mathbf{q}, \mathbf{h}$ frame rotate with an angular velocity $\mathbf{U}$:

$$\dot{\mathbf{e}} = \mathbf{U} \times \mathbf{e}, \quad \dot{\mathbf{q}} = \mathbf{U} \times \mathbf{q}, \quad \dot{\mathbf{h}} = \mathbf{U} \times \mathbf{h},$$

$$\mathbf{U} \equiv X\bar{\mathbf{e}} + Y\bar{\mathbf{q}} + Z\bar{\mathbf{h}} = \frac{M_2 A}{2\mu\omega a^5 (1 - e^2)^2} \quad (3.46)$$

$$\left[\Omega_h \bar{\mathbf{h}} \times (\bar{\mathbf{h}} \times \mathbf{\Omega}) + \left\{ \Omega^2 - \frac{3}{2} (\Omega_e^2 + \Omega_q^2) \right\} \bar{\mathbf{h}} + \frac{15 G M_2}{a^3} \frac{1 + \frac{3}{2} e^2 + \frac{1}{8} e^4}{(1 - e^2)^3} \bar{\mathbf{h}} \right]. \quad (3.47)$$

$\Omega_e, \Omega_q, \Omega_h$ are the components of $\mathbf{\Omega}$ in the orbital frame, and $A \propto R^5$ is given by Eq. (3.31).

The effect on $\mathbf{e}$ of Z alone – the component of $\mathbf{U}$ in the $\bar{\mathbf{h}}$ direction, i.e. the last two of the three terms in brackets in Eq. (3.46) – is to produce what is usually called 'apsidal motion'. The line of apses, in other words the major axis, parallel to $\mathbf{e}$, turns about the $\mathbf{h}$ vector at rate Z (rad/s), provided terms in $\mathbf{U}$ perpendicular to $\bar{\mathbf{h}}$ are ignored:

$$\dot{\mathbf{e}} = Z\bar{\mathbf{h}} \times \mathbf{e},$$

$$Z = \frac{M_2 A}{2\mu\omega a^5 (1 - e^2)^2} \left[\Omega^2 - \frac{3}{2} (\Omega_e^2 + \Omega_q^2) + \frac{15 G M_2}{a^3} \frac{1 + \frac{3}{2} e^2 + \frac{1}{8} e^4}{(1 - e^2)^3} \right]. \quad (3.48)$$

In a close binary, normally $\Omega \sim \omega$ as a result of tidal friction. Then $\Omega_e, \Omega_q \sim 0$; and if $M_1 \sim M_2$ the third term (due to the distortion by the companion) dominates the first (due to rotation) by a modest factor. But in binaries with $P \gtrsim 5$ days (and shorter periods for very young binaries), non-corotation is possible, and the first two terms may dominate if the star rotates rapidly. The second term gives a *negative* contribution to apsidal motion, unless the spin is parallel to the orbit. If, in fact, the spin is oriented randomly relative to the orbit, as arguably it might be at age zero, the expectation value of the first two terms together is zero.

If both stars are extended objects, Z will be the sum of two similar terms, one for each star (as will also X and Y). We must also add in the apsidal motion due to GR, Appendix C(a),

$$Z_{GR} = \frac{3GM\omega}{ac^2(1-e^2)}. \tag{3.49}$$

This is by no means negligible, even for orbital periods of a few days, and since it drops much more slowly with separation than any term of Eq. (3.48) it dominates for periods over ~ 10 days, provided that both components are still on the main sequence.

The term in $\mathbf{U}$ that is perpendicular to $\overline{\mathbf{h}}$, i.e. the first term in the brackets in Eq. (3.47), or equivalently the combination of X and Y terms in Eq. (3.47), produces precession. The total angular momentum, $\mathbf{H} \equiv \mu h + I\mathbf{\Omega}$, is conserved, and so $\overline{\mathbf{H}}$ is a vector fixed in an inertial frame. Hence we can write

$$\dot{\mathbf{h}} = \left[\frac{M_2 A H \Omega_h}{2\mu\omega I a^5(1-e^2)^2} \right] \overline{\mathbf{H}} \times \mathbf{h}, \tag{3.50}$$

which shows that $\mathbf{h}$ precesses about the fixed direction $\overline{\mathbf{H}}$ at a rate given by the expression in brackets.

There appears to be a common mistake in the literature regarding apsidal motion, in circumstances where there is or may also be precession. The rate of turning of the line of apses is often directly measurable, and is generally compared to the quantity Z, which is the sum of Eqs (3.48) and (3.49). However, as shown in Section 4.8, the observed rate of apsidal motion depends on X and Y, as well as Z. There are not many known binaries where there is clear evidence of non-parallel rotation as well as measurable apsidal motion, but I shall briefly discuss one example in Section 4.2.

Processes (c) to (e) above are non-conservative, and so I shall leave the details to Chapter 4 and Appendices B, C. I will also leave process (f) to there, although it is conservative. It is sufficient for the present to note that process (d), the frictional dissipation of tidal motion, will tend to (i) make the intrinsic spin align itself parallel to the orbit, (ii) bring the spin into 'pseudo-synchronisation' with the initially eccentric orbit, and (iii) on a slower timescale, reduce the eccentricity of the orbit to zero. Consequently, it is reasonable to start by assuming that close binaries have circular orbits and corotating components, as is commonly (though not universally) observed.

The force (3.45) leads to a revision of the basic Keplerian relation (3.1) between period and separation. If we assume that stellar spin is parallel to orbital spin, and if in addition we add in GR (Appendix C), we obtain for circular orbits

$$\omega^2 = \frac{GM}{a^3} \left[1 + \frac{A\Omega^2}{2GM_1a^2} + \frac{3AM_2}{M_1a^5} + \frac{3GM}{c^2a} \right]. \tag{3.51}$$

The present accuracy of most orbital determinations does not make this correction worth while – except for radio pulsars where, in fact, several more corrections are necessary – but it will no doubt be more important in the future.

3.3   Conservative Roche-lobe overflow

When a star in a circular orbit expands to fill, and then overfill, its Roche lobe – Fig. 3.1c – it will start to lose mass to its companion. For the time being we make the 'conservative' assumption that total mass and orbital angular momentum are constant.

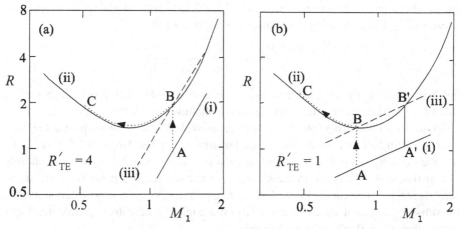

Figure 3.4 Schematic behaviour of Roche-lobe radius and stellar radius as functions of primary mass during evolution governed by the simplistic relation (3.52). Units are arbitrary, except that the total mass is 2 units. The curves are (i) the ZAMS radius, (ii) the Roche-lobe radius, (iii) the radius at time $t = t_{NE} \log 2$. The star starts on curve (i), at point A or A′, and evolves vertically until it reaches curve (ii) at B or B′. From B in either panel, it can proceed to evolve along curve (ii) to C, losing mass while still evolving on a nuclear timescale. In (b) it cannot do this from point B′, since curve (ii) is steeper than curve (iii) there.

Whether or not the mass transfer can proceed in a steady stable manner depends largely on the relative rates of change of stellar radius, and of Roche-lobe radius, with respect to changes of the loser's mass.

3.3.1 Effect of RLOF on the loser

Let us consider first a simple hypothetical case where *2, the 'gainer', is just a point mass, and where the internal structure and evolution of *1 imposes a radius–mass–age relation

$$\log R = \log R_0 + R'_{TE} \log \frac{M_1}{M_0} + \frac{t}{t_{NE}}. \tag{3.52}$$

Here R_0 and M_0 are the initial mass and radius of *1, and t_{NE} is a constant nuclear-evolution timescale, on which the star's radius increases with age t. The ZAMS expression for R is a power law with constant slope $R' = R'_{TE}$ (using a prime to denote a *logarithmic* derivative); the suffix TE stands for 'thermal equilibrium'. If the star's mass is changed at any non-zero age the $(\log R, \log M)$ relation has the same slope (for simplicity) as at zero age. Figure 3.4a shows, in the $(\log M_1, \log R)$ plane, a (very unrealistic) case where $R'_{TE} = 4$. The star starts at $t = 0$ from point A with a radius equal to half its Roche-lobe radius, so that when $t = t_{NE} \log 2$ it just fills its Roche lobe at point B. Curve (i) is the ZAMS, curve (ii) is the Roche-lobe radius relation, and curve (iii) is the stellar radius–mass relation at $t = t_{NE} \log 2$.

Table 3.1 shows that $R'_L = 2.61$ at $q = 2$. The fact that $R'_L < R'_{TE} = 4$ at B in Fig. 3.4a means that it is possible for *1 to evolve along (but actually very slightly above) the curve $R = R_L$ from B to C, decreasing M_1, i.e. transferring mass, on roughly the nuclear timescale t_{NE}. The mass-loss rate at time $t > t_N \log 2$ is found by solving the implicit equation

$$\log R_L(M_1) = \log R(M_1, t), \tag{3.53}$$

obtained by equating the lobe radius – Eqs (3.5 and 3.11) – to the stellar radius as approximated by Eq. (3.52). Differentiating this equation wrt time, we obtain

$$\frac{d \log M_1}{dt} = -\frac{1}{t_{NE}} \cdot \frac{1}{R'_{TE} - R'_{L}}. \tag{3.54}$$

This has a physically acceptable solution, i.e. one with $\dot{M}_1 < 0$, provided that $R'_{TE} > R'_{L}$, which is just the condition that curve (ii) has a shallower slope than curve (iii).

Figure 3.4b shows two (more realistic) cases with $R'_{TE} = 1$. In each case the star starts (A or A′) with half its Roche-lobe radius, but with $q \sim 0.8$ at A or $q \sim 1.5$ at A′. In the first case we have $R'_{L} < 0 < R'_{TE}$ at B, and so evolution can proceed A → B → C entirely on a nuclear timescale. However, in the second case the star reaches point B′ where $R'_{L} \sim 2.61$, which is *steeper* than curve (iii) with $R' = R'_{TE} = 1$, and so there is no solution in which the star subsequently evolves along curve (ii) on a nuclear timescale. Equivalently, Eq. (3.54) implies that $\dot{M}_1 > 0$, which is unphysical.

For $R'_{TE} = 1$ the condition $R'_{L} < R'_{TE}$ will be satisfied for any $q \lesssim 1.25$ (see Table 3.1), but for $R'_{TE} \lesssim 0.46$ the condition cannot be satisfied for any $q \gtrsim 1$. In fact, on a realistic ZAMS – Eq. (2.2) – $R'_{TE} \sim 0.5$ for $M \gtrsim M_\odot$, and increases to ~ 1 for $0.1\, M_\odot \lesssim M \lesssim M_\odot$. Thus, there is only a rather restricted range of initial masses and mass ratios in which Roche-lobe overflow (RLOF) *starts*, and continues, on a nuclear timescale. However, it does not follow that in all other cases something catastrophic must happen; it only follows that mass transfer must accelerate beyond the nuclear timescale.

Such a simplistic approximation as Eq. (3.52) does not allow for the fact that once the timescale of mass loss approaches the *thermal* timescale, which is ~ 1000 times shorter than the nuclear timescale – Eq. (2.43) – the luminosity and radius of the star can be significantly altered (Crawford 1955, Morton 1960). Equation (3.52) can be seen as giving the 'thermal equilibrium' radius of a star; but this must be modified if a star is out of thermal equilibrium.

The material of the star expands and cools off as it rises to the surface from deep within the gravitational potential well of the star, and this process will absorb or release heat that would otherwise diffuse down the internal temperature gradient. Since most of the temperature and pressure gradient is concentrated near to the surface of the star, it is reasonable to use a 'steady, thin-shell' approximation (Paczyński 1967) for the rate of energy release ϵ. If the star changes its mass M_1 at the rate $\dot{M}_1$, there is an effective thermal-energy source of strength

$$\epsilon_{th} = -T \left(\frac{\partial S}{\partial t}\right)_m \equiv T \left(\frac{\partial S}{\partial m}\right)_t \left(\frac{\partial m}{\partial t}\right)_S \tag{3.55}$$

$$\approx T \frac{\partial S}{\partial m} \dot{M}_1 \tag{3.56}$$

$$= -\frac{C_P T G m}{4\pi r^4 p}(\nabla - \nabla_a)\dot{M}_1, \tag{3.57}$$

from Eq. (2.28), to be added on top of other energy sources, i.e. nuclear reactions. Equation (3.57) assumes a uniform composition, which is usually the case in the outer layers where most of the thermal energy release takes place (Fig. 3.5).

The quantity m in Eq. (3.55) is a Lagrangian mass coordinate, specifically the mass contained within an equipotential surface of the family (3.2). Clearly if we think of mass as flowing steadily through an (almost) constant entropy profile that is very steep in the outer

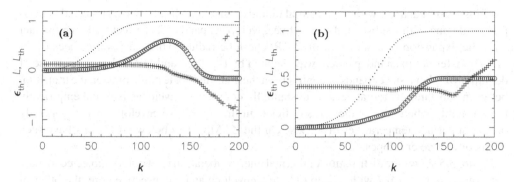

Figure 3.5 Dots: the luminosity L in a star losing mass at a slow, steady rate. Pluses: ϵ_{th}, the thermal energy generation rate from Eq. (3.55). Circles: the thermal luminosity L_{th}. The abscissa is k, the number of the mesh zone in the discretised model, $k = 0$ at the centre, $k = 200$ at the photosphere. Only alternate meshpoints are plotted. (a) $4\,M_\odot$, with radiative envelope and convective core; the negative contribution from the radiative envelope ($k \gtrsim 100$) ultimately dominates. (b) $0.4\,M_\odot$, with radiative core and convective envelope; the thermal term is positive throughout, the convective boundary being at $k \sim 108$. The vertical scales are quasi-logarithmic (actually, arcsinh), and allow positive and negative values to be distinguished; L, L_{th} are not on the same scale. In (a), some erratic behaviour near the surface is due to the presence of two narrow convection zones.

layers of the star, then we can expect that $(\partial m/\partial t)_S \sim \dot M_1$. As a first approximation it is helpful to think of $T(M_1 - m)$ and $S(M_1 - m)$ as given, having the values of the undisturbed, or mass-constant, star, provided that the ratio of the rate of energy release, Eq. (3.55) integrated over the star, to the unperturbed nuclear luminosity L_N is not large. This ratio can be estimated with the approximation (3.57), particularly if, for illustrative purposes, we assume that $\nabla - \nabla_a$ and $C_P T\rho/p$ are constants in the outer layers (where most of the contribution comes from):

$$\frac{L_{th}}{L_N} \approx -\frac{\dot M_1}{L_N}\frac{C_P T\rho}{p}(\nabla - \nabla_a)\int_{*1}\frac{Gm}{r^2}\,dr$$

$$\sim -\frac{5}{2}(\nabla - \nabla_a)\frac{GM_1\dot M_1}{RL_N}. \tag{3.58}$$

The last term on the right is the ratio of gravitational to nuclear luminosity, but it is also the ratio of the Kelvin–Helmholtz timescale,

$$t_{KH} = \frac{GM_1^2}{RL_N}, \tag{3.59}$$

to the mass-loss timescale $|M_1/\dot M_1|$. For main-sequence stars we can, from hydrostatic equilibrium, say that $GM_1/R \sim \Re T/\mu$ – cf. Eq. (2.13) – i.e. that the thermal and gravitational energies are comparable, and we refer to both timescales as the Kelvin–Helmholtz timescale $t_{KH} \sim \Re T M_1/\mu L_N$.

The sign of L_{th} is important. Stars with outer layers that are stable to convection necessarily have $\nabla < \nabla_a$, and so the outer layers absorb energy as the star loses mass; the star's luminosity is decreased. Stars with convective outer layers have $\nabla > \nabla_a$, and so in such stars mass loss will *increase* the luminosity (Paczyński 1967). The importance of the sign of L_{th} is that it is found in practice that the star's radius tends usually to vary in the same way as its luminosity, so that a star with a substantial radiative envelope shrinks when subjected to mass loss while one with a substantial convective envelope expands. This effect will then interact with the expansion or contraction of the Roche-lobe radius, which also varies because of mass transfer but in an independent way dictated by Eqs (3.5) and (3.13). For the purpose of deciding whether a star expands or contracts on thermal-timescale mass transfer, the transition between substantially radiative and substantially convective envelopes is found empirically to be roughly when 50% of the star's radius is in the convective envelope, which generally occurs at surface temperatures of $\sim$5 kK. On the ZAMS, stars below $\sim$0.75 $M_\odot$ have such deep convective envelopes.

Figure 3.5 shows the distribution of thermal energy liberation in two stars subjected to mass loss, one a star of 4 $M_\odot$ with a deep radiative envelope and a convective core, the other of 0.4 $M_\odot$ with a deep convective envelope and a radiative core. Because the quantities plotted can vary by large factors, and also change sign, they are shown on a quasi-logarithmic scale which nevertheless preserves sign; the units are arbitrary. In the 4 $M_\odot$ star there is energy *release* in the convective core (out to $k \sim 100$), and indeed some way beyond it (to $k \sim 125$), but the energy absorption in layers further out ($k \sim 125$–170) makes for a negative total contribution to the star's luminosity. Beyond $k \sim 170$ there is so little mass that although the energy generation rate is large, and fluctuating in sign because of two small convection zones, it contributes little to the luminosity. That the rate is positive in the radiative core, when one might expect it to be negative from Eq. (3.57), is because the steady thin-shell approximation only works in the surface layers.

In the other star (Fig. 3.5b) the energy generation rate is positive throughout. Although the rate is largest near the surface there is little mass there, and most of the thermal luminosity comes from $k \sim 80 - 130$, straddling the radiative/convective boundary at $k \sim 108$.

We can now generalise the simplistic example of Eq. (3.52) to include a term with a coefficient R'_{TD}, which allows for thermal disequilibrium:

$$\log R \sim \log R_0 + R'_{TE} \log \frac{M_1}{M_0} + \frac{t}{t_{NE}} + R'_{TD} \, t_{KH} \frac{d \log M_1}{dt}, \qquad (3.60)$$

where R'_{TD} is a coefficient of order unity and is positive for largely radiative stars and negative for largely convective stars. The apparent problem in Fig. 3.4b, that $*1$ wants to increase its radius as a result of nuclear evolution, but also wants to decrease it if it is to keep to its Roche-lobe radius, is solved, at least provisionally, by the introduction of the R'_{TD} term in Eq. (3.60), provided that $R'_{TD} > 0$. For as $|\dot{M}_1|$ increases, the extra term is negative and so can allow the star to have a radius less than its thermal-equilibrium radius. This shrinkage gives a degree of negative feedback, allowing the star to remain close to, but just above, its lobe radius, i.e. with $0 < \Delta R \ll R_L$. In Section 3.3.2 we analyse the stability a little more closely, and find a rather more general condition than the provisional one here, $R'_{TD} > 0$.

We can estimate the degree of overfill necessary by considering the hydrodynamics of the compressible flow near the L1 neutral point (Fig. 3.1c). The flow must pass through a transition, and this allows us to estimate the mass-loss rate from Bernoulli's equation (Jedrzejec, quoted by Paczyński and Sienkiewicz 1972). As usual, in such analyses we have to make a geometrical approximation, taking the flow to be nearly parallel to an axis (the z-axis in cylindrical polars, Fig. 3.6a). This is somewhat in conflict with the reality that the critical Roche equipotential through the L1 point crosses itself at an angle in excess of 60°. Nevertheless, the kind of estimate that emerges is usually wrong only by factors of order unity.

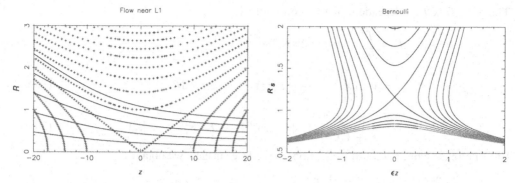

Figure 3.6 (a) Roche-lobe overflow approximated as a cylindrical stream of slowly varying cross-section. Equipotentials are shown by pluses, flow lines are solid. The radius of the outermost streamline is $R_s(z)$. To justify the approximation that the velocity is mainly in the z direction we have enlarged the z scale by a factor of 10, i.e. we have taken $\epsilon = 0.1$ in Eq. (3.61). (b) Solutions of Eq. (3.68), in units where $F = K = \omega = 1$, for various values of the upstream potential ϕ_s. We take $n = 3/2$. The only solution which has large R_s at the left and small R_s at the right is one which passes through the singular point, and has $\phi_s = 1.2(2.5)^{1/6}$, $R_s(0) = (2.5)^{1/6}$.

We assume that the potential near the L1 point can be approximated, apart from an additive constant, by

$$\phi \approx \omega^2(R^2 - \epsilon^2 z^2), \qquad (3.61)$$

where ω is comparable to the orbital frequency, and ϵ is small (but actually $\epsilon^2 > 2$). We then assume that $\mathbf{v}$ is approximately in the z direction and a function only of z, i.e. $\mathbf{v} \approx (0,\ 0,\ v(z))$, so that the motion is irrotational ($\nabla \times \mathbf{v} = 0$). We further assume that the fluid is isentropic and adiabatic, so that the pressure is directly related to the density:

$$P = \frac{C}{n+1}\,\rho^{1+1/n}, \quad \frac{1}{\rho}\nabla P = C\nabla\rho^{1/n}. \qquad (3.62)$$

Bernoulli's equation then says that throughout the flow

$$C\rho^{1/n} + \frac{1}{2}v^2 + \phi = \text{constant} = \phi_s, \quad \text{say}, \qquad (3.63)$$

where ϕ_s is the potential on the free surface far upstream ($v \sim 0$, $\rho = 0$); ϕ_s is a measure of the extent to which the star overfills its lobe. Then at a general z, both within the fluid and on the free surface $R = R_s(z)$,

$$C\rho^{1/n} + \frac{1}{2}v^2(z) + \omega^2\{R^2 - \epsilon^2 z^2\} = \phi_s = \frac{1}{2}v^2(z) + \omega^2\{R_s^2(z) - \epsilon^2 z^2\}. \qquad (3.64)$$

Hence

$$\rho = \frac{\omega^{2n}}{C^n}\{R_s^2(z) - R^2\}^n. \qquad (3.65)$$

The mass flux F, independent of z by continuity, is

$$F = \int_0^{R_s(z)} v(z)\rho \, 2\pi R \, dR$$

$$= v(z) \int_0^{R_s(z)} \frac{\omega^{2n}}{C^n} \{R_s^2(z) - R^2\} 2\pi R \, dR \tag{3.66}$$

$$= \frac{\pi}{n+1} \frac{\omega^{2n}}{C^n} v(z)\{R_s(z)\}^{2n+2} \equiv K v R_s^{2n+2}, \quad \text{say.} \tag{3.67}$$

Eliminating v from Bernoulli's equation on the free surface, we obtain

$$B(R_s, z) \equiv \frac{1}{2} \frac{F^2}{K^2} \frac{1}{R_s^{4n+4}} + \omega^2 (R_s^2 - \epsilon^2 z^2) = \phi_s. \tag{3.68}$$

This relation gives R_s as a function of z for any ϕ_s. Some curves of constant ϕ_s are plotted in Fig. 3.6b, in dimensionless form ($F = K = \omega = 1$).

As usual in such problems, all the solution curves except two are symmetrical about $z = 0$, i.e. they give the same solution downstream as upstream, which is not what we require. We are therefore restricted to a solution that passes through the singular point where $\partial B/\partial R_s = 0 = \partial B/\partial z$, from which we obtain

$$(n+1)v^2(0) = \omega^2 R_s^2(0) = \frac{2n+2}{2n+3} \phi_s. \tag{3.69}$$

Thus the relation between flux, or mass-loss rate, and the upstream potential excess ϕ_s is

$$-\dot{M}_1 = \dot{M}_2 = F = \frac{\pi}{(n+1)^{3/2}\omega^2 C^n} \left(\frac{2n+2}{2n+3} \phi_s\right)^{n+3/2} \sim (GM)^2 \frac{\rho}{v_{\text{sound}}^3} \left(\frac{\Delta R}{R_L}\right)^3$$

$$\equiv \frac{M_1}{t_{\text{HD}}} \left(\frac{\Delta R}{R_L}\right)^3, \quad \text{say} \quad (\Delta R \geq 0), \tag{3.70}$$

$$= 0 \qquad (\Delta R \leq 0).$$

We have taken $n = 3/2$, used $\phi_s \sim GM\Delta R/R_L^2$, $\omega^2 \sim GM/a^3$, and eliminated C in terms of ρ and v_{sound}. ΔR is the excess of stellar radius over lobe radius. Several factors of order unity, including R_L/a, are ignored.

This approximation can easily be used as a boundary condition in stellar evolution computations, if the model is discretised using an implicit adaptive non-Lagrangian mesh as set out in Appendix A. Equation (3.70) can be expected to be better for convective envelopes, which may be fairly closely isentropic (although n may be increased considerably above $3/2$ by hydrogen ionisation), than for radiative envelopes, which have a steep entropy gradient. However, for radiative stars we expect only a slight degree of overfill, because of the negative feedback described earlier, and so the variation of entropy with depth may not be very significant.

The hydrodynamical timescale t_{HD} defined in Eq. (3.70) is quite short, of much the same order as the usual dynamical or pulsational timescale $(R^3/GM_1)^{1/2}$; although if we take seriously several dimensionless factors that we have ignored above the result might be two orders of magnitude longer. In circumstances where $\dot{M}_1$ is on a thermal timescale, say 1 megayear, we can estimate from Eq. (3.70) that $\Delta R/R_L \lesssim 0.01$. However, if mass loss is on the nuclear timescale, $\sim 10^3$ slower, the corresponding degree of overfill would be $\lesssim 0.001$.

A convective star, on the other hand, will suffer from *positive* feedback in similar circumstances, and we can expect the overfill to grow until $\Delta R/R_L \sim 1$ ultimately, giving mass transfer on the timescale t_{HD}; although most of our assumptions will have broken down before such rates are reached. It is important to note, however, that these high rates are *not* demanded, even if the loser's atmosphere is convective, *unless* the mass ratio is such as to require the Roche lobe either to shrink, or to expand less rapidly than the star, as the mass transfer proceeds.

A fully convective star responds to rapid mass loss almost adiabatically. This means that it is approximately an $n = 3/2$ polytrope, with an effective equation of state $p = K\rho^{5/3}$, K being a constant that is given by the constant entropy in the star. Homology shows that such a star has $R \propto M^{-1/3}$. So we see that R'_{TD} is -0.33 in the limit of a fully convective star subject to very rapid mass loss. Note that whereas white dwarfs have $R'_{TE} = R'_{TD} = -0.33$ (in the low-mass, non-relativistic regime), low-mass largely convective main-sequence stars have $R'_{TE} \sim 1$, $R'_{TD} = -0.33$. White dwarves all have the same entropy (zero, apart from the modest contribution of the non-degenerate ions), whereas the entropy of a red dwarf is a function of its mass unless it is gaining or losing mass so rapidly that the process is approximately adiabatic.

3.3.2 Modes 1, 2 and 3; cases A, B, C and D

We find it convenient to define three 'modes' of mass transfer during RLOF, according to the three timescales expected so far: nuclear – mode 1; thermal – mode 2; and hydrodynamic – mode 3. In fact there is at least a fourth timescale that can be important, namely the timescale of angular momentum loss. Since this is of course non-conservative, I will defer a discussion till Section 4.3. But it is usually a slow process, and so we will include it under mode 1. Which mode will operate is determined principally by (a) the mass ratio, (b) the response of thermal-equilibrium models to loss of mass, and (c) the response of thermal-disequilibrium models to loss of mass. The last two are summed up, in our simplistic model (3.60), by the coefficients R'_{TE} and R'_{TD}. In Section (3.3.3) I present a simple linearised model which may help to clarify the nature of the onset of RLOF, but for the present we continue with a more qualitative analysis.

By the 'thermal-equilibrium' models for a potentially mass-losing star, at an arbitrary stage of evolution, we mean the sequence of models that would be obtained if the rate of mass loss is fast compared with the nuclear timescale, but slow compared with the thermal timescale. For a ZAMS star, the thermal-equilibrium model sequence is simply the ZAMS itself, and so the quantity R'_{TE} used above is simply the slope of the ZAMS in the $\log M$–$\log R$ plane. We have already argued that since $0.5 \lesssim R'_{TE} \lesssim 1$ on the ZAMS, there is only a rather limited range of mass and mass ratio where mode 1 can prevail at the onset of RLOF. So for most systems the mass loss has to speed up, and we might anticipate in the first instance that it would speed up on the short timescale t_{HD} – Eq. (3.70). But in mode 2, where the back reaction of the mass-loss rate on the star's structure can no longer be neglected, the situation is mitigated provided that the star has a predominantly radiative envelope ($R'_{TD} > 0$; but see Section 3.3.3). The timescale for growth becomes t_{KH} rather than t_{HD}. This will continue until and unless the Roche-lobe radius, after decreasing and then increasing with continued mass transfer, gets back to the thermal-equilibrium radius at a smaller value of the mass ratio. Thus in Fig. 3.4b, the true evolution of a radiative star which first fills its Roche lobe at point B$'$ is along but slightly above curve (ii) between B$'$ and B, losing mass on a thermal timescale, and then from B to C and beyond on a nuclear timescale. But if the loser at point B' were predominantly convective ($R'_{TD} < 0$), the star would expand faster, up to the hydrodynamic timescale, beyond B'.

The amount of mass transfer that must proceed in mode 2, before $*1$ can stabilise in thermal equilibrium, clearly depends on the initial mass ratio q_0. In Fig. 3.4b, the initial detached evolution A'B' will be further to the right for larger q_0, and so the point B where the thermal-equilibrium curve (iii) intersects the Roche-lobe radius curve (ii) again will be further to the *left*. There is a very approximate symmetry about $q \sim 1$, suggesting that at B $q \sim 1/q_0$, i.e. the mass ratio is approximately reversed; further decrease of q takes place in mode 1 rather than mode 2. A slightly more precise condition for this transition is given below – Eq. (3.71).

For red giants, it is clear that changes rapid compared with nuclear evolution leave the core mass M_c constant, and hence from Eq. (2.50) the nuclear luminosity is also constant since this is almost entirely dictated by the core mass. So Eq. (2.47), describing the location of the Hayashi track as a function of total mass, simply tells us that $R'_{TE} = -0.31$. This only breaks down at a late evolutionary phase when the star turns a corner at the top right of the HRD and starts to shrink rapidly towards a white dwarf, and R'_{TE} becomes positive. The adiabatic response rate R'_{TD} for red giants was estimated by Hjellming and Webbink (1987), using semi-analytical models with an $n = 3/2$, $\gamma = 5/3$ envelope and a point-mass core. An interpolation formula that fits their tabular values reasonably well is

$$R'_{TD} = -\frac{1}{3} + \frac{14M_1}{8M_1 + 13M_c} \frac{M_c}{M_1 - M_c}. \tag{3.71}$$

This starts at -0.33 for a negligible core, but becomes positive once the core mass grows above $\sim 0.2M_1$.

The situation for red giants is rather like Fig. 3.4b, except that line (iii), the thermal-equilibrium line, will actually slope *up* to the left. Curve (ii) has slope -0.31 at $q \sim 0.66$ (Table 3.1), so that mode 1 evolution is still possible in a situation that starts at a point like A with $q < 0.66$. But if the star starts at a point like A', mode 3 is expected. In mode 3 the mass transfer is so rapid that probably our conservative assumptions break down. I will consider this further in Section 5.2.

A relatively easy situation to analyse, though one which may at first sight seem academic, is the case of a white dwarf filling its Roche lobe. For this to be feasible, the companion must be an even smaller entity, such as a more massive white dwarf, or a neutron star or black hole. The orbital period must be very short, ~ 1 min. Since white dwarfs are basically inert, there is no nuclear evolution to drive the white dwarf towards its Roche lobe; but instead, angular momentum loss by gravitational radiation (Section 4.1) will drive the Roche lobe towards the white dwarf. The radius–mass relation, Eq. (2.51), of a white dwarf is almost entirely determined by the degenerate equation of state, and depends very little on the thermal structure. Hence, for low-mass white dwarfs with $R'_{TE} \sim -0.33$, if $q \gtrsim 0.63$ the situation is unstable, requiring mass transfer on a hydrodynamical timescale. This is the value of q at which $R'_L \sim R'_{TE}$ (Table 3.1, Eq. 3.16). For more massive white dwarfs R'_{TE} is more negative, and the situation is unstable at still lower q. But if the 'initial' q is low enough, mode 3 can be avoided, and the evolution can in principle continue on the same timescale as the angular momentum loss (which however is likely to be very rapid at the short periods required).

Evolution driven by angular momentum loss (AML), either GR or a combination of magnetic braking and tidal friction (MB), as in Chapter 4, is probably important in many other types of binary – novae, low-mass X-ray binaries, and W UMa binaries for example. We can still determine the rate of mass transfer from Eq. (3.53), bearing in mind that R_L will also have an explicit time-dependence via the orbital angular momentum H_0 – Eqs (3.6 and 3.11). For semidetached systems this gives, analogously to Eq. (3.54),

$$\frac{\mathrm{d}\log M_1}{\mathrm{d}t} = -\left(\frac{1}{t_{NE}} + \frac{2}{t_{AML}}\right)\frac{1}{R'_{TE} - R'_L},$$

$$\frac{1}{t_{AML}} \equiv \frac{|\dot{H}_0|}{H_0} = \frac{1}{t_{GR}} + \frac{1}{t_{MB}}. \tag{3.72}$$

The GR timescale, and an estimate of the MB timescale, are given by Eqs (4.1) and (4.30) in Chapter 4. Although t_{AML} is not related to t_{NE} it is usually much longer than t_{KH}, and so it is reasonable (in the case $R'_{TE} > R'_L$) to consider this as an extension of mode 1.

Modes 1–3 are different from the more traditional cases A, B, C of Kippenhahn and Weigert (1967). The definition of the latter relates to the state of evolution of the interior: in case A, the loser is still in the main sequence band, in case B it is beyond the main sequence but before helium ignition, and in case C it is beyond helium ignition. But the behaviour of the mass-losing primary is more closely related to whether the envelope is radiative or convective than to whether helium has ignited (case C) or not (case B). The refinements 'early case B' and 'early case C' make this distinction: they are much the same as mode 2 followed by mode 1, and the alternatives 'late case B' and 'late case C' are much the same as mode 3. Case A should probably also be divided into 'early' and 'late', since low-mass main-sequence stars ($\lesssim 0.75\,M_\odot$) are likely to give mode 3 RLOF as a result of angular momentum loss rather than nuclear evolution (Section 4.5). We see in Section 3.5 that when we consider not just the onset but also the continuation of RLOF, there are at least eight subtypes just within *conservative* early case A.

To emphasise the significance of radiative or convective envelopes, we will in effect redefine the cases: case $\mathcal{B}$ and case $\mathcal{C}$. Case $\mathcal{B}$ is the situation where the loser is in the Hertzsprung gap, and therefore has a mainly radiative envelope, at the onset of RLOF, and case $\mathcal{C}$ is the situation where the loser is on the giant branch, and therefore has a mainly convective envelope. Furthermore, because massive stars increase their radii by a large factor while crossing the Hertzsprung gap, and because we perceive in later discussion some potentially significant differences depending on whether a star is in the left-hand or right-hand portion of the Hertzsprung gap at the onset of RLOF, in case $\mathcal{B}$ we will sometimes wish to distinguish case $\mathcal{B}_1$ and case $\mathcal{B}_2$. Anticipating later discussion, we place the boundary provisionally at a period of ~ 100 days. Stars less massive than $\sim 8\,M_\odot$ will not encounter case $\mathcal{B}_2$ (Table 3.2).

We also find it convenient to define case D, the situation where the binary is too wide for RLOF to occur at all. In the conservative approximation here, this is of little interest; except that the conservative approximation obviously cannot hold, since most stars lose considerable mass at a late stage in evolution, and some very massive ones at a relatively early stage. I will return to this in Chapter 4.

3.3.3 A simple linearised model for the onset of RLOF

It is possible to write down a simple linearised model for the onset of RLOF, that is surprisingly helpful in explaining why RLOF develops at different speeds in different circumstances. This model is a generalisation of Eqs (3.52)–(3.54) and (3.60) and (3.72). It consists of four parts:

(a) The thermal equilibrium radius R_{TE} and luminosity L_{TE} of *1 are approximated as functions of the current mass M_1 and the mass M_c of burnt fuel. We write

$$\frac{d \log R_{TE}}{dt} = \frac{\partial \log R_{TE}}{\partial \log M_1} \frac{d \log M_1}{dt} + \frac{\partial \log R_{TE}}{\partial M_c} \frac{dM_c}{dt}$$

$$\equiv R'_{TE} \frac{d \log M_1}{dt} + \frac{1}{t_{NE}}, \tag{3.73}$$

where R'_{TE} and t_{NE} are generalisations of the corresponding quantities in Eq. (3.52).

(b) The radius of a perturbed star relaxes on a timescale t_{KH} to the thermal-equilibrium radius, but departs from that radius as a result of mass loss. We write

$$\log R + t_{KH} \frac{d \log R}{dt} = \log R_{TE} + R'_{TD} t_{KH} \frac{d \log M_1}{dt}. \tag{3.74}$$

This is an improvement on Eq. (3.60), because it says that if, for example, mass transfer abruptly ceases R does not abruptly return to R_{TE}, but approaches it on a timescale which we in effect *define* as the Kelvin–Helmholtz timescale. It says, as before, that mass loss affects R significantly only if it happens on a timescale comparable to, or faster than, this timescale. The coefficient R'_{TD} is positive for radiative stars and negative for convective. Fully convective stars, and also white dwarfs, respond to rapid mass loss as $n = 3/2$ polytropes, and so $R'_{TD} \sim -0.33$ in these cases.

(c) A star whose radius exceeds its Roche-lobe radius R_L by some fraction f, where we take

$$f \equiv \log(R/R_L), \tag{3.75}$$

loses mass at a rate proportional to f. We write

$$\frac{d \log M_1}{dt} = -\frac{f}{t_{HD}}, \quad f \geq 0, \tag{3.76}$$

where t_{HD} is the dynamical timescale estimated in Eq. (3.70). Note that, for simplicity, we take a linear dependence on f rather than a cubic dependence as implied by Eq. (3.70).

(d) The radius of the Roche lobe changes in response to mass transfer, or to angular momentum loss, at a rate

$$\frac{d \log R_L}{dt} = \frac{\partial \log R_L}{\partial \log M_1} \frac{d \log M_1}{dt} + \frac{\partial \log R_L}{\partial \log H} \frac{d \log H}{dt}$$

$$\equiv R'_L \frac{d \log M_1}{dt} - \frac{2}{t_{AML}}, \tag{3.77}$$

where R'_L is the coefficient of Eq. (3.16) and Table 3.1, and t_{AML} is the timescale on which angular momentum is lost from the system.

Equations (3.74)–(3.76) combine to give a first-order differential equation for f:

$$f + t_{KH} \dot{f} = \log \frac{R_{TE}}{R_L} + (R'_L - R'_{TD}) \frac{t_{KH}}{t_{HD}} f. \tag{3.78}$$

Differentiating this wrt time, on the assumption that the dimensionless parameters R'_{TD}, R'_L and the timescales t_{HD}, t_{KH} are all constants, and using Eqs (3.73) and (3.77), we derive a second-order inhomogeneous linear differential equation for f:

$$\ddot{f} + \left(\frac{R'_{TD} - R'_{L}}{t_{HD}} + \frac{1}{t_{KH}} \right) \dot{f} + \frac{R'_{TE} - R'_{L}}{t_{KH} t_{HD}} f = \frac{1}{t_{KH}} \left(\frac{1}{t_{NE}} + \frac{2}{t_{AML}} \right) . \qquad (3.79)$$

Bearing in mind that the hierarchy of timescales is normally $t_{HD} \ll t_{KH} \ll t_{NE}, t_{AML}$, we can solve the corresponding characteristic equation to obtain, for the complementary function:

$$f \propto e^{\lambda t}, \quad \lambda \approx -\frac{R'_{TD} - R'_{L}}{t_{HD}} \quad \text{or} \quad -\frac{R'_{TE} - R'_{L}}{R'_{TD} - R'_{L}} \frac{1}{t_{KH}}. \qquad (3.80)$$

We see that

 (a) if $R'_{L} < R'_{TD}$ and $R'_{L} < R'_{TE}$ both roots are negative
 (b) if $R'_{TE} < R'_{L} < R'_{TD}$ the only positive root is on the Kelvin–Helmholtz timescale
 (c) if $R'_{TD} < R'_{L}$ the larger root is positive and on the hydrodynamic timescale.

So we obtain the following conditions:

$$\text{mode 1}: \ R'_{L} < R'_{TD}, \ R'_{L} < R'_{TE}; \quad \text{mode 2}: \ R'_{TE} < R'_{L} < R'_{TD};$$

$$\text{mode 3}: \ R'_{TD} < R'_{L}. \qquad (3.81)$$

In the case that the solution is stable (mode 1), it tends to the particular integral

$$f = \frac{t_{HD}}{R'_{TE} - R'_{L}} \left(\frac{1}{t_{NE}} + \frac{2}{t_{AML}} \right), \qquad (3.82)$$

which is just Eq. (3.72), combined with Eq. (3.76). Thus the star overfills its Roche lobe by a very small amount; this would actually be somewhat larger if we kept in the more realistic cubic dependence of Eq. (3.70), but not so large as to be measurable.

The condition for mode 2 is normally satisfied for a star with a radiative envelope, and with mass larger but not considerably larger than the companion. The condition for mode 3 is normally satisfied if the star has a convective envelope ($R'_{TD} < 0$). But it should be noted firstly that mode 3 can also apply to mainly radiative stars if q is large enough initially, and secondly that mode 1 can still apply to convective stars if R'_{L} is sufficiently negative, i.e. if the loser is substantially less massive than the gainer. Firstly, even if $R'_{TD} \sim 4$, which is not untypical of radiative envelopes, we can have $R'_{L} \gtrsim R'_{TD}$ when $q \gtrsim 2.5$ (Table 3.1, Eq. 3.16). Secondly, even if $R'_{TE} \sim -0.31$, as is typical for red giant losers, R'_{L} is less than this if $q < 0.65$ (Table 3.1).

Although I suggested provisionally, in Section 3.3.1, that for mass transfer to be stabilised on the thermal timescale it was desirable to have R'_{TD} positive, closer examination shows that the condition is rather more complex: R'_{TD} can be negative if R'_{L} is more negative, and R'_{TD} positive is not enough if R'_{L} is more positive.

This discussion of modes 1–3 can be summarised as follows. During various long-lived stages of a star's life (e.g. MS, RG, WD) there is a parameter R'_{TE} that measures the (logarithmic) sensitivity of the radius to the mass when the mass is thought of as varying slowly compared with the thermal timescale but rapidly compared with the evolutionary timescale. There is also a parameter R'_{TD} which measures the sensitivity of radius to mass when mass is added (or subtracted) rapidly, on or about the thermal timescale. From R'_{TE} we get a critical mass ratio $q_{cr}(R'_{TE})$ by solving for q the Roche-lobe radius relations (3.5), (3.13) and (3.17):

$$R'_{TE} = R'_L(q_{cr}) \approx 2.13q_{cr} - 1.67. \tag{3.83}$$

Similarly we get a $q_{cr}(R'_{TD})$ from R'_{TD}. Then the mass ratio q_b at the beginning of Roche-lobe overflow determines the initial mode of mass transfer thus:

mode 1 – if $q_b < q_{cr}(R'_{TE})$, $q_{cr}(R'_{TD})$
mode 2 – if $q_{cr}(R'_{TE}) < q_b < q_{cr}(R'_{TD})$
mode 3 – if $q_b > q_{cr}(R'_{TD})$.

If mode 2 is what is indicated, it continues until a transition mass ratio q_t, say, is reached, and then settles into mode 1. A rough estimate of q_t, based on the approximate symmetry of the Roche-lobe-radius curve and the thermal-equilibrium-radius curve about the point $q = q_{cr}(R'_{TE})$ where they have the same gradient, is

$$q_t \sim q_{cr}^2/q_b. \tag{3.84}$$

For the upper ZAMS, where $R'_{TE} \sim 0.5$ and so $q_{cr} \sim 1$, this means that the mass ratio is approximately reversed ($q_t \sim 1/q_b$), but for terminal-MS and post-MS stars we usually find $R'_{TE} \sim 0$ to -0.33, so that $q_{cr} \sim 0.8$ to 0.65. The mass ratio therefore is substantially more-than-reversed before mode 2 gives way to mode 1.

However, these conditions are based on the assumption that RLOF, at least at its onset (i.e. before it becomes rapid, if that is what is indicated), is approximately conservative of both mass and angular momentum. In fact some angular momentum can be transferred from orbit to stellar spin, and so in effect 'lost': the Roche lobe will shrink faster for a given amount of mass transferred, making the process more unstable. This is a modest correction if q_b is not large. At a rather extreme mass ratio ($q_b \gtrsim 12$), however, it is possible for the loser's spin angular momentum to be comparable to that of the orbit. We must expect the Darwin instability (Section 5.1), in which the orbit may rapidly shrink and $*2$ crashes into $*1$.

It might be supposed that q_b will always be greater than unity, since the more massive star initially is always the one to fill its Roche lobe first. However, mass loss by stellar wind, accelerated by rotation, may cause $*1$ to lose significant mass, i.e. on a nuclear timescale, *before* $*1$ fills its lobe (Sections 4.4–6), so that it is not impossible that a convective loser will experience relatively mild mode 1 RLOF. I will argue in later chapters that some observed systems support this possibility.

3.3.4   *Effect of RLOF on the gainer*

Figure 3.4 did not include the behaviour of the gainer, which I will now discuss. As a first approximation, we can expect that the response of a star to the gain of mass is the inverse of its response to loss: a star with a predominantly radiative envelope expands, while one with a predominantly convective envelope contracts. However, the situation is not quite symmetrical. Material leaving the surface of the loser can reasonably be assumed to have little velocity, and the same temperature and density as the photosphere, but material added to the gainer will have different kinetic and thermal energy from the gainer's surface, so that some extra energy (positive or negative) may have to be allowed for.

The main point to note at present is that the gainer may well swell up in response to accretion, and in fairly close binaries this can easily lead it to fill its own Roche lobe (Yungelson 1973, Webbink 1976). This is all the more likely because the thermal timescale of the gainer,

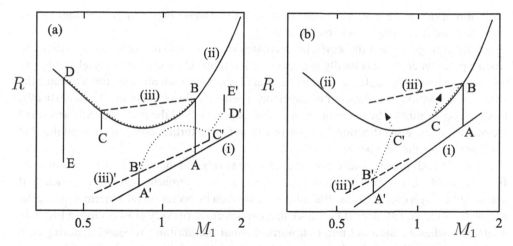

Figure 3.7 Schematic behaviour of radius and mass for both loser and gainer, in situations (a) where the gainer does not expand to fill its own lobe, and (b) where it does. Curve (i) – ZAMS; curve (ii) – Roche-lobe radius; curve (iii) – thermal-equilibrium radius. Primed letters indicated ∗2's position corresponding to the unprimed letter for ∗1. In (a), after Pennington (1986), the transition from mode 2 to mode 1 was interrupted at point C by a brief detached phase (see text). In (b), after Robertson and Eggleton (1977), evolution beyond contact was followed using a prescription like Eqs (3.85) and (3.86).

which is normally the less massive and luminous component at the *beginning* of Roche-lobe overflow, will be substantially longer than for the loser. Thus supposing that both components have radiative envelopes, the gainer's increase in radius in response to a given $|\dot{M}|$ will be proportionately larger than the loser's decrease. The situation is illustrated in Fig. 3.7, where the left-hand panel shows a situation in which contact was avoided and the right-hand panel shows contact being reached.

Contact can be avoided with some choices of initial mass ratios and periods (Section 3.5). In Fig. 3.7a, ∗1 starts at A and evolves with no loss of mass to point B, at which it fills its Roche lobe. Curve (iii) is the path along which ∗1 would evolve if it lost mass but somehow remained in thermal equilibrium. This line is no longer parallel to the ZAMS, curve (i), but instead, if continued indefinitely to *higher* masses, would approach (i) asymptotically. In effect, ∗1 has a core of given mass, developed during the phase AB. The effect of this core diminishes if the envelope becomes more and more massive, but makes the structure rather red-giant-like as the envelope decreases so that the radius increases relative to the ZAMS radius. If the core is substantial enough the radius might even increase *absolutely* as the envelope loses mass, although the situation in Fig. 3.7a is not envisaged as quite that extreme.

However, as indicated in the previous section, the star does not in fact follow curve (iii), because this requires a large degree of overfill and, therefore, rapid mass loss. When mass loss reaches the thermal timescale the envelope shrinks below curve (iii), and instead follows curve (ii), the Roche-lobe curve. Strictly speaking, the path lies very slightly above curve (ii). This continues until curves (ii) and (iii) intersect again, at which point it is possible for the star to return to thermal equilibrium. Ignoring for the moment the small spur at C, ∗1 can now continue to lose mass on a *nuclear* timescale. It grows *absolutely* because of nuclear evolution, trying to become a red giant as it would if it were single; but it continues to lose mass because it is trying to get above curve (ii).

As the core grows and the envelope decreases in mass, but still increases in radius, we reach a point where there is hardly any envelope left (point D), and then the envelope shrinks abruptly – much as a single star climbs the giant branch but then abruptly turns towards the white-dwarf region as it runs out of envelope. Then ∗1 follows the path DE. Realistically, this may be terminated by the ignition of helium at point E, and the star may settle as a small He main-sequence star (Section 2.5). There will in fact be further evolution beyond this, but we ignore it for the time being.

During the AB phase, ∗2 also evolves, but more slowly and therefore by a smaller amount from A′ to B′. Curve (iii)′ is analogous to curve (iii), approaching (i) asymptotically if continued to high enough mass. But now ∗2 is swollen by accretion on a thermal timescale, and follows the dotted curve B′C′. This can approach curve (ii) very closely (and in Fig. 3.7b actually reaches it). But as ∗1 returns towards thermal equilibrium so does ∗2. Ignoring once again the spur at C and C′, ∗2's further evolution is on a nuclear timescale, but is now quite rapid since ∗2 is substantially more massive than ∗1 was originally. It therefore traces C′D′ while ∗1 traces CD. Once ∗1 shrinks away from its Roche lobe at D, ∗2 evolves upwards rather rapidly (D′E′).

The spur at C during which the binary returns to being detached sometimes occurs, for one of two reasons. Firstly, ∗1 may exhaust hydrogen in its core. If single, it could shrink temporarily as is normal at the end of the main sequence (Fig. 2.1), leading to a short, detached phase. Secondly, it may happen as the star approaches thermal equilibrium, because the centre of the star and its envelope respond not only at different rates but also in different directions. The core, being convective, may actually *expand* while the envelope shrinks, and when the mass loss slows down first the envelope expands rapidly back to thermal equilibrium and then the core contracts less rapidly back to thermal equilibrium. This last phase may cause temporary detachment from the lobe, as at C, with a small corresponding spur in ∗2's evolution at C′.

There are two places in the evolution of Fig. 3.7a where, with slightly different parameters, contact might be reached. One is on the stretch B′C′, and the other on the stretch C′D′. The first possibility is shown in Fig. 3.7b. I have not illustrated the second, but it is easy to see that if C′D′ in Fig. 3.7a slopes up more steeply, and extends further, it may reach curve (ii) before ∗1 becomes detached at D.

In Fig. 3.7b, contact occurs at points C and C′. For the present, it is not at all clear what the outcome should be, although in the calculation on which Fig. 3.7b is based a particular model for mass and energy transport was used, which predicted that the contact would become deeper, temporarily, but that the direction of mass transfer would reverse. I discuss various evolutionary possibilities in Section 3.4. But there certainly exist binaries with 'contact' geometry (Fig. 3.1b), and in sufficient numbers that it is unlikely that they are all evolving on a thermal timescale (unless the evolution is somehow cyclic). Thus some consideration of how stars evolve once contact is established has to be undertaken (Section 3.3.5).

In binaries that are not particularly close, where ∗1 has to expand to well beyond its MS radius before filling its Roche lobe, ∗2 will usually be much smaller than its own Roche lobe. Then material falling into ∗2's lobe can be sufficiently deflected by the Coriolis force that, instead of impacting nearly directly on to ∗2 (as implied in Fig. 3.1c), it settles into a ring around ∗2. In the absence of a dissipative agency such as viscosity, this ring would simply accumulate all the transferred mass. But dissipation can, at least in principle, lead to something like a steady state, with gas flowing from L1 to the outer edge of a disc around *2, working its way through the disc, and finally flowing from the inner edge of the disc through some boundary layer on to the surface of *2. Provided that *1 loses mass in modes 1 or 2 (but not the very rapid mode 3), and that all or at least most of the mass lost by *1 is gained by *2, the behaviour of the gainer may still be more-or-less the inverse of the behaviour of the loser. But it becomes more likely, of course, that contact will be avoided, the larger is the lobe around *2 relative to the unperturbed radius of *2.

In some observed systems the gainer is a strongly magnetic star, and in that case the accretion flow may be dominated by magnetic forces, rather than by viscosity. Magnetically dominated accretion appears to be in the form of a stream of gas down the magnetic field lines on to a magnetic polar cap. This may still, however, have the net result that most or all of the material lost by *1 is accreted by *2. In Chapter 6 we discuss further the viscous and magnetic accretion processes.

I will discuss later some computational results obtained by converting the single-star evolution code of Appendix A into a 'conservative' binary-star code. At a first level of approximation this is very easily done. Firstly, *1 is evolved with the boundary condition (3.70) instead of the usual condition $M_1 = $ constant. The only information about *2 that is needed during this calculation is its mass, and that is known from the conservative hypothesis. This evolution gives the mass-loss history $M_1(t)$, among other things. Subsequently, *2 is evolved with the boundary condition that its mass at time t is $M - M_1(t)$, M being the constant total mass. This procedure generally works well, until either (a) *2 evolves to fill its own Roche lobe, or (b) *1 or *2 goes supernova. In principle, if (a) happens, we might follow the reverse RLOF by reversing the above procedure, but in practice, either (a1) *1 is still filling its lobe, so that we have a contact system, with the possibility of *luminosity* transfer – see next section – or (a2) the mass ratio is very extreme at this point because of previous mass transfer, and so dynamical-timescale reverse mass transfer is expected. A simple hydrostatic code will not do: see Section 5.2.

The above procedure can in fact be generalised to include some non-conservative processes, at least at a simplistic level. However, throughout much of this book we are seeking to learn about non-conservative processes by comparing observed stars with conservative expectations. Unfortunately there are sufficiently many unknown factors in most non-conservative models that it is difficult to constrain them by computer models. Common sense is probably more helpful at present.

Three effects of accretion on the gainer which we have not discussed at any length are

(a) the fact that material about to be accreted has a different temperature and density from the material on the surface of the gainer
(b) the accreted material may also have a different composition, if the gainer has been stripped down as far as its core
(c) the accreted material may also have a different angular momentum, especially if it has passed through an accretion disc instead of travelling by a fairly direct trajectory from the L1 point.

It is somewhat difficult to come up with an unequivocal model of the first process, but one could add some ad hoc term to the energy equation of the gainer in the photospheric layer. The second process is likely to lead to mixing in the surface layers, because the newly added material will usually have a higher molecular weight than the average in the outer layers. This will lead to Rayleigh–Taylor instability; but mixing with the deeper layers will rapidly dilute the adverse molecular weight gradient so that it is unlikely to have a major effect. The third process can probably be modelled fairly easily, provided that one's stellar models incorporate a model for the rate of rotation of the star. One would of course have to incorporate several other processes that should influence the distribution of angular momentum in a star: tidal friction, for example, and magnetic fields. The last is likely to be the most problematic. I will adhere in this book to the argument by Spruit (1998) that even a very weak magnetic field is likely to enforce a fairly uniform angular velocity, at least in layers stable to convection.

3.4   Evolution in contact

Figure 3.7b illustrated a situation where ∗2 expanded to fill its own Roche lobe (at point C′) shortly after ∗1 had begun RLOF at point B. This is a rather common situation, particularly for short-period binaries with markedly unequal initial masses. The result will be a 'contact binary'. I have also mentioned that contact may alternatively be reached on the stretch C′D′ in Fig. 3.7a, supposing the parameters are slightly different. In either case we expect a contact binary to be formed, but in the first case it is likely that contact is reached at a quite early stage of evolution, with little prior exchange of mass, while in the second case it happens quite late in evolution (though still usually within the main-sequence band) and after considerable exchange of mass.

When both stars overfill their Roche lobes simultaneously, i.e. are in contact, it is evident that not only can mass flow in either direction (in principle, at least) between them, but also that energy can flow between the components, even without a net flow of mass. Furthermore, although we leave most of our discussion of observational material till later (Section 5.3), observation appears to be telling us quite unequivocally that heat *is* somehow being transported between the components, leading to surface temperatures that are equal to within two or three per cent even when the masses, in some cases, are different by as much as a factor of ten.

Briefly, the salient observational facts that a theoretical model has to explain are:

(a)   Contact systems are quite common, and rather stable in the sense that the orbital periods do not change significantly on timescales less than ∼1 megayear. Recall (Section 1.2) that we do not have to wait a megayear to determine a rate of period change on this timescale. Thus mass transfer has to be slow: possibly on a thermal timescale but certainly not on a dynamical timescale.

(b)   Periods, and the masses, radii and luminosities of the more massive component, are roughly consistent with a main-sequence structure.

(c)   Mass ratios are typically ≳2, and can be as extreme as 10 or more. This is very different from short-period *detached* binaries (Section 1.5.4), where even after allowing for selection effects, which favour near-equal masses, mass ratios ≲ 2 preponderate.

(d)   The temperatures of the two components differ typically by ≲ 2–3%, so that the lower-mass component is *considerably* overluminous for its mass. This suggests that a substantial fraction of the luminosity of the more massive component is being transferred, presumably within the contact envelope, and radiated from the surface of the less massive component.

A mathematical prescription for the rates of mass and heat flow is required in order to have a closed set of equations capable of solution. No such prescription has been widely accepted, but most attempts rely on the concept that mass and energy flow will be related to (a) the difference in surface Roche potential and (b) the difference in temperature or entropy or enthalpy between the surface layers of the two components. Guided mainly by dimensional arguments, and an attempt to generalise Eq. (3.70), we might try to approximate both $\dot{M}$ and ΔL as functions of the mass coordinate m as it varies through the outer (contact envelope) layers:

$$\frac{\mathrm{d}\dot{M}}{\mathrm{d}m} = \pm\frac{v}{r}, \quad v^2 \sim 2|\phi_{s2} - \phi_{s1}|, \tag{3.85}$$

$$\frac{\mathrm{d}\Delta L}{\mathrm{d}m} = \lambda(m)\,(h_2 - h_1), \tag{3.86}$$

with an inner boundary condition that both are zero on (as well as below) the L_1 surface. The ϕ_s are the surface potentials of the two stars, the h are the enthalpies. The relative thickness of the contact envelope appears to be typically $\sim 2\%$ in radius, or $\sim 10^{-5}$ in mass. Both $\dot{M}$ and ΔL will have opposite signs in $*2$ and $*1$. The velocity v is from analogy with Eq. (3.69) for Bernoulli flow, but with a sign determined presumably by which surface potential is the higher; r is some mean radius, the same for both components. The coefficient v/r, dimensions (time)$^{-1}$, must be quite small, since the orbital period, and so the stellar masses, do not change appreciably on timescales of less than ~ 1 megayear. This suggests that $v/r \lesssim 10^{-6}\,\mathrm{s}^{-1}$. In the heat-transfer equation we suggest that the heat flux is proportional to the difference in enthalpies h; the factor λ, also with dimensions (time)$^{-1}$, might also be of order v/r, but we believe that a model which gives as much heat transfer as observation suggests, combined with the rather small difference in temperatures observed, would need a *considerably* larger value of λ, say $10^{-4}\,\mathrm{s}^{-1}$.

One way to achieve a higher value of λ than of v/r is to postulate a closed circulation pattern in the outer layers, so that λ has one sign deep in the contact envelope and the opposite sign further up. But attempts to achieve this tend to fall short of the amount of heat transfer that seems to be required by observation.

One possible model that has not yet been considered in detail is that the main agency for transferring heat is *differential rotation*, of the same character that is observed in the equatorial region of the Sun's convective envelope (Section 2.2.4). We have already noted that the origin of this differential rotation is by no means clear, although it is difficult to see what else it can be, other than the interaction of the Coriolis force with turbulent convection. But it is clear that the equatorial region of the Sun is travelling about 10% faster than the main body. If such a flow with the same *relative* difference were travelling round the entire envelope of a contact binary, with an orbital period 100 times shorter than the Sun's rotation period, it might *just* be capable of transferring the amount of energy required. The parameter λ in this case would be $\sim \Delta\Omega$. It is to be hoped that three-dimensional simulations, such as are just becoming feasible, might cast light on this or other possibilities.

It seems likely that Eqs (3.85) and (3.86), whatever values within reason are assigned to v/r and λ, usually provide some degree of negative feedback, as does Eq. (3.70) for a semidetached system, so that $\phi_{s1} \approx \phi_{s2}$ and $h_1 \approx h_2$. However, although Eq. (3.70) will clearly give negative feedback in the semidetached evolution of, for instance, mode 2 of

Fig. 3.4 (and positive feedback in mode 3), it is by no means clear what circumstances, if any, would do the same in contact evolution. Consequently we will confine ourselves throughout the later chapters to two very basic types of binary evolution in contact, viz:

(a)   mass flows from ∗1 to ∗2 : forward mass transfer in contact (mode CF)
(b)   mass flows from ∗2 to ∗1 : reverse mass transfer in contact (mode CR).

The letter C in CF distinguishes contact mass transfer from semidetached RLOF, which can also be forward or reverse: modes SF and SR. But whereas in the semidetached case it will be obvious which of ∗1 and ∗2 is the loser, it is by no means obvious in the contact case and we will hve to rely on a posteriori arguments rather than a priori arguments. Note that we return here to our more general definition of ∗1 as the component that was *initially* more massive (Section 1.4). In the present chapter, up to this point, ∗1 has meant the star that is under discussion, usually the star that is filling, or about to fill, its Roche lobe.

We can further qualify the mode of mass transfer, in contact as well as semidetached geometry, by the numbers 1–3, representing successively faster timescales. However, for the contact modes the rate as well as the direction is more a matter of speculation than of calculation. We hope that in the not too distant future it will be possible to model binary stars in a fully three-dimensional way, including both the thermodynamics and the hydrodynamics, and this should lead to a clarification of the direction and rate of mass transfer in contact geometry. Whatever the mass flow, we assume here that the heat flow will be whatever is required to almost equalise the two surface temperatures.

We believe it is likely that, as in Fig. 3.7b, if a system evolves rapidly to contact its direction of mass transfer will reverse. However, if it does then the binary will widen, which should lead quickly to the breaking of contact. We can expect a cyclic behaviour, 'thermal relaxation oscillations' or TROs: a limit cycle about an unstable equilibrium in which the system is slightly in contact and transferring less luminosity than would be enough to equalise the temperatures as observed. This unstable equilibrium will itself evolve slowly, in response to nuclear evolution or (more probably) magnetic braking – Section 4.5 – moving presumably towards more unequal masses since very few contact systems are known with nearly equal masses and very many are known with strongly *unequal* masses.

In some circumstances there is the possibility that after two stars come into contact they will merge into a single star. If after contact is reached *rapidly*, as in Fig. 3.7b, the direction of mass transfer reverses while the rate decreases to a low value (mode SF2 → mode CR1), we would have a possible explanation for the facts (a) and (c) above. Continuation of this evolution will obviously lead to an end-point where ∗2 is entirely entirely eaten up by ∗1.

Many binaries of short period can be expected to evolve into contact; in fact if RLOF begins while ∗1 is still in the main sequence band (case A) there is only a small region in the space of initial period and initial mass ratio where it does *not* happen. It is very unfortunate that evolution during this important phase is poorly understood. Those systems that do avoid contact will normally evolve through a semidetached forward phase (mode SF2 → mode SF1) until at a late stage ∗2 expands to fill *its* Roche lobe and initiate reverse mass transfer, which is probably mode SR3 because the mass ratio is typically quite extreme at this stage.

How contact binaries evolve is one of the most important unsolved problems of stellar astrophysics. That it is not yet solved may be a consequence largely of the numerical difficulty of implementing physical models in computer codes, but is also because of the difficulty in understanding what physical processes are most important.

Table 3.5. *Abbreviations for evolutionary states*

Evolutionary state	Sub-type	
P pre-main-sequence	*TT*	T Tau
	Be/Ae	Herbig emission-line stars
	BD	Brown dwarf ~0.01–0.08 $M_\odot$
	JMP	Jupiter-mass planet ≲0.01 $M_\odot$
M main sequence	*UMS*	Upper main sequence ≳8 $M_\odot$
	IMS	Intermediate main sequence ~2–8 $M_\odot$
	LMS	Lower main sequence ~0.08–2 $M_\odot$
H Hertzsprung gap	*HG*	He not yet ignited; star expanding on thermal timescale
	CHeB	Core He-burning
	HB	Horizontal branch
	BL	Blue loop
	δC	Cepheid
	GKGC	G/K-giant clump: core He-burning, shallow convective envelope
	post-AGB	post-asymptotic-giant-branch
G red giant	*FGB*	First giant branch: non-burning He core, deep convective envelope
S red supergiant	*GKGC*	G/K-giant clump: core He-burning, deep convective envelope
	AGB	Asymptotic giant branch
	TP-AGB	Thermally-pulsating AGB
	TZO	Thorne–Żytkow object, red supergiant with NS/BH core
R hot remnant	*WR*	Wolf–Rayet (WN, WC, WO)
	UHeMS	Upper He main sequence ($M \gtrsim 1.4 M_\odot$)
C hot core	*SDB*	pre-He-WD
	SDO	pre-C/O-WD
	PNN	Planetary nebula nucleus
E He-burning star	*LHeMS*	Lower He main sequence ($M \lesssim 1.4 M_\odot$)
	EHB	Extreme horizontal branch
	SDOB	Sub-dwarf OB
	SDB	Possibly the same as *EHB* or *SDOB*
W white dwarf	*HeWD*	He white dwarf
	COWD	C/O white dwarf
	NeWD	Ne white dwarf
N neutron star	*NS*	normally-rotating neutron star
	XRP	X-Ray pulsar
	MSP	Millisecond pulsar; rapidly-rotating neutron star
B black hole	*BH*	Black hole

3.5 Evolutionary routes

I will summarise this chapter using a compact notation defined in Tables 3.5–3.7. I define eleven broad evolutionary states for each component, and four geometrical states for a pair of components. Thus *MMD* means a basic binary where each component is on the main sequence and the system is detached; *GMS* means an Algol-like system where *1 is a red giant, *2 is still on the main sequence, and the system is semidetached. The evolution of

Table 3.6. *Abbreviations for geometrical states*

Type	Sub-type	Geometrical state
D		Detached; circular orbit; at least one star not much smaller than Roche lobe
S	SF, SR	Semidetached; mass transfer in forward ($*1 \rightarrow *2$) or reverse ($*2 \rightarrow *1$) direction
C	CF, CR	Contact, both stars exceed Roche radii; forward or reverse; includes common envelope detached in eccentric orbit; typically wider than D, but we do not
E		always discriminate

Table 3.7. *Some major modes of evolution*

0 – NE – Nuclear evolution
1 – F1, R1 – RLOF: mass transfer, forward (F) or reverse (R), slow (Nuclear or MB) timescale; Section 3.3
2 – F2, R2 – RLOF: ditto, fast (thermal) timescale; Section 3.3
3 – F3, R3 – RLOF: ditto, very fast (dynamical) timescale; Section 3.3
All six modes above apply to semidetached evolution (SF, SR) and also to evolution in contact (CF, CR).
The following modes are non-conservative: see later
4 – GR – gravitational radiation; Section 4.1
5 – TF – tidal friction; Section 4.2
6 – NW, PC, SW – normal (single-star) wind; Sections 2.4, 4.3: copious subtypes P Cyg, superwind
7 – MB – orbital angular momentum loss by stellar wind, magnetic braking and tidal friction; Section 4.5
8 – PA – partial accretion from stellar wind; Sections 4.3, 6.4
9 – EW – companion-enhanced stellar wind; Section 4.6
10 – BP – bi-polar re-emission; Section 4.7
11 – TB – influence of a third body; Section 4.8
12 – DI – tidal friction with Darwin instability; Section 5.1
13 – CE – common envelope evolution with spiral-in; Section 5.2
14 – EJ – rapid envelope ejection, common envelope without spiral-in; Section 5.2
15 – SN – supernova explosion; Section 5.3
16 – DE – dynamical encounters in dense clusters; Section 5.4
17 – IR – irradiation of the loser by accretion luminosity from the gainer; Section 6.2
We sometimes use 1, 2, 3 to qualify Modes GR–DE, indicating roughly the timescale, e.g. TF1, PC2, CE3.

a system can be written as something resembling a Markov chain:

$$MMD \rightarrow MMS \rightarrow HMS \rightarrow GMS \rightarrow SMS \rightarrow EMD \rightarrow EHD \rightarrow EGD \rightarrow EGSR \rightarrow \cdots \tag{3.87}$$

The extra R in the last step emphasises *reverse* RLOF; I might have said $MMSF$ at an earlier step, to emphasise *forward* RLOF, but I take that as the default option. The above route may be roughly appropriate to initial parameters ($4 + 3 M_\odot$, 2.5 days). The route is still far from complete, but what follows from the $EGSR$ state probably involves non-conservative evolution, as discussed in the next three chapters.

Table 3.7 lists a number of modes of evolution, of which only the first three are actually conservative; the others will be discussed in more detail later. In our notation we sometimes append these to states such as GMS. For example, GMS; NE, MB is an Algol in which it is claimed that magnetic braking (mode MB) is about as significant in modifying the binary as nuclear evolution (mode NE).

For case A, experience (Nelson and Eggleton 2001) suggests at least eight fairly distinct routes, depending on starting parameters. We call them sub-cases AD, AR, AS, AN, AB, AG, AE and AL. We shall have to add some non-conservative sub-cases later. Their definitions are as follows, where X is the ratio of the initial orbital period to the period such that $*1$ would exactly fill its Roche lobe while on the ZAMS:

(AD) 'dynamic RLOF': when $*1$ is low on the main sequence and so possesses a largely convective envelope, or when the mass ratio is fairly extreme, we can expect $*1$ virtually to explode very shortly after it overfills its Roche lobe, and engulf $*2$ – route (3.88).

(AR) 'rapid to contact': the stars come into contact very rapidly, before much mass is exchanged. This happens when q_0 is roughly in excess of 1.5 to 2 but not large enough for case AD. It also depends on X and can happen at low q_0, $q_0 \sim$ 1–1.5, if $X \lesssim 1.2$. We anticipate thermal relaxation oscillations, Section 3.4 – route (3.89).

(AS) 'slow to contact': the stars come into contact slowly, on a nuclear timescale, after a considerable exchange of mass. This happens for q_0 between 1 and $\sim$1.5, and for $X \sim$ 1.2–2 – route (3.90).

(AN) 'normal': $*2$ never fills its Roche lobe, at least until $*1$ has reached a long-lived compact remnant, white dwarf, neutron star or black hole. This can happen for $q_0 \lesssim$ 1.5–2, and $X \sim$ 2–4 – route (3.91).

(AB): in a limited range of M_1, $\sim$5–12 $M_\odot$, $*1$ has two distinct episodes of RLOF. The first leaves a helium-burning core of 0.8–2 $M_\odot$, but this is able to expand back to supergiant size (Fig. 2.18) and lose further mass, ending as either a C/O white dwarf or a supernova and neutron star – route (3.92).

Note that in AR and AS 'rapid' and 'slow' refer to the evolution, *before* contact, and not necessarily to evolution *during* contact. It is not clear how fast is evolution in contact, but it can hardly be a great deal faster than mode NE or we would not see many such systems.

Sandwiched between cases AS and AN are three further alternatives:

(AG) '(sub)giant contact': for low-mass stars, it is possible for one or both components to develop a deep convective envelope before coming into contact; $*1$ or $*2$ may still be a main-sequence star in terms of central hydrogen – route (3.93).

(AE) 'early overtaking': $*2$ gains so much mass that its evolution is accelerated beyond that of $*1$. It may reach the Hertzsprung gap first, and evolve into contact fairly soon afterwards, perhaps after 'reverse' mass transfer, with $*1$ shrunk temporarily inside its lobe and $*2$ filling its lobe – route (3.94).

(AL) 'late overtaking': $*2$ gains enough mass to catch up with and overtake $*1$, *before* $*1$ becomes a compact remnant, but *after* it has detached from its Roche lobe; for example, when it is a helium-burning star. For massive stars, it may be $*2$ that supernovas first (Pols 1992), in a rather limited range of initial conditions, because the helium core in $*2$ is much more massive than the helium-star remnant of $*1$, and so evolves much faster – route (3.95).

These three options, along with sub-case AS, are consequences of the fact that if ∗1 loses mass while it is still on the main sequence its evolution can be substantially slowed down, while the evolution of ∗2 subject to mass gain can be substantially speeded up. But the nearer ∗1 is to the TMS when it begins RLOF, i.e. the larger X is, the harder it is for ∗2 to catch up, let alone overtake.

Within the framework of conservative RLOF, the number of case $\mathcal{B}$ options is probably smaller than for case A. This is because the evolution of ∗1 speeds up considerably after the TMS, and so even if ∗2 gains considerable mass it is less likely to catch up with or overtake ∗1. We can however identify the following sub-cases: $\mathcal{B}$N, $\mathcal{B}$L, $\mathcal{B}$B, $\mathcal{B}$R and $\mathcal{B}$D. These are analogous to sub-cases AN, AL, AB, AR and AD. For case $\mathcal{C}$, there is, in principle, only one conservative option, sub-case $\mathcal{C}$D, since in the conservative paradigm ∗1 would expand, not contract, on the onset of RLOF, and the rate of mass transfer would rapidly become catastrophic. It is something of a strain to call this, and also sub-cases AD and $\mathcal{B}$D, conservative, because they can be expected to evolve very rapidly into a situation ('common envelope' evolution, Section 5.2) where conservative assumptions are no longer realistic.

In our concise notation, the case A options are

$$\text{AD}: \quad MMD \to MMS; F3 \to MMC \to (MMC; CE \to M \to H \to G \to \ldots \to W) \tag{3.88}$$

$$\text{AR}: \quad MMD \to MMS; F2 \to MMC \to (MMC; R2 \leftrightarrow MMS; F2 \to HMS; F2 \leftrightarrow$$
$$\leftrightarrow HMC; R2 \to HMC; DI \to HMC; CE \to H \to G \to \ldots \to W) \tag{3.89}$$

$$\text{AS}: \quad MMD \to MMS; F2 \to MMS; F1 \to MMC \to (MMC; F1 \to MHC; F2 \to$$
$$\to MHC; F1 \to MHC; DI \to MHC; CE \to H \to G \to \ldots \to W) \tag{3.90}$$

$$\text{AN}: \quad MMD \to MMS; F2 \to MMS; F1 \to MMD \to HMD \to HMS; F2 \to HMD \to$$
$$\to RMD \to RMD; SN \to (NME \to NHE; TF \to NHD \to$$
$$\to NHD; DI \to NHC; CE \to NRD \to NRD; SN \to NNE) \tag{3.91}$$

$$\text{AB}: \quad MMD \to MMS; F2 \to MMS; F1 \to MMD \to HMD \to HMS; F2 \to HMD \to$$
$$\to EMD \to HMD \to SMS; F1 \to HMD \to CMD \to WMD \to$$
$$\to WHD \to (WHS; R3 \to WHC; CE \to WRD \to WRD; SN \to WNE) \tag{3.92}$$

$$\text{AG}: \quad MMD \to MMS; F1 \to GMS; F1 \to GGS; F1 \to GGC \to (G \to \ldots \to W) \tag{3.93}$$

$$\text{AE}: \quad MMD \to MMS; F2 \to MMS; F1 \to MHS; F1 \to MHC \to \tag{3.94}$$
$$\to (MHC; F2 \to H \to \ldots \to N)$$

$$\text{AL}: \quad MMD \to MMS; F2 \to MMD \to MMS; F1 \to MMD \to HMD \to HMS; F2 \to$$
$$\to HMD \to RMD \to RHD \to RHS; R3 \to (RHC; CE \to RHD \to$$
$$\to RRD \to RRD; SN \to RNE \to RNE; SN \to NNE) \tag{3.95}$$

Portions of these routes in parentheses are speculative concluding stages, often involving non-conservative modes to be discussed later (but defined briefly in Table 3.7). The earlier portions are drawn from computed conservative models. These models were computed until either (a) ∗2 reached (reverse) RLOF – which often happened while ∗1 still filled its own

Roche lobe, so that the two stars came into contact – or (b) ∗1 reached an immediately pre-supernova state, defined as carbon burning reaching $100 L_\odot$.

There is a presumption in routes AN and AL as described above that ∗1 is massive enough to explode and leave a neutron star (and not to disrupt the binary – see Section 5.3). Lower mass systems would tend to have evolutionary states E,W instead of R,N, but are otherwise fairly similar. In fact there are ranges of M_{10} and q_0 where the the final products might be two white dwarfs, two neutron stars, or one of each (with the neutron star descended from either ∗1 or ∗2). We continue to use cases AN and AL to describe these lower-mass variants. We also do not discriminate between neutron stars and black holes, for the moment.

In addition to such variants, there are also minor variations to be found if we compute a large number of conservative models. Occasionally there is a detached portion of evolution interrupting a semidetached stage: $MMS;F1 \to MMD \to MMS;F1$. Among about 50 computed pairs with a range of masses and periods, at least 25 tracks were found that could be distinguished in minor ways, but only the above eight seemed importantly different. Of these eight, only three seem reliably to avoid ending up as a merged single star (AN, AB, AL), although it must be borne in mind that the progress within parentheses above is very tentative.

The reason for these several possibilities AD–AL is partly the acceleration of ∗2's evolution by its accretion from ∗1, along with the deceleration of ∗1's evolution, and partly the possibility of a wide range of initial mass ratios. If q_0 is larger than ∼2, not only is the evolution rapid (thermal) because of the initially decreasing Roche lobe around ∗1, but there is also the possibility that as ∗2 becomes less luminous and cooler because of thermal RLOF its surface develops a deep convection zone which can then result in even faster (dynamic) RLOF.

Figure 3.8 shows, on the basis of several computed evolutionary runs, the expected sub-case as a function of initial parameters. The initial period is implied by X: $X = 3$, for instance, means that the initial period was three times longer than the period at which the system would have experienced RLOF at zero age. Conservative evolution implies some constraints on the *current* mass ratio as a function of X. For example, systems which evolve by case AS usually reach contact before q is reduced below ∼0.35; and thus a system with a current $q < 0.3$, say, cannot be case AS, even though its X might lie in the right range in Fig. 3.8. If a system is found that appears to violate this, we would look for some non-conservative process that might explain it.

Of course Fig. 3.8 ought really to be three-dimensional, since all three initial parameters M_{10}, q_0 and X_0 influence the outcome. I have illustrated the entire space with only two two-dimensional cuts through it.

I will have occasion later to introduce four *non-conservative* sub-cases: AA, AM, AW and AU. The first involves substantial angular momentum loss and the second substantial mass loss, as a result of rotation-enhanced dynamo activity on the lower main sequence. The last two involve substantial mass loss in massive OB stars. I will also have to introduce some non-conservative analogues in case $\mathcal{B}$ ($\mathcal{B}$A, $\mathcal{B}$W and $\mathcal{B}$U) and in case $\mathcal{C}$ ($\mathcal{C}$W and $\mathcal{C}$U).

Table 3.8 collates data for a number of systems that might be supposed a priori to fit within case A. The two quantities X and Y (columns 9 and 10) are intended to help with classification. For interacting, or by hypothesis *formerly* interacting, binaries, X is the ratio of the present period of the system to the period that the system would have had initially, if (a) it has evolved conservatively, (b) the initial mass ratio was $q_0 = 4/3$, and (c) ∗1 just filled its Roche lobe on the ZAMS. The reference value q_0 is an arbitrary choice, but in fact X is

Table 3.8. *Some close detached, semidetached and contact binaries related to case A*

Name	Spectra	State	P	M_1	M_2	R_1	R_2	X^a	Y^a	Case	Reference
Y Cyg	O9.8 + O9.8	*MME*	3.00[b]	17.5	17.3	6.0	5.7	2.91	1.04	AN	Hill and Holmgren 1995
VV Ori[c]	B1 + B5	*MMD*	1.49	10.8	4.5	5.0	2.5	1.68	1.01	AD	Popper 1993
XZ Cep	BIII + O9.5V	*mMS*	5.10	6.4	15.8	10.5	7	3.2	1.35	AL	Harries et al. 1998
LY Aur	BOIII + O9.5III	*MMs*[d]	4.00	13	24	13	16	2.94	2.38		Stickland et al. 1994
V Pup	B2 + B1V	*MMS*	1.45	9	17	5.3	6.3	1.17	1.16	AS	Popper 1980
TT Aur	B6 + B3	*MMS*	1.33	5.4	8.1	4.2	3.9	1.59	1.12	AS	Popper and Hill 1991
SV Cen	B3-4 + B1V	*MMs*[d]	1.66	11	9.3	7.2	6.9	1.91	1.83	AE	Wilson and Starr 1976
u Her	B8-9 + B2V-III	*MMS*	2.05	2.9	7.6	4.4	5.8	1.53	1.73	AS	Hilditch 1984
Z Vul	A2III + B3V	*MMS*	2.45	2.3	5.4	4.5	4.7	2.34	1.71	AE	Popper 1980
λ Tau[c]	A5III + B3V	*MMS*	3.95	1.9	7.2	5.3	6.4	1.74	1.97	AE	Fekel and Tomkin 1982
DM Per[c]	A6III + B5	*MMS*	2.73	1.83	5.82	4.59	3.96	1.71	1.38		Hilditch et al. 1986
AT Peg	G9IV + A4V	*mMS*	1.15	1.05	2.22	2.15	1.86	1.74	1.10	AS	Maxted et al. 1994b
TV Cas	G5IV + B9V	*mMS*	1.81	1.53	3.78	3.29	3.15	1.85	1.40	AS	Khalasseh and Hill 1992
AF Gem	G0IV + B9.5V	*mMS*	1.24	1.16	3.37	2.32	2.61	1.09	1.24		Maxted and Hilditch 1995
δ Lib[c]	G0-5 + B9.5	*mMS*	2.33	1.7	4.9	4.4	4.1	1.78	1.57		Tomkin 1978, Worek 2001
U CrB	G0III + B5.5V	*mMS*	3.45	1.46	4.98	4.94	2.73	2.05	1.04	AE	Heintze and van Gent 1988
TX UMa	G1III + B8V	*mMS*	3.06	1.18	4.76	4.24	2.83	1.40	1.10		Maxted et al. 1995a
HU Tau	G2IV + B8V	*mMS*	2.06	1.14	4.43	3.21	2.57	1.03	1.04		Maxted et al. 1995b
TU Mus	O7.5V + O9.5V	*MMC*	1.39	23.5	13.3	7.5	6.2	1.06	1.33	AR	Terrell et al. 2004
LZ Cep	O9Vn + O9Vn	*MMC*	3.07	7	18	8	11	1.77	1.96	AS	Howarth et al. 1991
RZ Pyx	B4 + B4	*MMd*[d]	0.66	5.8	4.7	2.7	2.5	0.93	0.98	AR	Hilditch and Bell 1987
V499 Sco	B6 + B5	*MMD*	2.33	2.2	7.1	4	6	1.33	1.86	AL	Wilson and Rafert 1981
V640 Mon	O7.5If + O6If	*MMD*	14.4	43	51	22	18:	4.7	1.65	AU	Bagnuolo et al. 1992
RT And	F8V + K0V	*MMD*	0.63	1.24	0.91	1.26	0.90	2.01	1.42	AA	Popper 1994
V361 Lyr	F8-G0 + K4	*MMS*	0.31	1.26	0.87	1.02	0.72	0.84	0.92	AA	Hilditch et al. 1997
ε CrA	A8 + F0	*MMC*	0.59	1.5	0.15	2.2	0.7	0.08	3.75	AA	Tapia and Whelan 1975
W UMa	F8 + F8	*MMC*	0.33	1.35	0.7	1.2	0.85	0.75	1.30	AA	Hilditch et al. 1988
AH Vir	G8IV + G8IV	*ggC*	0.41	0.45:	1.4:	0.8:	1.3:	0.55	0.95	AA	Hilditch 1981
AS Eri	K0 + A3	*GMS*	2.66	0.2	1.9	2.2	1.8	0.33	1.14	AA	Popper 1980
R CMa	G8IV + F1	*GMS*	1.14	0.17	1.07	1.15	1.5	0.59	1.53	AA	Sarma et al. 1996

[a] Parameters relating to hypothetical evolution: see text.

[b] $e = 0.15$; all other systems have zero eccentricity.

[c] Member of a close triple system: outer period ~0.1–3 year.

[d] The reference cited suggests that this is a contact system. Nevertheless, in this book we interpret it otherwise. Under 'state', a letter in lower case indicates a substantial degree of uncertainty. Note that for contact binaries, and some semidetached and even detached binaries, the identification of the *originally* more massive component (i.e. *1) is arguable – see text.

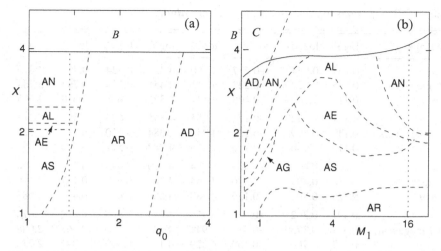

Figure 3.8 (a) Schematic division of the X versus q_0 plane, for fixed initial M_{10} ($16\,M_\odot$). X is the ratio of the period to the period P_{ZAMS} at which $*1$ would fill its Roche lobe on the ZAMS, and q_0 is the initial mass ratio. $X \sim 1\text{–}4$ for case A RLOF. Regions are shown in which the sub-cases AD–AN, routes (3.88)–(3.94), can be expected to take place; see text. The location of these boundaries is only qualitative. They depend quite strongly on M_{10}. Contact is likely to be avoided only in the regions AL, AN. Adapted from Pols (1994). (b) Schematic division of the X versus M_{10} plane, for fixed q_0 (1.33). AD – dynamic RLOF; AR – rapid to contact; AS – slow to contact; AE and AG – early catch-up; AL – late catch-up; AN – normal, no catch-up. Panels (a) and (b) intersect roughly along the dotted line; but both figures are only qualitative.

not very sensitive to q_0 provided $1 < q_0 < 1.5$. For detached binaries X is just the ratio of present period to the period where $*1$ would fill its Roche lobe at the ZAMS. Y is the ratio of the radius of $*2$ to the radius of a ZAMS star of the same mass. Both X and Y ought to be greater than unity, and for case A evolution X ought to be less than ~4.

Among the more massive systems in the upper part of Table 3.8, it is possible to assign plausible sub-cases to the majority (column 11). Table 3.9 shows the result of a least-squares fit for five of these 'hot Algols' to the grid of theoretical models by Nelson and Eggleton (2001). For those systems without an assignment in Table 3.8, I have the following comments:

LY Aur: X is too large for case AS, and yet $*2$ is so large that it is almost in contact (or perhaps has already reached it). As a long shot, I suggest that the components are non-coeval, and ended up in the same binary because of dynamical interaction in a young, dense cluster. Such a model is fairly convincing for ι Ori, Section 5.4.

λ Tau, DM Per: q is too small for case AS, and more appropriate to cases AL or AN; yet X is too small for cases AL or AN, and more appropriate to case AS. I suspect that in both examples the influence of an unusually close third body is significant; the periods are 33 days and 100 days. The close third body may have removed a modest amount of angular momentum from the inner orbit (Section 4.8), so that the *initial X* was large enough for case AL.

AF Gem–HU Tau: the same problem as with λ Tau and DM Per. But here it might be magnetic braking, in cool systems with G-type components and relatively deep convective envelopes, that has removed a modest fraction of the original angular momentum (Section 4.5).

Table 3.9. *Best-fit conservative models for five hot algols*

Name/age/χ^2	log P	log M_1	log q	log T_1	log T_2	log R_1	log R_2	log L_1	log L_2
V Pup	0.163	0.954	−0.277	4.360	4.420	0.724	0.799	3.850	4.200
10 megayears	0.185	0.927	−0.223	4.345	4.451	0.724	0.793	3.784	4.344
1.311	0.117	1.100	0.100	4.444	4.395	0.668	0.610	4.065	3.755
TT Aur	0.124	0.732	−0.175	4.255	4.395	0.623	0.591	3.210	3.710
16 megayears	0.149	0.769	−0.201	4.242	2.384	0.650	0.615	3.220	3.720
1.775	0.119	0.950	0.150	4.369	4.287	0.581	0.493	3.593	3.088
u Her	0.312	0.462	−0.409	4.064	4.300	0.643	0.763	2.490	3.680
64 megayears	0.330	0.497	−0.386	4.054	4.286	0.673	0.757	2.516	3.612
0.949	0.120	0.800	0.150	4.287	4.200	0.493	0.402	3.088	2.554
Z Vul	0.391	0.362	−0.367	3.955	4.255	0.653	0.672	2.070	3.300
107 megayears	0.387	0.375	−0.417	3.949	4.245	0.670	0.668	2.088	3.268
0.776	0.137	0.700	0.150	4.229	4.138	0.432	0.341	2.735	2.185
U CrB	0.538	0.164	−0.420	3.767	4.170	0.694	0.436	1.430	2.510
218 megayears	0.547	0.158	−0.481	3.761	4.187	0.703	0.433	1.403	2.567
1.604	0.235	0.550	0.200	4.137	4.003	0.341	0.227	2.185	1.417

For each star the first line gives observational data, the second gives the best-fit conservative model from the grid of Nelson and Eggleton (2001), and the third gives the corresponding zero-age parameters. Age and χ^2 are in the LH column.

Note that $X \propto P \propto H^3$, i.e. a 10% loss of angular momentum allows X at age zero to have been 30% larger.

Among the systems in the bottom lines of Table 3.8, evidence of mass loss or angular momentum loss is overwhelming. V640 Mon (Plaskett's star) is arguably the most massive binary known. It is too wide for RLOF, and yet the larger, presumably more evolved, component is significantly the less massive. We attribute this to stellar wind (Section 4.3), probably enhanced above what it would have been if ∗1 were single. Among the remaining low-mass systems, ϵ CrA has $X \sim 0.08$, which requires that the angular momentum has more than halved since age zero. W UMa, the prototype contact binary and AH Vir, are not quite so extreme. V361 Lyr is a semidetached system arguably in sub-case AR, but it also has too little angular momentum. AS Eri has much too little angular momentum to have evolved through the state of equal masses that obviously ought to be passed through by any semidetached system; it would have overflowed its *outer* Roche lobe by a substantial factor at that stage, if it evolved conservatively. R CMa is not quite so extreme in regard to angular momentum, but its *total* mass is so low that it cannot have started evolution except with a rather extreme mass ratio in the opposite sense. Then we would expect case AD, and dramatic evolution into a rapid merger.

Although there exist low-mass detached systems like RT And with $X > 1$, there is no point in assuming that they have *not* lost angular momentum and mass when many similarly cool systems clearly have. I will discuss magnetic braking and binary-enhanced stellar wind at some length in the next chapter.

Table 3.10 gives observed parameters for a small selection of systems arguably related to case $\mathcal{B}$. The sub-cases of case $\mathcal{B}$ that are analogous to AD, AR, AN, AB and AL start with

Table 3.10. *Possible case B systems*

Name	Spectra	State	P	M_1	M_2	R_1	R_2	X	Y	Case	Reference
V356 Sgr	A2II + B4V	*HMS*	3.90	4.7	12.1	14	6	5.8	1.36	*BN*	Popper 1980
RZ Sct	F5 + B3II	*HhS*	15.2	2.5	11.7	15.9	15.8	3.9	3.7		Olson and Etzel 1994
RZ Oph	MIII + B7: + F5Ie	*GMs*	262	0.7:	5.7:	60:	3:	27:	1.1:	*BN*	Knee et al. 1986, Zola 1991
ϕ Per	HeI em + B1IIIpe	*EMD*	127	1.15	9.3	1.3:	5.5–8[a]	11		*BB*	Gies et al. 1998
3 Pup	? + A2Iabe	*eHD*	161	?	0.006[b]					*BL*	Plets et al. 1995
HD51956	B2-3e + F8Ib:	*eHD*	107		0.0016[b]					*BL*	Burki and Mayor 1983, Ake and Parsons 1990
υ Sgr	AI + ?	*HMs*	138	2.5	4.0					*BU*	Dudley and Jeffery 1990
V379 Cep	B2III + B2III	*hhD*	99.7	1.9	2.9	5.2	7.4		3.8	*AL*[c]	Gordon et al. 1998
δ Ori A	B0.5III + O9.5II	*hhD*	5.73	5.6	11.2	5	13				Harvin et al. 2002
V505 Mon	B5Ib + ?	*hmd*	53.8	4.55[b]						*BU*	Chochol and Mayer 2002
V2174 Cyg	BN2.5Ib:e + ?	*hme*	225	5.9[a]						*BU*	Bolton and Rogers 1978

[a] Polar–equatorial radii of rapid rotator
[b] Mass function
[c] Followed by reverse case *BU*: see text
All eccentricities are low or zero.
Uncertain or guessed evolutionary states are in lower case.

$MMD \rightarrow HMD \rightarrow HMS \rightarrow \ldots$ V356 Sgr is reasonably well modelled with starting parameters $(9.4 + 7.4\,M_\odot; 5\,\text{days})$. However, RZ Sct is much more difficult. The large size of its gainer, 3.7 times its ZAMS radius, argues for something like case AS, but the long period, characterised by $X \sim 3.9$, argues for case AN or its analogue case $\mathcal{B}$N, in which $*2$ does not grow to anything like its Roche lobe size until well after $*1$ has detached itself. We do not have a good model for this system; nor is it obvious how some of the non-conservative processes of the next two chapters would help. Perhaps the least implausible suggestion is that the gainer has been so much spun up in the accretion process that it is in a state of *differential* rotation, and largely centrifugally supported (Section 6.2).

On the other hand RZ Oph and ϕ Per agree reasonably with conservative case $\mathcal{B}$. The FI spectrum seen in RZ Oph is interpreted as the accretion cloud around $*2$. There is not much scope for mass loss in either system. The minimum initial mass for $*1$ is half the present *total* mass, and a maximum is given by requiring that the current mass of $*1$ is greater than (or equal to, in the case of ϕ Per) the core mass at the terminal main sequence. These estimates do not conflict, and in fact agree rather well. We can model RZ Oph with starting parameters $(3.4 + 3.0\,M_\odot; 16.5\,\text{days}; \text{case } \mathcal{B}\text{N})$, and ϕ Per with $(5.9 + 4.6\,M_\odot; 8\,\text{days}; \text{case } \mathcal{B}\text{B})$. The latter system is part way between its first episode of RLOF and its second, when $*1$ will reexpand back to its Roche lobe as a helium red giant.

3 Pup and HD51956 are A/F supergiants in single-lined orbits, with very small mass functions. In one of them a hot companion, sub-luminous for a main sequence star of its early type, is detected. They are probably similar to ϕ Per except that (a) $*1$ was less massive originally, and so its remaining core is less luminous, and (b) $*2$ has evolved further, and is approaching reverse RLOF.

It is very unlikely that the reverse RLOF will be conservative, given the extreme mass ratios to be expected, and seen in both ϕ Per and RZ Oph. There do, however, exist a few binaries that are arguably in the next stage: we look at υ Sgr and V379 Cep. In both of these $*1$ can be interpreted as either a helium star or a star with a substantial helium core and hydrogen-rich envelope. In both, $*2$ is *severely* undermassive compared with what we would expect as a result of conservative RLOF. We suggest (Section 5.2) that $*2$ in V379 Cep and $*1$ in υ Sgr have lost substantial envelopes (perhaps $10\,M_\odot$), but without any substantial orbital shrinkage and without any substantial transfer to the companion. I identify this process later as mode EJ, leading (in case $\mathcal{B}$) to sub-case $\mathcal{B}$U. The ESB2 system δ Ori A is somewhat similar to V379 Cyg. Each component is undermassive by a factor of ~ 2–3 relative to what would be expected.

Chochol and Mayer (2002) suggest that something similar may happen rather generally if the initial period is several tens of days. They point to V505 Mon and V2174 Cyg (and several others) as systems where, because the mass function is very large, the companion should be of high mass, and yet in fact is not seen (except that in V505 Mon an envelope around it gives eclipses). This suggests rather that M_1 is now rather small because of mass loss, despite the BI spectrum, and that the unseen star did not accrete much of this lost mass. We believe a case can be made that (a) if $*1$ is in the Hertzsprung gap but to the hot side of roughly BI when it fills its Roche lobe, then RLOF may be reasonably conservative, but (b) if $*1$ is to the cool side of this boundary – without yet being at the Hayashi track – then RLOF, if that is the right term, may be largely or wholly non-conservative of mass, although the orbital period does not shrink by a large factor. In the case of V379 Cep we are talking about *reverse* RLOF, and so (a) hardly applies – to ∗2 – because an earlier conservative *forward* phase tends to lengthen the period so that only (b) applies.

It is difficult to identify any example of a binary that contains a fairly massive component ($\gtrsim 10\,M_\odot$), has a period ($\sim$50–500 days) corresponding to the onset of RLOF in the right-hand half of the Hertzsprung gap, and that can be reasonably accounted for by *conservative* RLOF. This is the reason why I feel that such systems need the non-conservative model of Section 5.2.

Broadly, I conclude that conservative RLOF gives a reasonable model of case A binary evolution for systems both of whose components are (and always have been) in the range late O to F. Cooler systems seem to be subject to significant magnetic braking, or mass loss. Early, massive systems, often containing Wolf–Rayet components, generally show clear evidence of mass loss, and are discussed in Section 4.3. Case $\mathcal{B}$ presents a more complex picture, with possibly conservative behaviour for the shorter periods and highly non-conservative behaviour for the longer periods. The transition may occur near the middle of the Hertzsprung gap, roughly on a sloping line that corresponds to initial periods of $\sim$50–100 days.

4

Slow non-conservative processes

I now consider a number of processes by which either angular momentum or mass, or both, may be lost from the system. Such 'non-conservative' processes can modify the orbit very substantially. Some operate on a long timescale, and some on a short – indeed, very short – timescale. I will deal with the latter in the next chapter. Firstly I consider some slow processes.

4.1   Gravitational radiation: mode GR

One slow but inevitable process is gravitational radiation, a general relativistic effect which can become significant in binaries with $P \lesssim 0.6$ days. Formulae for this (Peters 1964, Shapiro and Teukolsky 1983) are obtained – Appendix C(d) – by averaging the rates of energy loss and angular momentum loss over the approximately Keplerian orbit:

$$\frac{\dot{h}}{h} = -\frac{1}{t_{GR}} \frac{1 + \frac{7}{8}e^2}{(1 - e^2)^{5/2}}, \tag{4.1}$$

$$\frac{\dot{P}}{P} = \frac{3}{2}\frac{\dot{a}}{a} = -\frac{3}{2}\frac{\dot{\mathcal{E}}}{\mathcal{E}} = -\frac{3}{t_{GR}} \frac{1 + \frac{73}{24}e^2 + \frac{37}{96}e^4}{(1 - e^2)^{7/2}}, \tag{4.2}$$

and

$$\frac{\dot{e}}{e} = -\frac{1}{t_{GR}} \frac{\frac{19}{6} + \frac{121}{96}e^2}{(1 - e^2)^{5/2}}, \tag{4.3}$$

where

$$t_{GR}(P) = \frac{5}{32} \frac{c^5 a^4}{G^3 M^2 \mu} = \frac{5}{32} \frac{M^2}{M_1 M_2} \left(\frac{cP}{2\pi a}\right)^5 \frac{P}{2\pi}$$

$$= 376.8 \frac{(1 + q)^2}{q} P^{8/3} M^{-5/3} \text{ (gigayears).} \tag{4.4}$$

M, as usual, is the total mass (in Solar units) and P is the period in days. For a circular orbit of initial period P_0, the period decreases to zero in a time $t_{GR}(P_0)/8$. At $q = 1$ and $M = 2.8\,M_\odot$ (two neutron stars), this time is less than ~ 10 gigayears if $P \lesssim 0.63$ days. For two white dwarfs of $\sim 0.6\,M_\odot$ the period required is 0.37 days, and for two black holes of $10\,M_\odot$ it is 2.1 days.

Gravitational radiation tends to circularise the orbit, on much the same timescale as the orbital shrinkage. We can integrate the ratio of Eqs (4.2) and (4.3) to obtain period as a function of eccentricity:

$$\log P = \frac{18}{19}\log e - \frac{3}{2}\log(1-e^2) + \frac{1305}{2299}\log\left(\frac{19}{6} + \frac{121}{96}e^2\right) + \text{constant}, \quad (4.5)$$

where the arbitrary constant is determined by the initial P_0, e_0. The time $T_{\text{GR}}(P_0, e_0)$ taken to shrink to zero can then be found by integrating Eq. (4.3), with P in the factor t_{GR} taken from Eq. (4.5). The resulting function of e can be integrated numerically from an initial e_0 to zero. We can approximate the result by the interpolation formula

$$T_{\text{GR}}(P_0, e_0) = \frac{1}{8} t_{\text{GR}}(P_0)X(e_0), \quad \text{where} \quad X(e) \approx (1-e^2)^{3.689-0.243e-0.058e^2}, \quad (4.6)$$

which is accurate to about 1% for $e \leq 0.99$. Thus if the initial eccentricity is 0.7 the time taken to shrink to zero is about 10% of the time required if the initial eccentricity were zero, for the same initial period.

Pulsar J1915 + 1606 has parameters $(1.387 + 1.441\,M_\odot, 0.323\,\text{days}, e = 0.617$; Thorsett and Chakrabarty 1999). The present timescale of period change, from Eq. (4.2), is 0.368 gigayears, which is in good agreement the measured value 0.364 gigayears (Taylor and Weisberg 1989). The time until the two components merge, from Eq. (4.6), is 0.302 gigayears. Since pulsars 'die', i.e. stop pulsing detectably, in perhaps 3 megayears, the system cannot have been 'born' (in its present form) with e in excess of about 0.62. A recently discovered *pair* of pulsars (J0737-3039; Lyne *et al.* 2004) has $P = 0.102$ days, $e = 0.088$. One of the pulsars is of very short spin period, 0.022 s, and is presumably the older pulsar, which has been spun up by accretion during an earlier phase as a massive X-ray binary with an OB or WR companion. The other pulsar has spin period 2.7 s, and is presumably the remnant of the companion which exploded within the last few megayears. Although the GR merger timescale for this system is substantially shorter (70 megayears), this system also cannot have been much different when it 'started' from what it is now. It would have taken about 12 megayears to reduce its eccentricity from 0.085 to 0.08. The precursor system could have been like V1521 Cyg (Cyg X-3; Table 5.3), where a neutron star is accreting from a Wolf–Rayet-like star in a 0.2 day orbit (van Kerkwijk *et al.* 1996b). Although such an orbit is small, it is quite large enough to contain a *helium* ZAMS star of 2.5 $M_\odot$ or somewhat more – Section 2.5 – which could evolve to a supernova without overflowing its Roche lobe and totally engulfing the NS companion. The helium main sequence component is presumably the remnant of an OB star from an earlier wider binary which may have undergone mode CE – Section 5.2.

The binary of shortest known period so far is RX J0806 + 15, with a period of 321.5 s (Hakala *et al.* 2003). It appears to consist of two white dwarfs. There is a measured period decrease on a timescale $P/\dot{P} = -0.16$ megayears. This may be due entirely to GR, and suggests masses (if $q = 1$) of 0.51 $M_\odot$ each.

4.2 Tidal friction: mode TF

Tidal friction is a process which operates in the Earth–Moon system, slowing down the Earth's rotation and (to conserve angular momentum) driving the Moon outwards; most of this friction is due to the turbulent dissipation of tidal motion in shallow parts of the oceans (Taylor 1919). Tidal friction is also what keeps Jupiter's moon Io in a permanently molten volcanic state, while Jupiter's other moons are cold; although in Jupiter's satellite system the other moons near Io, and not just Jupiter itself, help provide the time-dependent tidal distortion that generates the heat released. Tidal friction is a dissipative process converting the kinetic energy of time-dependent distortions into heat, while conserving angular momentum. I mentioned in Section 3.1 that it can be expected to drive a binary towards a state of uniform rotation, implying both a circular orbit, and corotation of both stars with the orbit.

When a body is solid, as is the Earth (approximately), it is reasonable to assume that it is in uniform rotation that can be represented by an angular velocity Ω. For gaseous bodies, this seems a rather bold assumption, but let us make it nevertheless. We consider the case where *2 is a point mass rather than an extended body, so that its own angular velocity can be ignored. The effect of tidal friction on the orbit, and on the rotation of *1, can be modelled by a dissipative force (Darwin 1880, Kopal 1959, Jeffreys 1959, Alexander 1973, Hut 1981), in addition to the gravitational force (itself a combination of point-mass gravity plus a quadrupole term). The dissipative force can be determined – Appendices B and C(c); Eggleton *et al.* (1998) – by the assumption that the rate of dissipation of energy is proportional to the square of the time rate of change of the quadrupole tensor of *1, viewed in the frame that rotates with the star. Evidently this is zero if and only if (a) the orbit is circular, (b) the stellar rotation is parallel to the orbital rotation, and (c) the star corotates with the orbit. The model leads to a perturbative acceleration

$$\mathbf{f} = -\frac{9\sigma M_2^2 A^2}{2\mu d^{10}}[3\,\mathbf{d}\,(\mathbf{d}\cdot\dot{\mathbf{d}}) + (\mathbf{d}\times\dot{\mathbf{d}} - \Omega d^2)\times\mathbf{d}]. \tag{4.7}$$

As usual μ is the reduced mass; A is the same as in Eq. (3.31), and depends only on the radius R and an internal structure constant Q – Section 3.2.1. The dissipation coefficient σ (dimensions $m^{-1}l^{-2}t^{-1}$) can be related to the turbulent viscosity within *1 – Eqs (B72), (B73), (C54) and (C55) – by

$$\sigma = \frac{2}{M_1^2 R^4 Q^2}\int_0^{M_1} w\,l\,\gamma(r)\,\mathrm{d}m \equiv \frac{2}{M_1 R^2 Q^2 t_{\mathrm{visc}}}, \quad \text{say}, \tag{4.8}$$

where w, l are estimates of the mean velocity and mean free path of turbulent eddies and $\gamma(r)$, of order unity in the outer layers and dropping to ~ 0.002–0.01 in the core of an MS star, is a dimensionless function of position in the star that depends only on its zero-order structure – Appendix B(xi), Eqs (B65), (B69). The quantity t_{visc} is a dissipative timescale intrinsic to the star: see Eq. (4.32) below. The behaviour of $\gamma(r)$ in some models was shown in Fig. 3.2.

To conserve total angular momentum $\mathbf{H} \equiv \mathbf{H}_o + I\Omega = \mu\mathbf{d}\times\dot{\mathbf{d}} + I\Omega$, there must be a corresponding couple on the star, so that

$$\frac{\mathrm{d}}{\mathrm{d}t}\,I\Omega = -\mu\mathbf{d}\times\mathbf{f}, \tag{4.9}$$

where I is the moment of inertia of *1.

Let us define a tidal-friction timescale t_{TF} by

$$t_{\mathrm{TF}} = \frac{2\mu a^8}{9\sigma M_2^2 A^2} = \frac{t_{\mathrm{visc}}}{9}\left(\frac{a}{R}\right)^8\frac{M_1^2}{M_2 M}(1-Q)^2. \tag{4.10}$$

Equation (4.7) leads, by averaging over the zero-order Keplerian orbit – Hut 1981, and Appendix C(c) – to

$$\frac{\dot{\omega}}{\omega} = -\frac{3\dot{a}}{2a} = \frac{3\dot{\mathcal{E}}}{2\mathcal{E}}$$

$$= \frac{3}{t_{\text{TF}}} \left[\frac{1 + \frac{31}{2}e^2 + \frac{255}{8}e^4 + \frac{185}{16}e^6 + \frac{25}{64}e^8}{(1 - e^2)^{15/2}} - \frac{\Omega}{\omega} \frac{1 + \frac{15}{2}e^2 + \frac{45}{8}e^4 + \frac{5}{16}e^6}{(1 - e^2)^6} \right], \quad (4.11)$$

where ω is the *mean* orbital angular velocity, i.e. $2\pi/P$, rather than the variable instantaneous angular velocity, and Ω is the spin, assumed for the time being to be parallel to the orbit, i.e. to $\mathbf{h} = \mathbf{d} \times \dot{\mathbf{d}}$. We similarly obtain rates of change of eccentricity and of orbital angular momentum (per unit reduced mass):

$$\dot{e} = -\frac{9e}{t_{\text{TF}}} \left[\frac{1 + \frac{15}{4}e^2 + \frac{15}{8}e^4 + \frac{5}{64}e^6}{(1 - e^2)^{13/2}} - \frac{11\Omega}{18\omega} \frac{1 + \frac{3}{2}e^2 + \frac{1}{8}e^4}{(1 - e^2)^5} \right], \quad (4.12)$$

$$\dot{h} = -\frac{h}{t_{\text{TF}}} \left[\frac{1 + \frac{15}{2}e^2 + \frac{45}{8}e^4 + \frac{5}{16}e^6}{(1 - e^2)^{13/2}} - \frac{\Omega}{\omega} \frac{1 + 3e^2 + \frac{3}{8}e^4}{(1 - e^2)^5} \right]. \quad (4.13)$$

For the intrinsic spin, we obtain

$$\frac{\dot{\Omega}}{\Omega} = -\frac{\dot{h}}{\lambda h}, \quad (4.14)$$

where

$$\lambda \equiv \frac{I\Omega}{\mu h} = \frac{Mk^2}{M_2 R^2} \frac{R^2 \Omega}{a^2 \omega}. \quad (4.15)$$

The factor λ is the ratio of spin to orbital angular momentum, with k the radius of gyration of *1, as in Eq. (3.18).

Each of Eqs (4.11)–(4.13) can be written in the form

$$\frac{\dot{x}}{x} = \pm \frac{1}{t_{\text{TF}}} \left[f_{x1}(e) - \frac{\Omega}{\omega} f_{x2}(e) \right], \quad (4.16)$$

where x is ω, e or h; plus applies to ω, and minus to e, h. The functions f_{x1}, f_{x2} are tabulated in Table 4.1. It can be seen that even for a modest e, for example 0.4, the rates of variation of a, h and e are *considerably* larger than for a nearly circular orbit of the same period (or equivalently of the same semimajor axis).

By subtracting $\dot{\omega}/\omega$ – Eq. (4.11) – from $\dot{\Omega}/\Omega$ – Eq. (4.14) – in the case that $\Omega \parallel \mathbf{h}$, we obtain

$$\frac{\mathrm{d}}{\mathrm{d}t} \log \Omega/\omega = \frac{1}{\lambda t_{\text{TF}}} \left[f_{h1}(e) - \lambda f_{\omega 1}(e) - \frac{\Omega}{\omega} \{ f_{h2}(e) - \lambda f_{\omega 2}(e) \} \right]. \quad (4.17)$$

Since λ is normally small, we see that even if Ω is initially several times greater or smaller than ω, *1 spins down or up rather rapidly at first towards 'pseudo-synchronism' (Hut 1981), i.e. towards a value

$$\frac{\Omega}{\omega} = f(e, \lambda) \equiv \frac{f_{h1}(e) - \lambda f_{\omega 1}(e)}{f_{h2}(e) - \lambda f_{\omega 2}(e)} \approx \frac{f_{h1}(e)}{f_{h2}(e)} \quad \text{if} \quad \lambda \ll 1, \quad (4.18)$$

Table 4.1. *Functions of eccentricity involved in tidal friction*

e	$f_{\omega1}(e)$	$f_{\omega2}(e)$	$f_{e1}(e)$	$f_{e2}(e)$	$f_{h1}(e)$	$f_{h2}(e)$	Ω/ω PS	Ω/ω e-stable	λ D-stable
0.0	3.000	3.000	9.000	5.500	1.000	1.000	1.000	1.636	0.333
0.1	3.747	3.427	9.970	5.870	1.148	1.083	1.060	1.698	0.309
0.2	6.812	5.017	13.53	7.152	1.707	1.374	1.242	1.892	0.248
0.3	16.20	9.091	22.47	10.01	3.177	2.040	1.557	2.245	0.171
0.4	48.18	20.03	46.08	16.35	7.284	3.562	2.045	2.818	0.102
0.5	182.9	54.47	120.0	32.05	20.97	7.473	2.805	3.746	0.052
0.6	959.8	194.0	425.1	79.71	80.83	19.82	4.077	5.333	0.0210
0.7	8253	1034	2361	281.4	482.4	74.20	6.502	8.392	0.0063
0.8	1.73×10^5	11282	28862	1829	6268	508.3	12.33	15.78	0.0011
0.9	3.14×10^7	6.97×10^5	2.33×10^6	51022	5.33×10^5	14846	35.91	45.67	5×10^{-5}

Columns 2 and 3 relate to $d\omega/dt$, Eq. (4.11); columns 4 and 5 to de/dt, Eq. (4.12); columns 6 and 7 to dh/dt, and also $d\Omega/dt$, Eqs (4.13) and (4.14).
Of the last three columns two are Ω/ω ratios for pseudo-synchronism (PS) taking $\lambda \ll 1$, and for e-stability; and finally λ for D-stability, Eq. (4.21).

where $f(e, \lambda)$ is determined by the vanishing of the term in square brackets of Eq. (4.17). Subsequently e, Eq. (4.12), and ω, Eq. (4.11), decrease on a slower timescale, Ω/ω being in transient equilibrium with e (i.e. pseudo-synchronised) until the orbit is circularised. Both timescales, synchronisation and circularisation, depend strongly on the ratio of stellar radius to orbital semimajor axis. From Eqs (4.10), (4.14) and (4.15), the timescale of synchronisation (λt_{TF}) depends on the sixth power and of circularisation (t_{TF}) on the eighth power of a/R.

From Table 4.1, we see how the pseudo-synchronous spin rate departs quite rapidly from the synchronous rate: at $e = 0.4$, and small λ, the ratio is ~ 2.04. This, is of course, because much the greatest part of the effect comes from near periastron.

From Eq. (4.12) for e-evolution, if the star is spinning faster than a certain amount e *increases*, so that we have a kind of instability that we call the 'e-instability': the condition, also given in Table 4.1, is that

$$\frac{\Omega}{\omega} > \frac{f_{e1}(e)}{f_{e2}(e)} \geq \frac{18}{11}. \tag{4.19}$$

However, if λ is not small another instability, the Darwin or 'D instability' can come into play. When $e = 0$, Eq. (4.17) gives

$$\frac{d\log(\Omega/\omega)}{dt} = \frac{1 - 3\lambda}{\lambda t_{TF}}\left(1 - \frac{\Omega}{\omega}\right), \tag{4.20}$$

and for $\lambda \geq 1/3$, if Ω/ω departs slightly from unity the departure grows. When $e > 0$ the critical λ is smaller: the condition for D instability is

$$\lambda > \frac{f_{h2}(e)}{f_{\omega2}(e)}, \tag{4.21}$$

also shown in Table 4.1. I will discuss the D instability, starting from a more elementary viewpoint, in Section 5.1.

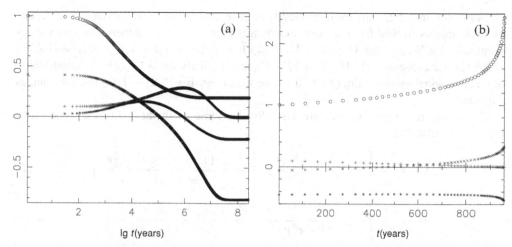

Figure 4.1 Orbital and spin evolution in a generic massive X-ray binary, with a B supergiant of $24\,M_\odot$, $30\,R_\odot$ and a neutron-star companion in an orbit with $P = 9$ days, $e = 0.1$ initially. The initial stellar rotation rate is (a) supersynchronous ($\times 1.8$), or (b) sub-synchronous ($\times 0.6$). Eccentricity (pluses), orbital frequency ω relative to its initial value (circles), the degree of asynchronism $\log(\Omega/\omega)$ (asterisks), and the ratio of spin to orbital angular momentum $\log(I\Omega/\mu h)$ (crosses) are plotted against time. In (a) the orbit starts both D unstable and e unstable. It decircularises at first (e instability). Once the orbit has widened slightly it becomes stable to both processes, and settles down as a much wider binary ($P \sim 45$ days). However, nuclear evolution (neglected) would cause problems well before 10 megayears. In (b) the orbit is e stable and slightly D stable to start with. The Darwin instability occurs after a small degree of spin-up. This causes the orbit to shrink catastrophically in ~ 1 kiloyears.

It is not difficult to integrate Eqs (4.10)–(4.15) numerically, to investigate the variation of the orbit under tidal friction in more detail. We must, of course, remember that λ is itself a function of ω and Ω, and t_{TF} of a, or equivalently of ω. Some illustrations of these processes are given in Fig. 4.1. They are loosely based on the massive X-ray binary GP Vel, and start with a neutron star companion in a 9 days, slightly eccentric, orbit with a massive OB supergiant. Because the mass ratio is very different from unity, the synchronisation timescale is not necessarily shorter than the circularisation timescale. We therefore suppose that the rotation rate of the OB star is somewhat different from pseudo-synchronous, and is either larger (Fig. 4.1a) or smaller (Fig. 4.1b). The orbital evolution is interestingly different in the two cases.

We now consider the more general case, where both stars are extended objects that may be rotating in independent directions non-parallel to $\mathbf{h}$. The equations governing the four vectors $\mathbf{e}, \mathbf{h}, \boldsymbol{\Omega}_1, \boldsymbol{\Omega}_2$ are

$$\dot{\mathbf{e}} = \mathbf{U} \times \mathbf{e} - V\mathbf{e}, \tag{4.22}$$

$$\dot{\mathbf{h}} = \mathbf{U} \times \mathbf{h} - W\mathbf{h}, \tag{4.23}$$

$$I_1\dot{\boldsymbol{\Omega}}_1 = -\mu\mathbf{U}_1 \times \mathbf{h} + W_1\mathbf{h}, \tag{4.24}$$

$$I_2\dot{\boldsymbol{\Omega}}_2 = -\mu\mathbf{U}_2 \times \mathbf{h} + W_2\mathbf{h}. \tag{4.25}$$

$\mathbf{U}$, as in Section 3.2.2, is the angular velocity of the $\mathbf{e}$, $\mathbf{q}$, $\mathbf{h}$ frame relative to an inertial frame, but now thanks to tidal friction there are changes of $\mathbf{e}$, $\mathbf{h}$ parallel to themselves as well as perpendicular. We find that $\mathbf{U} = \mathbf{U}_1 + \mathbf{U}_2 + Z_{GR}\overline{\mathbf{h}}$, with $\mathbf{U}_1 = X_1\overline{\mathbf{e}} + Y_1\overline{\mathbf{q}} + Z_1\overline{\mathbf{h}}$ (see below) and a similar expression for $\mathbf{U}_2$. $V = V_1 + V_2$, and similarly for W. Z_{GR} is the contribution of GR to apsidal motion – Eq. (3.50). It can be seen that $\mu\mathbf{h} + I_1\mathbf{\Omega}_1 + I_2\mathbf{\Omega}_2$ is a constant, as expected.

The dissipative terms V_1, W_1 are virtually the same as in the parallel case above – Eqs (4.12 and 4.13):

$$V_1 = \frac{9}{t_{\mathrm{TF1}}}\left[\frac{1 + \frac{15}{4}e^2 + \frac{15}{8}e^4 + \frac{5}{64}e^6}{(1-e^2)^{13/2}} - \frac{11\Omega_{1h}}{18\omega}\frac{1 + \frac{3}{2}e^2 + \frac{1}{8}e^4}{(1-e^2)^5}\right], \tag{4.26}$$

$$W_1 = \frac{1}{t_{\mathrm{TF1}}}\left[\frac{1 + \frac{15}{2}e^2 + \frac{45}{8}e^4 + \frac{5}{16}e^6}{(1-e^2)^{13/2}} - \frac{\Omega_{1h}}{\omega}\frac{1 + 3e^2 + \frac{3}{8}e^4}{(1-e^2)^5}\right]. \tag{4.27}$$

The contributions X_1, Y_1, Z_1 to the rotation of the axes caused by rotational and tidal distortion of $*1$ (including the small contribution of tidal friction), are given by

$$X_1 = -\frac{M_2 A_1}{2\mu\omega a^5}\frac{\Omega_{1h}\Omega_{1e}}{(1-e^2)^2} - \frac{\Omega_{1q}}{2\omega t_{\mathrm{TF1}}}\frac{1 + \frac{9}{2}e^2 + \frac{5}{8}e^4}{(1-e^2)^5}, \tag{4.28}$$

$$Y_1 = -\frac{M_2 A_1}{2\mu\omega a^5}\frac{\Omega_{1h}\Omega_{1q}}{(1-e^2)^2} + \frac{\Omega_{1e}}{2\omega t_{\mathrm{TF1}}}\frac{1 + \frac{3}{2}e^2 + \frac{1}{8}e^4}{(1-e^2)^5}, \tag{4.29}$$

$$Z_1 = \frac{M_2 A_1}{2\mu\omega a^5}\left[\frac{2\Omega_{1h}^2 - \Omega_{1e}^2 - \Omega_{1q}^2}{2(1-e^2)^2} + \frac{15 G M_2}{a^3}\frac{1 + \frac{3}{2}e^2 + \frac{1}{8}e^4}{(1-e^2)^5}\right]. \tag{4.30}$$

$V_2 \ldots Z_2$ are the same, with suffices 1,2 interchanged. Ω_e, Ω_q, Ω_h are the components of the appropriate $\mathbf{\Omega}$ in the orbital frame (Section 3.2.2). X, Y, apart from the terms due to tidal friction, give the same precession rate as in Section 3.2.2. Z is the same as in Section 3.2.2: tidal friction does not contribute to apsidal motion. The tidal friction terms in X, Y tend to parallelise the spins on much the same timescale as synchronisation.

In Appendices B and C(c) I derive the force law due to tidal friction – Eq. (4.7) – in two apparently different ways: in B, we determine the tidal velocity field and work out its rate of dissipation if turbulent viscosity is the main dissipative agent; and in C, we start from the more general principle that the rate of dissipation should be a positive semidefinite function of the rate of change (as seen in the frame that rotates with the star) of the quadrupole tensor. That the two approaches lead to the same dependence on $\mathbf{d}, \dot{\mathbf{d}}$ is presumably confirmation that the tidal velocity field of Eqs (B63)–(B65) is correct, at least in its $\mathbf{d}$ dependence. Identifying Eq. (B71) with Eq. (C53) gives the otherwise indeterminate coefficient σ of Appendix C in terms of the specific dissipative model of Appendix B. The result is Eq. (4.8); but this in turn has a coefficient t_{visc} which is a dissipative timescale intrinsic to the star – Eq. (4.31) below. However, it is gratifying that the more general approach of Appendix C leads unequivocally to a specific dependence on $\mathbf{d}$ of the frictional force, and we might hope that observation might ultimately pin down the value of the coefficient σ, or equivalently t_{visc}, even if theoretical models are discordant.

The rate of dissipation can be estimated from the tidal velocity field $\mathbf{u}$ of Eq. (3.36) by (a) calculating its shear, (b) squaring it and (c) multiplying by the viscosity and integrating over the star. This is done in Appendix B(xi). We assume that the main viscosity is due to turbulent convection, with a coefficient of viscosity $\sim wl$, where w and l (Section 2.2.3) are the mean turbulent velocity and mixing length. For the velocity field of Eq. (3.36) the dissipation has a timescale t_{visc} where

$$\frac{1}{t_{\text{visc}}} \equiv \frac{1}{M_1 R_1^2} \int_0^{M_1} wl\, \gamma(r)\,\mathrm{d}m, \quad \gamma(r) \equiv \beta^2 + \frac{2}{3}r\beta\beta' + \frac{7}{30}r^2\beta'^2. \tag{4.31}$$

The factor $\gamma(r)$, along with α, β, is illustrated in Fig. 3.2.

To estimate this rate of dissipation in a star, Table 3.4 lists $\overline{\gamma}$, which is γ averaged over the inner 25% of the mass of a polytrope. Typically this fraction is convective in upper main sequence stars. The core-averaged $\overline{\gamma}$ is ~ 0.01 for near-main-sequence stars (roughly, $n \sim 3$ polytropes), but starts to drop rapidly as stars become more centrally condensed. A first estimate for wl/R_1^2 is that it is roughly the reciprocal of the global convective timescale t_G of Eq. (2.32), and so

$$\frac{1}{t_{\text{visc}}} \sim \frac{\overline{\gamma}}{t_G} = \overline{\gamma}\left(\frac{L}{3M_1 R^2}\right)^{1/3}. \tag{4.32}$$

If M, R, L are in Solar units, t_G is approximately in years.

In principle $\overline{\gamma}$ can be computed by solving Clairault's equation – Eq. (3.30) – for the factor $\alpha(r)$ which measures the departure from sphericity throughout a stellar interior, and then summing up two integrals, Eqs (3.38) and (4.31). However, weighing the computational effort against the uncertainty, particularly in the rate of turbulent dissipation, let us settle here for a rough interpolation formula. On the main sequence and in the Hertzsprung gap, this relates the quantity $\overline{\gamma}$ of Table 3.4 to k^2/R^2, the dimensionless gyration radius (squared) for simple polytropes with different degrees of central condensation; while on the Hayashi track with its fully convective envelopes ($R \sim R_{\text{HT}}$) the formula gives $\overline{\gamma} \sim 1$:

$$\overline{\gamma} \sim \frac{2}{2R^2/5k^2 + (2R^2/5k^2)^{3.2}} + \left(\frac{R}{R_{\text{HT}}}\right)^8. \tag{4.33}$$

This only requires in our stellar evolution code that the moment of inertia ($I = Mk^2$) be integrated along with the structure equations; I is also needed in other equations for orbital change.

I believe that in the past the factor which we call $\overline{\gamma}$ has been substantially underestimated. Early estimates were based on the assumption that the velocity field of time-dependent tidal motion was either incompressible or irrotational, and appeared to lead to $\gamma \sim (r/R)^7$. For a convective core with $r \sim 0.3R$, this implies $\gamma \sim 10^{-3.5}$. However, the tidal velocity field determined in Appendix B, the *exact* solution of the conservation equation to first order, is neither incompressible nor irrotational, and leads to a finite value of v/r even at the centre (Fig. 3.2), which is at least an order of magnitude greater. The value of t_{visc} arrived at will be found, at several points in later discussion, to be reasonably in accord with the rather weak constraints that observation imposes.

In the SMC there is a radio pulsar, 0045-7319, which has a very eccentric orbit ($e = 0.808$) and period 51.2 days (Kaspi *et al.* 1994a). The companion is an early B star. Unusually, there appears to be negligible stellar wind from the B star (which is presumably not in rapid rotation), and consequently the pulsar rotates unusually steadily, without erratic spin-up or spin-down due to accretion from a stellar wind. As a result, the measured slow spin-up of

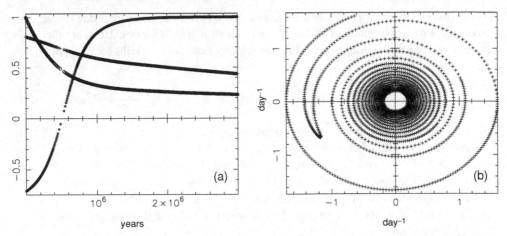

Figure 4.2 A model of tidal friction in 0045-7319. An obliquely counter-rotating B star has an NS companion in a wide eccentric orbit. (a) Circles – orbital period relative to initial value (51 days); pluses – eccentricity; asterisks – cosine of the angle between spin and orbit. The rotation was assumed retrograde initially wrt the orbit, perhaps as a result of a supernova kick. The rotation becomes aligned in about 1 megayears. (b) The component of stellar spin in the direction of the orbital major axis (horizontal) plotted against the component in the direction of the latus rectum (vertical). Starting near the upper left, the spin axis precessed counter-clockwise around the orbital axis, until the third component of spin (not shown) passed through zero. Then the precession reversed, the spin axis rotating clockwise about the orbital axis while both components plotted here gradually decrease to zero. The coefficient of viscosity used in (b) was artificially large relative to (a), so that the spiral pattern is less tightly wound by about a factor of 220 than it would really be.

the *orbit*, on a timescale of 0.45 megayears, can reasonably be attributed to the influence of tidal friction alone. Figure 4.2a shows the expected long-term behaviour of the orbit if we start, somewhat arbitrarily, with the B star's rotation axis inclined at 135° to the pulsar orbit. The rotation is parallelised and synchronised in about 1 megayear, and the orbital eccentricity is reduced from 0.8 to 0.4 in about 3 megayears. Figure 4.2b shows the two components of the B star's rotation perpendicular to the orbit, plotted against each other. There is counter-clockwise precession until the inclination is reduced from 135° to 90°, and then clockwise precession until parallelism is reached. In Fig. 4.2b (but *not* in 4.2a) viscosity was increased artificially by 220, so that the spiral is less tightly wound than it would normally be.

There are several different and sometimes strongly conflicting estimates of, in effect, the parameter t_{visc}: Alexander (1973), Campbell and Papaloizou (1983), Savonije and Papaloizou (1984), Scharlemann (1982), Tassoul and Tassoul (1992), Zahn (1977), Zahn and Bouchet (1989). This no doubt partly reflects the inherent difficulty of dealing with fluid (as compared with solid but elastic or slightly inelastic) bodies, where the interior motions may in principle be very complex. In the Sun, for example, one might imagine that the 'turbulent viscosity' in the surface convection zone (the outer $\sim 30\%$ by radius) would enforce rigid-body rotation there, and yet there is a $\gtrsim 25\%$ increase in rotational angular velocity between the poles and the equator (Fig. 2.7b). It is unlikely that the fluid-dynamical and MHD problems inherent in modelling the interior motion in general and tidal friction in particular will be solved soon. They will presumably require fully three-dimensional modelling.

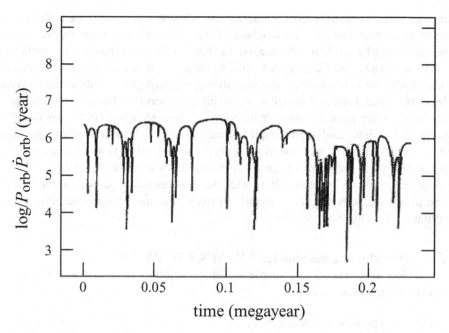

Figure 4.3 Timescale for orbital period change, as a function of time, for a 10 $M_\odot$ star in a binary orbit with a neutron star: initial period 55 days, initial eccentricity 0.81. There are changes in either direction, but the overall trend is decay, and the orbit spins up. After Witte and Savonije (1999).

Witte and Savonije (1999) have modelled 'dynamic tides', and their effect on the orbit of a non-corotating star of 10 $M_\odot$ in an eccentric orbit. For a perturbing $*2$ in a Keplerian orbit, the perturbing potential within $*1$ can be decomposed, by classical procedures, into a sum of products of Legendre polynomials in polar angle, Fourier terms in azimuthal angle, and functions of radial distance from the stellar centre. For moderate to large eccentricity there is a considerable number of terms that contribute comparably strongly. For each term in the decomposition, Witte and Savonije (1999) calculate the rate of dissipation within the star, using an implicit two-dimensional (r, θ) numerical hydrodynamics code which includes the Coriolis term. There is a rich spectrum of normal modes, whose frequencies gradually change as the star evolves; the lines of the spectrum have widths determined by the rate of dissipation. Some of these lines, as they move, will pass through resonances with the tidal forcing terms, so that the rate of dissipation can fluctuate considerably. Figure 4.3 shows how the period-change timescale $|P/\dot{P}|$ varies with evolution in a particular case: a rapidly counter-rotating B star within an NS companion, an initial orbital period of 55 days, and an initial eccentricity of 0.81. It is, therefore, a candidate to evolve into the the pulsar 0045-7319 discussed above. The decay timescale of the orbital period fluctuates by over two orders of magnitude, but averages, over an interval of ~ 0.25 megayears, to ~ 0.6 megayears. For the same star ($R \sim 7\,R_\odot$, $L \sim 10^4\,L_\odot$), and much the same binary ($P = 51$ days, $e = 0.8$, $P_{\rm rot} = -3.1$ days), Eq. (4.32) gives $t_{\rm visc} \sim 17$ years, Eq. (4.10) gives $t_{\rm TF} \sim 160$ gigayears and Eq. (4.11) gives $|P/\dot{P}| \sim 0.45$ megayears.

Clearly an approximation as bland as the equilibrium-tide model cannot be relied on for highly eccentric orbits. Equation (4.7) is based on the concept that the bodies are continually adapting their shapes to be in near equilibrium, while they rotate relative to this equilibrium tide (if they are not already synchronised). This is fairly close to what happens on Earth, as a result of the Moon's tidal influence: but the Earth is small compared to its Roche lobe, and the Moon's orbit is only slightly eccentric. In the much more extreme situation modelled by Witte and Savonije (1999) there is a very strong, and strongly time-dependent, perturbation which stimulates a range of normal modes of differing periods: after a brief intense periastron passage the stars would be very far from equilibrium. Mardling (1995) has found that the oscillations, rather than dissipating quickly, may persist till the next periastron passage, when they are just as likely to increase as decrease the eccentricity and asynchronicity. The result may be chaotic in some circumstances. A purely dissipative model like Eq. (4.7) is unlikely to apply to markedly eccentric orbits with close encounters at periastron, but it still seems to be a reasonable first-order dissipative correction to the conservative zero-order purely gravitational problem.

4.3   Wind processes: modes NW, MB, EW, PA, BP

Let us now consider a simple model for the effect on orbital period and separation of the following processes:

(a)   normal single-star winds (NW)
(b)   magnetic braking with tidal friction (MB)
(c)   binary-enhanced stellar wind (EW)
(d)   partial accretion of stellar wind (PA)
(e)   bipolar re-emission (BP).

All of these have to do with stellar winds, which can remove mass and angular momentum from the system, as well as transfer mass from one component to the other. In this section we ignore GR, but it is not difficult in principle to add it as well.

Single stars can lose mass by stellar wind. In a binary, some of this wind may be accreted by a companion and some may be lost to infinity. The latter portion can be expected to carry angular momentum from the system. The general problem may be quite complicated, and requires a detailed treatment of the flow of gas between and around the stars. For example, in detached as well as semidetached binaries, the portion of the wind recently captured by the gainer may accumulate in a disc around the gainer before being accreted by the star itself; and some of the material in the disc may be expelled in jets rather than accreted at all. Some aspects of this are discussed further in Chapters 5 and 6.

In Appendix C(e), a formulation is given for the effect on the orbit of the combination of (a) isotropic wind from either or both components and (b) the transfer of mass, either by accretion from the wind or RLOF, from one star to the other. This formulation treats the loss and the transfer of momentum and angular momentum in a consistent way, and can be applied to situations where either the loss and transfer rates are steady, i.e. do not change significantly in the course of one orbit, or where they depend on orbital phase, perhaps quite strongly. To start with, we content ourselves with a more intuitive approach. If the wind is isotropic, and fast compared with the orbital speed, a preliminary expectation is that it carries off the same angular momentum per unit mass as resides in the orbital motion of the mass-losing star, i.e. $H_0 M_2/M M_1$, where H_0 is the orbital angular momentum. If all the wind from $*1$ goes to infinity, we can model the effect on the orbit by

$$-\frac{1}{H_0}\frac{dH_0}{dt} = \frac{M_2}{MM_1}\zeta, \quad \frac{dM}{dt} = \frac{dM_1}{dt} = -\zeta, \quad \frac{dM_2}{dt} = 0. \tag{4.34}$$

The mass-loss rate ζ might be given by Eq. (2.73) for a massive OB component, and by Eq. (2.74), or the theoretical estimate of Eq. (4.84) below, for a red giant. We do not (yet) include the *additional* loss of angular momentum due to magnetic braking in Section 4.4. The above system of equations is easily solved analytically, at least if we are only interested in the way that the period changes with the masses. In that case, we can cancel ζ; we only need a numerical value for ζ if we want to know period or masses as a function of time. The solutions of Eq. (4.34) for H_0, a and P, using Eqs (3.13) and (3.14) for the last two, are easily found to be

$$H_0 \propto \frac{M_1}{M}, \quad a \propto \frac{1}{M}, \quad P \propto \frac{1}{M^2}. \tag{4.35}$$

Thus the period *increases* as the mass and mass ratio decrease, in contrast to RLOF where the period first decreases and only increases again once the mass ratio has passed through unity. Our assumptions about wind ensure that it does not change the velocities of the stars (instantaneously), but it weakens the gravity of the remaining mass, thus causing the stars to spiral out.

We now consider a more general model for the influence on orbital period of stellar winds, which might originate from either component and which might be magnetically linked to that component (magnetic braking, mode MB). The winds might be either 'normal' (mode NW), i.e. what the component would experience even if single, or 'enhanced', i.e. larger than normal by virtue of the tidal disturbance due to the other component (enhanced wind, mode EW): the present model does not distinguish these possibilities. We also include the possibility that a fraction of the wind lost by $*1$ may be accreted by the companion (partial accretion, mode PA).

We further include the possibility that a portion of the material from $*1$, temporarily accreted by $*2$, is expelled from the neighbourhood of $*2$; so that in effect some of the wind escaping from $*1$ and leaving the system carries with it the specific angular momentum of the orbit of $*2$ rather than of $*1$ (bipolar reemission, mode BP). Such a process may be particularly important in those semidetached systems where the gainer is a compact star (white dwarf, neutron star or black hole). There is clear evidence in some such systems of outflowing bipolar jets that originate near the compact gainer, and that are presumably fuelled by the energy released in the accretion process.

Suppose that

(a) $*1$ loses mass isotropically to infinity at a rate ζ_1
(b) $*2$ does the same at rate ζ_2
(c) $*1$ also loses mass to $*2$, either by RLOF, or by accretion of a portion of the wind from $*1$, or both, at a rate ξ
(d) the wind to infinity from $*1$ carries specific angular momentum K_1 times the specific orbital angular momentum of $*1$; we expect $K_1 \sim 1$ if there is no magnetic linkage of the wind to the star, but otherwise we might have $K_1 > 1$
(e) a similar factor K_2 applies for the wind escaping from $*2$.

To be pedantic, it is difficult to see how a star can lose mass isotropically while simultaneously accreting from a companion. However, we can think of accretion as confined to a plane, and perhaps even to a point, with the mass loss being *nearly* isotropic in the remaining solid angle, or in a cone with axis perpendicular to the orbital plane ('bipolar').

Then we can write

$$\dot{M}_1 = -\xi - \zeta_1, \quad \dot{M}_2 = \xi - \zeta_2, \quad \dot{M} = -\zeta_1 - \zeta_2, \tag{4.36}$$

and the model for angular momentum loss has

$$\frac{1}{H_o}\frac{dH_o}{dt} = -K_1\frac{M_2}{MM_1}\zeta_1 - K_2\frac{M_1}{MM_2}\zeta_2. \tag{4.37}$$

Equation (4.34) was just the special case of this, which has $\xi = \zeta_2 = K_2 = 0$, $K_1 = 1$. Note that Eq. (4.37) with $K_1 = K_2 = 1$ is also what is obtained from a slightly more rigorous treatment in Appendix C(e), where $h = H_o/\mu$ is the orbital angular momentum per unit reduced mass. That treatment also shows that the eccentricity remains constant, even if non-zero, provided that the wind – including the fraction $\xi/(\xi + \zeta_1)$ accreted by $*2$ – is independent of orbital phase.

We consider later some more detailed expressions for K_1, K_2. For the present, we take the Ks to be constants, and also the ratios $\xi : \zeta_1 : \zeta_2$. There are therefore four independent constant parameters in the model; in Table 4.2 we normalise ξ and the ζs by taking the largest to be unity. Then integration of Eqs (4.36), (4.37) gives

$$\log H_o = \text{constant} + \frac{\zeta_1 K_1}{\zeta_1 + \xi}\log M_1 + \frac{\zeta_2 K_2}{\zeta_2 - \xi}\log M_2 - \frac{\zeta_1 K_1 + \zeta_2 K_2}{\zeta_1 + \zeta_2}\log M. \tag{4.38}$$

This determines how a and P will vary as the masses vary, using Eqs (3.13) and (3.14) respectively. The assumption that the four parameters are constant is not in fact a very good one, but is made simply because it allows the elementary integral (4.38) to be extracted.

In Section 3.3 we saw that the nature of RLOF at its onset (nuclear, thermal or dynamic timescale) depends largely on a comparison of R'_L, the logarithmic rate of change of lobe radius against mass of loser, with various coefficients intrinsic to the star itself. We can also calculate R'_L in our simplistic non-conservative model here. Using Eq. (3.13) for the orbital radius a as a function of H_o and the masses, and Eq. (3.6) for the lobe radius as a function of a and mass ratio q, we obtain, after some manipulation

$$R'_L \equiv \frac{d\log R_L}{d\log M_1} = \frac{(0.33 + 0.13q)\{\xi(1+q) + \zeta_1\} + 2\xi(q^2 - 1) + \zeta_1\{2K_1 - 2 - q\}}{(\xi + \zeta_1)(1 + q)}. \tag{4.39}$$

Equation (3.81) shows that large positive values of R'_L contribute to instability, and negative values to stability. Of course the above result neglects the usually small contribution of spin to the total angular momentum; also we have ignored ζ_2 for simplicity, but it can be included with a little extra difficulty.

Table 4.2 shows the variation of P, M and R'_L (including $\zeta_2 \neq 0$) with mass-ratio q, for various values of ξ, ζ_1, ζ_2, K_1 and K_2. The solutions are normalised so that $P = 1$, $M = 2$ at $q = 1$. Note that although Eq. (4.38) is formally undefined when any of the linear combinations of ξ, ζ_1, ζ_2, that appear in the denominators vanish, the singularities are removable: we need only vary some of the parameters by tiny amounts from their singular values to obtain sufficiently accurate answers, as was done for certain rows in Table 4.2. Equation (4.39) does *not* depend on the constancy of these parameters, since it comes directly from the differential

Table 4.2. *Period, total mass, and stability factor R'_L as functions of mass-ratio in some non-conservative models*

Row	ξ	ζ_1	ζ_2	K_1	K_2	$q = 4$	2	1.33	0.5	0.25	0.125
1	1	0.001	0.001	1	1	3.81	1.42	1.06	1.43	3.82	16.2
RLOF						2.00	2.00	2.00	2.00	2.00	2.00
						6.84	2.59	1.17	−0.61	−1.14	−1.40
2	1	0.1	0.001	3	1	3.83	1.46	1.08	1.26	2.62	7.66
RLOF,						2.06	2.03	2.01	1.97	1.94	1.93
EW, MB						6.24	2.43	1.19	−0.31	−0.74	−0.94
3	0.001	1	0.001	1	1	0.161	0.445	0.735	1.78	2.56	3.17
NW						5.00	3.00	2.33	1.50	1.25	1.13
						−0.624	−0.468	−0.355	−0.071	0.089	0.195
4	0.001	1	0.001	1.5	1	0.325	0.685	0.898	0.969	0.650	0.331
NW, MB						5.00	3.00	2.33	1.50	1.25	1.13
						−0.424	−0.135	0.073	0.595	0.888	1.08
5	0.2	1	0.001	2	1	1.39	1.17	1.09	0.676	0.332	0.125
NW, PA,						3.50	2.63	2.23	1.61	1.40	1.28
MB						0.949	0.595	0.612	0.952	1.22	1.41
6	0.05	1	0.001	2	1	0.865	1.09	1.09	0.569	0.203	0.051
NW, PA,						4.41	2.87	2.30	1.53	1.29	1.17
MB						0.106	0.310	0.533	1.17	1.56	1.81
7	1	1	0.001	1	1	1.81	1.02	0.946	1.54	3.16	7.61
NW, PA						2.51	2.25	2.10	1.80	1.66	1.58
						3.10	1.05	0.402	−0.340	−0.525	−0.604
8	1	0.1	0.001	15	1	6.90	2.22	1.34	0.559	0.328	0.183
RLOF,						2.06	2.03	2.01	1.97	1.94	1.93
EW, MB						6.68	3.16	2.12	1.14	1.01	1.00
9	1	0.1	1	15	1	79.0	3.09	1.39	0.644	0.435	0.278
RLOF,						5.00	3.00	2.33	1.50	1.25	1.13
BP, MB						5.33	2.20	1.34	0.718	0.763	0.869
10	0.5	1	0.001	1	1	1.15	0.848	0.891	1.60	2.95	5.75
RLOF,						2.86	2.40	2.15	1.71	1.54	1.44
NW, PA						1.86	0.550	0.153	−0.25	−0.32	−0.34
11	1	0.001	0.5	1	1	5.11	1.31	0.988	1.91	6.91	36.4
RLOF,						2.50	2.25	2.10	1.80	1.67	1.59
BP						6.10	2.06	0.739	−0.834	−1.27	−1.48

For each combination of parameters, periods are on the first row, total mass on the second, and R'_L, Eq. (4.39), on the third; periods are normalised to 1 and total mass to 2, for $q = 1$. The principal modes involved are indicated at the left; modes NW and EW are not distinguished in these models.

form of Eqs. (4.36) and (4.37), rather than the integrated form Eq. (4.38). We should also note that the spin angular momenta of the components have been ignored in comparison with the orbit; but they can be included by using a more complicated K_1, K_2, as indicated below. It can be seen that there are choices for the parameters that can keep the period constant to within a factor of 3 as q varies all the way from 4 to 0.125 (e.g. row 4), as well as choices that allow P either to increase or to decrease by substantial factors.

The above model is essentially the same as that of Soberman *et al.* (1997), with the proviso that they reversed the roles of $*1$, $*2$, i.e. their model has $\xi < 0$. This does not affect the mathematics, but it slightly complicates the comparison. Subject to this proviso, our $\zeta_1/(\xi + \zeta_1)$, $\zeta_2/(\xi + \zeta_1)$ and K_1 are equivalent to their parameters α, β and A respectively, and they adopt $K_2 = 1$.

The estimate $K_1 \sim 1$ assumes that $*1$ is small compared with the binary separation, and also that magnetic linkage of the wind to the star is negligible. We can estimate K_1 a little better by allowing for $*1$'s finite radius, while still assuming that $*1$ is locked into corotation with the orbit as a result of tidal friction; and we can also allow for the possibility that the wind is forced magnetically to corotate out to an Alfvén radius R_A – Eqs (4.56) and (4.85) below. Simplifying to the case where $\xi = \zeta_2 = 0$, we can write

$$\frac{d}{dt}(H_o + I\omega) = -\zeta_1 \left[\frac{M_2}{MM_1} H_o + \left(R_A^2 + \frac{2}{3}R^2 \right)\omega \right],$$

$$\omega = \frac{G^2 M_1^3 M_2^3}{H_o^3 M} = \frac{H_o M}{M_1 M_2 a^2}, \tag{4.40}$$

using Eq. (3.14) for ω as a function of H_o. After some manipulation, this can be written in the form of Eq. (4.37) – still with $\zeta_2 = 0$ – provided that

$$K_1 \sim \left[1 + \frac{M^2}{M_2^2} \frac{R_A^2 + \frac{2}{3}R^2}{a^2} - \lambda\left(\frac{2M}{M_2} + 1 \right) \right](1 - 3\lambda)^{-1}, \quad \lambda = \frac{M}{M_2}\frac{k^2}{a^2}, \tag{4.41}$$

λ being the usual ratio $I\Omega/H_o$ of spin to orbital angular momentum. The denominator of K_1 is usually not much different from unity; nevertheless, it will approach zero at fairly large mass ratios for nearly lobe-filling components as we approach the Darwin instability (Sections 4.2 and 5.1). Take $q \sim 1$, $R_A \sim 0$, $R \sim R_L \sim 0.38a$, and $k^2 \sim 0.075$ from Eq. (3.18). Then we have $K_1 \sim 1.4$. If q is moderately large, say 2, and if R_A is still zero while $R \sim R_L$, then $K_1 \sim 2$; thus we should not assume that only magnetic braking will increase K_1 significantly above unity. However, as q drops below unity, the effect becomes fairly insignificant. A minor term, involving the change of moment of inertia of the star as the mass changes, has been ignored.

Returning to the case where *both* stars may have winds, but assuming that λ, and hence $I\omega$, can be neglected, we can see from Eq. (4.40) that the rate of change of period $(2\pi/\omega)$ is

$$\frac{\dot{P}}{P} = \frac{2(\zeta_1 + \zeta_2)}{M} + \frac{3\xi}{M}\left(\frac{1}{q} - q \right) - \frac{3M}{M_1 M_2}\left(\zeta_1 \frac{R_{A1}^2 + \frac{2}{3}R_1^2}{a^2} + \zeta_2 \frac{R_{A2}^2 + \frac{2}{3}R_2^2}{a^2} \right). \tag{4.42}$$

In Sections 4.4 and 4.5 I outline a procedure for estimating R_A, which can be incorporated into Eqs (4.40) and (4.41). Magnetic linkage of $*1$ to its wind, possibly enforcing corotation of the wind out to an Alfvén radius several times the stellar radius, may amplify K_1 very considerably (Table 4.4 below), and allow, for example, a significant value of $K_1\zeta_1$ even when $\zeta_1 \ll \xi$, as expected in semidetached binaries. In Chapter 6 I will consider a simple model for $\xi/(\zeta_1 + \xi)$, the fraction of wind from *1 that is accreted by *2, in detached binaries.

If mass transfer is by accretion of a part of the wind from one star by the other, rather than by RLOF, there is no reason to suppose that the orbit will be circular, and in that case ξ may well depend on the phase in the orbit. Appendix C(e) and Section 6.5 show how, nevertheless, the effect of such a variable ξ on the parameters of the orbit can be determined by averaging over a Keplerian orbit.

In some circumstances, the gainer may use part of the accretion energy to blow off a fraction of the transferred mass. If the *nuclear* energy of the transferred gas can be so used, as in classical novae, then possibly all, or even more than all, of the transferred mass may be ejected (episodically). This process of 'bipolar reemission' or mode BP can also be modelled crudely by the above formulation, taking $\zeta_2 \geq \xi$, along with $\zeta_1 \approx 0$ and $K_2 = 1$. We assume that the material leaves *2 isotropically, or at any rate with bipolar symmetry, and with the specific angular momentum of the orbit of *2. Rows 9 and 11 of Table 4.2 give such models, with row 9 also having some magnetic braking from its wind. Possibly the bipolar flow from *2 might be linked magnetically to *2, or its accretion disc, and this might increase K_2 above unity.

Equation (4.39) can be read in an alternative way, as giving the rate of mass transfer ξ when the wind parameters ζ_1, K_1 are known. Suppose that

(a) *1 fills its Roche lobe, transferring mass, while also losing mass to infinity by stellar wind
(b) its radius R relates to M in some definite way, as on the ZAMS – Eq. (2.2) – so that R' is known
(c) $\zeta_2 = 0$, and so K_2 is irrelevant.
     Then by equating R_L' to the known R', Eq. (4.39) at a given q determines the ratio ξ/ζ_1, i.e. the ratio of mass transfer to mass loss by wind. In row 8 of Table 4.2, for example, we see that $R_L' \sim 1$ for $q \sim 0.5$–0.125. On the lower main sequence, $R \propto M_1$ is a reasonable approximation to the ZAMS relation, and so for the assumed $K_1 = 15$, we see that $\xi \sim 10\zeta_1$. This particular case has a small non-zero ζ_2, allowing a moderate amount of bipolar reemission.

Clearly for serious study of long-term evolution of systems subject to a combination of winds, RLOF, MB, etc., it would be necessary to formulate credible expressions for $\zeta_1, \zeta_2, \xi, K_1$ and K_2, rather than treat them (or their ratios) as constants. The equations for H_0, M_1 and M_2 can then be integrated specifically. In such a calculation we could also include GR. With a little further elaboration we could also include the synchronisation and circularisation of Section 4.2. But the values in Table 4.2 give an impression of how the evolution might go in a reasonably representative variety of cases, and are easily calculated.

Massive binaries seem particularly prone to stellar wind; not surprisingly, since massive single stars are (Section 2.4). Table 4.3 gives parameters for a few massive binaries. I have as usual attempted to nominate as *1 the component which I believe was *initially* the more massive. For V640 Mon this is not the *currently* more massive; it seems more likely to be the *larger* component. Although the components are large, neither is close to its Roche lobe. V729 Cyg is much more extreme, but there it seems reasonable to suppose that RLOF is

Table 4.3. *Some massive binaries*

Name	Spectra	State	P	M_1	M_2	R_1	R_2	X^a	Y	Reference
V640 Mon	O7.5If + O6If	M M D	14.4	43	51	22	18:	4.7	1.65	Bagnuolo et al. 1992
V429 Car	WN7 + O6.5–8.5	M M E	80.3	55	21			27		Schweikhardt et al. 1999
0534-69	O3If + O6:V	M M C	1.40	41	27	10	8	1.03	1.1	Ostrov 2001
V729 Cyg	O6f + O7f	m M s	6.60	11:	39:	17:	33:	1.2	3.6	Leung and Schneider 1978
CQ Cep	WN7 + O6	R M d	1.64	30	24	8.8	7.9	1.4	1.18	Harries and Hilditch 1997
CV Ser	WC8 + O8–9III–IV	R M D	29.7	11	22			10		Massey and Niemelä 1981
V398 Car	WN4 + O4-6	R M D	8.26	13	25			4		Niemelä and Moffat 1982
V444 Cyg	WN5 + O6	R M d	4.21	9.3	28	2.9, 15:[b]	8.5	1.5	1.15	St-Louis et al. 1993
V348 Car	B0 + B1	M M d	5.56	35	35	20.5	20.5	4.3	2.4	Hilditch and Bell 1987
V448 Cyg	B1III-Ib + O9.5V	M M S	6.52	14	25	16.3	6.7	5.0	1.0	Harries et al. 1998
V382 Cyg	O7.7 + O7.3	M M C	1.89	19.3	26	8.4	9.6	1.6	1.4	Harries et al. 1998

[a] Includes some assumptions about winds: see Table 4.4.
[b] Radii of star, envelope.

All eccentricities are small or zero, except V429 Car ($e = 0.6$).

Under 'state', a letter in lower case indicates a more-than-usual degree of uncertainty. For many of these systems, it is uncertain which component is actually *1, the *originally* more massive component; see text.

Table 4.4. *Possible starting conditions for 11 massive binaries*

No.	Name	Case	Initial parameters masses; period		Current parameters masses; period	ξ	ζ_1	ζ_2
1	V640 Mon	AU	$72 + 66$; 6.7 days	to	$43 + 51$; 14 days	0	2	1
2	V429 Car	$\mathcal{B}$W	$84 + 21$; 42 days	to	$55 + 21$; 80 days	0	1	0
3	0534-69	AUR	$41 + 27$; 1.4 days					
4	V729 Cyg	AUS	$43 + 39$; 1.4 days	to	$11 + 39$; 6.6 days	1	2	1
5	CQ Cep	AUN	$58 + 15$; 2.1 days	to	$30 + 24$; 1.6 days	1	2	0
6	CV Ser	$\mathcal{B}$U	$33 + 22$; 11 days	to	$11 + 22$; 30 days	0	1	0
7	V398 Car	AUN	$29 + 17$; 5.4 days	to	$13 + 25$; 8.3 days	1	1	0
8	V444 Cyg	AUN	$33 + 16$; 1.9 days	to	$9 + 28$; 4.2 days	1	1	0
9	V348 Car	AUN	$35 + 35$; 5.6 days					
10	V448 Cyg	AN	$24 + 16$; 5.7 days	to	$14 + 25$; 6.5 days	1	0	0
11	V382 Cyg	AS	$25 + 21$; 1.8 days	to	$19 + 26$; 1.9 days	1	0	0

going on, in addition to wind from *both* components. Several of these systems seem to be in the awkward situation that both winds and RLOF have shaped their history.

If components lose enough mass by stellar wind they may avoid RLOF altogether, partly because the orbit gets wider as the total mass drops, and partly because one or more components may be stripped down to their helium cores, which are normally quite small. Alternatively, if they do not avoid RLOF altogether, the effect of RLOF may, nevertheless, be substantially modified from the conservative picture described in Section 3.5. To supplement our evolutionary notation (cases AR, ..., AN) there, we add two more cases, AW and AU. In case AW we suppose that the *normal* single-star stellar wind is enough to prevent RLOF. It is not clear that there is any system in the Galaxy that qualifies, but two components of $\sim 100 \, M_\odot$, in a 20-day orbit, might do. In case AU, we suppose that the normal wind is not enough, but the wind is enhanced by binarity so that there is enough wind from one or both components to modify the outcome severely. We can expect analogues of cases AR, AS, AN, but with a somewhat different outcome. We refer to these analogues as cases AUR, AUS and AUN, and attempt to describe their expected evolutionary progress in an extension of the notation of Section 3.5:

$$\text{AW:} \quad MME; TF1, NW \to MMD; NW1 \to MMD; PC2 \to RMD; NW1 \to RMD; SN \to \dots \quad (4.43)$$

$$\text{AU:} \quad MME; TF1 \to MMD \to MMD; EW1 \to MMD; EJ2 \to RMD; NW1 \to \dots \quad (4.44)$$

$$\text{AUN:} \quad " \qquad " \quad " \to MMS; F1, EW1 \to MMS; EJ2 \to RMD; NW1 \to \dots \quad (4.45)$$

$$\text{AUS:} \quad " \qquad " \quad " \to MMS; F1, EW1 \to MMC; EJ2 \to RMD; NW1 \to \dots \quad (4.46)$$

$$\text{AUR:} \quad " \qquad " \quad " \to MMS; F2, EW \leftrightarrow MMC; R2, EW \to M \to \dots \quad (4.47)$$

We hypothesise that there are case $\mathcal{B}$ alternatives at high mass similar to the above:

$$\mathcal{B}\text{W:} \quad MME \to HME \to HME; PC2 \to RME; NW1 \to RME; SN \to \dots \quad (4.48)$$

$$\mathcal{B}\text{U:} \quad MME \to HME; TF2 \to HMD; EW2 \to HMD; EJ2 \to RMD; NW1 \to \dots \quad (4.49)$$

$$\mathcal{B}\text{UN:} \quad " \qquad " \quad " \to HMS; F1, EW2 \to HMS; EJ2 \to \dots \quad (4.50)$$

$$\mathcal{B}\text{UR:} \quad " \qquad " \quad " \to HMS; F2 \to HMC; EJ2 \to \dots \quad (4.51)$$

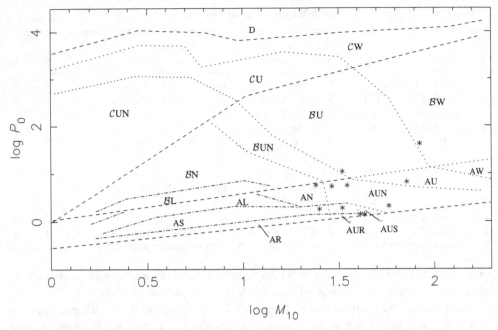

Figure 4.4 The P_0/M_{10} plane, with schematic boundaries between different routes; q_0 is assumed to be in the range $\sim$1–1.5. Dashed lines indicate, approximately, (a) the ZAMS lower boundary, (b) the TMS boundary between case A and case $\mathcal{B}$, (c) the beginning of the giant branch, i.e. the boundary between cases $\mathcal{B}$ and $\mathcal{C}$, and (d) the lower boundary of case D (no RLOF, and no mass loss). Dash-dotted lines are some of the boundaries within case A (Fig. 3.8); a number of cases (AD, AG, AE, AB, $\mathcal{B}$B) have been ignored. Estimated initial models (asterisks) for 11 massive systems are taken from Table 4.4: some have q_0 outside the range hypothesised for this diagram. Dotted boundaries are particularly uncertain. Several features of this diagram are not discussed till later.

We might note that if mass loss does prevent a binary from reaching RLOF, then the distinction between case A and case $\mathcal{B}$ becomes meaningless; but rather than labour this point we will continue to think of case A as fairly short initial periods and case $\mathcal{B}$ as longer initial periods.

In Table 4.4, we make very tentative estimates of starting parameters, and of the non-conservative coefficients ζ_1 and ζ_2 (noting that only their *ratios*, to each other or to the conservative coefficient ξ, matter). Two systems have probably not evolved enough to have lost or transferred much mass. One is a very massive contact binary in the LMC. The other, V348 Car, is a system of surprisingly high total mass yet with rather little evidence of either current or former mass loss. Presumably this is because (a) the mass is very evenly split between two almost equal components and (b) 35 $M_\odot$ may be more-or-less the upper limit below which winds are unimportant, at least within the main-sequence band, and at least until some outer layers have already been removed by RLOF. Guided partly by this very tentative insight, partly also by the probability that the initial period should not have been uncomfortably small, and further that a Wolf–Rayet remnant of known mass implies a precursor massive enough to have contained it within its He rich core, we will attempt to determine plausible initial parameters.

In Fig. 4.4 I attempt to locate some provisional boundaries in the plane of *initial* period and *initial* mass of *1. Of course, initial mass ratio must also play a part, but I have assumed here that initial mass ratios are not large, or do not matter in the case that RLOF is avoided.

The term 'partial accretion' might be used to describe two rather different physical processes: (a) while *1 loses mass by stellar wind, *2 is able to accrete a portion of it, or (b) during RLOF only some of the matter lost by *1 ends up on *2, while the remainder is somehow driven from the system. However, here we distinguish between these, calling the second 'bipolar reemission' (mode BP) and only the first 'partial accretion' (mode PA). They have in common that both may involve non-zero ξ *and* ζ_1, but they are likely to differ mainly in the amount of angular momentum that is carried to infinity by the escaping gas. We assume as a starting point that mass lost to the system in mode BP first changes its specific angular momentum from the orbital value of *1 to the orbital value of *2, while in mode PA it does not. These can lead to very different behaviours of the orbit, since if the mass ratio is well away from unity the specific angular momenta of the two stars are fairly different. This difference is seen in rows 7 and 11 of Table 4.2, where in both cases half the gas lost by *1 is accreted by *2.

Although something like mode BP is seen in several mass-transferring binaries, it is hard to judge what fraction is expelled and what retained. Equally, there is no doubt observationally that something like mode PA takes place, but with great uncertainty about the fraction of wind that is accreted.

Suppose that *1 is subject to a wind of strength ζ_1', and at the same time to RLOF of strength ξ'. Say that *2 accretes (temporarily) the stream ξ' as well as a fraction α_2 of ζ_1'; but then reemits, from the near neighbourhood of *2, a fraction β_2 of all the matter temporarily accreted. We can continue to model the effect on the masses, and on the orbit, by Eqs (4.36) and (4.37), provided that we write

$$\zeta_1 = (1 - \alpha_2)\,\zeta_1', \quad \zeta_2 = \beta_2\,(\xi' + \alpha_2\,\zeta_1'), \quad \xi = \xi' + \alpha_2\,\zeta_1'. \tag{4.52}$$

We will return to these processes in Sections 6.3 and 6.4. Note that in Eq. (4.52) we have ignored any intrinsic wind ζ_2' from *2, but this can in fact easily be added into ζ_2.

If we attempt to follow the evolution of a binary in some detail, with a non-conservative model for orbital evolution, and using a stellar evolution code for the interiors, it is difficult to avoid the necessity for solving for both components *simultaneously*. This is because the behaviour of *1 may be influenced not just by M_2, as in the conservative case, but also by such parameters as α_2 and β_2, which themselves will at the least depend on R_2, L_2 as well as M_2. Of course, at a very crude level of approximation, we might start by assuming α_2, β_2 to be given constants, in which case we can still evolve *1 without direct knowledge of the structure of *2.

A process of envelope loss that we do not consider here in detail is seen in some massive binaries, say containing a Wolf–Rayet component and an O star. Both stars have winds, and a region where they collide is sometimes observed, particularly in X-rays. Even if the region is not directly evident, it is obvious that winds from both components must have some collision front. However, in default of a detailed model, we assume here that what goes on in the collision region does not react back on the orbit, and that the effect on the orbit of two independent winds is given by the same simple mathematical model, Eq. (4.38).

4.4   Magnetic braking and tidal friction: mode MB

Rotating single stars that lose mass by winds may be subject to magnetic braking, with magnetic fields possibly linking the star to the outflowing wind (Schatzman 1962) and forcing the wind to corotate out to an Alfvén radius of several stellar radii. This applies particularly to relatively late stars, ~F2 or later. The fact that such stars are generally slow rotators, whereas earlier stars are rapid rotators, suggests that magnetic braking can operate effectively only in stars with convective envelopes. This is not conclusive evidence that the radiative/convective transition in the envelope is the major cause, since if *all* stars were subject to magnetic braking on a timescale of say 3 gigayears, stars earlier than ~F2 would be little affected and most of those later would be strongly affected. However, it seems a plausible starting point.

Even without magnetic linkage, we expect some spin-down as a consequence of stellar wind. If the star rotates roughly uniformly, the mass leaving the surface has more specific angular momentum than the average in the star, by a factor of about $2R^2/3k^2$. R, k are the radius and radius of gyration of the star. The factor $2/3$ assumes that the mass-loss is uniform over the surface. We will however continue to use the term 'magnetic braking' (MB) to describe the combined effect. For the Sun the Alfvén radius is $\sim 12\,R_\odot$, and the gyration radius $\sim 0.26\,R_\odot$.

If a star is in a close binary, close enough that tidal friction keeps it locked in corotation with the system, then MB drains angular momentum not just from the stellar spin, but from the orbit. Hence this mechanism can alter the fundamental orbital separation, and for example make RLOF occur earlier than might otherwise be the case. Although the process is normally slow, it can in some cases be more rapid than nuclear evolution or gravitational radiation. We refer to the combined effect on a binary as MB; we rely on the context to determine whether we are talking about magnetic braking of single stars, of stars in widish binaries which might spin down without exchanging angular momentum with the orbit, or of closish binaries where tidal friction leads to exchange of spin and orbital angular momentum.

A rather detailed theoretical model of axisymmetric, stationary winds with 'frozen-in' fields (i.e. the limit of infinite conductivity) can be developed (Mestel 1968, Mestel and Spruit 1987). The mathematics of this model is outlined in Appendix D. The stationary, axisymmetric assumptions mean that there are five functions to be solved for: the density, the toroidal components of both the velocity and the magnetic fields, and the stream functions of the poloidal components of both fields. The five equations determining them are the two independent components of the steady dynamo equation

$$\nabla \times (\mathbf{v} \times \mathbf{B}) = 0, \tag{4.53}$$

and the three of the steady momentum equation

$$\rho \mathbf{v} \cdot \nabla \mathbf{v} = -\nabla p - \rho \nabla \Phi + \frac{1}{\mu_0}(\nabla \times \mathbf{B}) \times \mathbf{B}. \tag{4.54}$$

Three first integrals can be extracted rather generally, and also a fourth if the wind is assumed to be adiabatic or isothermal; one of these integrals tells us that field lines and stream lines in a plane through the rotation axis coincide. The remaining equation, which can be viewed as determining the poloidal part of the magnetic field, is unfortunately rather complicated. The model is sketched in Fig. 4.5. If it is assumed that the magnetic field of the star, in the absence of wind, is roughly dipolar, then the model shows that field lines originating on polar

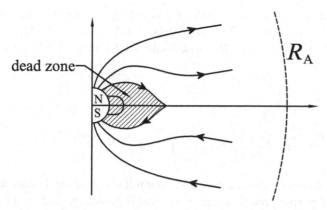

dead zone

R_A

Figure 4.5 An artist's impression of the field lines and stream lines (which coincide) in an expanding magnetic stellar wind, assuming axial symmetry. The magnetic field (arrows) is outwards on the northern hemisphere and inwards on the southern hemisphere. The gas flow is outwards on both hemispheres, but is zero in the 'dead zone', a toroidal belt separating the flows. There is a cusp in the critical field line that separates the wind zone from the dead zone. In the simplest model, a current sheet is required in the equatorial plane, to support discontinuities in the tangential magnetic field.

caps will be stretched by escaping wind to reach infinity roughly radially (the 'wind zone'), while field lines originating in an equatorial belt will cross the equatorial plane normally, and trap a region of hot gas (the 'dead zone') in which the gas flow is purely toroidal. On field lines within the wind zone there will be an 'Alfvénic point' at a distance R_A, say, such that at smaller distances the wind is obliged by magnetic stress to corotate with the star, while at larger distances the wind expands freely conserving its specific angular momentum. If the Alfvénic point is at several times the radius of the star, the escaping gas will remove a much larger amount of angular momentum per unit mass than is contained in the body of the star, and thus the rotation will be braked.

In general, we expect the torque to depend on both the mass-loss rate $|\dot{M}_1|$ and the surface dipolar magnetic field B_P. Using the analysis of Mestel and Spruit (1987), outlined in Appendix D, but simplifying to a very considerable extent as summarised in Eqs (D27)–(D34), we can estimate the dependence as follows. The wind is presumed to be corotating with the star from the stellar surface to an Alfvén radius R_A at which it attains escape velocity. We can then write

$$|\dot{M}_1| \sim 4\pi R_A^2 \rho_A v_A, \quad \rho_A v_A^2 \sim \frac{B_A^2}{\mu_0}, \quad v_A^2 \sim \frac{2GM_1}{R_A}, \quad \frac{B_A}{B_P} \sim \left(\frac{R}{R_A}\right)^2, \quad (4.55)$$

where suffix A refers to the Alfvén radius. The magnetic field is to be identified very loosely with the poloidal field B_P of Appendix D. Its assumed decrease as $B \propto r^{-2}$ is based on a 'split monopole' model of the field. The field lines are assumed to be dragged out almost radially by wind, with a northern monopole in one hemisphere, a southern monopole in the other, and a toroidal current sheet in the equatorial plane separating them (Fig. 4.5).

From estimates (4.55), we easily obtain

$$\left(\frac{R_A}{R}\right)^{3/2} = C_1 \left(\frac{R^5}{2GM_1}\right)^{1/2} |\dot{M}_1|^{-1} \frac{4\pi B_P^2}{\mu_0}, \quad (4.56)$$

and corresponding dependences for v_A, ρ_A, B_A. C_1 is a fudge factor that we calibrate from Solar data. For the Sun, $B_P \sim 1.25 \times 10^{-4}$ tesla (~ 1.25 gauss) and $|\dot{M}_1| \sim 10^{-7.6}$ $M_\odot$/ megayears. Then the observed value $R_A \sim 12\,R_\odot$ means that we should take $C_1 \sim 0.6$. The braking rate should be

$$\frac{\dot{P}_1}{P_1} = -\frac{1}{I\Omega}\left(\frac{dI\Omega}{dt}\right)_{MB}$$

$$\sim \frac{|\dot{M}_1|R_A^2}{I} \sim C_1^{4/3}\frac{R^2}{M_1 k^2}\left(\frac{R^5}{2GM_1}\right)^{2/3}|\dot{M}_1|^{-1/3}\left(\frac{4\pi B_P^2}{\mu_0}\right)^{4/3}. \qquad (4.57)$$

It can be seen that the spin-down timescale depends rather weakly on $\dot{M}_1$ and rather strongly on B_P. We can imagine that spin-down becomes very rapid if B_P is large *and* $|\dot{M}_1|$ is small, but at least in stars with active surface dynamos it is likely that they correlate positively (see Section 4.5).

For notational purposes, we call the *rotational* period P_1, to distinguish it from the *orbital* period P – although for the moment we are only talking about single stars. But we use Ω rather than Ω_1 because (a) we use ω rather than Ω for the mean orbital angular velocity, and (b) we assume throughout most of this chapter, just for clarity, that only *1 is active. This latter assumption also allows us to drop the suffix unity on R and L; yet we keep a suffix on M_1, because in a binary M_2 and the *combined* mass $M \equiv M_1 + M_2$ are still relevant, even if *2 is inert.

Unfortunately, the above model does not by itself predict either the magnitude of the wind or the strength at the stellar surface of the magnetic field; these have to be fed in as boundary conditions. Presumably they are determined by processes inside and at the surface of the star. I will attempt to model these in the next section. Nor can the magnetic-braking model readily incorporate the fact that most magnetically active stars show non-axisymmetric and non-stationary behaviour. So, although we rely on the concept of corotation out to an Alfvénic radius, and free expansion beyond, we discuss here first some more empirical determinations of the braking rate.

Observations of the rotation rates of single, roughly Solar, stars (Kraft 1967), in clusters of different ages, suggest that these stars slow down on a timescale of about 10^3–10^4 megayears, which is much less than their mass-loss timescales of 10^7 megayears – Eq. (2.73). This is roughly consistent with the fact that the Solar wind is observed to corotate out to $\sim 12\,R_\odot$ (Pizzo *et al.* 1983). Skumanich (1972) estimated, from Kraft's data, a formula for rotation period as a function of age: his result can be written

$$P_1 \sim 0.4t^{\frac{1}{2}}, \qquad (4.58)$$

with P_1 in days, t in megayears. This was based on rotation velocities of stars in three clusters of known ages. Equation (4.58) is consistent with a magnetic braking rate for Solar-type stars given by

$$\frac{\dot{P}_1}{P_1} \sim \frac{1}{t_0}\left(\frac{4}{P_1}\right)^2, \qquad t_0 \sim 200\,\text{megayears}, \qquad (4.59)$$

in the same units. However, for the Sun ($P_1 = 26$ days) this gives a braking timescale too short by a factor of ~ 2.5: it should be ~ 20 gigayears (Pizzo *et al.* 1983).

Stępień (1995) proposed a braking law for Solar-type stars, based partly on observation and partly on theory, which roughly agrees with Eq. (4.59) at $P_1 \sim 4\text{--}8$ days, but gives a smaller torque at both larger and smaller period. We adopt here the essence of Stępień's result, but use a different mathematical form for his expression, a form that integrates analytically but that agrees to within about 50% for the period range $0 \lesssim P_1 \lesssim 27$ days:

$$\frac{\dot{P}_1}{P_1} \sim \frac{1}{t_0} \left[1 + \left(\frac{P_1}{9} \right)^2 \right]^{-2}, \qquad t_0 \sim 200 \, \text{megayears}. \qquad (4.60)$$

Stępień's formula has $e^{-0.2P_1}$ instead of the expression to the right of t_0^{-1}. Equation (4.60) gives a value of spin-down timescale for the Sun of ~ 18 gigayears, in better agreement than Eq. (4.59) with the fairly direct observational determination of Pizzo *et al.* (1983). We can integrate the resulting formula to give an estimate of the time taken for a single star to spin down from P_a to P_b:

$$\frac{t}{t_0} \sim \left[\ln \left(\frac{P_1}{9} \right) + \left(\frac{P_1}{9} \right)^2 + \frac{1}{4} \left(\frac{P_1}{9} \right)^4 \right]_{P_a}^{P_b}. \qquad (4.61)$$

For the Sun, this means that ~ 5.4 gigayears would be required for spin-down from an initial value of a few days to the present, only $\sim 20\%$ longer than it should be.

Stępień (1995) suggested that Eq. (4.60) can perhaps be generalised to stars other than Solar-like if

(a) The ratio $P_1/9$ is replaced by a multiple of the Rossby number $\sigma \equiv P_1/t_{\text{ET}}$. This is the ratio of rotation period to the convective envelope turnover time t_{ET}, which is ~ 15 days for the Sun, Eq. (2.33).

(b) A multiplicative factor, $f(M_1, R, L)$, say, is introduced, which is unity for $M_1 \sim M_\odot, R \sim R_\odot, L \sim L_\odot$. Then we might hope that a more general expression could be something like

$$\frac{\dot{P}_1}{P_1} \sim \frac{1}{t_0} \frac{f(M_1, R, L)}{(1 + 2.8\sigma^2)^2}, \qquad \sigma \equiv \frac{P_1}{t_{\text{ET}}}. \qquad (4.62)$$

In the next section, I will attempt a simplistic model that gives a functional form for f.

In a wide binary, angular momentum loss by winds would simply slow down one or other star (or both) independently of the orbital motion, but in a *close* binary tidal friction may keep the stars corotating so that the magnetically coupled winds may drain the orbital angular momentum (Huang 1966, Mestel 1968, Eggleton 1976, Verbunt and Zwaan 1981, Mestel and Spruit 1987). Because the orbital moment of inertia is much greater than a single star's, this might seem like a small effect, but it is balanced by the fact that the star could be rotating much more rapidly than if it were single. Indeed, as the orbit loses angular momentum the star spins up, not down, because of tidal coupling. As a consequence binaries with $P \lesssim 1.5$ days, or thereabouts, may (if they contain a Solar-type star) be forced appreciably closer, even to RLOF, in the course of the Hubble time (van't Veer 1976, Vilhu 1981, Rucinski 1983). If tidal friction is strong enough to ensure corotation, so that the angular momentum lost in the wind is drained ultimately from the orbit, Eq. (4.62) along with Eq. (3.14) tells us that the rate of orbital period change, in the absence of significant *mass* loss, is

$$-\frac{1}{t_{\text{MBb}}} \equiv \frac{1}{H_o} \left(\frac{dH_o}{dt} \right) = \frac{\dot{P}}{3P} \approx \frac{1}{H_o} \left(\frac{dI\Omega}{dt} \right)_{\text{MB}}$$

$$= -\frac{\lambda}{t_{\text{MBs}}} \sim -\frac{\lambda}{t_0} \frac{f(M_1, R, L)}{(1 + 2.8\sigma^2)^2}, \tag{4.63}$$

where λ, as in Eq. (4.15), is the ratio of spin to orbital angular momentum. The letters b and s in the subscripts refer to binary and single-star timescales. The factor λ as well as σ contains a P-dependence. From Eqs (4.15), (3.5), (3.11) and (3.18),

$$\lambda = \lambda_0 \left(\frac{P_{\text{cr}}}{P} \right)^{4/3}, \quad \lambda_0 = \frac{M}{M_2} \frac{R_{\text{L}}^2}{a^2} \frac{k^2}{R^2}. \tag{4.64}$$

If we take $q \sim 1$, and polytropic index ~ 3, we get a representative value of $\lambda_0 \sim 0.023$. P_{cr} is the period at which the star fills its lobe (Table 3.2). Hence

$$\frac{\dot{P}}{P} = -\frac{3\lambda_0}{t_0} \left(\frac{P_{\text{cr}}}{P} \right)^{4/3} \frac{f(M_1, R, L)}{(1 + 2.8\sigma^2)^2}. \tag{4.65}$$

If initially $P \ll t_{\text{ET}}$ (e.g. $P \lesssim 2$ days), then $\sigma \sim 0$, although the P-dependence implicit in σ can be integrated easily enough. Then a binary containing a dynamo-active star spins up from period P_0 to P_{cr} in a time

$$\frac{t}{t_0} \sim \frac{1}{4\lambda_0 f(M_1, R, L)} \left[\left(\frac{P_0}{P_{\text{cr}}} \right)^{4/3} - 1 \right]. \tag{4.66}$$

With $t_0 \sim 200$ megayears – Eq. (4.60) – and $\lambda_0 \sim 0.023$ – Eq. (4.64) – and assuming Solar parameters, the period could decrease from a few days to the point where contact is reached ($P \sim P_{\text{cr}} \sim 0.3$ d) in something like the Hubble time.

It is regrettable that in using either Eq. (4.60), as here, or Eq. (4.59) to estimate magnetic braking in short-period binaries we are extrapolating well outside the range of validity of either. The two estimates differ by a very large factor at, say, 0.1–0.3 days, which is the kind of period relevant to contact binaries and cataclysmic binaries.

Mass loss will cause a single star to spin down, even in the absence of magnetic field, because the specific angular momentum at the surface is greater than the mean. We replace Eq. (4.57) by

$$-\frac{1}{I\Omega} \frac{dI\Omega}{dt} \sim |\dot{M}_1| \frac{\frac{2}{3}R^2 + R_{\text{A}}^2}{I} \equiv \frac{1}{t_{\text{MBs}}}. \tag{4.67}$$

Although t_{MBs} defined in this equation is not wholly magnetic in origin, we refer here to the *combination* as mode MB.

In a binary that is sufficiently close for tidal friction to enforce corotation, we have yet another term for angular momentum loss, because even if $R = R_{\text{A}} = 0$, mass leaving $*1$ will carry off *orbital* angular momentum. Following Eq. (4.34), we write

$$-\frac{1}{H_o}\left(\frac{dH_o}{dt}\right) = \frac{|\dot{M}_1|M_2}{MM_1}\left[1 + \frac{\frac{2}{3}R^2 + R_A^2}{a^2}\frac{M^2}{M_2^2}\right]$$

$$\equiv K\frac{|\dot{M}_1|M_2}{MM_1} \equiv \frac{1}{t_{MBb}},$$ (4.68)

where H_o is the orbital angular momentum, a is the orbital radius, and $M \equiv M_1 + M_2$. K is a factor giving the excess of actual angular momentum loss to its minimal value. For binaries, we use the term MB to cover all *three* terms, for brevity: usually it is obvious by context whether we are referring to single stars or binaries, and where it is that we use MBs and MBb. We define a third timescale, for binary *mass* loss, by

$$\frac{|\dot{M}_1|}{M_1 + M_2} \equiv \frac{1}{t_{ML}}.$$ (4.69)

Presumably a complete theory of winds, determining $\dot{M}_1$, and of dynamo activity, determining B_P, would provide values for these quantities which when substituted into Eq. (4.57) give something close to the semi-empirical Eq. (4.62), for Solar-type stars. These two theories are probably not independent, since we can see in the Sun that it is largely or wholly magnetic energy dissipation that drives the wind. In the next section we shall attempt to quantify this in a very crude way.

4.5 Stellar dynamos

Much excellent data on stellar rotation and stellar activity exists, but it is not easy to translate this into a usable mathematical formulation of magnetic braking. Mainly, this is because what are usually measured are parameters such as the strength of the emission cores in the HK lines of calcium, or the X-ray or radio flux. These themselves need a comprehensive theoretical model to be translated into such quantities as magnetic field and mass-loss rate. Equivalently, although much excellent theoretical work has been done on stellar dynamos, this work usually shows chaotic behaviour; and it normally starts by *assuming* a given law of differential rotation. Massive three-dimensional computational effort will be required to build a self-consistent MHD model of a rotating convection zone, that one hopes would generate for itself the necessary differential rotation as well as the magnetic field and the mass loss.

I present here an elemetary recipe for dynamo activity, which is necessarily ad hoc in the absence of a detailed theory. Such a theory would no doubt be very complicated. The recipe's purpose is to fill in the two missing links of the previous analysis, i.e. to determine as far as possible from first principles the two parameters $|\dot{M}_1|$ and B_P that themselves determine the braking rate. The present analysis is largely dimensional, but also makes use of some observational relationships such as Stępień's (1995) correlation as approximated in Eq. (4.60). It is based on the following assumptions:

(a) The velocity field in the star is determined by its overall rotation, by the extent and strength of its turbulent convection, and by their interaction. In particular, we suppose that there is *differential* rotation, which is driven solely by the combination of these two influences. Then if $\Delta\Omega$ is some measure of the differential rotation (mainly concentrated at the base of the convection zone, in the Sun), we expect that

$$\Delta\Omega = \frac{1}{t_{\text{ET}}} E(\sigma), \quad \sigma \equiv \frac{P_1}{t_{\text{ET}}}, \quad t_{\text{ET}} \equiv \frac{l}{w}, \quad (4.70)$$

where w and l are the velocity and mean free path of turbulent convective eddies at some reference point, say half way in radius between the top and the bottom of the zone – this is one way that t_{ET} was estimated in Section 2.2.3. E is some dimensionless function to be determined or guessed at, and σ, as in Eq. (4.62), is the Rossby number, also dimensionless. On the Solar surface, rotation is seen to vary with latitude as $\Omega \propto 1 - 0.08\, P_2(\cos\theta)$. Consequently we take $\Delta\Omega = 0.08\Omega$, which gives $E \sim 0.3$ for the Sun. We also have $\sigma \sim 1.7$ for the Sun.

(b)   The magnetic field, driven by an $\alpha\Omega$ dynamo, can be represented by two values: B_{P}, an overall poloidal field, and B_ϕ, a toroidal field. This type of dynamo, which might better be called the α, Ω, $\Delta\Omega$ dynamo, relies on differential rotation $\Delta\Omega$ to wind up the internal poloidal field and produce a toroidal field. In a purely axisymmetric situation it would not be possible for this toroidal field to be converted back into a poloidal field, thus completing a feedback loop, and so the poloidal field would ultimately decay by ohmic dissipation. But small-scale non-axisymmetric perturbations due to turbulent convection can be allowed for at least crudely. Using a Fourier transform analysis (Appendix E), they lead to a small complex coefficient α which allows a poloidal field to be regenerated. When the equations of the $\alpha\Omega$ dynamo are reduced to their barest essentials – Eqs (E19) – they emerge as

$$\dot{B}_\phi \sim \frac{\Delta\Omega}{l} R B_{\text{P}}, \quad \dot{B}_{\text{P}} \sim \frac{\alpha B_\phi}{R}. \quad (4.71)$$

The (complex) coefficient α relates to the 'helicity' of the turbulent motion, i.e. the mean value of $\mathbf{v} \cdot \nabla \times \mathbf{v}$. In a non-rotating situation this can be expected to average to zero, but Coriolis force leads to cyclonic turbulence with a non-zero mean helicity. The process can be observed fairly directly on the face of the Sun. The toroidal flux loops deep in the convection zone rise to the surface because of magnetic buoyancy. They become 'kinked' by the turbulent convection, so that they emerge at the surface as pairs, rather than all at once. With no helicity, these pairs would, on average, be aligned east–west, but cyclonic turbulence gives them on average a slight north–south tilt. The pairs drift polewards in a large-scale meridional circulation current, which presumably, like the differential rotation, is a consequence of the interaction of convection with rotation. As they drift and decay (not so much by ohmic diffusion, but by the highly non-linear process of field-line reconnection above the photosphere), the tilt increases, giving a small contribution to the large-scale poloidal field.

Equations (4.71), with complex α, give exponential growth as well as cyclic behaviour, both on a timescale of $\sqrt{l/|\alpha\Delta\Omega|}$. However, we can expect that the neglected non-linear dissipation terms will prevent growth beyond some amplitude, and so lead to a limit-cycle, whose frequency Ω_{c} is likely to be comparable to the growth rate and cycle frequency of the linear regime. We identify this with the Solar cycle frequency, which allows us to estimate α, at least for the Sun and a number of Solar-type stars for which cyclic activity is observed. Although Ω_{c} might depend on field-strength, for example, we assume for simplicity that, like $\Delta\Omega$, Ω_{c} is some Rossby-number-dependent function

Table 4.5. *Solar parameters connected with dynamo activity*

Observed		Stellar model		Dynamo model	
M_1	2.0×10^{30}	R_{HT}	1.27×10^9	σ	1.7
R	7.0×10^8	D	2.0×10^8	$|\alpha|$	0.029
L	3.8×10^{26}	l	6.6×10^7	B_ϕ	0.030
Ω	2.8×10^{-6}	w	51	E	0.29
	2.2×10^{-7}	t_{ET}	1.30×10^6	F	0.0129
Ω_c	10^{-8}	ρ	11.7	H	6.4×10^{-4}
B_P	1.25×10^{-4}	$w\sqrt{\mu_0\rho}$	0.58	C_1	0.58
$|\dot{M}_1|$	1.7×10^9	k^2	3.4×10^{16}	C_2	0.155
$\bar{\rho}$	1.4×10^3				
w_G	36				

All quantities are in SI units.

(with dimensions of time^{-1}) of the turbulent velocity field:

$$\Omega_c \sim \frac{1}{t_{ET}} F(\sigma). \tag{4.72}$$

Then, from Eqs (4.71), α and the ratio B_P/B_ϕ are clearly given by

$$\Omega_c^2 \sim \frac{|\alpha|\Delta\Omega}{l}, \quad |\alpha| = \frac{l}{t_{ET}} \frac{F^2(\sigma)}{E(\sigma)}, \quad B_\phi \sim \frac{R}{l\Omega_c} B_P \sim \frac{R}{l} \frac{E}{F} B_P. \tag{4.73}$$

We assume that both B_P and B_ϕ scale like $w\sqrt{\mu_0\rho}$, where w is the mean velocity of convection, since this has the appropriate dimensions. Consequently we write

$$B_P = w\sqrt{\mu_0\rho}\, II(\sigma), \quad B_\phi = w\sqrt{\mu_0\rho}\, \frac{R}{l} \frac{EH}{F}, \tag{4.74a,b}$$

taking the density ρ, like w, l, to be to be a mean value somewhere in the convection zone.

(c) The toroidal field B_ϕ is produced near the base of the convection zone, where $\Delta\Omega$ is concentrated, but levitates to the surface, being shredded in the process by the turbulent convection. It emerges chaotically at the surface, and largely dissipates above the photosphere, driving a wind which carries away part of the field not dissipated. Arguably the wind is driven by the rate of dissipation of magnetic energy in the course of a magnetic cycle:

$$\frac{2GM_1}{R} |\dot{M}_1| \sim C_2\, 4\pi R^2 D\, \Omega_c\, \frac{B_\phi^2}{2\mu_0}, \quad \text{taking} \quad B_\phi \gg B_P, \tag{4.75}$$

where D is the depth of the convection zone (so that $4\pi R^2 D$ is approximately its volume). C_2 is another fudge factor, which represents the fraction of magnetic energy going into escape. Other fractions might be radiated away, or used to drive the wind to higher than escape velocity. Numerical estimates suggest that $C_2 \sim 0.16$ for the Sun, and since there seems no obvious reason why this should depend on σ we take it to be constant. Table 4.5 gives Solar quantities estimated by a combination of direct observation, stellar-structure modelling, and the dynamo model proposed here.

For a given M_1, R, L, the extent of the convection zone can be estimated reasonably well as a function of the ratio of the stellar radius R to the radius R_{HT} that the star would have if it were on the Hayashi track. R_{HT} is a function of L, M_1, given by Eq. (2.47). In SI units

$$R_{HT} = R_\odot \left[1.65 \left(\frac{L}{L_\odot} \right)^{0.47} + 0.17 \left(\frac{L}{L_\odot} \right)^{0.8} \right] \left(\frac{M_\odot}{M_1} \right)^{0.31}. \tag{4.76}$$

For the Sun, $R/R_{HT} \sim 0.55$, and this ratio drops rapidly further up the main sequence. We reintroduce the two global quantities $w_G, \bar{\rho}$, i.e. the mean convective velocity and the mean density – Eqs (2.29) –

$$w_G = \left(\frac{LR}{3M_1} \right)^{1/3}, \quad \bar{\rho} = \frac{3M_1}{4\pi R^3}. \tag{4.77}$$

The velocity w_G is a convenient dimensional quantity even when the star, or a part of it, is not actually convecting. Then an appropriate D, l, w, ρ in the surface convection zone can be approximated by empirical power-law depences on R/R_{HT}:

$$D \sim R \left(\frac{R}{R_{HT}} \right)^{2.1} \sim 3l, \quad w \sim w_G \left(\frac{R}{R_{HT}} \right)^{-0.6}, \quad \rho \sim \bar{\rho} \left(\frac{R}{R_{HT}} \right)^{8.0}. \tag{4.78}$$

The powers were estimated from ZAMS models of 0.8 and 1.2 $M_\odot$. We see for instance that t_{ET} of Eq. (4.70) and the magnetic field B_P of Eq. (4.74a) are

$$t_{ET} = 0.33 \left(\frac{3M_1 R^2}{L} \right)^{1/3} \left(\frac{R}{R_{HT}} \right)^{2.7},$$

$$B_P = \left(\frac{3\mu_0}{4\pi} \right)^{1/2} \left(\frac{L^2 M_1}{9R^7} \right)^{1/6} \left(\frac{R}{R_{HT}} \right)^{3.4} H. \tag{4.79a,b}$$

All quantities here are in SI units, not Solar units. These relations give B_ϕ from Eq. (4.74b); substituting into Eq. (4.75) we obtain $|\dot{M}_1|$ and the binary mass-loss timescale t_{ML}:

$$\frac{M_1 + M_2}{t_{ML}} \equiv |\dot{M}_1| = C_2 \frac{RL}{4GM_1} \left(\frac{D}{l} \right)^3 \left(\frac{R}{R_{HT}} \right)^{2.0} \frac{E^2 H^2}{F}. \tag{4.80}$$

Then we can use B_P and $|\dot{M}_1|$ in Eq. (4.56) to obtain the Alfvén radius,

$$\left(\frac{R_A}{R} \right)^{3/2} = \frac{2C_1}{C_2} \left(\frac{l}{D} \right)^3 \left(\frac{2GM_1}{R} \right)^{1/2} \left(\frac{3M_1}{LR} \right)^{1/3} \left(\frac{R}{R_{HT}} \right)^{4.8} \frac{F}{E^2}, \tag{4.81}$$

and use this in Eq. (4.57) to obtain the braking rate for single and also binary stars:

$$\frac{1}{\lambda t_{MBb}} \sim \frac{1}{t_{MBs}} \sim \frac{3C_1^{4/3}}{C_2^{1/3}} \frac{R^2}{k^2} \frac{l}{D} \left(\frac{R^3}{2GM_1} \right)^{1/3} \left(\frac{L}{3M_1 R^2} \right)^{5/9} \left(\frac{R}{R_{HT}} \right)^{8.4} \frac{F^{1/3} H^2}{E^{2/3}}. \tag{4.82}$$

Comparing this result with the semi-empirical Eq. (4.62), we can identify the factor $f(M_1, R, L)$ with the first few factors of this expression, and the σ-dependent factor $(1 + 2.8\sigma^2)^{-2}$ with the last factor, $F^{1/3} H^2 / E^{2/3}$, apart from constant factors in each case, which can be put together and identified with the empirical timescale t_0. The approximation signs in these equations allow for the fact that we have not yet applied the corrections suggested in Eqs (4.67) and (4.68).

Although we have identified one combination of $E(\sigma)$, $F(\sigma)$ and $H(\sigma)$ with an empirical function of σ, we need two more. Unfortunately we have a direct measure of $\Delta\Omega$ or B_P for few stars other than the Sun. However, in a few cases a magnetic activity cycle can be seen and the frequency Ω_c estimated. Brandenburg *et al.* (1998) found $\Omega_c \propto \Omega^{1.5} t_{ET}^{0.5}$ ($F \propto \sigma^{-1.5}$), for six active stars with rotation periods of $\sim$10–20 days and $\sigma \sim 0.5$–1.4. They also found that for 15 less active stars with rotational periods of about 25–50 days and $\sigma \sim 1.2$–1.8 there was a similar slope but with Ω_c larger by a factor of $\sim$5; at $\sigma \sim 1.2$–1.4 there were one or two stars on both branches. They interpret this behaviour as suggesting a mode change in the dynamo at an intermediate σ.

If, then, we put together two approximate pieces of observational information with one theoretical postulate, we can estimate each of E, F, H as functions of σ. They are

(a) $F^{1/3}H^2/E^{2/3} \propto (1+2.8\sigma^2)^{-2}$, by comparing Eqs (4.82) and (4.59).
(b) $F \propto \sigma^{-1.5}$, as in the paragraph above; except that on the supposition that there should be saturation at small σ rather than a divergence we replace this by $F \propto (1+2.8\sigma^2)^{-0.75}$.
(c) $E/F^2 \propto \sigma$, on the basis that α in Eq. (4.72) should be proportional to $\Omega \propto 1/\sigma$ (Appendix E); but we also assume here that there is saturation at small σ (rapid rotation), so that $E/F^2 \propto (1+2.8\sigma^2)^{0.5}$.

Then our model requires

$$E \sim \frac{2.7}{1+2.8\sigma^2}; \quad H \sim \frac{0.0096}{(1+2.8\sigma^2)^{1.21}};$$

$$F \sim \frac{0.014}{(1+2.8\sigma^2)^{0.75}} \text{ if } \sigma \lesssim 1.3; \quad F \sim \frac{0.07}{(1+2.8\sigma^2)^{0.75}} \text{ if } \sigma \gtrsim 1.3. \quad (4.83)$$

The jump in F at $\sigma \sim 1.3$ takes account of the suggestion by Brandenburg *et al.* (1998) that there are two distinct modes on either side of $\sigma \sim 1.3$.

Note that Eq. (4.82) did *not* include the extra spin-down terms, shown in Eq. (4.68), which occur even when $R_A = 0$. However, it is easy to add in these extra terms, since we know both $|\dot{M}_1|$ and R_A separately.

Numerically, our final results for the mass-loss rate and the Alfvén radius as functions of M_1, L, R and P_1 are, in SI units,

$$|\dot{M}_1| \sim 0.050 \frac{RL}{GM_1} \left(\frac{R}{R_{HT}}\right)^{2.0} \frac{1}{(1+2.8\sigma^2)^{3.67}}, \quad (4.84)$$

and

$$\left(\frac{R_A}{R}\right)^{3/2} = 0.0039 \left(\frac{2GM_1}{R}\right)^{1/2} \left(\frac{3M_1}{LR}\right)^{1/3} \left(\frac{R}{R_{HT}}\right)^{4.8} (1+2.8\sigma^2)^{1.25}, \quad (4.85)$$

with

$$\sigma = \frac{P_1}{t_{ET}}, \quad t_{ET} = 0.33 \left(\frac{3M_1 R^2}{L}\right)^{1/3} \left(\frac{R}{R_{HT}}\right)^{2.7}, \quad (4.86)$$

and with R_{HT} given in terms of M_1, L by Eq. (4.76). The above are valid for $\sigma \lesssim 1.3$; if $\sigma \gtrsim 1.3$, $|\dot{M}_1|$ should be multiplied by five, and $(R_A/R)^{3/2}$ divided by five.

Table 4.6. *Some timescales: nuclear evolution, Kelvin–Helmholtz, magnetic braking (single), gravitational radiation, circularisation by tidal friction, synchronisation, mass loss, and magnetic braking (binary)*

	Type	M_1	R	L	M_2	P, P_1	t_{NE}	t_{KH}	t_{MBs}	$\frac{1}{3}t_{GR}$	$\frac{2}{7}t_{TF}$	t_{syn}	t_{ML}	$\frac{1}{3}t_{MBb}$	K_1
1	MMD	1.00	1.00	1.00	1.00	12.0	7400	31	1110	–	8200	5.1	–	–	1.09
2	MMD	1.00	1.00	1.00	1.00	4.0	7400	31	290	–	230	.	–	11600	1.12
3a	MMD	1.00	1.00	1.00	0.75	1.5	7400	31	210	–	0.13	.	19400	5600	1.53
3b	MMD	0.75	0.70	0.20	1.0	1.5	–	126	22	–	1.2	.	–	3400	3.5
4a	MMD	0.80	0.73	0.24	0.60	3.0	–	114	27	–	44	0.14	–	5600	2.8
4b	MMD	0.60	0.57	0.09	0.80	3.0	–	220	10.8	–	145	0.21	–	5200	3.2
5	MMD	0.5	0.46	0.038	0.5	0.8	–	450	4.2	–	0.48	.	–	540	53
6a	MMC	1.35	1.2	1.5	0.7	0.33	6600	32	120	7500	.	.	13400	710	12.1
6b	MMC	0.7	0.87	0.75	1.35	0.33	–	23	340	7500	.	.	–	2900	1.60
7	GMD	1.3	3.85	8.0	1.6	4.0	1200	1.72	4.5	–	.	.	630	82	2.1
8	GMD	4.0	35.0	400	3.0	110	74	0.036	0.24	–	.	.	28	6.1	2.0
9	GMD	1.4	10.0	50.0	1.6	12.0	210	0.123	1.73	–	.	.	67	13.0	1.50
10	GMS	0.80	3.60	6.00	3.70	2.87	980	0.93	5.2	–	.	.	970	56	1.26
11	GMS	0.50	2.00	1.00	2.00	1.50	3700	3.9	2.3	–	.	.	2400	71	2.8
12	HMS	1.10	3.20	10.0	4.50	2.00	810	1.18	32	–	.	.	2200	161	1.12
13	MWD	1.00	0.89	0.70	0.60	0.56	10500	50	153	–	.	.	23000	2700	4.8
14	MWD	0.50	0.46	0.038	0.60	0.35	–	450	4.2	–	.	.	–	210	123
15	MWD	0.25	0.27	0.009	0.60	0.21	–	800	7.8	–	.	.	–	620	66
16	MWS	1.00	0.89	0.70	0.60	0.28	10500	50	152	7000	.	.	23000	1200	10.5
17	MWS	0.50	0.46	0.038	0.60	0.175	–	450	4.2	3600	.	.	–	84	310
18	MWS	0.25	0.27	0.009	0.60	0.103	–	800	7.8	1590	.	.	–	240	168
19	MND	0.50	0.46	0.038	1.40	0.30	–	450	4.2	11600	.	.	–	400	48

M, R, L in Solar units, P in days, all timescales in megayears. Timescales over 20 gigayears (–) and under 0.1 megayears (.) are omitted. Timescales are for period, mass or eccentricity decrease. K_1 is as defined in Eqs (4.37) and (4.41) with $\xi = 0$. Letters M, H, G, W, N under 'type' refer to main sequence, Hertzsprung gap, red giant, white dwarf and neutron star, respectively; D, S, C refer to detached, semidetached and contact, respectively.

For a red giant $R \sim R_{HT}$, and in that case Eq. (4.84) bears close comparison with Reimers' (1975) empirical formula (2.74), which, also in SI units, is

$$|\dot{M}_1| \sim 1.3 \times 10^{-5} \frac{RL}{GM_1}. \tag{4.87}$$

These agree provided that $\sigma \sim 1.2$–1.8 is reasonably typical for red giants; but clearly we should expect a substantial spread. It is no doubt a coincidence, but a rather interesting one, that although stars slow down as they become red giants the convective turn-over time goes up to the extent that the Rossby number does not change as much as one might expect. For those red giants rotating sufficiently rapidly that $\sigma \lesssim 0.3$ (which will usually only be those that are in relatively close binaries where they are forced by tidal friction to corotate) we can expect mass-loss rates larger than Reimers' formula by $\gtrsim 10^3$. Note that if the enhancement of Eq. (4.87) is by only a factor of 10^2 it puts the mass loss on a nuclear timescale. We will return to this in Section 4.6.

The above formulation represents a 'complete' theory of magnetic braking, to the extent that it predicts the braking rate of a single star, and hence of a binary assuming tidally-induced synchronism, as a function of mass, radius, luminosity and rotation period only. As by-products, the model predicts also such quantities as Ω_c, $\Delta\Omega$, B_P, B_ϕ. For example, the relative differential rotation is

$$\frac{\Delta\Omega}{\Omega} = \frac{\sigma E(\sigma)}{2\pi} \sim \frac{0.43\sigma}{1 + 2.8\sigma^2}, \tag{4.88}$$

which has a maximum value of 13% at $\sigma \sim 0.6$, but which is very small both for rapidly rotating K dwarfs ($\sigma \sim 0.01$) and slowly rotating red giants ($\sigma \sim 10$). Of course, the model is extremely tentative, but we shall use it as a reference point for discussion.

To clarify a point that might seem confusing, we should emphasise that the B_ϕ of dynamo theory (above, and Appendix E) is different from and independent of the B_ϕ of magnetic-braking theory (Appendix D); but on the other hand B_P is taken to be the same. Both B_ϕ can be seen to be consequence of differential rotation, but in the very different environments of the stellar interior (specifically, the base of the convection zone) and of the stellar wind. It is not yet clear what drives the interior differential rotation, although it can be measured in some detail by helioseismology. It is probably caused by a combination of Coriolis force with turbulent convection. What drives differential rotation in the wind is more simply the fact that the poloidal field cannnot be strong enough to enforce corotation indefinitely, but only as far as the Alfvén radius. In our analysis of the wind – Eqs (4.55) to (4.57) – we make no reference to the external B_ϕ because we have already eliminated it (crudely) using the precepts of Appendix D.

Some timescales expected in a few cases are shown in Table 4.6: nuclear evolution, thermal (Kelvin–Helmholtz) evolution, magnetic braking (for a single star, i.e. for $*1$ assumed single but rotating with the period listed), gravitational radiation – Eq. (4.4), circularisation by tidal friction – Eqs (4.10 and 4.31), synchronisation by tidal friction – Eq. (4.15), mass loss – Eq. (4.80), and magnetic braking with tidal friction (for a binary star). The last two are based on Eq. (4.85), but t_{MBs} includes the additional non-Alfvénic term of Eq. (4.67) and t_{MBb} also includes the further term in Eq. (4.68). Note that the circularisation, GR and MBb timescales are for e-folding of eccentricity and *period*, not angular momentum. The timescales involving tidal friction assume $e \sim 0$, $\omega \sim \Omega_\parallel$.

Table 4.6 also gives the parameter K_1 of Eqs (4.37) and (4.41). This is an estimate of the ratio of actual to orbital specific angular momentum carried off in the wind. The stellar activity produces both mass loss and angular momentum loss, and whether it is the mass loss, which tends to increase the separation, or the angular momentum loss, which tends to decrease it, that dominates is approximately decided by K_1 and the mass ratio: angular momentum loss wins if $K_1 \gtrsim 1 + q$.

In Table 4.6 only *1 is considered active, *2 being supposed inert, although in some cases (rows 3a, 3b, etc.) we interchange them to see which is the more active. For example, rows 4a and 4b imply that the timescale for period change by mode MB in a binary with parameters $(0.8 + 0.6\,M_\odot,\ 3\,d)$ is 2.7 gigayears, the harmonic sum of the two values of timescale. Various evolutionary states (M, H, G, W, N; Table 3.5) are hypothesised for each component. Some systems are assumed detached, some semidetached, and one system is in contact. These data show that:

(a) a 12-day orbit containing two Solar-type stars can circularise in substantially less than a Hubble time; half the time listed, since the components contribute equally. A 4-day orbit can shrink its period significantly by mode MB in a similar time.

(b) two M dwarfs in an 0.8-day binary (row 5) can shrink their orbit by mode MB in much less than a Hubble time

(c) in a contact binary of roughly Solar temperature (row 6) the magnetic-braking timescale may be substantially shorter than the evolutionary timescale, due mainly (at least in the case tabulated) to the more massive component

(d) red giants in close binaries (rows 7–12) can lose both mass and angular momentum on roughly the nuclear timescale

(e) short-period binaries containing a late main-sequence dwarf can shorten their periods (rows 13–19) on much less than a Hubble timescale, with mode MB dominating mode GR by an order of magnitude or more

(f) binaries with nearly lobe-filling components have very short tidal-friction timescales, and so can generally be assumed to be circular and corotating

(g) for short-period binaries containing a Solar-type star, mode MB may be more effective (row 13) than Mode NE in bringing the system towards RLOF.

The model presented here for magnetic braking is similar to that of Tout and Pringle (1992), who however restricted their discussion to fully convective pre-main-sequence stars. They also used somewhat different approximations regarding conditions at the Alfvén radius. Even if one accepts the general concept that simple equations may be sufficient to model the dynamo and it consequences, there remains a fair amount of choice about the nature of the formulae to be used.

Although the estimates (4.84) and (4.85) above may apply at some level of approximation to *single* stars, it is by no means obvious that they can be applied directly to components of *binaries*. For example, it would not be surprising if tidal friction in a binary, as one potentially active component evolves towards filling its Roche lobe, brings the surface into corotation more quickly than the interior, thus possibly enhancing the differential rotation causing the dynamo. But equally, once near-uniform rotation is achieved, tidal friction might *diminish* the differential rotation of the sort observed on the Solar surface.

I have incorporated the above detailed yet speculative model into codes which follow either the orbital evolution alone (taking the interior evolution to be negligible) or the

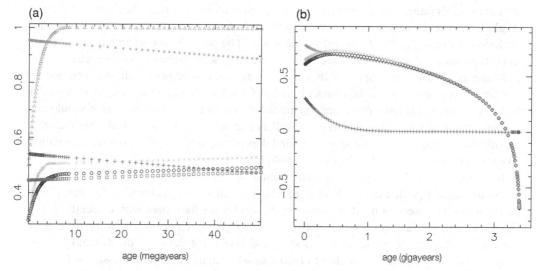

Figure 4.6 Orbital and spin evolution of a model for BY Dra. (a) the short term, starting 200 megayears ago and ending 150 megayears ago. The initial period and eccentricity were chosen to give the present period and eccentricity after 200 megayears. (b) the long term, starting from present conditions. Eccentricity, pluses; orbital period (log days), asterisks; rotational periods (log days) of ∗1 and ∗2, circles and crosses. For (a) only: pseudo-synchronous period (log days), squares; cosine(inclination), for ∗1 only, triangles. 'Inclination' means the angle between the stellar spin and the orbital spin.

combination of stellar and orbital evolution. Figure 4.6 shows results for the active K/M dwarf binary BY Dra (Boden and Lane 2001). This is a double-lined binary, which does not eclipse but which has a remarkably small *interferometric* orbit (4.4 mas). Although the inclination is in principle measured, it is somewhat uncertain (151.8 + 3.5°) and allows a substantial range of masses. We assume initial parameters ($0.73 + 0.64\,M_\odot$, 9 days, $e = 0.54$). After ∼200 megayears it reaches its present parameters (6 days, $e = 0.3$). We choose this age because the system has a cpm companion, M5V at $17''$ (Zuckerman *et al.* 1997), which is presumably at least this old. Because we cannot be clear how the binary formed in the first place, we cannot be sure that evolution 'started' with synchronous, parallel rotation in both components, and so we assume arbitrarily that the initial spin periods were both 2 days, and that the two axes were at 60° to the orbital axis, and 120° to each other in the same plane. In Fig. 4.6a, we see that pseudo-synchronisation and parallelisation take ∼6 megayears. Our model gives the *spin* period of ∗1 as 4 days, not very different from the observed value of 3.83 days. In Fig. 4.6b we follow the evolution much further, and find that the orbit shrinks to RLOF at ∼3.4 gigayears, by which time the masses have dropped 15% and 10% respectively. They might form a contact binary (case AR, Section 3.5), but also might merge quickly in a hydrodynamic burst of RLOF (case AD). However, conservative evolution would not lead to any interaction at all (within a Hubble time), and so we define a new sub-case: case AA. Yet another sub-case occurs if mass loss is relatively *stronger* than angular momentum loss. In that case the binary widens and can avoid interaction. We call this case AM. These are discussed further in the next section.

Unfortunately we cannot draw as strong conclusions as we would like from BY Dra, because (a) the inclination and, therefore, the masses are rather uncertain, and the dynamo model depends rather sensitively on mass on the lower main sequence; and (b) the third body might influence the close pair's eccentricity through Kozai cycles (Section 4.8). These were included in Fig. 4.6 and had little effect, but it would be possible to start with quite different conditions and end up with much the same system as is seen, thanks to Kozai cycles.

Magnetic braking, and therefore dynamo activity, is a crucial process in the evolution of certain types of binary (Algols, contact binaries, CVs, LMXBs), and at some stage a model has to be included in attempts to understand the course of their evolution. But it should not be forgotten that winds carry off mass as well as angular momentum, and whether the orbit shrinks or expands in response to stellar wind depends mainly on the ratio of Alfvén radius to *orbital* radius, not *stellar* radius – Eq. (4.68).

It is often asserted that dynamo activity should vanish if a star becomes fully convective, as on the lowest portion of the main sequence. We can see no justification for that, either in theory or in observation. Many very late M dwarfs are flare stars with evidently active dynamos, and yet fully convective. Active low-mass young red or brown dwarfs have been seen in the Pleiades, with rotational periods of 2–3 hours. The details of the dynamo process may well be very different from those in more slowly-rotating and more massive Solar-type stars, but it is clear that considerable dynamo activity takes place in rapidly-rotating stars at the bottom of the main sequence and beyond.

4.6   Binary-enhanced stellar winds: modes EW, MB

Isolated stars that evolve to large radii will rotate very slowly, partly because the moment of inertia increases and partly because even a modest stellar wind, and even without the extra effect of magnetic braking, tends to remove a disproportionate amount of angular momentum. For example the 4 $M_\odot$ star of Table 3.2 increases its radius by ~30 between the ZAMS and helium ignition, and if it started with a rotational period of 1 day would have slowed to more than 1000 days. However, at its temporary maximum radius of over 70 $R_\odot$ it could still (just) fit within a binary of period 110 days. Thus, tidal friction might spin such a star to 10 times the rotation rate that it would experience if single. Since the convective turnover time of red giants is ~ 100–200 days, this can be expected to have an effect on its mass-loss rate. Our specific model of the previous section predicts a specific increase, but even if the model is not correct some increase is to be expected.

Cool giants and supergiants usually show evidence of winds, but there may be a dichotomy between those with relatively tenuous, hot fast winds and those with more copious, cool and slow winds (Linsky and Haisch 1979). The former may be driven partly at least by dynamo activity, while arguably in the latter it is the more direct effect of high luminosity, with radiation pressure acting on grains that form in the cool superphotospheric region. Possibly rotation plays a more minor role in the latter. But any star which is close to filling its Roche lobe is also rotating within a factor of three of its break-up velocity – Table 3.1 – and this seems very likely to be a cause of 'enhanced wind', whatever the detailed mechanism of the 'normal wind'. It is not easy to demonstrate either observationally that such a process is taking place, or theoretically that such a process must take place, but there are several pointers from individual systems, some of which we discuss shortly.

For very massive stars, which also have winds, there is little direct evidence for magnetic braking. Indeed, because massive stars in general rotate rapidly, while also losing mass at a much greater rate than lower-main-sequence stars, it seems likely that magnetic stresses cannot make the wind corotate to any great distance. But even a wind not linked to the star

Table 4.7. *Some RS CVn and possibly related systems*

Name	Spectra	State	P	e	M_1	M_2	R_1	R_2	X^a	Y^a	Reference
RS CVn	K0IV + F4IV-V	GMD	4.80		1.44	1.41	4.0	2.0	12	1.45	Popper 1988a
Z Her	K0IV + F4	GMD	3.99		1.31	1.6	2.7:	1.9:	11	1.25	Popper 1988a
RW UMa	K4IV-V + F5V	GMD	7.33		1.45	1.5	3.8:	2:	18	1.4	Popper 1980
RZ Eri	K2III + F5m	GMD	39.3	0.35	1.62	1.68	7.0	2.8	90	1.9	Popper 1988b
SZ Psc	K1IV + F8V	GMD	3.97		1.6	1.3	5.1	1.5	9	1.2	Popper 1988b
λ And	G8III-IV + ?	gmD	20.5	0.04	0.0006^b						Walker 1944
AR Lac	K0 + G2	GhD	1.98		1.3	1.3	3.1	1.8	5.33	1.42	Popper 1980
WW Dra	G8IV + G2IV	GhD	4.63		1.34	1.36	3.9	2.1	12.2	1.57	Popper 1988b
α Aur	G8III + G0III	SHD	104		2.61	2.49	11.4	8.8	192	4.9	Barlow *et al.* 1993
V643 Ori	K7III + K2III	ssD	52.4	0.014	2.0	3.4	22	16	109	7.5	Imbert 1987
OW Gem	G8Ib + F2Ib-II	HHE	1259	0.52	3.9	5.8	32	30		10.4	Griffin and Duquennoy 1993

[a] $X \gtrsim 4$ implies case $\mathcal{B}$ or $\mathcal{C}$; Y is the ratio of R_2 to its unevolved value; see Section 3.5.
[b] Mass function, or if two values $M \sin^3 i$.

magnetically should cause some braking: Eq. (4.67) shows that the single-star spin-down timescale is about a tenth of the mass-loss timescale. Thus a massive star that rotates rapidly cannot have lost more than, say, 10% of its mass so far, and less than that if R_A is significant. Equation (2.71) suggests that massive stars can indeed be expected to lose a few per cent of their masses. But it seems quite possible that convective *cores* might be just as effective as envelopes in sponsoring dynamo activity.

Table 4.7 contains a number of systems, mostly 'RS CVn binaries', in which the larger and cooler star shows evidence of considerable activity, much more than one sees in isolated stars of the same spectral type. Among them Z Her shows clearly that the more evolved star is substantially the *less* massive, and yet is well short of filling its Roche lobe. Two or three other systems show a marginal mass deficit.

It is much more difficult to construct serious models of non-conservative binary evolution to fit such systems, than it is to fit conservative models as in Table 3.9. In a conservative model we know the total mass and angular momentum from the presently observed parameters, and only the initial mass ratio has to be varied in the hope of getting a good fit. It is obvious that no conservative model will give Z Her, but it is not obvious whether we should assume minimal mass loss, or whether perhaps *both* components have lost mass, though presumably more from ∗1. In practice, our recipe in Section 4.5 gives some mass loss even from ∗2. A further problem is that we would like the same recipe to hold for RS CVn as for Z Her. These systems have rather similar masses *and* periods, and yet show rather different effects of mass loss: arguably slight or even non-existent in RS CVn, while unarguably very substantial in Z Her. We may be faced with the unattractive possibility that mass loss is far from deterministic, but instead rather chaotic, so that it acts very differently in otherwise similar systems.

Two systems in Table 4.7 that show something of a similar contrast are α Aur and V643 Ori. The former appears to be quite normal – although the fact that the orbit is circular suggests that ∗1 came quite near to filling its Roche lobe at its peak radius during He ignition, and we might have expected some enhanced activity then. By contrast, the almost equally wide system V643 Ori appears to have suffered considerably, despite the fact that the orbit is still very slightly eccentric. I would like to suggest that the *original* masses in V643 Ori were somewhat larger than those in α Aur. Coupled with the probability that the initial period was somewhat smaller, say $\sim$30 days, so that *1 would have been very near its Roche lobe at helium ignition, this might have implied substantial mode EW. This could mean that when *2 in turn reached helium ignition the binary was wider and so *2 suffered less mode EW than *1. We expect both giants to be post-helium-ignition because their masses, assuming one or both were more massive originally, would lead to non-degenerate helium-core ignition very quickly after crossing the Hertzsprung gap.

Another system that shows surprising masses, to much the same extent as in V643 Ori, is OW Gem. Here we can hardly appeal to mass loss, whether intrinsic, or enhanced by binarity, since the orbit is much larger and also eccentric. The G8II star is not luminous enough to be the remains of a star that was initially the *more* massive. I suggest here a quite different explanation: that the FI–II star is the merged remnant of a former *close* sub-binary, and that the initial triple system had parameters something like $(3.9\,M_\odot + (3.9 + 1.9\,M_\odot; 2\,\text{days})$; 1260 days). It seems reasonable to assume that the mass ratio in the sub-binary was sufficiently large ($\sim$2) to trigger a rapid merger (case AD) rather than normal RLOF. However, it may be stretching credulity to suggest that V643 Ori had a similar history, even though a few adequately close triples are known to exist (VV Ori, λ Tau, DM Per, Table 3.8; HD 109648, Fig. 1.7b).

Systems such as those in Table 4.7 are probably a gold mine for researching the kind of non-conservative processes described here. For example, RZ Eri shows a similar but smaller mass deficit than Z Her, and also has substantial eccentricity remaining in its orbit. One might expect pseudo-synchronism here, but the observed rotation period of *1, determined from its spottedness, is 31.4 days, in between the pseudo-synchronous rate (23 days) and the synchronous rate. This could mean that magnetic braking is keeping it at a pseudo-equilibrium which is slower, or it could mean that tidal friction is not quite strong enough to keep up with evolutionary expansion.

However, the gold will be difficult to extract, because as noted above, in non-conservative systems it is much harder to guess the initial parameters. Attempts that we have pursued so far suggest that quite substantial amounts of mass and angular momentum can be lost, and so there is a considerable range of initial parameter space to explore. In addition, since it is unlikely that the non-conservative model of Section 4.5 is exactly right, we would have to treat several 'constants' in it as unknown variables; how many depends on taste.

Figure 4.7 is an attempt to model RZ Eri. We started somewhat arbitrarily with $e = 0.5$, but this led to a satisfactory pair of present masses, as well as eccentricity and radii, after *1's mass was reduced by about 12% at age $\sim$1.7 gigayears. However, the rotation rate of *1 was determined to be nearly pseudo-synchronous, in contrast to what is observed. It is not yet clear whether some other starting point in parameter space might do better, or whether the 'constants' in the non-conservative model need fine tuning.

Note that M_1 was reduced in this calculation from 1.75 to $0.7\,M_\odot$ *before* RLOF was reached. This had the effect of removing the possibility of hydrodynamic RLOF, that one would otherwise expect in conservative case $\mathcal{C}$. We call this kind of evolution sub-case $\mathcal{C}$UN of case $\mathcal{C}$U, the U referring to 'unusually' strong wind – unusual in comparison to the rather weak but detectable wind of a normal (single) red giant that is nowhere near the top of the giant branch.

More precisely, and in analogy with case $\mathcal{B}$ (Sections 3.5 and 4.3), we will define sub-cases of case $\mathcal{C}$ where different amounts of envelope of *1 are blown away by enhanced wind,

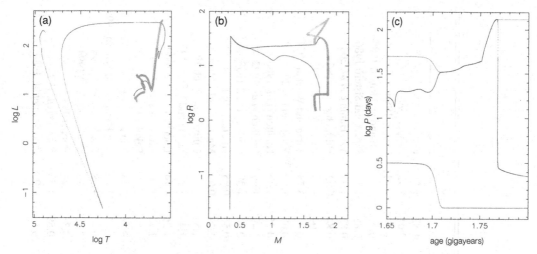

Figure 4.7 Possible evolution of RZ Eri, taking account of enhanced wind, magnetic braking and tidal friction in both components. The system was started with parameters $(1.75 + 1.68 \, M_\odot; \; e = 0.5; \; 49 \, \text{days})$; both stars were started with an initial rotation period of 2 days. (a) The theoretical HRD, $*1$ being the thinner line. At the end of the evolution $*1$ is starting to undergo a nova-like outburst, due to accretion from the wind of $*2$. (b) Stellar radii (lower curves) and Roche-lobe radii (upper curves) for both components as functions of their masses. (c) The evolution of eccentricity (lowest curve), orbital period (uppermost curve) and rotational period of $*1$, during the time interval when they were varying most rapidly.

thus:

$$\mathcal{C}W: \quad MME \rightarrow HME \rightarrow GME \rightarrow SME \rightarrow SME;SW \rightarrow CME \rightarrow WME \rightarrow \dots \qquad (4.89)$$

$$\mathcal{C}U: \quad MME \rightarrow MME;TF2 \rightarrow MMD \rightarrow HMD \rightarrow GMD;EW1,MB1 \rightarrow CMD \rightarrow \dots \qquad (4.90)$$

$$\mathcal{C}UN: \quad " \qquad " \qquad " \qquad " \rightarrow GMS;F1,EW1,MB1 \rightarrow CMD \rightarrow \dots \qquad (4.91)$$

$$\mathcal{C}UD: \quad " \qquad " \qquad " \qquad " \rightarrow GMS;F3 \rightarrow GMC;CE \rightarrow CMD \rightarrow \dots \qquad (4.92)$$

In the first case, the *normal* wind (superwind) that terminates the evolution of an AGB star prevents the star from ever filling its Roche lobe. In the second, the same effect may be produced by *enhanced* wind, at an earlier stage of evolution. In the third, RLOF is not entirely avoided, but the mass ratio is so reduced by mode EW that the subsequent RLOF is on a nuclear rather than dynamical timescale. In the fourth, either because mode EW was less severe or because the initial mass ratio was larger in the first place, the RLOF is at a dynamical rate, and leads to common-envelope evolution (mode CE, Section 5.2). The final letters N and D imply some similarity with cases AN and AD of Section 3.5. RZ Eri is a system which, I suggest, belongs to case $\mathcal{C}UN$.

There are, of course, several possible variants of these cases. Case $\mathcal{C}W$ might happen to stars of $\sim 1 \, M_\odot$ even on the FGB, rather than the AGB. The other cases might happen in sufficiently wide binaries that $*1$ reaches the AGB before modes TF and EW becomes important. The initial mass ratio may be sufficiently close to unity that *both* components are giants either before or after RLOF begins, as was presumably the case for RT Lac to AR Mon in Table 4.8.

Table 4.8. *Some low-mass binaries: before, during and after RLOF*

Name	Spectra	State	P	e	M_1	M_2	R_1	R_2	X^a	Y^a	Reference
UV Leo	G2V + G2V	MMD	0.60		1.11	1.11	1.13	1.13	1.82	1.11	Popper 1998
RT And	F8V + K0V	MMD	0.63		1.24	0.91	1.26	0.90	2.01	1.42	Popper 1994
YY Gem	M1Ve + M1Ve	MMD	0.81		0.59	0.59	0.62	0.62	4.60	1.13	Popper 1980
V388 Cyg	A3 + G2	MMS	0.86		2.2	0.8	2.6	1.5	0.99	2.06	Milano and Russo 1983
FT Lup	F2V + K5	MMS	0.47		1.4	0.6	1.4	0.95	0.88	1.71	Lipari and Sisterò 1986
CN And	F5V + G5	MMS	0.46		1.30	0.51	1.43	0.92	0.83	1.97	Van Hamme et al. 2001
V361 Lyr	F8-G0 + K4	MMS	0.31		1.26	0.87	1.02	0.72	0.84	0.92	Hilditch et al. 1997
VZ Psc	K3 + K7	MMS	0.26		0.81	0.65	0.78	0.70	1.04	1.16	Hrivnak et al. 1995
ε CrA	A8 + F0	MMC	0.59		1.5	0.15	2.2	0.7	0.08	3.75	Tapia and Whelan 1975
EQ Tau	G2 + G2	MMC	0.34		1.32	0.59	1.16	0.82	0.69	1.50	Yang and Liu 2002
W UMa	F8 + F8	MMC	0.33		1.35	0.7	1.2	0.85	0.75	1.30	Hilditch et al. 1988
RW Dor	K3 + K0	MMC	0.29		0.64	0.43	0.79	0.67	1.49	1.69	Hilditch et al. 1992
CC Com	K7V + K7V	MMC	0.22		0.79	0.43	0.74	0.55	0.84	1.38	Hilditch et al. 1988
AH Vir	G8IV + G8IV	ggC	0.41		1.4:	0.45:	1.3:	0.8:	0.55	0.95	Hilditch 1981
RZ Cas	K0IV + A3V	gMS	1.20		0.73	2.21	1.94	1.67	1.21	0.99	Maxted et al. 1994a
ZZ Cyg	K5 + F7	gMS	0.63		0.6:	1.2:	1.2:	1.3:	1.52	1.14	Guinan and Koch 1977
V1010 Oph	G6 + A5	gMs	0.66		0.65	1.35	1.3	1.8	1.41	1.36	Leung and Wilson 1977
W Crv	K2 + G5	gMs	0.39		0.68	1.00	0.92	1.01	1.26	1.14	Rucinski and Lu 2000
β Per	G8III + B8	GMS	2.87		0.8	3.7	3.6	3.1	1.16	1.30	Richards et al. 1996
β Per[b]	(...) + F1m	(...)ME	679	0.23	4.5	1.7					Fekel 1981
U Cep	G8IV + B7-8	GMS	2.49		2.8	4.2	4.9	2.7	3.69	1.13	Tomkin 1981
TT Hya	K1IV + B9.5V	GMS	6.95		0.6	2.6	5.9	2.0	3.45	1.06	Van Hamme and Wilson 1993
DN Ori	G5III + A2	GMs	13.0		0.34	2.8	6.7	2.4	0.44	1.08	Etzel and Olson 1995
S Cnc	G8III + B9.5V	GMs	9.49		0.23	2.4	5.0	2.2	0.70	1.20	Olson and Etzel 1994
AS Eri	K0 + A3	GMS	2.66		0.2	1.9	2.2	1.8	0.33	1.14	Popper 1980
R CMa	G8IV + F1	GMS	1.14		0.17	1.07	1.15	1.5	0.59	1.53	Sarma et al. 1996
RX Cas	K1III + Ash	gMS	32.3		1.8	5.8	24:	2.5:	20	0.9:	Andersen et al. 1989

Name	Sp. type	State									Reference
SX Cas	K3III + A6sh	gMS	36.6		1.5	5.1	23.5	3.0	21.7	1.13	Andersen et al. 1988
RT Lac	G9IV + G5IV:	GGS	5.07		0.63	1.57	4.6	4.3	8.12	2.9	Popper 1980
RZ Cnc	K4 + K1	GGS	21.6		0.54	3.2	12.2	10.2	5.61	5.0	Popper 1980
AR Mon	K3 + K0	GGS	21.2		0.8	2.7	14.2	10.8	16.5	8.8	Popper 1980
DL Vir	K0-2 + A3V	GMS	1.32		1.1:	2.2:	2.4:	1.8:	2.1	1.1	Schöffel 1977
DL Vir[b]	(...) + G8III	$(...)GE$	2260:	0.44	3.3:	1.9:		10:			Schöffel 1977
θ Tuc	F: + A7IV	HMD	7.10		0.063[c]	0.7[c]					De May et al. 1998
V1379 Aql	SDB + K0III	CGE	20.7	0.09	0.30	2.27	0.05	9.0	3.82	5.2	Jeffery and Simon 1997
FF Aqr	SDOB + G8III	CGD	9.2		0.35:	1.4:	0.15:	6:	8.5	4.4	Vaccaro and Wilson 2002
AY Cet	WD + G5IIIe	WhD	56.8	0.09:	0.55:	2.1:	0.012	6.8	42	4	Simon et al. 1985
V651 Mon	SDO + A5V	CMD	16.1	0.07:		0.007[c]					Méndez and Niemelä 1981
AA Dor	SDO + 4kK	CMD	0.26		0.3:	0.05:	0.16:	0.09:			Włodarczyk 1984

[a] $X \sim 1$–4 implies case A, in conservative evolution; $X < 1$ implies non-conservative evolution; see Section 3.5. Y is the ratio of R_2 to the ZAMS radius at the observed mass.

[b] Outer orbit of triple system.

[c] Mass function, or if two values $M \sin^3 i$.

Under 'State', a lower-case letter denotes considerable uncertainty.

We might note that among low-mass stars, where the effects of dynamo activity are more likely to be significant than at intermediate masses, there is rather little scope for case $\mathcal{B}$, because at $M \lesssim 1.5\,M_\odot$ the main sequence terminates very close to where the giant branch, with substantial convective envelopes, begins. We have chosen to define the $\mathcal{B}/\mathcal{C}$ boundary as where the convective envelope is sufficiently deep that the star expands (case $\mathcal{C}$) rather than contracts (case $\mathcal{B}$) in response to mass loss on a thermal timescale. This will not coincide with the boundary between where dynamo activity is negligible and where it is important, but we hope, for simplicity, that the two boundaries are not very far apart. On the main sequence they *are* quite far apart, perhaps early F for dynamo activity and late K for mass-loss-driven expansion, but they may be closer together, say early G and late G, for (sub)giants. In this section some binaries that we discuss may be technically case $\mathcal{B}$, but we largely ignore this. For the *massive* binaries discussed in Section 4.3, case $\mathcal{B}$ covered a much wider range of initial parameters.

Table 4.8 contains a selection of binaries with one or more cool components, where we can expect some binary-enhanced activity. Some are pre-RLOF, some are undergoing semidetached RLOF, some are in contact, and some are arguably post-RLOF systems. We might have evolution from a detached system like UV Leo through a semidetached state like V361 Lyr to a contact state like W Uma. In this hypothetical sequence the total mass decreases modestly from 2.22 to 2.05 $M_\odot$, and the angular momentum decreases also modestly; but it could easily be the case that relatively more mass is lost and the contact system might be more like EQ Tau, or even RW Dor.

We have argued in Section 3.4 that contact might not be a stable configuration, but that instead the system pursues a relaxation cycle about an unstable equilibrium of marginal contact with poor thermal contact and markedly unequal temperatures. We anticipate that the unstable equilibrium itself gradually changes, because of progressive loss of angular momentum. This means that the system cycles between contact and semidetached states, with the mass ratio oscillating but growing slowly larger in the mean. During an oscillation the contact phase lasts for something like the thermal timescale of the less massive component, and during the semidetached phase of the more massive. Thus we may expect the semidetached phase to be short compared with the detached phase. Four systems in Table 4.8, FT Lup to VZ Psc, may arguably be in the semidetached phase of the oscillation, but several hundred contact systems are known, and so the ratio of timescales may indeed be quite small.

An alternative view might be that these semidetached systems are approaching contact for the first time, and once in contact will remain in contact. However, in that case we expect only about one semidetached system per thousand contact systems, which does not appear to be the case. Nevertheless, the nature of the evolution in contact binaries remains one of the least understood processes in binary-star evolution.

It is likely that both nuclear evolution and magnetic braking contribute about equally to the long-term evolution of some contact binaries. But the balance will no doubt itself change in the course of evolution, and will also depend on the initial mass. Nuclear evolution will only be significant if at least one component is $\gtrsim M_\odot$, while magnetic braking may only be significant if the system is of spectral type G/K/M, and perhaps (for these *very* rapid rotators) type F as well. These ranges overlap in stars of ~ 1–$1.5\,M_\odot$. In ϵ CrA it may be nuclear evolution that dominates now: *1 appears to be significantly evolved. But at an earlier stage M_1 was probably less, and it may have been mainly magnetic braking that transformed it from a detached system like UV Leo to a contact binary like EQ Tau, followed by incipient nuclear evolution that took it to something like its present form.

We can anticipate that there may be a mild dichotomy between (a) systems of somewhat low total mass, where magnetic braking dominates nuclear evolution and the ultimate merger produces a main-sequence star, and (b) more massive systems where nuclear evolution dominates and the ultimate merger produces a red subgiant instead. Probably ϵ CrA is already close to a merger that will leave it as a partially evolved main-sequence star. On the other hand AH Vir seems part way to merging as a cool subgiant.

I may have biased this discussion by representing in Table 4.8 that $*1$ of AH Vir is the *currently* more massive component. This is just a hypothesis; there are also some ultra-short-period systems like ZZ Cyg and W Crv that may have arisen by fairly normal (mode NE) RLOF, although I suspect that there has been considerable magnetic braking as well. Even more magnetic braking in the future may bring such systems into a contact configuration like AH Vir, but with $*1$ and $*2$ interchanged. All of the systems above β Per in Table 4.8 look as though they may come from a rather small volume of initial-parameter space, where nevertheless considerable diversity is achieved because modes NE, MB and EW are rather finely balanced there.

We attempt to categorise some scenarios where mode MB or mode EW play a significant role. Because of the above expectation of diversity we restrict ourselves to two main sub-cases of case A, although each has several possible variants. We call them cases AA and AM, the second letter in each standing for 'angular momemtum loss' and 'mass loss', respectively. Both processes happen in both cases, but the relative importance is perceived as varying. Case AA is likely to lead to progressively shorter periods, although if the mass ratio increases as the angular momentum decreases (as is likely in contact binaries) it is possible for the period actually to increase modestly. Case AM would lead to longer periods. In the former case, we expect contact binaries as for conservative case AR (route 3.89), except that the components merge while still on the main sequence, instead of after $*1$ has evolved into the Hertzsprung gap. In the latter case, it is possible that RLOF is entirely avoided, as the stars decrease in mass to the point where there is no longer nuclear evolution in a Hubble time. This could be quite common in old clusters. Binaries with initial masses that would put them near the current turnoff may be pushed down the main sequence at much the same rate as the turnoff itself moves downward, instead of evolving into the sub-giant region.

$$AA: \quad MMD; TF, MB \to MMS; F2 \to MMC \to MMC; R2, MB1 \leftrightarrow MMS; F2, MB1 \to$$

$$\to MMC; DI \to MMC; CE \to M; MB \to G \to \dots \to W. \tag{4.93}$$

$$AM: \quad MME; TF, EW, NE \to MMD; no NE. \tag{4.94}$$

Table 4.8, as well as listing a number of short-period Algols like V361 Lyr that may be related to case AA, lists some more normal (i.e. longer-period) Algols, and a few possible post-Algols. I have already argued (Section 3.5) that AS Eri and R CMa show signs of substantial angular-momentum and mass loss, respectively. DN Ori and S Cnc are similar to AS Eri, though not quite so extreme. I feel fairly sure that all of the Algols (β Per to DL Vir) are subject to all of modes NE, MB and EW, but with mode NE probably the dominant one, by a modest margin, at least in the early evolution. This is the opposite of our interpretation of contact binaries, where I have suggested that mode MB dominates earlier, and (possibly) mode NE later. The Algol systems must have started with substantially longer periods than those listed above β Per, and also with substantially greater total masses, and both of these are likely to give mode NE an initial advantage; but it is clearly not a big advantage.

Some Algols (RT Lac to AR Mon) contain two giants or sub-giants. This is to be expected in conservative case AL, but once again it is likely that the non-conservative modes MB and EW have played a part here. In fact it is especially likely, since two cool stars may be more effective than just one.

DL Vir is particularly interesting as a system where there is not only a third body, but a substantially evolved third body. We can be reasonably confident that both cool stars in this triple started with much the same mass ($1.9\,M_\odot$), since they are both well evolved. Thus here we have some handle on the *initial* parameters of the Algol, independent of conservation or the lack of it. If we take the quoted numbers at face value, then we also have an upper limit on the initial masses of the Algol pair ($\sim 1.9 + 1.9\,M_\odot$) and a lower limit ($\sim 1.9 + 1.4\,M_\odot$). Thus the upper limit of the amount of mass lost is $\sim 0.5\,M_\odot$. This is only $\sim 14\%$ of the initial mass, but the Algol is also not very evolved (i.e. has quite a large mass ratio) compared with the others. Unfortunately the masses are not well determined, though they would repay an analysis with modern instrumentation.

θ Tuc is an excellent example of a probable post-Algol. It is not eclipsing, so that the inclination can only be guessed, but likely current masses are ($0.2 + 1.8\,M_\odot$), not unlike AS Eri, which is at a slightly earlier stage of evolution (and slightly less wide). V1379 Aql is at first glance an even later stage, and of a somewhat wider binary. However, V1379 Aql has two problems: (a) the hot sub-dwarf is more massive than we would expect at this period (~ 0.23–$0.25\,M_\odot$) and (b) the orbit is significantly eccentric ($e = 0.09 \pm 0.01$). Possibly the answer to the first is that the system has less metallicity than the Solar system (Jeffery, private communication 1998). I suspect that the eccentricity can only be caused by the presence of a third body, so far undetected (Section 4.8). Obviously, we expect the orbit of a post-Algol to be highly circular, if unperturbed. A third body like that in β Per or DL Vir, but perhaps half the mass, and in a substantially inclined orbit, could easily have this effect.

The last four systems in Table 4.8 are also combinations of a hot sub-dwarf and a main-sequence or giant star. They might be post-Algols, but either the estimated mass of the hot sub-dwarf is a little too high, or that of the companion too low, for such a scenario. I will discuss them again in the context of common-envelope evolution (mode CE, Section 5.2).

4.7   Effects of a third body: mode TB

If the binary is part of a triple system, with a third body of mass M_3 in a wider outer orbit, there can be an appreciable effect on the inner orbit. This is more pronounced the closer the third body, and also the more inclined is its orbit to the inner orbit; but in the case that the inclination of the two orbits is greater than $39°$ ($\sin^{-1}\sqrt{2/5}$) the effect ('Kozai cycles') can be surprisingly large even if the outer orbit is quite wide.

I give here a model – Appendix C(f) – based on the quadrupole level of approximation. At this level there is an extra acceleration $\mathbf{f}$ in the inner binary given by

$$f_i = S_{ij}(\mathbf{D})d_j, \quad S_{ij} = \frac{GM_3}{D^5}(3D_i D_j - D^2\delta_{ij}), \tag{4.95}$$

where $\mathbf{d}$ is, as before, the separation of the inner pair and $\mathbf{D}$ is the separation of the outer pair, i.e. the vector from the center of gravity of the inner pair to the third body. At the same level of approximation, there is (somewhat surprisingly) *no* extra acceleration within the outer orbit.

Since there is a couple on the inner orbit, and not on the outer, angular momentum is not conserved. However, it is conserved *approximately*, since it is implicit in the approximation that the angular momentum of the inner orbit is small compared to the outer.

If we average the effect of $\mathbf{f}$ over the inner orbit, according to the precepts of Appendix C, we find that the energy of the inner orbit is unaffected – because a potential force does no work around a closed curve. Thus a and P are constant, at this level of approximation. But the vectors $\mathbf{e}, \mathbf{q}, \mathbf{h}$ defining the 'orbital frame', as in Sections 3.2.2 and 3.4.2, can vary, in direction as well as magnitude, and their variation is given once again by

$$\dot{\mathbf{e}} = \mathbf{U} \times \mathbf{e} - V\mathbf{e}, \quad \dot{\mathbf{h}} = \mathbf{U} \times \mathbf{h} - W\mathbf{h}, \quad (4.96a,b)$$

where $\mathbf{U}$ is the angular velocity of the orbital frame relative to an inertial frame. The equation for $\mathbf{q}$ is easily obtained from these, since by definition $\mathbf{q} = \mathbf{h} \times \mathbf{e}$. Unlike the V, W terms that we introduced in Eqs (4.22 and 4.23), V, W here are not dissipative: they can be negative as well as positive. The orbital-averaging technique, which assumes, of course, that the tensor S, and therefore $\mathbf{D}$, does not vary significantly during one inner orbit, gives $\mathbf{U}$, V and W as

$$\mathbf{U} \equiv X\overline{\mathbf{e}} + Y\overline{\mathbf{q}} + Z\overline{\mathbf{h}}$$

$$= \frac{a^2}{2h} [(1 + 4e^2)S_{13}\,\overline{\mathbf{e}} + (1 - e^2)S_{23}\,\overline{\mathbf{q}} + (1 - e^2)(4S_{11} - S_{22})\overline{\mathbf{h}}], \quad (4.97)$$

and

$$V = \frac{5a^2}{2h} (1 - e^2)\,S_{12}, \quad W = -\frac{5a^2 e^2}{2h}\,S_{12}. \quad (4.98a,b)$$

As before, X, Y give precession and Z gives apsidal motion; but see the last paragraph of this section.

We now average the S_{ij} over an outer orbit, which is in fact exactly Keplerian at this level of approximation. We obtain

$$\langle S_{ij} \rangle = C\,(\delta_{ij} - 3\overline{H}_i\overline{H}_j), \quad C \equiv \frac{GM_3}{2A^3(1 - E^2)^{3/2}}, \quad (4.99)$$

where $A, E, \overline{\mathbf{H}}$ are for the outer orbit the equivalent of $a, e, \overline{\mathbf{h}}$ for the inner orbit, i.e. the semimajor axis, the eccentricity and the unit vector in the direction of the orbital angular momentum. Surprisingly but conveniently, the tensor S is symmetrical about the $\overline{\mathbf{H}}$-axis even although the outer orbit, if eccentric, is not.

In the inner-orbital frame, with the 1, 2, 3-directions parallel to $\mathbf{e}, \mathbf{q}, \mathbf{h}$ respectively, S_{12} means $S_{ij}\overline{e}_i\overline{q}_j$, etc., and so

$$\langle S_{11} \rangle = C\,\{1 - 3(\overline{\mathbf{H}} \cdot \overline{\mathbf{e}})^2\}, \quad \langle S_{12} \rangle = -3C\,\overline{\mathbf{H}} \cdot \overline{\mathbf{e}}\,\overline{\mathbf{H}} \cdot \overline{\mathbf{q}}, \quad \text{etc.} \quad (4.100)$$

These expressions allow us to replace the S_{ij} in Eqs (4.98) for $\dot{\mathbf{e}}$, $\dot{\mathbf{h}}$ by their outer orbital averages, which are now known functions of the constant $\overline{\mathbf{H}}$ and the basis vectors $\mathbf{e}, \mathbf{h}$ (and $\mathbf{q} \equiv \mathbf{h} \times \mathbf{e}$). Thus we have a closed set of equations, which can be integrated by a stepwise procedure such as Runge–Kutta. In fact, two first integrals can be extracted analytically – Appendix C(f) – leaving a first-order ordinary differential equation for say e, whose solution is an elliptic integral.

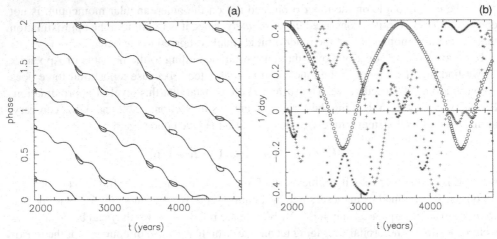

Figure 4.8 (a) Eclipse history of SS Lac, 1912–4900, subject to precession driven by a third body in an orbit inclined at 29° to the inner orbit. Phase is plotted vertically, covering two complete inner orbits. Eclipses occur within the isolated leaf-shaped patches. Eclipses typically last for just over a century, several centuries apart. Also shown are the phases of inferior and superior conjunction. (b) Rotational history of ∗1 in SS Lac, starting (arbitrarily) with corotation in 1912. The three components of the spin vector are plotted: circles are the component of spin parallel to the orbit. The spin is partially retrograde for ∼25% of the time. However, the spin history, unlike the eclipse history, depends quite sensitively on the dimensionless quadrupole moment Q of Section 3.2.2.

Two effects of the third-body force are precession and apsidal motion, both of which come from the rotation rate $\mathbf{U}$ of the orbital frame. These effects are both on a timescale

$$t_{\text{TB}} \equiv \frac{h}{a^2 C} \sim \frac{M + M_3}{M_3} \frac{P_{\text{out}}^2}{2\pi P} (1 - E^2)^{3/2}, \qquad (4.101)$$

where P_{out} is the period of the outer orbit. Figure 4.8 illustrates the effect of precession and apsidal motion on the inner orbit (14.4 d) of SS Lac (Table 2.1), a binary system which showed eclipses up till about 1950, and not subsequently. A triple companion in a 679 day orbit was detected by Torres and Stefanik (2000). This third body can account for the eclipse history provided its orbit is inclined at 29° to the 14.4 day orbit (Eggleton and Kiseleva 2001).

However, a third effect is that h and e can also change on the same timescale. Fluctuations in these quantities are periodic, although if calculated with a more exact three-body code there tend to be modest departures from strict periodicity. In the case that the inclination of the two orbits is in the range 39–141° ($|\bar{\mathbf{H}} \cdot \bar{\mathbf{h}}| < \sqrt{3/5}$) the fluctuations can be very large, with a range that is *independent* of t_{TB}, and hence of P_{out}. Table 4.9 shows the relation between inclination on the one hand and minimum and maximum eccentricity on the other. It can be seen that an inclination of only 60°, about the average to be expected if triple orbits are the result of random encounters, can lead to eccentricities in excess of 0.75. However, although the eccentricity and angular momentum fluctuate, perhaps by quite a large amount, the period and semimajor axis are constant (in the lowest approximation), because the the third body's force is still a potential force, and so does no work in total around an orbit of the inner pair. P and a are determined purely by the energy of the orbit, and not by its angular momentum.

Table 4.9. *Limits of Kozai cycles*

η_a	Probability	e_a	e_b	e_a	e_b	e_a	e_b
0	0.000	0	0	0.3	0.3	0.5	0.5
10	0.015	0	0	0.3	0.309	0.5	0.510
20	0.060	0	0	0.3	0.341	0.5	0.543
30	0.124	0	0	0.3	0.407	0.5	0.600
40	0.224	0	0.149	0.3	0.521	0.5	0.679
50	0.357	0	0.558	0.3	0.669	0.5	0.772
60	0.500	0	0.764	0.3	0.808	0.5	0.863
70	0.658	0	0.897	0.3	0.914	0.5	0.937
80	0.826	0	0.974	0.3	0.978	0.5	0.984
90	1.00	0	1.00	0.3	1.00	0.5	1.00

The first column is the initial inclination and the second the cumulative probability of this inclination. Remaining pairs of columns are the initial (minimum, subscript a) and maximum (subscript b) inner eccentricity.

Neither precession nor apsidal motion is expected to have much effect on the long-term evolution of a binary. But the substantial cyclic variations in e possible at high inclination (Kozai cycles; Kozai 1962), coupled with the approximate constancy of a, means that tidal friction can become important (at periastron) during some part of the cycle, even if it is unimportant during that part of the cycle when the eccentricity is small. The effect means that over many Kozai cycles the inner orbit will shrink as well as become circularised, the final period being roughly the period when the stars are close enough for apsidal motion due to their distortion to dominate over apsidal motion due to the third body. Of course, the possible importance of this process depends on (a) the frequency of triples, relative to binaries, and (b) the frequency of high inclinations relative to low. Neither frequency is well known at present.

We must not, however, let ourselves be carried away. Kozai cycles can be quenched by perturbations, apart from the third-body perturbation, that cause apsidal motion at much the same rate as the third body does. Rotation, mutual distortion and GR are all capable of doing this, if strong enough; but they all drop off fairly rapidly with the inner separation. Therefore, for a given inner binary, there will be a maximum size of outer orbit that can generate Kozai cycles, but this may still be several thousand times larger than the inner orbit.

Figure 4.9 relates to one or even both sub-components of the remarkable quadruple sytem ADS 11061 (Tokovinin *et al.* 2003). This consists of four rather similar late-F dwarfs, marking out the turnoff region of an isochrone of $\sim$2.5 gigayears. One orbit is long and thin (1274 days, $e = 0.9754$), the other smaller and rounder (10.5 days, $e = 0.374$). The outer orbit is not known, though it can reasonably be estimated to be $\sim 10^4$ years. Its inclination is almost certainly different from those of the two sub-binaries, which differ from each other. Assuming a range of initial parameters for all three orbits, we get a wide variety of possible scenarios. In some, neither orbit is much affected by Kozai cycles, but in others one or both orbits are affected, perhaps seriously. Figure 4.9 is a possible model of the long, thin orbit. Starting with parameters as listed in the caption, the eccentricity cycles powerfully with e fluctuating between $\sim$0 and $\sim$0.98, and the inclination between $70°$ and $86°$. But in the close periastra at

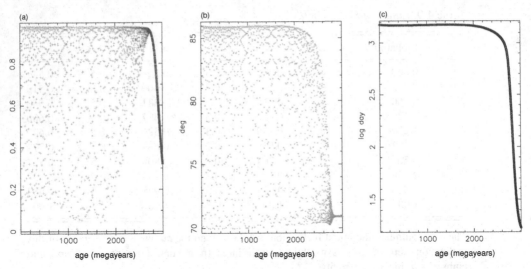

Figure 4.9 Possible orbital evolution for 41 Dra, the more eccentric sub-binary of ADS 11061. Initial conditions: $P_1 = 1500$ days, $e_1 = 0.01$ (inner orbit), $P = 15\,000$ years, $e = 0.73$ (outer orbit), mutual inclination $86°$. (a) Eccentricity. (b) Inclination relative to outer orbit. (c) Period. Individual Kozai cycles are about 20 megayears long, and are severely undersampled by the plotting processss.

the peaks of eccentricity, tidal friction after ~ 2 gigayears reduces the range considerably (but not its *upper* limit), and by 2.7 gigayears the eccentricity though still large starts to diminish rapidly. By 3 gigayears the orbit is much smaller and only moderately eccentric, as is observed in the *other* sub-binary. The inclination (Fig. 4.9b) cycles intimately with the eccentricity, and the period (Fig. 4.9c) drops, though to 20 days rather than 10 days as seen in the other system. Either or both sub-binaries may have suffered, or be suffering, such evolution, for all one can tell at present.

A further effect of a third body, which is smaller but can have longer-term importance, comes from a combination of third-body perturbation and tidal friction even in the case that the orbits are coplanar. In an unperturbed binary, tidal friction simply circularises the orbit, which then remains circular so that the frictional dissipation goes to zero. But in an orbit that is continually being perturbed by a third body, circularisation cannot be completed because fresh eccentricity (though probably of small magnitude) is always being added. Tidal friction tries to remove eccentricity, because it tries to dissipate time-dependent tides such as are raised by eccentric orbits. Thus the conversion of mechanical energy into heat continues unabated, though no doubt rather slowly, leading to a secular decrease in the orbital period. Since, in the triple system as a whole, angular momentum should be conserved, loss of orbital energy and consequential loss of angular momentum, from the close pair will result in a widening of the wide pair, but on a still smaller scale since the wide pair will have greater angular momentum to start with.

Unfortunately, this process is not well modelled by the equations above; nor would going to a higher order (say, octupole) approximation help. If, for example, we consider an outer orbit exactly parallel to the inner orbit, the off-diagonal components of the S-tensor are zero: hence $V = 0$ – Eq. (4.100) – and so $\mathbf{e}$ is unchanged in magnitude – Eq. (4.100). But in such a triple there will in fact be small fluctuations in e, on the timescale of the inner orbit as well as on longer timescales. The basis of the above approximation, and several others in this chapter, is that we determine the effect of a small perturbation by integrating around an *exactly* Keplerian orbit, assuming the orbit changes insignificantly on the orbital timescale. To model the process of the previous paragraph we need a full three-body code, rather than an orbitally-averaged approximation; but we also need to include the quadrupolar gravity perturbation of intrinsic spin and mutual distortion.

Using a three-body code, we can investigate tentatively the root-mean-square eccentricity fluctuation introduced into the inner orbit, on a timescale that is comparable to the inner orbit. A preliminary estimate is

$$e_{TB} \sim \frac{M_3}{M + M_3} \frac{1}{X^2}, \qquad X \equiv \frac{P_{out}}{P}. \tag{4.102}$$

The effect on the binary can then be modelled by adding to Eq. (4.12), in the limit $e \ll 1$, a source term on the right:

$$e\dot{e} \approx -\frac{7e^2}{2t_{TF}} + \frac{1}{P} \left(e_{TB}^2 - e^2 \right). \tag{4.103}$$

In the absence of tidal friction this means, as we require, that $e^2 \to e_{TB}^2$ on an orbital timescale. The equilibrium value will be little affected by tidal friction, in fact, because typically $t_{TF} \gg P$. The effect of the small $e^2 \sim e_{TB}^2$ propagates through Eq. (4.17), in transient equilibrium, to give Ω/ω – Eq. (4.18) – and hence to give $\dot{h}/h$ from Eq. (4.13) as

$$\frac{\dot{h}}{h} \sim -\frac{15\lambda e_{TB}^2}{2t_{TF}}, \tag{4.104}$$

where λ is as usual the ratio of spin to orbital angular momentum. For semidetached or contact binaries, where t_{TF} is about as small as possible, since $a \sim 2R$ in Eq. (4.10), this effect can be significant if $P_{out} \lesssim 100P$.

Georgakarakos (2003) has determined a much more substantial estimate than (4.102). For a general outer eccentricity his expression is some ten lines long, and so for illustration we specialise here to circular outer orbits. He gives

$$e_{TB}^2 \sim \left(\frac{M_3}{M + M_3} \right)^2 \left(\frac{225}{128} \frac{M_*^2}{X^{10/3}} + \frac{43}{4} \frac{1}{X^4} + \frac{122}{3} \frac{1}{X^5} + \frac{20961}{4096} \frac{M_*^2}{X^{16/3}} + \frac{365}{9} \frac{1}{X^6} \right), \tag{4.105}$$

where X is the period ratio, as above, and

$$M_* \equiv \frac{M_2 - M_1}{M^{2/3} (M + M_3)^{1/3}}. \tag{4.106}$$

The angular momentum loss rate is still given by Eq. (4.104).

We therefore have 'third-body' modes of secular orbital change, which we label mode TB. There are at least two rather different modes, one involving Kozai cycles with tidal friction operating only near their peaks in eccentricity, and one involving possibly coplanar (or nearly coplanar) systems where the third body is unusually close. We believe that the Kozai mechanism might be quite common, and the other less common, and we do not attempt for the present to distinguish them.

However, precession and apsidal motion, although conservative, can be important in relation to observed properties of binaries. A handful of binaries is known where eclipses have been seen over some stretch of time and not seen over some other stretch of time (Fig. 4.8). This is presumably because of precession, and although precession can also be caused by oblique rotation of one or both components (Section 3.2.2) it is more likely to be caused by a third body, because spin angular momentum is normally small compared to orbital whereas the third-body effect can be large. There is also a handful of close binaries where apsidal motion can be measured and is found to be discrepant with estimates based on Eq. (3.48). Here it may also be the effect of a third body that is causing the discrepancy.

The effect of $\mathbf{U}$, the rotation rate of the orbital frame, on the inclination and apsidal motion of the orbit can be seen as follows, and independently of whatever mechanism is causing the frame to rotate. Let $\bar{\mathbf{J}}$ be a unit vector pointing from the centre of gravity of the orbit to the observer. This is a fixed vector in an inertial frame (apart from a small contribution from the acceleration of the binary around the centre of gravity of the triple, which can however also be allowed for), but it is a variable vector in the orbital frame $\bar{\mathbf{e}}, \bar{\mathbf{q}}, \bar{\mathbf{h}}$. If i is the inclination of the orbit to the line of sight and if ω_{lp} is the longitude of periastron – we use ω_{lp} rather than the more conventional ω since the latter is used here for the mean angular velocity of the orbit – then these are given generally by

$$\cos i = \bar{\mathbf{h}} \cdot \bar{\mathbf{J}} \quad \text{and} \quad \tan \omega_{\mathrm{lp}} = \frac{\bar{\mathbf{e}} \cdot \bar{\mathbf{J}}}{\bar{\mathbf{q}} \cdot \bar{\mathbf{J}}}. \tag{4.107a,b}$$

Differentiating with respect to time in the inertial frame, we obtain, after some manipulation,

$$\frac{di}{dt} = \frac{\mathbf{U} \times \bar{\mathbf{h}} \cdot \bar{\mathbf{J}}}{|\bar{\mathbf{h}} \times \bar{\mathbf{J}}|}, \quad \frac{d\omega_{\mathrm{lp}}}{dt} = \frac{\mathbf{U} \cdot \bar{\mathbf{h}} - \bar{\mathbf{J}} \cdot \bar{\mathbf{h}} \, \bar{\mathbf{J}} \cdot \mathbf{U}}{|\bar{\mathbf{h}} \times \bar{\mathbf{J}}|^2}. \tag{4.108a,b}$$

I have used the fact that $\bar{\mathbf{e}}, \bar{\mathbf{h}}$ satisfy the same equations (Eqs (4.96)), as $\mathbf{e}, \mathbf{h}$, except that the V, W terms are absent; and $\bar{\mathbf{q}}$ satisfies a similar equation.

In the event that the rotation rate $\mathbf{U}$ is purely in the $\bar{\mathbf{h}}$ direction, i.e. that $\mathbf{U} = Z\bar{\mathbf{h}}$, Eq. (4.110) gives $d\omega_{\mathrm{lp}}/dt = Z$, in other words the rate of rotation of the line of apses is just Z. However, it is instructive to note that if $\mathbf{U}$ has $\bar{\mathbf{e}}, \bar{\mathbf{q}}$ components as well, corresponding to precession, then $d\omega_{\mathrm{lp}}/dt \neq Z$. Thus the quantity Z, which is often referred to as 'apsidal motion', is not, in fact, the only contributor to this effect. I believe therefore that it is inappropriate to describe Z as 'apsidal motion', although this is commonly done. Equations (4.109 and 4.110), and this comment, apply irrespective of whether $\mathbf{U}$ is caused by a third body or to any other process, such as oblique rotation.

Table 4.10 shows some of the rich variety of triples that can be found; two are quadruples. All have been selected from the minority of systems in which the *outer* period is $\lesssim 30$ years, so that there is some probability that not only the inner pair but even the outer pair may interact in the course of evolution. Six of the systems, those with Greek letters, as well as three more similar triples (ξ Tau, κ Peg and p Vel), are among the brightest 500 stars, a set that may be reasonably representative of stars with masses $\gtrsim 2\, M_\odot$. Thus it seems possible that such compact multiple systems may represent $\sim 2\%$ of systems (including single stars as 'systems'), although of course multiples may be somewhat overrepresented in a magnitude-limited sample. It would not be surprising if a further few systems of these 500 are similarly multiple, given particularly the difficulty of recognising small third bodies such as M dwarfs at separations of a few AU from binary B/A companions.

Table 4.10. *Some close triple and quadruple systems*

β Per	((K0-3IV + B8V; 0.8 + 3.7 $M_\odot$; SD, 2.87 days) + F1V; 4.5 + 1.7 $M_\odot$; 1.86 years, $e = 0.23$)
β Cap	((B8V + ?; 3.3 + 0.9 $M_\odot$; 8.68 days) + K0II-III; 4.2 + 3.7 $M_\odot$; 3.76 years, $e = 0.42$)
λ Tau	((A4IV + B3V; 1.9 + 7.2 $M_\odot$, SD, 3.97 days) + ?; 9.1 + 0.7: $M_\odot$; 0.09 years = 33 days, $e = 0.15$:)
η Ori	((B1V + B3V, 15 + 12 $M_\odot$; 7.98 days) + B1V; 27 + 14 $M_\odot$; 9.5 years, $e = 0.43$)
DL Vir	((? + A3V, 1.1: + 2.2 $M_\odot$; SD, 1.32 days) + G8III, 3.9 + 1.9: $M_\odot$; 6.2 years, $e = 0.44$)
CQ Dra	((WD + ?; SD, 0.16 days) + M3III; 4.7 years, $e = 0.3$, $f = 0.0076 M_\odot$)
VV Ori	((B1V + B5V, 10.8 + 4.5 $M_\odot$; 1.49 days) + A3:, 15.3 + 2.3: $M_\odot$; 0.33 years, $e = 0.3$)
DM Per	((A6III + B5; 1.8 + 5.8 $M_\odot$; SD, 2.73 days) + B7:; 7.6 + 3.6: $M_\odot$; 0.27 years)
SU Cyg	(F2-G0I-II, δ Cep + (B7.5HgMn + A0:;3.2 + 2.6 $M_\odot$; 4.675 days); 6.2 + 5.8 $M_\odot$; 549 days, $e = 0.34$)
τ CMa	(O9II + (B: + B:;1.28 days); 0.42 years, $e = 0.29$, $f = 6.1 M_\odot$)
V907 Sco	((B9.5V + B9.5V; $P_1 = 3.78$ days) + ?; $P = 99.3$ days, $f = .004 M_\odot$)
μ Ori	((A7m + ?;1.8 + 1 $M_\odot$; 4.45 days) + (F3V + F3V, 1.4 + 1.4 $M_\odot$; 4.78 days); 2.8 + 2.8 $M_\odot$; 18.8 years, $e = 0.76$)
QZ Car	((O9.7Ib + B2V:; 40: + 9: $M_\odot$; 20.7 days, $e = 0.34$) + (O9V + B0Ib; 28 + 17 $M_\odot$; 6.00 days); 49: + 45 $M_\odot$; $\lesssim 25.4$ years, $\lesssim 0.012''$)

References: DL Vir – Schöffel (1977); CQ Dra – Reimers *et al.* (1988); DM Per – Hilditch *et al.* (1986); SU Cyg – Evans and Bolton (1990); τ CMa – Stickland *et al.* (1998), van Leeuwen and van Genderen (1997); V907 Sco – Lacy *et al.* (1999); QZ Car – Morrison and Conti (1980); others from Fekel (1981). Where the eccentricity is not given, it is zero to observational accuracy; SD stands for semidetached.

Note that outer periods can be as small as 33 days (λ Tau), that β Cap and DL Vir contain *two* red giants, that SU Cyg contains a Cepheid pulsator and that CQ Dra contains a possible cataclysmic binary. SU Cyg and τ CMa have a third body more massive than the combined mass of the close pair, but it seems more normal that the third body is the least massive of the three, or else that all three are of comparable mass. The last may be a selection effect, since such systems are easiest to recognise. From the point of view of dynamical interaction by Kozai cycles, what matters most is the inclination between the outer orbit and the inner orbit or orbits. This is usually not well known, although in β Per it is a well-determined $100°$. However, Kozai cycles can also be important in the many more systems where the outer period is up to $\sim 10^3$–10^4 years.

4.8 Old Uncle Tom Cobley and all

We can now put together most of the various slow orbital perturbations discussed in this chapter. We have seen that many of them have a somewhat similar mathematical form, to the extent that they cause a rotation rate $\mathbf{U} \equiv X\overline{\mathbf{e}} + Y\overline{\mathbf{q}} + Z\overline{\mathbf{h}}$ of the orbital ($\overline{\mathbf{e}}$, $\overline{\mathbf{q}}$, $\overline{\mathbf{h}}$) frame, and also variations of the orbital triad $\mathbf{e}$, $\mathbf{q}$, $\mathbf{h}$ parallel to themselves. We include specifically GR (both as a conservative and a non-conservative process), quadrupolar distortion due to both rotation and a companion, tidal friction, mass loss (possibly with magnetic braking), mass exchange and a third body. It is somewhat unlikely that all these processes are operating significantly in the same binary at the same time, but it is convenient to have to hand a computer code that can include any of them as necessary.

We can write the combination as

$$\frac{d\mathbf{e}}{dt} = \mathbf{U} \times \mathbf{e} - V\mathbf{e}, \tag{4.109}$$

$$\frac{d}{dt}\mathbf{H}_o = \mathbf{U} \times \mathbf{H}_o - W\mathbf{H}_o - \left[\frac{\zeta_1 M_2}{M M_1} + \frac{\zeta_2 M_1}{M M_2}\right]\mathbf{H}_o, \quad \mathbf{H}_o \equiv \mu\mathbf{h}, \tag{4.110}$$

$$\frac{d}{dt}I_1\mathbf{\Omega}_1 = -\mathbf{U}_1 \times \mathbf{H}_o + W_1\mathbf{H}_o - \zeta_1\left[R_{A1}^2 + \frac{2}{3}R_1^2\right]\mathbf{\Omega}_1, \tag{4.111}$$

$$\frac{d}{dt}I_2\mathbf{\Omega}_2 = -\mathbf{U}_2 \times \mathbf{H}_o + W_2\mathbf{H}_o - \zeta_2\left[R_{A2}^2 + \frac{2}{3}R_2^2\right]\mathbf{\Omega}_2. \tag{4.112}$$

$\mathbf{U} \equiv X\overline{\mathbf{e}} + Y\overline{\mathbf{q}} + Z\overline{\mathbf{h}}$ is the sum of $\mathbf{U}_1, \mathbf{U}_2$ in Section 4.2 (tidal friction and distortion), and also of terms due to GR – Eq. (3.50) – and to a third body – Eq. (4.97). V, W are similar sums, although the GR terms now come from Section 4.1. A term can be added into V to give the additional third-body effect of Eq. (4.104). The mass loss rates (to infinity) ζ_1, ζ_2 can be obtained from Section 4.5, or from some other model. The prescription of Section 4.5 includes a model for binary-enhanced wind (mode EW), because it makes the ζ depend on the rotation rates, which are often much faster for components of binaries than for single stars if tidal friction is significant. The Alfvén radii R_{A1}, R_{A2} can also be obtained from Section 4.5, or from some other model. The masses M_1, M_2, M, μ are all now possibly variable, so that we have to add to the ensemble Eqs (4.36), which allow for mass transfer as well as mass loss to infinity. The moments of inertia are also possibly variable, not only because of varying masses but also because of evolution. The entire ensemble of differential equations can be integrated by a Runge–Kutta procedure, as is done for Figs. 5.2 and 6.3, although care should be taken because in some circumstances (e.g. when tidal friction is so strong that it enforces very-near corotation) the equations can become quite 'stiff'.

For some examples computed in this book we use two different codes. One consists only of the above equations, supplemented by very simplistic approximations to stellar evolution that give radius and luminosity explicitly as functions of mass and time; for example Eqs (2.49) and (2.50) for red giants. The other is a full stellar evolution code, in which, for example, a differential equation for the moment of inertia is added to the set for pressure balance, heat transport, etc. In this code Eqs (4.109)–(4.112) are added as boundary conditions; for the present they are only scalar, i.e. $\mathbf{U}$ is ignored and all the vectors are assumed parallel.

In Eqs (4.109)–(4.112) the term for angular momentum loss due to winds has been split into two parts, one because winds, even without magnetic braking, carry off *orbital* angular momentum (Eq. 4.110), and the other because magnetic braking carries off *spin* rather than orbital angular momentum Eq. (4.111). In the case that tidal friction is strong, Eq. (4.111) leads to a transient equilibrium between the tidal friction term $W_1\mathbf{H}_o$ and the magnetic-braking term (the rotational term $\mathbf{U}$ being usually unimportant in such a case). Substituting this, and the equivalent from *2, into Eq. (4.110), we see that in effect the angular momentum is drained directly from the orbit. An example was given in Fig. 4.6.

5

Rapid non-conservative processes

There are at least four processes that might change an orbit radically on a short timescale. These are (a) the Darwin instability, in binaries with a rather extreme mass ratio, (b) hydrodynamic mass transfer in a semidetached configuration, e.g. where the loser is convective and more massive than the gainer (modes SF3 or SR3 of RLOF), (c) a supernova explosion in one component and (d) a dynamical encounter with a previously independent star or system, during a chance fly-by in a dense cluster of stars. In fact, I feel that there is need for at least a fifth process, which I shall tentatively identify in the section that deals with (b).

5.1   Tidal friction and the Darwin instability: mode DI

Although tidal friction attempts to dissipate relative motion and so lead to a state of uniform rotation, with the components corotating with the binary in circular orbits, there is no guarantee that such a state is actually attainable. Consider a detached binary with a circular orbit, where *2 is sufficiently small that its moment of inertia can be neglected but *1 is not. In the absence of winds, the total angular momentum H, if *1 corotates, is given by

$$\frac{H}{M_1} = \omega k^2 + \frac{(GM)^{2/3}}{\omega^{1/3}(1+q)} = \text{constant}, \tag{5.1}$$

where k is the radius of gyration of *1. The second term is the orbital contribution: it comes from Eq. (3.14). Through evolution, k may grow with time. However, Eq. (5.1) when differentiated wrt ω at constant H, M_1, M_2 shows that k, considered as a function of ω, has a maximum value, say k_0, which occurs when

$$\omega k_0^2 = \frac{1}{3}\frac{(GM)^{2/3}}{\omega^{1/3}(1+q)}, \tag{5.2}$$

i.e. when the ratio λ – Eq. (4.15) – of spin to orbital angular momentum is 1/3. Hence if k grows beyond k_0, corotation must break down (Darwin 1879, Counselman 1973, Pringle 1974). The star, by growing, takes angular momentum from the orbit, which thereby shrinks and rotates faster, making it impossible, beyond some point, for the star to keep up. Of course if *1 fills its Roche lobe before $k = k_0$, the conventional picture of corotation until Roche-lobe overflow can be sustained. Using Eq. (3.18), the critical condition for an $n = 3$ polytrope is

$$\frac{1}{3}\frac{a^2}{1+q} = k_0^2 \approx 0.076\, a^2 x_L^2(q). \tag{5.3}$$

With $x_L(q)$ given by Eq. (3.5), this reduces to $q \equiv q_D \approx 12$ (Table 3.4). For $q > q_D$, tidal friction cannot maintain corotation as the star expands all the way to its Roche lobe. As shown in Section 4.2, we expect the system to move out of corotation in this case, which means also that the eccentricity will depart from zero. The likely outcome is that the smaller body will plunge into its companion, and experience a variant of common envelope evolution (next section): either the system will merge, ∗2 becoming smeared out in the envelope of ∗1, or enough orbital energy may be released for the envelope to be blown away leaving a much closer binary. For $n = 1.5$, as might be more appropriate for red giants, $q_D \sim 5$; for $n \sim 3.5$–4, more appropriate to substantially evolved main-sequence stars, $q_D \sim 18$–30.

If the orbit is already eccentric, then Eq. (4.17) shows that the instability will set in when the ratio λ of spin to orbital angular momentum is a function of e. For $e = 0.4$ we obtain $\lambda = 0.18$. In principle, we can integrate the system of Eqs (4.11)–(4.14) in order to follow the way in which the instability develops, as for the two examples given in Fig. 4.2. However, the model presented here and in Section 4.2 assumes that the star stays in *uniform* rotation, and it is not clear that the whole star will in fact spin up uniformly as a result of tidal friction: another possibility is that the surface spins up relatively rapidly while the interior spins up more slowly. Non-uniform rotation is thought to be unstable on something like a thermal timescale, but the development of the Darwin instability may be on a faster timescale.

5.2   Common envelopes and ejection: modes CE, EJ

We have already seen that rapid (hydrodynamic) mass loss, i.e. mode 3, is expected if the loser is a red giant with a convective envelope, and has more than ~ 0.66 of the gainer's initial mass. It is very unlikely that the gainer could accrete at anything like this rate, so that much or all of the envelope's mass may end up in a halo round the binary, and possibly, though not certainly, be expelled. It will be hard to make this process more precise. Paczyński (1976) suggested a 'common-envelope' process, in which the material lost very rapidly by ∗1, too rapidly to settle easily on ∗2, will collect in a differentially rotating envelope around *both* stars. Unlike the envelope of a normal contact binary, which is expected to be in hydrostatic equilibrium and uniform corotation, and therefore limited by the *outer* Roche-lobe radius – Eq. (3.13) – the common envelope hypothesised following mode 3 RLOF is differentially rotating, and therefore not restricted by the outer Roche-lobe radius. The formation of the envelope is envisaged as so rapid that tidal friction will be irrelevant. Ordinary dynamical friction will occur between ∗2 and the common envelope through which it moves, and this will transfer angular momentum from the orbit to the envelope, but not so efficiently as to bring the envelope into corotation with the orbit. Thus the mutual orbit of ∗2 and the *core* of the red giant or supergiant loser (∗1) can shrink, perhaps by a large factor, while the common envelope remains at about the same size as the binary was at the onset of RLOF. The timescale of this orbital shrinkage, though short compared with even the thermal, let alone the nuclear timescale of ∗1, will probably be long compared with the orbital period, so that ∗2 will 'spiral in' towards the core of ∗1 in a fairly tight spiral.

An order-of-magnitude estimate is that when ∗2 has spiralled in to distance a from the centre of ∗1, the drag force on ∗2 will be

$$F_{\text{drag}} \sim \rho v^2 R_L^2, \tag{5.4}$$

with v the orbital velocity, and ρ the ambient density, which we take to be the unperturbed density at radius $r \sim a$ within a red-giant envelope. R_L is the Roche-lobe radius (shrinking, along with the orbit), and R_L^2 represents roughly the cross-sectional area, since the fluid motion will be seriously affected by *2's gravity out to distance R_L. The drag luminosity will be

$$L_{\text{drag}} \sim v F_{\text{drag}},\qquad(5.5)$$

and we equate this roughly to the rate of loss of energy from the orbit:

$$L_{\text{drag}} \sim \rho v^3 R_L^2 \equiv \frac{GM^2}{at_{\text{CE}}},\qquad(5.6)$$

where t_{CE} is an estimate of the common-envelope timescale. Taking $R_L \sim a$, $v^2 \sim GM/a$, and $P \sim 2\pi a/v$, we obtain

$$\frac{t_{\text{CE}}}{P} \sim \frac{M}{2\pi\rho a^3}.\qquad(5.7)$$

Equation (2.44) shows that in the extensive radiative inner portion of a red-giant envelope,

$$\rho r^3 \sim \text{constant} \sim M_{\text{shell}},\qquad(5.8)$$

where M_{shell} is the mass in the nuclear-burning shell. This decreases as the red giant evolves, being $\sim 10^{-2}M_c$ near the base of the giant branch and $\sim 10^{-6}M_c$ near the AGB tip. Since we are assuming as the crudest approximation that $r \sim a$, i.e. that when the companion has moved to a separation a it encounters material with a density comparable to that in the unperturbed red giant at radius $r \sim a$, we expect a tight spiral in a highly evolved red giant, but a much less tight one lower down the giant branch. In the convective outer portion of a red-giant envelope, the spiral should also be less tight.

It is not clear a priori where this process will end. It is possible that *2 will spiral in so close to *1's core that it will simply be smeared out, adding to the envelope's mass but losing its separate identity; this process will be called a 'merger'. This is likely to produce a single star that is very rapidly rotating, at least to start with. But there may be several types of merger, since although one component is probably a red giant the other might be either a main-sequence star or a white dwarf or neutron star.

Alternatively, if the gravitational energy released by the orbit's shrinkage is dumped sufficiently rapidly into the common envelope, the envelope may be blown away. A condition which is presumably necessary, though it can hardly be sufficient, is that the energy released as the orbit shrinks is greater than the binding energy of the envelope. We can write this roughly as

$$\alpha_{\text{CE}}\frac{GM_2}{2}\left(\frac{M_c}{a'} - \frac{M_1}{a}\right) = E_B \equiv \int_{M_c}^{M_1}\left(\frac{Gm}{r} - U\right)\,dm,\qquad(5.9)$$

with the prime referring to the final state, when M_1 has been reduced to M_c, E_B being the binding energy of the envelope – Eq. (2.62) – and α_{CE} being a factor to allow for the fact that some of the energy will be radiated away rather than channelled into unbinding the envelope. If this condition is satisfied, then it seems possible that the outcome of mode CE is a binary of shorter, perhaps much shorter, period than the original, surrounded temporarily by a planetary nebula which consists of the ejected common envelope glowing in the UV radiation of the hot core of *1. However, it is quite possible that not enough energy is liberated, and the outcome is a merger: *2 becomes smeared out in the deeper regions of *1.

We have noted that if stars are evolved without mass loss the binding energy E_B may become negative at some point on the AGB. We interpreted that in Sections 2.3 and 2.4 as a crude indication that the envelope would be lost spontaneously at about that stage in the evolution of a single star. Evidently it would also imply that at a somewhat earlier stage the envelope could be rather easily blown away by the common-envelope interaction, with very little contraction of the binary.

E_B is often modelled as

$$E_B = \frac{GM_1 M_e}{\lambda R_1},$$
(5.10)

where the envelope mass is $M_e = M_1 - M_c$ and λ is some coefficient of order 0.5 (de Kool 1990). However, this may be too simple: a precise definition of E_B is very unclear (Section 2.4), partly because it is uncertain where to define the boundary between core and envelope, and partly because it is unclear whether to include all, some, or none of the *thermal* energy with the gravitational energy – Eq. (5.9) assumes that all the thermal energy (U) is to be included.

Dewi and Tauris (2001) list five possible definitions of the core. They are

(a)   the point where the H-burning energy generation rate has a maximum
(b)   the point where $X_H = 0.1$
(c)   the location of an inflection in the $\rho(m)$ distribution
(d)   the place where E_B (viewed as a function of M_c), after varying slowly with M_c in the outer layers starts to increase rapidly as M_c decreases towards the He-burning shell
(e)   the base of the convective envelope.

The first of these usually gives the smallest, and the last the largest, core mass M_c. We can probably rule out (e) because for stars that have not yet reached the Hayashi track, even if only by a very narrow margin, it gives hardly any envelope mass. But the others can still give a range of about 10% in core mass.

To elaborate, consider an envelope (Fig. 5.1) that can be approximated as two layers with different $\rho(r)$ power laws:

$$\rho = \rho_0 \left(\frac{R_c}{r}\right)^\alpha, \quad R_c \le r \le R_{ab}, \quad \text{and} \quad \rho = \rho_0 \left(\frac{R_c}{R_{ab}}\right)^\alpha \left(\frac{R_{ab}}{r}\right)^\beta, \quad R_{ab} \le r \le R_1.$$
(5.11)

The gravitational part of the binding energy of this envelope, defined as positive, has the form

$$E_G = \frac{GM_\alpha^2}{R_c} W_1(R_{ab}/R_c, \alpha) + \frac{GM_\alpha M_c}{R_c} W_2(R_{ab}/R_c, \alpha)$$

$$+ \frac{GM_\beta^2}{R_{ab}} W_1(R_1/R_{ab}, \beta) + \frac{GM_\beta(M_c + M_\alpha)}{R_{ab}} W_2(R_1/R_{ab}, \beta),$$
(5.12)

where

$$W_1(x, \alpha) \equiv \frac{2F(x, 5 - 2\alpha) - x^{2-\alpha} F(x, 3 - \alpha) - F(x, 2 - \alpha)}{F^2(x, 3 - \alpha)},$$
(5.13)

$$W_2(x, \alpha) \equiv \frac{F(x, 2 - \alpha)}{F(x, 3 - \alpha)},$$
(5.14)

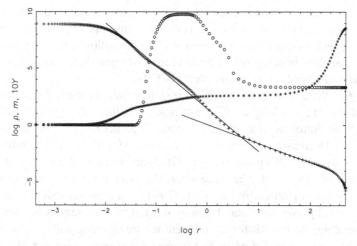

Figure 5.1 The distribution of $\log \rho$ (kg/m^3, pluses), m (Solar units, asterisks) and $10Y$ (circles) in a red supergiant of $\sim 9 M_\odot$, as a function of radius (Solar units). The helium burning shell is fairly well-defined in the region 0.06–$0.15 R_\odot$, but the hydrogen/helium interface, with no burning, is spread out over 0.6–$5 R_\odot$. Straight lines of slope -3.8 and -1.7 have been drawn by hand.

and

$$F(x, \gamma) \equiv \frac{x^\gamma - 1}{\gamma} \quad (\gamma \neq 0), \quad \equiv \ln x \quad (\gamma = 0). \tag{5.15}$$

The masses M_α, M_β in the two parts of the envelope (of total mass M_e) satisfy

$$\frac{M_\alpha}{M_\beta} = \left(\frac{R_c}{R_{ab}}\right)^{3-\alpha} \frac{F(R_{ab}/R_c, 3 - \alpha)}{F(R_1/R_{ab}, 3 - \beta)}, \quad M_e = M_\alpha + M_\beta. \tag{5.16}$$

It is not especially easy to see which terms dominate, since this depends rather critically on the values of α, β relative to the (removable) singularities at 2, $5/2$ and 3. But probably in most circumstances we can assume that $\alpha > 3$, $\beta < 2$, and we can also usually assume that $R_c \ll R_{ab} \ll R_1$. This leads to $M_\alpha \ll M_\beta \sim M_e$, i.e. to

$$\frac{M_\alpha}{M_e} \sim (R_{ab}/R_c)^{\alpha-3}(R_{ab}/R_1)^{3-\beta} \frac{3 - \beta}{\alpha - 3} \ll 1, \tag{5.17}$$

and in addition the core is comparable in mass to the envelope so that $M_\alpha \ll M_c$. Then the leading terms in E_G are

$$E_G \sim \frac{GM_\alpha M_c}{R_c} (\alpha - 3) + \frac{GM_e^2}{R_1} \frac{3 - \beta}{5 - 2\beta} + \frac{GM_e M_c}{R_1} \frac{3 - \beta}{2 - \beta}. \tag{5.18}$$

Although M_α is normally small, so is R_c, and therefore it is not clear that the first term in Eq. (5.17) can be neglected relative to the second and third (the last two being comparable to each other). In fact the crux of the matter is where one thinks the core ends and the envelope begins.

In the star shown in Fig. 5.1, an $8.6 M_\odot$ red supergiant on the verge of non-degenerate carbon ignition, we can fairly clearly see that $\alpha \sim 3.8$, $\beta \sim 1.7$ and that $R_{ab} \sim 5 R_\odot$, $R_1 \sim 400 R_\odot$.

But we might see the boundary of the core as anywhere between the outer edge of the helium-burning shell ($R_c \sim 0.15\,R_\odot$) and the base of the convective envelope ($R_c \sim 5\,R_\odot$). There is no hydrogen-burning shell in this model because the H/He transition is too cool, and there is not even a sharp boundary in composition because the hydrogen shell, though covering a narrow range of mass, is spread out in radius over 0.6–$5\,R_\odot$.

We can think of R_{ab}, R_1, β, α as given, while altering the depth of the envelope by varying R_c and consequentially M_α. As long as R_c is not much less than $R_{ab} \sim 0.01 R_1$, the last two terms in Eq. (5.18) dominate, and are comparable. Together they can be estimated to correspond to $\lambda \sim 0.6$. Decreasing R_1, which also increases M_α, will have little effect until $R_c \sim 0.3 R_{ab} \sim 0.003 R_1$; but at this point $M_\alpha \sim 0.001 M_e$ and so as R_c decreases further the first term rapidly becomes dominant. The place where the first term starts to be important – definition (d) above – was suggested by Han *et al.* (1994) as a reasonable definition of the boundary, and is the one I prefer here; but obviously it is not the only reasonable definition. By definition the resulting E_G (which for the moment we are equating with E_B, i.e. we are ignoring the thermal energy contribution to E_B) is sensitive to using a smaller M_c, though not a larger.

Dewi and Tauris (2001) show that λ can be several times larger than ~ 0.5 if the thermal energy is included. In fact $\lambda \to \infty$, obviously, if we include all of the thermal energy, and allow the star to evolve to the point where the binding energy becomes zero. I showed in Section 2.3 that if single stars lose their envelopes at about this point we get a reasonable initial/final mass relation. Clearly if a single star can lose its envelope at this point, a binary companion will have little difficulty in removing the envelope slightly before this point. We can view λ as a function not only of the uncertain M_c, but also of an uncertain factor α_{th} such that our definition of E_B is revised to

$$E_B \equiv \int_{M_c}^{M_1} \left(\frac{Gm}{r} - \alpha_{th} U \right) \, \mathrm{d}m, \tag{5.19}$$

with $0 \le \alpha_{th} \le 1$.

Supposing for the moment that we ignore these uncertainties, and consider M_c, α_{th}, λ and α_{CE} as known, we can rewrite Eqs (5.9) and (5.10) to give the final value a' of the separation as

$$\frac{a'}{a} = \frac{M_c}{M_1} \frac{M_2}{M_2 + 2M_e/(\alpha_{CE}\lambda x_L)}, \qquad x_L(M_1/M_2) = \frac{R_1}{a}. \tag{5.20}$$

We are assuming that ∗1 just fills its Roche lobe at the start of this process: the function x_L is given by Eq. (3.5). At least in this formula the uncertainties are compounded into a single parameter, the quantity $\lambda \alpha_{CE}$ (λ itself incorporating uncertainties in M_c and α_{th}). Until these uncertainties can be reduced by fully three-dimensional modelling of the complicated gas dynamics (which may well be influenced by MHD), it is probably necessary, for the present, to treat $\lambda \alpha_{CE}$ as a free parameter that can best be estimated by seeing what values will give reasonable agreement, statistically, with the observed distribution of post-common-envelope systems.

That the energetics of ejecting gas from a binary is over-simplified here is seen by the stark contradiction between the result (5.20) in the case $\lambda \to \infty$, i.e. when the envelope is not in fact bound at all, and Eq. (4.35), which estimates the final period when one star loses all its envelope to infinity as a result of a spherically symmetrical wind. Equation (4.35) predicts that the separation *increases*, inversely proportional to the decreased total mass, on the basis (ultimately) of an angular momentum consideration; on the other hand Eq. (5.20) predicts that it *decreases* on the basis of an energy consideration, that the orbit conserves its energy (while remaining circular), since no energy is required to remove the envelope in this case. In reality, we should consider what happens to both angular momentum and energy, and should remember that a lot of energy is available, in principle, from the nuclear supply of the star. It is absolutely clear that the physics here is far from complete.

There have been several numerical attempts to model mode CE: see Taam and Sandquist (2000). Those that are three-dimensional appear to give $\alpha_{CE} \sim 0.3$–0.5. But our understanding at present can only be very tentative. Most attempts so far start somewhere in the middle of the spiral-in process, and it is not clear that an actual spiral in will pass through this state. We will need fully three-dimensional modelling of the entire process, starting from just before RLOF and ending when the envelope clears away; we cannot expect such modelling soon.

Table 5.1 is a collection of binaries that appear to be related to mode CE. Not all have short periods, because we wish to emphasise the fact that arguably rather similar *initial* systems seem to produce wide as well as close binaries. Mode CE probably produces mergers as well, but it is difficult to see how one would clearly recognise a single star as the merged remnant of a binary. Let us concentrate here on *detached* binaries, thus excluding CVs and LMXBs, on the grounds that if no mass transfer has yet taken place (*after* the mode CE interaction, of course) then we might hope for a clearer picture of the mode CE transition.

The first group of systems in the table are located in planetary nebulae, which suggests that they have only recently experienced mode CE. One component in each system is indeed very hot (SDO), and presumably powers the nebula. The next group is somewhat heterogeneous, with one component being apparently hot and undermassive (relative to the main sequence), and the other arguably a fairly normal unevolved or moderately evolved star. The third group contains a white dwarf and a relatively normal star; and the fourth group contains *two* highly evolved objects, and might be perceived as the outcome of two successive mode CE steps.

Because these systems have not (yet) begun to exchange mass in the course of post-CE evolution, we can make an estimate, but only a very tentative one, of parameters in the pre-CE state. Even this is ambiguous, however, since it is quite likely that the pre-CE red giant was losing mass by stellar wind. Our best attempts, for a subset of systems, are given in Table 5.2; they are still very subjective.

For the CMD and WMD systems of shortest period in Table 5.1 it is perhaps a little surprising how many have a very low value of M_2. The typical $*2$ appears to be a late M dwarf of 0.1–$0.2\,M_\odot$. One does see some more substantial stars (A5 in V651 Mon, G8III in FF Aqr), but they are found in longer-period systems. This leads us to suggest that it is the mass *ratio* – well before mode CE begins – which is most important in determining the final period. A massive companion may manage to blow away the envelope with rather little orbital shrinkage, while a low-mass companion has to spiral in much further, and perhaps in many cases merge.

This is illustrated in Fig. 5.2, where the estimates of Table 5.2 are plotted in a P_0, q_0 plane. In our estimates of precursor parameters we have gone for the possibility that has the lowest reasonable M_{10}, on the grounds that lower masses are more probable than higher masses, but we still find that virtually all the systems that have shrunk their orbits by large factors (marked by an asterisk in Fig. 5.2) had precursor mass ratios of $\gtrsim 4$, while conversely most

Table 5.1. *Some detached evolved binaries possibly related to common-envelope evolution*

Name	Spectra[a]	State	P	e	M_1	M_2	R_1	R_2	Reference
KV Vel	77kK + CIIIem	*CMD*	0.357		0.63	0.23	0.16	0.40	Hilditch *et al.* 1996
UU Sge	87kK + 6.3kK	*CMD*	0.465		0.63	0.29	0.33	0.54	Bell *et al.* 1994
V477 Lyr	60kK + 6.5kK	*CMD*	0.472		0.51	0.15	0.17	0.46	Pollacco and Bell 1994
BE UMa	105kK + 5.8kK	*CMD*	2.29		0.7:	0.36:	0.08:	0.7:	Ferguson *et al.* 1999
V651 Mon	100kK + A5Vm:	*CMD*	16.0	0.07:		0.0073[b]			Smalley 1997
IN Com	200kK: + (G + G)	*C(...)D*	41:			0.0016:[b]			Jasniewicz *et al.* 1987
IN Com	G: + G5III	*ggc*	1.99		0.004:[b]	0.016:[b]			Jasniewicz *et al.* 1987
HW Vir	SDB + 4.5kK	*EMD*	0.1167		0.48	0.14	0.18	0.18	Wood and Saffer 1999
AA Dor	40kK + 4kK	*CMD*	0.262		0.3:	0.05:	0.16:	0.09:	Włodarczyk 1984
FF Aqr	SDOB + G8III	*cGD*	9.2		0.35	1.4	0.16	7.2	Vaccaro and Wison 2002
V1379 Aql	SDB + K0III-IV	*EGE*	20.7	0.09	0.30	2.27	0.05	9.0	Jeffery and Simon 1997
HD 137569	B5III + ?	*hmE*	530	0.11	0.21[b]				Bolton and Thomson 1980
V652 Her	B2IIIp + ?	*hme*	3000:		0.7:	0.015:	1.7:		Kilkenny *et al.* 1996
HR Cam	19kK + M	*WMD*	0.103		0.41	0.10	0.018	0.125	Maxted *et al.* 1998
13471-1258	14.2K + M3.5/4	*WMd*	0.151		0.78	0.43	0.011	0.42	O'Donoghue *et al.* 2003
NN Ser	55kK + M5-6	*WMD*	0.130		0.57	0.12	0.019	0.17	Catalán *et al.* 1994
LM Com	29kK + M4.5	*WMD*	0.259		0.45	0.28			Orosz *et al.* 1999
CC Cet	WDA2 + M4.5e	*WMD*	0.284		0.39	0.18		0.21	Saffer *et al.* 1993
GK Vir	WDAO + M3-5V	*WMD*	0.344		0.51	0.10		0.15	Fulbright *et al.* 1993
V471 Tau	35kK + K2V	*WMD*	0.521		0.84	0.93	0.011	0.96	O'Brien *et al.* 2002
EG UMa	13kK + ?	*WMD*	0.668		0.38	0.26			O'Brien *et al.* 2002
Feige 24	55kK + M1.5V	*WMD*	4.23		0.47	0.30	0.032		Vennes and Thorstensen 1994
G203-47	? + M3.5V	*wME*	14.7	0.07		0.2[b]			Delfosse *et al.* 1999
IK Peg	35kK + A8p	*WMD*	21.7		1.1:	1.7:			Landsman *et al.* 1993
AY Cet	WD + G5IIIe	*WhD*	56.8	0.09:	0.55:	2.1:	0.012	15	Simon *et al.* 1985
HD121447	? + K7Ba5	*wGD*	186	0.02		0.025[b]			Jorissen *et al.* 1998
G77-61	? + MVp	*wMD*	246			0.173[b]			Dearborn *et al.* 1986
AG Dra	SDOe+ K3pIIIBa	*wgd*	549			0.006[b]			Smith *et al.* 1996
DR Dra	WD + K0III	*WGE*	904	0.07		0.0035[b]			Fekel *et al.* 1993
HD17817	? + K4IIIBa5	*wGE*	2866	0.43		0.0056[b]			Jorissen *et al.* 1998
α CMi	WDF + F5IV-V	*wME*	14910	0.41	0.60	1.50	0.0096	2.0	Girard *et al.* 2000
α CMa	WDA2 + A0Vm:	*wME*	18300	0.59	1.00	2.0	0.0084	1.7	Provencal *et al.* 1998
0957-666	WDA + WDA	*WWD*	0.061		0.32	0.37			Maxted *et al.* 2002
1101 + 364	WDA3 + WDA	*WWD*	0.145		0.33	0.29			Marsh 1995
1704 + 481.2	WDA4 + WD	*WWD*	0.145		0.39	0.54			Maxted *et al.* 2002
1704 + 481	(W+ W) + WDA5	*(...)WE*	4.5"		0.93	0.55:			Greenstein *et al.* 1983
1414-0848	8.9kK + 10.8kK	*WWD*	0.518		0.55	0.71	0.012	0.01	Napiwotzki *et al.* 2002
IQ Cam	? + SDB	*wED*	0.090			0.126[b]	0.01:		Koen *et al.* 1998
V2214 Cyg	? + SDB	*wED*	0.095			0.42[b]		0.18:	Maxted *et al.* 2000
HD 49798	XR + SDO6	*weD*	1.55			0.263[b]		1.5	Bisscheroux *et al.* 1997
UX CVn	? + BV	*whD*	0.574		0.42:	0.39:		1.1:	Schönberner 1978
V379 Cep	B2III + B2III	*hhE*	99.7	0.15	1.9	2.9	5.2	7.4	Gordon *et al.* 1998
EG52	DC9 + DC9	*WWE*	7500	0.18	0.65:	0.65:			Borgman and Lippincott 1983

[a] In several cases an effective temperature is listed.

[b] Mass function, or if two values, $M \sin^3 i$.

Table 5.2. *Estimated parameters well before interaction*

Name	M_{10}	M_{20}	P_0	q_0	M_{1m}	M_{2m}	P_m	M_{1f}	M_{2f}	P_f	Case
KV Vel	2.55	0.23	146	11.1	1.26	0.23	1518	0.63	0.23	0.36	$\mathcal{C}$UD
UU Sge	2.55	0.29	166	8.78	1.26	0.29	1576	0.63	0.29	0.47	$\mathcal{C}$UD
BE UMa	3.18	0.35	191	9.09	1.40	0.35	2161	0.70	0.35	2.29	$\mathcal{C}$UD
V651 Mon	2.04	1.00	185	2.04	1.20	1.80	1676	0.60	1.80	16.0	$\mathcal{C}$UD
HW Vir	1.20	0.14	274	8.57	0.62	0.14	853	0.48	0.14	0.12	$\mathcal{C}$UD
AA Dor	1.20	0.05	12	24.0	0.35	0.05	113	0.30	0.05	0.26	$\mathcal{C}$UD
FF Aqr	2.60	1.00	2.0	2.60				0.35	1.40	9.2	$\mathcal{C}$UN
HD137569	1.10	0.60	393	1.83	1.05	0.60	417	0.45	0.60	530	$\mathcal{C}$U
HR Cam	1.50	0.10	63	15.0	0.51	0.10	436	0.41	0.10	0.10	$\mathcal{C}$UD
NN Ser	2.04	0.12	126	17.0	1.14	0.12	1235	0.57	0.12	0.13	$\mathcal{C}$UD
LM Com	1.50	0.28	181	5.36	0.73	0.28	562	0.45	0.28	0.26	$\mathcal{C}$UD
CC Cet	1.50	0.18	64	8.33	0.57	0.18	322	0.39	0.18	0.28	$\mathcal{C}$UD
V471 Tau	4.20	0.93	415	4.52	1.68	0.93	3486	0.84	0.93	0.52	$\mathcal{C}$UD
EG UMa	1.50	0.26	68	5.77	0.64	0.26	260	0.38	0.26	0.67	$\mathcal{C}$UD
Feige 24	1.90	0.30	159	6.33	0.77	0.30	673	0.47	0.30	4.23	$\mathcal{C}$UD
G203-47	2.04	0.25	242	8.16	1.12	0.28	1451	0.56	0.28	14.7	$\mathcal{C}$UD
IK Peg	5.50	1.50	819	3.67	2.20	1.70	4710	1.10	1.70	21.7	$\mathcal{C}$UD
AY Cet	2.55	1.20	301	2.12	1.26	2.10	1986	0.63	2.10	56.8	$\mathcal{C}$UD
G77-61	2.04	0.30	252	6.80	1.10	0.33	1518	0.55	0.33	245	$\mathcal{C}$U
α CMi	2.04	1.48	2373	1.38				0.60	1.48	6800	D
α CMa	5.00	2.00	918	2.50				1.00	2.00	5000	D
0957-666	2.43	1.00	13	2.43				0.32	1.60	59	$\mathcal{B}$UN
0957-666	0.32	1.60	59	5.00	0.32	0.69	213	0.32	0.37	0.06	reverse $\mathcal{C}$UD
1101+364	2.49	1.00	3.5	2.49				0.33	1.70	15	$\mathcal{B}$UN
1101+364	0.33	1.70	15	5.15	0.33	0.62	66	0.33	0.29	0.14	reverse $\mathcal{C}$UD
1704+481	2.82	1.00	62	2.82				0.39	2.04	230	$\mathcal{C}$UN
1704+481	0.39	2.04	233	5.23	0.39	1.12	1525	0.39	0.56	0.14	reverse $\mathcal{C}$UD
IQ Cam	3.87	1.50	75	2.58				0.60	2.45	270	$\mathcal{C}$UN
IQ Cam	0.60	2.45	267	3.40	0.60	1.24	1666	0.60	0.62	0.09	reverse $\mathcal{C}$UD
HD49798	5.00	3.00	117	1.67				1.00	4.00	423	$\mathcal{C}$UN
HD49798	1.00	4.00	423	4.00	1.00	1.60	3267	1.00	0.80	1.55	reverse $\mathcal{C}$UD
EG52	2.82	2.40	394	1.17				0.66	2.64	1200	$\mathcal{C}$UN
EG52	0.66	2.64	1166	3.99				0.66	0.64	7500	reverse D

Suffix 0 refers to the hypothetical initial configuration; suffix m to the maximum period reached, just before common-envelope evolution if it occurs; and suffix f to the final state, so far. For doubly-evolved systems the second line begins where the first ends.
'Case' means the appropriate variant of Cases $\mathcal{B}$, $\mathcal{C}$ or D.

that have emerged with rather longer periods typically had less extreme mass ratios. At the longest periods, those near the boundary where no interaction occurs at all, even mass ratios of ~ 10 may not guarantee drastic shrinkage.

A system to take particular note of is AA Dor, which has a remarkably high initial mass ratio, as well as an unusually low mass for the SDOB component. Although this system is only SB1, it is also doubly eclipsing, and so its low mass function seems to translate plausibly into the low masses quoted. It is hard to see how a companion of only $0.05\,M_\odot$ can have

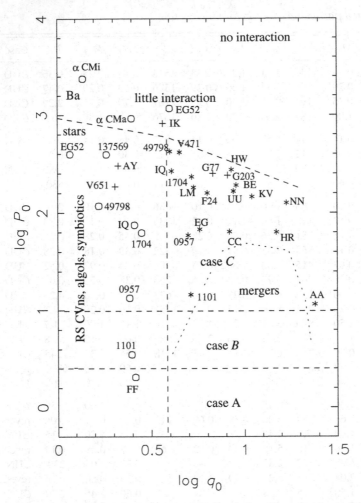

Figure 5.2 Possible locations in the initial period – mass ratio plane of the late case *B/C* precursor systems, from which some of the systems of Table 5.2 have formed. Each system is identified by some letters or numbers in its name. Asterisks: drastic shrinkage, current $P \lesssim 3$ days. Pluses: substantial shrinkage. Circles: slight shrinkage. Regions where probably no shrinkage occurred are labelled 'Ba Stars', 'RS CVns', 'Algols'. A region where we estimate that total shrinkage should occur is labelled 'merger'.

driven off the companion's envelope, which must have been of order 0.75 $M_\odot$ originally. We can imagine a two-stage process here, with mode EW playing an important part by driving off most of the envelope; then when the envelope is down to perhaps 0.1 $M_\odot$ mode CE drives off the remainder during a spiral-in episode. Mode EW could well be a prolonged and fairly efficient process, helped by the fact that as the total mass of the system drops the orbit widens, leaving *1 close to but not quite filling its Roche lobe during a substantial run up the first giant branch. Such a process may also have happened in other systems such as HW Vir. It is possible that mode DI played a role in systems like AA Dor and HW Vir, intermediating between modes EW and SF3, but it is not easy to be sure.

Mode 3 RLOF, and subsequent mode CE, is probably not limited to stars with convective envelopes. In a binary where the initial mass ratio is rather extreme (say $q_0 \sim 10$), it would be necessary under conservative assumptions for the orbit to shrink by a very large factor ($\sim$9, Table 3.1) before expanding again. Even if the loser has a radiative envelope, it seems difficult to imagine that thermal-timescale mass loss could succeed in contracting the stellar radius by such a large factor. A more likely outcome would be that $*2$ becomes swallowed by the envelope of $*1$, with the same kind of results as above. A further way in which mode CE evolution might be precipitated is by mode DI, as described in the previous section.

There is evidence, generally of a rather indirect character, that some binaries that might be expected to have undergone a common-envelope phase have not in fact had their orbits shrunk by any substantial amount. This could be a consequence of a number of factors. But one possibility that I believe may be important, for moderately massive stars (say $\gtrsim 10$–$30\,M_\odot$), and moderately long periods (say 50–1000 days), goes as follows. The evolved $*1$, still in the Hertzsprung gap, approaches its Roche lobe. Because it is quite massive and luminous, it is not very far below the Humphreys–Davidson limit, at which rather more massive *single* stars apparently become unstable, probably because they are also rather close to the Eddington limit. Massive single stars ($M \gtrsim 30\,M_\odot$) seem to eject almost their entire envelope at this stage, but the somewhat less massive stars discussed here presumably do not. Nevertheless, it may be that the presence of a binary companion in a suitable orbit somehow lowers the threshold, so that the evolved component ejects most of its envelope in a more-or-less spherical manner, as if its envelope were unstable to much the same extent that the more massive stars' envelopes are unstable. The loss of the envelope, if it happens in a roughly spherical manner, may lengthen rather than shorten the separation, so that there may not be the opportunity for the companion star to get caught up in it and spiral in to any significant extent.

Although stars that have an active nuclear burning region within them are producing energy at a rate that is capable, in principle, of driving off the star's envelope on a thermal timescale, it is clear that the conversion of radiant energy flux into outwardly directed mass flow is not usually efficient – otherwise stars would hardly evolve beyond the ZAMS. But single stars with a high L/M ratio, near the Eddington limit, apparently do achieve efficient energy conversion of this sort. I suggest that, in circumstances where one might expect a common envelope to be set up by rapid RLOF, the effect of a high L/M ratio may be to drive the common envelope away to infinity. In classic mode CE evolution, the energy to drive the envelope away is thought to come from the binary orbit, which necessarily means that angular momentum is also extracted at much the same rate; but if the energy comes from the radiation field, assumed to be near the Eddington limit, then little or no angular momentum per unit reduced mass need be extracted from the orbit, which therefore remains wide.

There may be at least two ways in which such evolution may come about. On the one hand it may be that, shortly *before* RLOF would be attained, the envelope becomes unstable on account of the lowered gravitational potential in the outer layers, and blows away; or, on the other hand, it might be that slightly *after* RLOF, the disturbance to the outer layers is sufficiently strong that the envelope blows away. The first way might be considered a variant of mode EW; but we cannot say that the normal wind is enhanced since normally (i.e. in a single star with $M \lesssim 30\,M_\odot$) there might be no wind at all at this stage. The second way, which in practice may be little different, is what we call 'envelope ejection': mode EJ.

I have already suggested – Section 3.5 – that V379 Cep in the fourth part of Table 5.1 (and also in Table 3.10) is a product of such a process, but in the second (reverse) stage of RLOF. V379 Cep is an ESB2, and so the masses, though not *very* secure, are not small by virtue of low inclination, as one might expect. Each is only perhaps one quarter of the mass to be expected for normal stars with their spectra. We imagine that this system may have started in case AL with parameters $(7 + 6.3 \, M_\odot; 3 \, \text{days})$. After a major episode of forward RLOF, this can be expected, on a conservative basis, to become detached again at parameters $(1.1 + 12.2 \, M_\odot; 100 \, \text{days})$. In this state *1 burns helium for some considerable time, but during this period *2 evolves rapidly to reverse RLOF. I hypothesise that either at the onset of this RLOF, or perhaps slightly earlier (at which point *2 would just be entering the Cepheid strip), *2 blew off most of its envelope very rapidly, and to infinity rather than to *1. Perhaps a few per cent of the envelope was accreted, in order (a) to raise the mass of *1 from 1.1 to 1.9 $M_\odot$, and (b) to turn *1 from a He main sequence star to something that is morphologically like a horizontal-branch star. The remnant of *2 must also hold on to some portion of its envelope, so that it can also resemble a horizontal branch star in structure, though not in mass.

The binary υ Sgr (Table 3.10) is a possible second example of this behaviour. It is a surprisingly bright member of the rather rare class of HdC stars (Section 2.5), which appear to have almost no hydrogen but high helium, and high carbon, a product of helium burning, as well. It has long been known as an SB1, but Dudley and Jeffery (1990) were able to detect a weak secondary spectrum in the UV. Although the system does not eclipse, there are faint indications of variable Hα absorption round the orbit (Nariai 1967), suggestive of an accretion flow and therefore of a fairly high inclination. Thus we may accept that the masses are not very different from the mass-functions listed in Table 3.10. A helium star of the mass of *1 can be expected to come from a star of $\sim 10 \, M_\odot$ initially, but little or none of the envelope has evidently been accreted by *2. I suggest that the envelope was largely blown to infinity, but with little change of orbit, as *1 approached both RLOF and the Cepheid strip simultaneously. Thus we suspect initial parameters of $(\sim 10 + 3 \, M_\odot; 150 \, \text{days})$. The conditions that I feel might be conducive to mode EJ are (a) *1, or in some cases *2, more massive than $\sim 10 \, M_\odot$, (b) a period that puts the star somewhere in the right-hand half of the Hertzsprung gap (and perhaps rather close to the Cepheid strip) as it also approaches RLOF. Unlike in mode CE, I feel that the mass ratio may be something of an irrelevance; although some binary companion is no doubt necessary since otherwise no star above $\sim 10 \, M_\odot$ would reach the giant branch.

I suggested at the beginning of this section that mode CE requires either a deep convective envelope, or a severe mass ratio (or both). The mode EJ that I describe is different at least to the extent that we require it to happen if the envelope is radiative, though very extended, as in a massive star near the middle or right-hand edge of the Hertzsprung gap, and it also appears to be required if the mass ratio is not particularly severe. Although we can identify only a handful of binaries where we feel that mode EJ is called for, I feel that the need for it is sufficiently pressing to define it as another mode, so that we have two common-envelope modes: a somewhat less dramatic mode EJ, and the more dramatic classical mode CE. In both modes, much mass is lost, but only in mode CE is a large fraction of the angular momentum also lost.

Some other binaries that we can tentatively identify with Mode EJ are: δ Ori A, V505 Mon and V2174 Cyg (Table 3.10) and PSR 0045-7319 (Table 5.3). In all of these, at least one component is severely undermassive for a credible conservative RLOF scenario. In δ Ori A *both* are undermassive for their spectral types, rather as in V379 Cep, though perhaps by factors of 2–3 rather than 4–5. This milder factor, and the substantially shorter period, may relate to considerably greater original masses than in V379 Cyg, $\sim 30\,M_\odot$ each. The period boundary for mode EJ perhaps slopes to periods as short as ~ 6 days at these high masses; this region ('$\mathcal{B}U$') is illustrated in Fig. 4.4.

It is clear that our mode EJ, operating more-or-less at the boundary between mode EW (in detached systems) and mode CE (in semidetached systems), is in just the contradictory regime described earlier in this section where the orbital separation a might be expected to either increase (mode EW) or decrease (mode CE). It is therefore particularly unlikely that we could predict what does in fact fact happen to the separation and period. I suggest a rather banal compromise: the period remains much the same.

5.3 Supernova explosion: mode SN

The effect on an orbit of a supernova explosion in one component of a binary can be readily estimated under the following very simple assumptions (Blaauw 1961, Brosche 1962). Suppose that two stars are in an elliptic orbit (eccentricity e, semimajor axis a, total mass M), and then one explodes instantaneously, sending to infinity a fraction $1 - F$ of the *total* binary mass. As a first approximation, which we will improve on later, we assume that the explosion is isotropic in the rest-frame of the supernova, and so the remaining objects continue instantaneously with the same separation $\mathbf{d}$ and relative velocity $\dot{\mathbf{d}}$ as immediately before. But because the total mass was changed instantaneously from M to $M' = FM$ they will now pursue a new Keplerian orbit described by e', a'. We can calculate the energy and angular momentum of the new orbit very simply, and hence a', e', from the initial conditions and F.

The semimajor axes before and after the supernova are related to the corresponding energies by

$$\frac{\dot{\mathbf{d}}^2}{2} - \frac{GM}{d} = -\frac{GM}{2a}, \quad \frac{\dot{\mathbf{d}}^2}{2} - \frac{GM'}{d} = -\frac{GM'}{2a'}, \tag{5.21}$$

and so eliminating $\dot{\mathbf{d}}$ we have

$$\frac{M'}{a'} = \frac{M}{a} - \frac{2(M - M')}{d}. \tag{5.22}$$

For Keplerian ellipses it is straightforward (Appendix C) to average the reciprocal separation $1/d$ over time, assuming constant probability of explosion per unit time. The average is $1/a$, so that the expectation value of $1/a'$ after the supernova is given by

$$\left\langle \frac{1}{a'} \right\rangle = \frac{2 - M/M'}{a} = \frac{1}{a}\left(2 - \frac{1}{F}\right). \tag{5.23}$$

Since $\mathbf{d}, \dot{\mathbf{d}}$ are instantaneously unchanged, the angular momentum $\mathbf{d} \times \dot{\mathbf{d}}$ per unit reduced mass will be the same after as before the supernova, and so the new eccentricity can be found from

$$GMa(1 - e^2) = |\mathbf{d} \times \dot{\mathbf{d}}|^2 = GM'a'(1 - e'^2). \tag{5.24}$$

Hence, we arrive at the results

$$\left\langle\frac{a}{a'}\right\rangle = 2 - \frac{1}{F}; \quad \frac{\langle e'^2\rangle - e^2}{1 - e^2} = \left(\frac{1}{F} - 1\right)^2; \quad \left\langle\left(\frac{P}{P'}\right)^{2/3}\right\rangle = \frac{2F - 1}{F^{2/3}}. \tag{5.25}$$

We see that if $F < \frac{1}{2}$, i.e. if less than half of the *total* mass of the binary is retained, the new orbit is (on average) unbound, and the binary is disrupted. That the orbit in this situation generally gets larger rather than smaller reflects the fact that most of the orbital angular momentum is in the motion of the less massive star. In the particular case that the orbit was circular before the supernova, we see that the final eccentricity is given simply by $e' = 1/F - 1$.

The above formulae suppose that the supernova explosion is isotropic in the rest frame of the exploding star. Quite a modest degree of anisotropy can make a considerable difference (Shklovskii 1970), since the material is ejected typically at something like a tenth of the speed of light, or several hundred times the orbital speed. We can make a somewhat more elaborate estimate of the effect of a 'kick' velocity $\mathbf{u}$, assuming that this velocity has a certain magnitude but random direction. We replace the second parts of Eqs (5.21) and (5.24) by

$$\frac{1}{2}|\dot{\mathbf{d}} + \mathbf{u}|^2 - \frac{GM'}{d} = -\frac{GM'}{2a'}, \tag{5.26}$$

$$|\mathbf{d} \times (\dot{\mathbf{d}} + \mathbf{u})|^2 = GM'a'(1 - e'^2). \tag{5.27}$$

Averaging over time (assuming that the supernova is equally likely at any time in the orbit, as before), and averaging also over solid angle for $\mathbf{u}$, we obtain

$$\left\langle\frac{a}{a'}\right\rangle = 2 - \frac{1}{F}(1 + K^2), \tag{5.28}$$

$$\frac{\langle e'^2\rangle - e^2}{1 - e^2} = \left(\frac{1}{F} - 1\right)^2 + \frac{K^2}{3F(1 - e^2)}\left[\frac{9 - 8e^2}{F} - 2(2 + e^2)\right] + \frac{K^4}{3F^2}\frac{2 + 3e^2}{1 - e^2}, \tag{5.29}$$

where K is a measure of the kick velocity in terms of the mean circular velocity before the supernova:

$$|\mathbf{u}| \equiv K\sqrt{\frac{GM}{a}} \sim 214K\left(\frac{M}{P}\right)^{1/3}, \tag{5.30}$$

with M in Solar units, P in days and u in km/s. For a kick of given magnitude, and (for simplicity) an initially circular orbit, the probability p of escape is

$$p = \max\{0, \min(\alpha, 1)\}, \quad \alpha \equiv \frac{(K + 1)^2 - 2F}{4K}. \tag{5.31}$$

For $K \sim 0$ this is either zero ($F > 1/2$) or unity ($F < 1/2$), and for $K > 1 + \sqrt{2}$ it is always unity, but for $0 < K < 1 + \sqrt{2}$ the probability has an intermediate value, because the orientation of the kick matters. An exceptionally well-placed kick can score a goal; the neutron star or black hole remnant colliding with the companion star. If the kick is not too strong the remnant may be trapped inside the companion, in a version of mode CE (Section 5.2). The outcome could then be a Thorne–Żytkow object (Thorne and Żytkow

1977), a red supergiant with a neutron star or black hole core (Leonard *et al.* 1994); or it might be a short-period orbit of the newly-formed neutron star with the core of the companion, the envelope being driven away.

Hansen and Phinney (1997) obtained a pulsar kick-velocity distribution

$$p(u) = \sqrt{\frac{2}{\pi}} \frac{u^2}{V^3} e^{-u^2/2V^2}, \qquad V = 190 \, \text{km/s}, \tag{5.32}$$

based on proper motions and distances to 86 pulsars. Their mean velocity V is equivalent to the circular velocity of a $20 \, M_\odot$ binary in a 30 day orbit. Lyne and Lorimer (1994) found a substantially larger value, 450 km/s; evidently the result is still fairly uncertain.

Table 5.3 lists a number of massive X-ray binaries. Clearly these have not been disrupted by a supernova, though it is likely that those that have been disrupted far outnumber them. If kick velocities of $\sim$450 km/s are typical, then it is rather surprising that any of them have survived, except perhaps the first three or four with the tightest orbits. But in these short-period systems it can be argued that a kick is *necessary*, although it must be a fairly well-placed one. The typical product of case A or $\mathcal{B}$ RLOF would normally be a much wider binary, with period $\sim$40–150 days (e.g. ϕ Per, Table 3.10, although $*1$ there is of somewhat too low a mass to explode). In such an orbit, it would be necessary for the kick to direct the neutron star towards the companion, generating an elliptical orbit with a periastron sufficiently close that tidal friction can then moderate the ellipse into a circle with the same angular momentum and semi-latus-rectum. It is not improbable that in some cases the neutron star is kicked right into the companion. Perhaps it could travel right through and emerge on the other side; but more probably it will be trapped inside, settle to the centre, and convert the companion into a Thorne–Żytkow object (TŻO), as above.

Among the longer-period examples, Pfahl *et al.* (2002) distinguish between a group with very eccentric orbits (e.g. V635 Cas, V725 Tau) and those with only mildly eccentric orbits (e.g. X Per, γ Cas). It seems unlikely that a kick played any substantial role in the latter group – they are much too wide for tidal friction to be important. Pfahl *et al.* suggest that the degree of anisotropy in the explosion may depend on the rotation rate of the pre-supernova core, and that this in turn depends on whether the preceding RLOF was case $\mathcal{B}$ or $\mathcal{C}$. Although Spruit (1998) suggested that internal magnetic stress would keep a core corotating with its envelope, this may be mitigated if the core is contracting on a thermal timescale, so that $*1$ remnants of case $\mathcal{B}$ might be rotating substantially more rapidly than remnants of case $\mathcal{C}$. Possibly rapid rotation reduces the anisotropy of the explosion when it occurs. Thus a bimodal distribution of kick velocities could be generated. A mean kick of only $\sim$20 km/s is required to account for the low-eccentricity systems.

The radio pulsar 0045-7319 in the SMC is one of the few 'HMXBs' which does *not* radiate in X-rays. The B star is, untypically, not a Be star with substantial though erratic wind. It probably rotates quite slowly. Consequently there appears to be nothing for the pulsar to accrete. Its orbital shrinkage subject to mode TF was discussed briefly in Section 4.2. Its previous evolution presents the problem that we expect $*2$ to be substantially more massive than it appears to be. We would expect, as a result of reasonably conservative RLOF, that $*2$ will become substantially more massive than $*1$ was *originally*, and $8.8 \, M_\odot$ (though tentative) seems too small. ϕ Per (Table 3.10) produced a slightly more massive $*2$, and yet its $*1$ is well short of becoming a supernova: I argued for an initial mass of $\sim$6 $M_\odot$. I suggest that 0045-7319 is a product of mode EJ (Section 5.2), by way of case $\mathcal{B}$U, as for some

Table 5.3. *Some high-mass X-ray binaries*

Name	Alias	Spectra[a]	State	P	e	M_1	M_2	R_2	Y	Reference
	LMC X-4	13.5s + O7III	NMd	1.41		1.5:	16	8	1.5	van Kerkwijk et al. 1995a
V779 Cen	Cen X-3	4.83s + O6.5II	NMD	2.09	0.0008	1.1:	19	11	1.9	van Kerkwijk et al. 1995a
QV Nor	1538-52	530s + B0Iabe	NME	3.73	0.08	1.3:	20	17	2.8	Reynolds et al. 1992
GP Vel	0900-40	283s + B0.5Ib	NhE	8.96	0.088	1.9:	24	30	4.5	van Kerkwijk et al. 1995b
BP Cru	GX301-2	699s + B1Iaa'	NhE	41.5	0.47	32[b]		80:		Kaper et al. 1995
	0535-668	0.069s + B2IIIe	NME	16.7	0.89	0.7:[b]				Skinner et al. 1982
V635 Cas	0115 + 63	3.61s + O-B3V	NMD	24.3	0.34	5[b]				Kelley et al. 1981
V725 Tau	0535 + 26	103s + B0IIe	NME	111	0.49	3.4[b]				Janot-Pacheco et al. 1987
X Per	0352 + 30	837s + O9.5IIIe	NMD	250	0.11	1.6[b]				Delgado-Marti et al. 2001
γ Cas	0053 + 604	XR + B0IVe	NMD	204	0.26		0.002:[b]			Harmanec et al. 2000
	0045-7319	0.93s + B1	NMD	51.2	0.808	1.4:	8.8:	6.4:		Kaspi et al. 1994a
V1357 Cyg	Cyg X-1	XR + O9.7Iab	BMD	5.60			0.25[b]			Bolton 1975, LaSala et al. 1998
V1343 Aql	SS433	H,He em + AI	Bhs	13.1		11:	19:			Margon 1984, Gies et al. 2002
V1521 Cyg	Cyg X-3	XR + WR:	NRd	0.200						van Kerkwijk et al. 1996b
J0737 + 3039		0.022s[c] + .28s[c]	NNE	0.102	0.088	1.34	1.25			Lyne et al. 2004
J1915 + 16	1913 + 16	0.059s[c] + ?	NNE	0.323	0.617	1.441	1.387			Thorsett and Chakrabarty 1999

[a] For a neutron-star component, spin period is given (s) where known.
[b] Mass function.
[c] Radio, not X-ray, pulsar.

other binaries that I propose were fairly wide and fairly massive originally. Initial parameters $(10 + 8 \, M_\odot; 50{-}100 \, \text{days})$ might produce such a system.

A handful of HMXBs contain a black hole, V1357 Cyg being the prototype. Masses of black holes are estimated to be in the range $7{-}12 \, M_\odot$ (Bailyn *et al.* 1998). Let us assume here that all black holes come from stars more massive than those that produce neutron stars; but this point deserves a much fuller discussion than there is room for here. Accepting this hypothesis, it is likely that the boundary is around $40 \, M_\odot$.

Evolution of *2 in HMXBs is likely to produce reverse RLOF. The reverse RLOF should be very hydrodynamic, and seems likely to lead to mode CE and either a merger or a very close binary of state NRD. However, I have already postulated mode EJ as a likely alternative to mode CE, in (reverse) case $\mathcal{B}$ systems that are moderately wide. Consequently we anticipate three different outcomes, possibly depending primarily on the period in the HMXB state:

(a) Shorter-period systems like V779 Cen, for example, might merge to form a single star, presumably a red-supergiant-like entity with a neutron-star core, i.e. a TŻO. The TŻO would probably be subject to extremely copious wind, which might remove the envelope in $\sim 10^4$ years leaving a bare neutron star, or black hole in the case of LMC X-4.

(b) Intermediate-period systems like GP Vel or V635 Cas, for example, where the binary is wide enough to allow reverse case $\mathcal{B}D$ rather than reverse case AD RLOF, might evolve through mode CE to become a short-period NRD system like V1521 Cyg. If the WR-like component of this system is $\gtrsim 2.5 \, M_\odot$, which is likely since *2 in the precursor NHD state is likely to be $\gtrsim 12 \, M_\odot$, then it can have a second supernova explosion. Even a fairly substantial supernova kick might fail to disrupt such a compact binary, and the result could be an NNE binary like PSR J1915 + 1606.

(c) Longer-period systems like X Per, for example, seem likely to evolve by reverse case $\mathcal{B}U$ to a comparably wide NRD, which would almost certainly be disrupted by a later supernova explosion.

At present no BNE or NBE system is known, but both types seem likely to exist. BP Cru has a sufficiently massive *2 that it might reasonably become a black hole (at least if the mass of the precursor is the major determinant). The system may have evolved to its present configuration by fairly conservative case AN from an initial binary such as V348 Car (Table 4.3) with both masses $\sim 35 \, M_\odot$. This makes *2 grow to a mass substantially greater than the initial *1. V1357 Cyg may have evolved more non-conservatively (case $\mathcal{B}UN$) from a *1 of initially much greater mass, but with *2 remaining arguably of low enough mass to leave a neutron star. We can also expect that some BBE systems exist, but they will be very hard to recognise; unless by good fortune they have a third body in a measurable orbit that yields a mass-function of $\gtrsim 20 \, M_\odot$, and an invisible companion.

5.4 Dynamical encounters in clusters: mode DE

If the space density of stars is n, and their mean velocity is v, then a binary with separation a will typically have a close encounter with another star after time t where

$$\pi a^2 n v t \sim 1. \tag{5.33}$$

In the Solar neighbourhood $n \sim 0.1 \, /\text{pc}^3$ and $v \sim 10 \, \text{km/s}$. Thus we need $a \sim 1000 \, \text{AU}$ ($\sim 0.005 \, \text{pc}$) if the time required for a close encounter is to be ~ 10 gigayears. This means that near collisions are unlikely for the kind of binary mainly considered here, with $a \lesssim 10 \, \text{AU}$.

However, in a dense cluster, such as a globular cluster or a young dense star-forming region, the space density can be a million times greater, and so systems with $a \sim 1$ AU are vulnerable.

Since most of this book is concerned with the relatively low-density Solar neighbourhood, say out to ~ 1 kpc and occasionally further, I will not attempt to discuss clusters and dynamical encounters in detail: see Heggie and Hut (2003). But when considering binary stars in globular clusters it is important to note that several interesting objects, such as low-mass X-ray binaries and radio-pulsar binaries that are found with surprising frequency there, may have been much influenced by dynamical encounters. In particular, one cannot assume that such binaries have always been binary, to the same extent that one probably *can* assume it in the Solar neighbourhood. Dense young star-forming regions also allow the possibility of dynamical encounters.

When considering the evolution of binary stars in dense clusters, therefore, it is important to include the N-body gravitational dynamics as well as the effects of RLOF, etc. (Aarseth 1996, 2001). We cannot ignore the fact that stars may loop many times into and out of the densest central core of the cluster. The number of stars in a cluster, $\sim 10^4$–10^6, is not large enough to allow simple statistical–mechanical arguments for estimating the degree of 'ionisation' of binaries. However, one aspect of thermodynamics that holds at least qualitatively is the tendency of more massive particles to diffuse towards the centre, as in selective diffusion in stars: I mentioned briefly the tendency of helium to diffuse inwards relative to hydrogen in Section 2.2.4. Close binaries on the one hand, and neutron stars on the other, tend to be more massive than the average globular-cluster star ($\sim 0.5\ M_\odot$), so they are somewhat more likely to interact near the centre. At the other end of the mass spectrum, light stars tend to be ejected to an outer halo, and some to escape velocity. Evaporation of the cluster is assisted by the fact that a cluster near to a galaxy is surrounded by a 'Roche lobe' whose radius can be estimated, as with binary stars, by Eq. (3.6). A crude estimate of the lobe radius is ~ 50 pc, for a cluster of 10^5 stars at a distance of 10 kpc from the Galactic centre.

It may seem a little odd that gravitational dynamics, being a strictly time-reversible process, can lead to long-term irreversible changes. It is true that (given a sufficiently exact code), we could evolve a cluster forward until, say, half the stars have escaped, and then reverse the evolution and watch the escaped stars being captured. But the (huge) volume of phase space occupied by the evolved cluster and its escapers is still a tiny fraction of the incomparably vaster total phase space involved, and a very special fraction. In general, we would not expect similarity between a cloud of escaped stars and a general cloud of stars, most of which would be non-capturable.

The internal distribution of stars in a globular cluster can be modelled in some respects like the internal structure of stars themselves, with a potential gradient balanced by a gradient of 'pressure' that is essentially the local velocity dispersion multiplied by the density. There is an outward 'heat' flux, with energy from the centre being transported outwards by gravitational interactions between stars; although unlike the atomic case a star may have to rotate many times around the cluster before undergoing a significant interaction. In addition there can even be a central 'nuclear' energy source, i.e. a binary at the centre that can grow more tightly bound while giving energy to neighbours. Lynden-Bell and Eggleton (1980) found that such a cluster (without a central energy source), contracting in a self-similar fashion, is rather like an $n \sim 11$ polytrope, i.e. fairly nearly but not quite isothermal. The slow self-similar contraction would in a long but finite time lead to the core's collapse to infinite density, the 'gravothermal catastrophe' (Antonov 1962, Lynden-Bell 1968), except that this is prevented (Bettwieser and Sugimoto 1984, Inagaki 1984) by the formation of a close binary at the centre. This is analogous to the ignition of nuclear fuel as in a star approaching the main sequence, or the tip of the giant branch.

Until relatively recently (McMillan *et al.* 1990, Heggie and Aarseth 1992) simulations of N-body gravitational dynamics tended to start with a large number of *single* stars – although, harking back to the first paragraph of this book, the long-range nature of gravity is such that it is not always clear whether sub-sytems are bound or not. But the introduction of a substantial fraction of 'primordial' binaries has been shown to be very significant for the evolution of a cluster, both dynamically and in terms of stellar evolution. The reason is that binaries are very difficult to form by two-body encounters among *single* stars. Energy and angular momentum conservation say that two bodies approaching each other on a hyperbolic orbit will depart on the same hyperbolic orbit – unless there happens to be a third star around close enough to interact at the same time and absorb some of the energy. Normally one or two binaries do form nevertheless, and they can dominate the later evolution: the binding energy of one fairly close binary (say 1 AU) is comparable to that of 10^5 stars within a sphere of ~ 1 pc. But single–binary, and a fortiori binary–binary, encounters are fairly rare in such a system, whereas if binaries are as common primordially in clusters as in the field such encounters can be very important. Furthermore, I will suggest below that primordial *triples* may be not just a luxury but a necessity for understanding such objects as the blue stragglers of M67 and other clusters.

Whether a primordial binary in a cluster survives for a long time or gets disrupted by encounters can be measured by its 'hardness'. A binary is 'hard' if the orbital velocity within the binary is large compared with the velocity dispersion of the cluster, and otherwise 'soft'. This leads rather easily to the condition that, in a cluster of N stars within radius R, hard binaries have $a \lesssim R/N$. With say 10^4 stars within 1 pc, this gives $a \lesssim 20$ AU. Dynamical encounters tend to make hard binaries harder, and soft binaries softer until they are disrupted (Heggie 1975).

Two kinds of three-body (or four-body) dynamical encounter may be particularly interesting: exchange reactions, and induced collisions. A neutron star encountering a binary of two K dwarfs may expel one dwarf and form a binary with the other. If this is close enough to have a period of only a few days, magnetic braking and tidal friction (mode MB, Table 3.8) may lead to interaction, to a low-mass X-ray binary and ultimately to a millisecond pulsar binary. Alternatively the neutron star (or another main-sequence star or binary) may perturb the binary to such an extent that the two K dwarfs crash into each other, and merge to form a single star. This single-star product could settle down into an apparently normal main-sequence star that could be substantially brighter and hotter than most main-sequence stars in these highly-evolved systems – a 'blue straggler'.

Figure 5.3 shows a colour-magnitude diagram for the old Galactic cluster M67 (~ 3.5 gigayears). The region to the left of the turn-off at colour ~ 1.0 is populated by 18 blue stragglers, six of which are known to be binary and one of which is known to be triple, as indicated. Even though binary evolution into Algols can, in principle, produce a blue straggler, this does not seem to be the case for the five SB1s noted: all have orbits that are either wide or eccentric or both, whereas Algols can be expected to have orbits that are compact and circular. The eclipsing blue straggler is a contact binary (EV Cnc), where presumably $*1$ is gaining mass at the expense of $*2$ as we expect in case AR. This is, in fact, the only binary blue straggler in M67 that is relatively well explained by its binarity. There may be

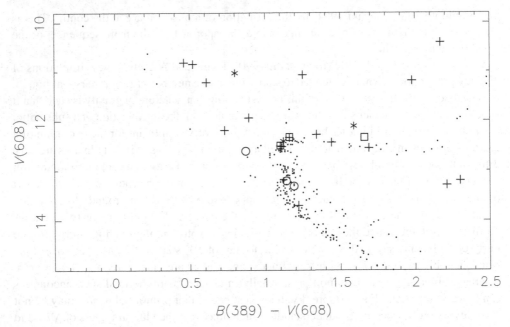

Figure 5.3 Colour-magnitude diagram of the turn-off region of M67. The magnitudes are from Fan *et al.* (1996), and are similar to but not the same as B, V. Central wavelengths (nm) are indicated. Membership is based on the proper-motion study of Girard *et al.* (1989) and on the radial velocity study of Mathieu *et al.* (1990): only candidates with a probability greater than 90% of being members by both criteria are plotted. We define the 18 systems bluer than $B - V = 1.0$ to be blue stragglers. Known SB1s (pluses), SB2s (squares), eclipsing systems (circles) and triples (asterisks) are marked: the blue-straggler triple (S1082) at $V \sim 11$ also has shallow eclipses. Stars that are moderately above the main sequence in the lower right are probably binary, and some may be triple.

some further binaries among the 18, since some have quite broad lines that are not amenable to accurate radial-velocity measurement.

The triple-star blue straggler S1082 in Fig. 5.3 is an extraordinary object (Sandquist *et al.* 2003). It actually consists of *two* blue stragglers, one of which is in a $\sim$1 day orbit with a component that is in the turn-off region, and the other of which is in a 1200 day eccentric orbit round the close pair. We believe that it may have required the near-collision of two primordial triples to produce such an outcome. Within each triple the close pair was perturbed into a merger, and one third body was kept (but somehow scattered into a *very* close orbit with one of the merged pairs) while the other was expelled.

Another blue straggler is so blue that it is difficult to account for, even as a merger of two turn-off stars: it is very near the top left in Fig. 5.3. Perhaps therefore it is a doubly-merged triple star. The close pair in S1082 appears to have only just missed such a fate.

Binary–binary encounters are quite likely in dense clusters with a substantial proportion of primordial binaries. For point masses a typical outcome is that the two lightest components are ejected and the two heaviest form an eccentric binary. On the evidence of M67 it seems that quite often, for non-point-mass stars, there is a merger instead of an eccentric binary. Another binary–binary outcome can be a hierarchical triple with a single star ejected. Several 'runaway' OB stars are seen with abnormally high space velocities. Some of these may be generated by supernova disruption (Section 5.3), and some by three- or four-body encounters. Hoogerwerf *et al.* (2000) have tracked the proper motions of OB runaways, and of neutron stars, backwards in time. They find several cases of apparent common origin, both for OB + OB pairs and OB + NS pairs, and they conclude that the two very different mechanisms occur about equally frequently.

Hoogerwerf *et al.* (2000) point in particular to three stars, two single (AE Aur and μ Col) and one triple (ι Ori A), which appear to have been scattered out of the Orion Nebula cluster (ONC) 2.5 megayears ago. This seems likely to have been a binary–binary (or binary–triple) collision in which one incoming system was broken up into two. The trajectories of AE Aur and μ Col are almost exactly in opposite directions from the ONC.

Bagnuolo *et al.* (2001) note that the two components of the spectroscopic sub-binary within ι Ori A are rather remarkable. The stellar and orbital parameters are (O9III + B0.8III–IV; 29.1 days; $e = 0.76$). Because the system does not eclipse, the inclination is not known, but the mass ratio is ~ 1.75 (Marchenko *et al.* 2000). The B star seems remarkably evolved, considering that it is much less massive, and it is difficult to believe that this is due to RLOF, for instance, since the orbit is so eccentric. Gualandris *et al.* (2004) suggest that the collision involved an exchange as well as a disruption, with the present ι Ori containing components that came from original binaries of substantially different ages – about 5 and 10 megayears are necessary. In fact ι Ori contains a speckle companion as well (Mason *et al.* 1998), in an orbit that might be ~ 40 years. It appears that a very complicated dance has taken place here.

There are quite a few massive binaries where one can question whether the two components are coeval; I mentioned LY Aur in Section 3.5. The quadruple star QZ Car in Table 4.10 contains one component (B0Ib; *22 as listed) that appears to be more evolved and yet less massive than its close companion (O9V; *21). If exchanges do occur among the massive singles, binaries and multiples of a region of star formation, it is probably only within the first ~ 10 megayears, while the region is still densely populated, and would only have a significant effect in those massive stars whose ages are of this order. Unfortunately early massive stars also tend to have the least certain parameters, and so it is difficult to be sure that two components are non-coeval. Since massive stars, and a fortiori binaries and triples, tend to congregate towards the centre of a cluster by gravitational settling, it would not be surprising if massive binaries were especially prone to show non-coeval components, or that primordial triples may play a major part in producing the blue stragglers of old clusters.

If single stars are treated as extended bodies rather than as point masses, a dynamical encounter of two single stars, leading to capture, may also occur (Fabian *et al.* 1975), though probably not often. As two stars approach close to each other on a hyperbolic orbit they can raise substantial time-dependent tides, which convert orbital energy into internal hydrodynamic energy. This may convert marginally hyperbolic orbits into marginally elliptic orbits, at least temporarily. In the longer term, viscosity may convert the hydrodynamical energy into heat that is radiated away, thus sealing the capture. In an extreme case the two stars may actually collide, and merge. In a less extreme case the highly eccentric orbit may be circularised over many orbits by tidal friction. It is a feature of circularisation by tidal friction that the apastron separation can decrease considerably but the periastron separation can only increase modestly. Tidal friction conserves angular momentum, and thus the semi-latus-rectum of the orbit: the ratio of this to the periastron separation is $1 + e$, which only decreases from 2 to 1 as the orbit circularises.

Induced collisions near the centres of young rich clusters may possibly be the origin of some stars at the high end of the mass spectrum. If two stars do merge, a great deal of energy is available, which may expand the merged star to red-giant proportions temporarily, making it all the more likely that it will merge with further stars. We might have a runaway process that continues until the density of stars is significantly reduced (Portegies Zwart and McMillan 2002). It does not seem impossible that *all* massive stars are produced in this way, rather than by an unusually large amount of accretion on to an initially low-mass core.

6

Accretion by the companion

Matter that leaves the surface of one component of a binary can be partly or wholly accreted by the companion. We have seen that the loser could be losing mass either by RLOF or by stellar wind, perhaps binary-enhanced; the accretion process has even more options, and these are modelelled with even less confidence. A major reason why the accretion process can be more complex than the mass-loss process is that gainers can have a very wide range of radii, from black holes and neutron stars (at $\sim$3–30 km) to white dwarfs ($\sim 10^4$ km) to normal dwarfs ($\sim$0.1–10 Gm), and even occasionally to sub-giants ($\sim$3–30 Gm) or giants ($\gtrsim$10–30 Gm); whereas the loser is usually only in the last three of these categories. Not only does the available energy of the accreted material vary (inversely) over the same range, but also different physical forces (magnetic, viscous, rotational, gravitational) may dominate at different radii from the gainer.

The study of accretion is one of the most active areas in stellar astrophysics. Phenomena, often dramatic, are observed to happen on timescales ranging upwards from milliseconds. This book will not attempt to cover the ground in detail – partly for lack of space, but also because this book is intended to concentrate on the long-term evolution of binaries rather than on their short-term behaviour. Naturally, to test long-term predictions observationally it would be helpful to be able first to model, and allow for, the observed short-term behaviour. For a fuller treatment the reader is referred to some standard works: Lewin and van den Heuvel (1983), Frank *et al.* (2002). The following few pages are an attempt to summarise the aspects of accretion that are most relevant to long-term evolution.

6.1 Critical radii

The character of an accretion flow depends importantly on the size of the gainer, particularly relative to the size of its Roche lobe. In a range of systems, one may see, or expect to see, at least four zones of different radii around the gainer, in which different physical processes are important. Proceeding outwards from the gainer, there may be some or all of the following:

(a) A magnetospheric zone, in which a magnetic field anchored in the rotating gainer dominates the flow, causing the (highly ionised) accreting gas to flow in along field lines, arriving on the gainer at its magnetic poles (Lamb *et al.* 1973). The magnetic field is normally assumed to be dipolar, oblique to the axis of rotation and not necessarily symmetrical about the centre. It has to be oblique to account for the observed rotational modulation; nevertheless in many analytic attempts to model magnetic-dominated accretion one often assumes axisymmetry for convenience.

(b) A Keplerian disc region, in which centrifugal force largely balances gravity in the radial direction, while viscosity drives an inward flux of gas superimposed on the almost-circular motion, and simultaneously drives an outward flux of angular momentum (Lüst 1952, Lynden-Bell 1969, Lin and Pringle 1976, Pringle 1981). The mechanism giving rise to the viscosity might be turbulent motion as a result of convective energy transport in the disc, but, more probably turbulent magnetic field in rough balance with the pressure supporting the disc against gravity in the direction perpendicular to the disc (Shakura and Sunyaev 1973, Balbus and Hawley 1991) – see Appendix F.

(c) A region of inward free-falling gas, where the specific angular momentum of the gas is small compared with Keplerian specific angular momentum. The material gains angular momentum while falling in, partly due to Coriolis force in the frame that corotates with the binary, and partly due to the non-central character of the forcefield (3.2) around ∗2.

(d) A region of wind flow from the loser. This flow might be fairly uniform and radial in the frame of the loser, if it were not for the perturbing gravitational field of the gainer.

The first region applies mainly to compact gainers like neutron stars and white dwarfs, although a magnetic Bp star accreting from a red supergiant might have a similar magnetospheric zone. The last region applies mainly where the loser underfills its Roche lobe but is losing mass by stellar wind.

Dictating the nature and extent (or existence) of these zones are several characteristic radii, some of which can be estimated only crudely:

(i) The radius R_2 of the gainer or, for a black hole, its Schwarzschild radius

$$R_S \equiv \frac{2GM_2}{c^2} . \tag{6.1}$$

This is the radius such that light cannot escape from within it. But circular orbits outside this, up to $3R_S$, are unstable to the extent that bodies in them will rapidly plunge inwards, to within R_S.

(ii) The corotation radius R_c, at which the angular velocity of material in Keplerian orbit is the same as the angular velocity Ω_2 of the gainer:

$$R_c = \left(\frac{GM_2}{\Omega_2^2} \right)^{1/3} . \tag{6.2}$$

If ∗2 is rotating close to break-up, then of course $R_c \sim R_2$.

(iii) An Alfvén radius R_A determined roughly by the balance of Alfvén speed v_A with Keplerian rotational speed v_K, as for Eq. (4.55):

$$v_A^2 \sim \frac{B_A^2}{\mu_0 \rho_A} \sim \frac{1}{\mu_0 \rho_A} \left(\frac{B_2 R_2^3}{R_A^3} \right)^2 \sim v_K^2 = \frac{GM_2}{R_A} . \tag{6.3}$$

B_2 is the dipole field strength at ∗2's surface; the field is assumed to diminish outwards like r^{-3}. Note that in Section 4.4 on magnetic braking, we took $B \propto r^{-2}$, appropriate for *outflowing* magnetically-coupled winds in the 'split-monopole' approximation; apart from this the analysis is the same. The Alfvén density ρ_A is estimated from the accretion rate by

$$\dot{M}_2 \sim 4\pi \rho_A R_A^2 v_A \sim 4\pi \rho_A R_A^2 v_K . \tag{6.4}$$

Combining these, we get

$$\left(\frac{R_A}{R_2}\right)^{7/2} \sim \frac{4\pi B_2^2}{\mu_0 \dot{M}_2}\left(\frac{R_2^5}{GM_2}\right)^{1/2}. \tag{6.5}$$

(iv) The radius R_l of the cylinder on which corotating material would have the speed of light:

$$R_l = \frac{c}{\Omega_2} = \frac{R_c^{3/2}}{(R_S/2)^{1/2}}. \tag{6.6}$$

(v) A characteristic disc radius R_D, where the specific angular momentum h of the material at the inner edge of zone (c) equals the Keplerian value. For particles moving in the plane of the orbit, Coriolis force, in the frame that rotates with the binary, ensures that $\mathbf{h} + \boldsymbol{\omega} r^2$, though not $\mathbf{h}$ itself, is conserved, except to the extent that the force within the lobe of *2 is not entirely central towards *2. This latter effect is fairly modest, for particles falling from rest at L1 into the vicinity of *2, and so for RLOF we can estimate h by saying that it is roughly the same as ωz^2, where z is the distance from the centre of *2 to L1. Then

$$\frac{R_D}{a} \equiv \frac{h^2}{GM_2 a} \sim \frac{(\omega z^2)^2}{GM_2 a} = \frac{M_1 + M_2}{M_2}\frac{z^4}{a^4} \sim \text{constant } (1+q)\,x_L^4(1/q). \tag{6.7}$$

Empirically, z is $\sim$20–35% greater than the Roche-lobe radius $R_{L2} = ax_L(1/q)$ – Eq. (3.5) – over a large range of mass ratio q. Hence the constant on the far right-hand side of Eq. (6.7) should be in the range 2.1–3.3. By comparing Eq. (6.7) with the more detailed calculations of Lubow and Shu (1975), we find that we can get very good agreement (to $\sim$3%) if we replace 'constant $(1+q)$' empirically by $1.9 + 2.2q$. Thus we arrive at the following expression for the ratio of disc radius to lobe radius:

$$\frac{R_D}{R_{L2}} \sim (1.9 + 2.2q)\,x_L^3(1/q). \tag{6.8}$$

This is not the radius to which the outer part of the disc would settle down in equilibrium, since the outward transport of angular momentum in a steady-state disc would push the boundary outwards. But if the radius of the gainer is greater than this radius we expect no substantial disc to form, because the stream will simply impact on to the trailing face of the gainer instead of forming a ring. Also, in the case of accretion from a wind rather than RLOF, the disc radius might be determined by the inhomogeneity in either the wind speed or the wind density in zone (d), rather than by the simple argument given above.

(vi) A Bondi–Hoyle accretion radius R_{acc}, where the kinetic energy in the outflowing wind from the loser balances the gravitational potential energy in the field of the gainer. If the wind has speed V_w relative to *1 in the radially outward direction when it reaches the orbit of *2, this balance gives (at a rather simplistic level of approximation)

$$R_{acc} \sim \frac{GM_2}{|\mathbf{V}_w - \mathbf{V}_{orb}|^2}, \quad \mathbf{V}_w = V_w\frac{\mathbf{d}}{d}, \quad \mathbf{V}_{orb} = \dot{\mathbf{d}}. \tag{6.9}$$

Although for circular orbits $\mathbf{d}$ and $\dot{\mathbf{d}}$ are perpendicular, we write this in a form which allows for the possibility of an eccentric orbit (Section 6.5).

(vii) The Roche-lobe radius of the gainer.

Table 6.1. *Estimates of critical radii and other parameters in the accretion process*

Loser (∗1)	OB star	B star	M dwarf	Red dwarf	Red supergiant
Gainer (∗2)	NS	NS	NS	WD	Bp dwarf
M_1	20	10	0.4	0.5	3
M_2	1.4	1.4	1.4	0.7	4
P (d)	3	100	0.3	0.15	1000
B_2 (T)	10^8	10^8	10^5	10^2	1
$\dot{M}_2$ ($M_\odot$/megayear)	10^{-3}	10^{-6}	10^{-3}	10^{-3}	1
P_2 (s)	0.3	3.0	0.05	10^3	1 d
V_w (km/s)	10^3	10^3	200	300	30
R_S (Schwarzschild)	5.9×10^{-6}	5.9×10^{-6}	5.9×10^{-6}		
R_2	1.4×10^{-5}	1.4×10^{-5}	1.4×10^{-5}	0.01	3
R_c (corotation)	1.1×10^{-3}	0.005	3.3×10^{-4}	0.19	6.7
R_A (Alfvén)	5.6×10^{-3}	4.0×10^{-2}	1.1×10^{-4}	0.17	24
R_l (light velocity)	0.021	0.21	3.4×10^{-3}	69	5.9×10^3
R_D (disc)	0.97	9.7	0.33	0.12	76
R_{acc} (Bondi–Hoyle)	0.23	0.26	1.61	0.49	300
R_L (Roche lobe)	4.5	46	1.12	0.52	325
$\dot{M}_{Edd}$ ($M_\odot$/megayear)	0.018	0.018	0.018	12	3.7×10^3
P_{eq} (s)	3.5	68	0.0095	870	5.8 d

Masses and radii in Solar units; periods in days or seconds as indicated; 1 tesla = 10^4 gauss.

Table 6.1 gives some typical values for these radii, and other parameters, in a variety of cases. As material falls on to the gainer, it generates a luminosity

$$L_{acc} \sim \frac{GM_2\dot{M}_2}{R_2}, \tag{6.10}$$

from which $\dot{M}_2$ may be estimated if we have a rough idea of the apparent brightness and the distance of the source. This luminosity might be very large if the gainer is compact. However, we expect that the luminosity should not be able to exceed the Eddington luminosity:

$$L_{acc} \leq L_{Edd} = \frac{4\pi cGM_2}{k_{th}}, \tag{6.11}$$

taking k_{th} to be the Thomson scattering opacity, $0.034 \, \mathrm{m^2/kg}$. According to Eq. (2.11) this is the maximum luminosity of a spherical star in hydrostatic equilibrium, the maximum being approached as $\zeta \equiv p_{rad}/p_{gas} \to \infty$. The Thomson scattering value is reasonably appropriate for hot luminous objects. Equation (6.11) means that the gainer may not be able to accept more than a fraction of the mass lost by the loser. By equating L_{acc} to L_{Edd} we obtain an upper limit $\dot{M}_{Edd}$ to the rate at which the gainer can accrete, except in a short-lived, unstable manner:

$$\dot{M}_{Edd} \sim \frac{4\pi cR_2}{k_{th}}. \tag{6.12}$$

If this is less than ∗1's mass-loss rate, the remainder of the mass is presumably either lost to the system as a whole, or else accumulates perhaps in the outer part of the gainer's

Roche lobe, or in a common envelope around the system. For white dwarfs, if we adopt a mass and radius of $0.7\ M_\odot$ and $7 \times 10^3\ \text{km}$, then we obtain $L_{\text{Edd}} \sim 2.8 \times 10^4\ L_\odot$, $\dot{M}_{\text{Edd}} \sim 12\ M_\odot/\text{megayear}$. For neutron stars, adopting $1.4\ M_\odot$ and $10\ \text{km}$, we have $L_{\text{Edd}} \sim 5.6 \times 10^4\ L_\odot$, $\dot{M}_{\text{Edd}} \sim 0.018\ M_\odot/\text{megayear}$.

We can rewrite Eq. (6.5) for the Alfvén radius in a slightly more transparent dimensionless form by using $\dot{M}_{\text{Edd}}$ as a reference value for $\dot{M}_2$, and by introducing a reference magnetic field B_0 defined by

$$\frac{B_0^2}{\mu_0} \equiv \frac{GM_2}{R_2^2 k_{\text{th}}}. \tag{6.13}$$

For a normal hot star, with photospheric boundary condition $\rho\kappa \sim g$ – Eq. (2.17) – the right-hand side is just the photospheric pressure, which represents an upper limit to the strength of the magnetic field in a starspot. However, for the reference value we continue to use the Thomson scattering opacity k_{th}, even although in the Sun, and most stars cooler than $\sim 10\ \text{kK}$, the *photospheric* opacity is substantially less, by two to four orders of magnitude. For a white dwarf we have $B_0 \sim 8.4\ \text{T}$. For a neutron star the physical picture is not very appropriate, but Eq. (6.13) nevertheless gives a reference field of $\sim 8.3 \times 10^3\ \text{T}$. Then Eq. (6.5) can be rewritten

$$\frac{R_A}{R_2} = \left(\frac{GM_2}{c^2 R_2}\right)^{1/7} \left(\frac{\dot{M}_{\text{Edd}}}{\dot{M}_2}\right)^{2/7} \left(\frac{B_2}{B_0}\right)^{4/7}. \tag{6.14}$$

The first factor on the right-hand side ~ 0.28 for a white dwarf, and ~ 0.80 for a neutron star.

When accreted material falls inside the Alfvén radius, it tries to corotate with the field and $*2$ as it follows the field lines to the surface, so that $*2$ is spun up or down depending on whether $\Omega_2 R_A$ is less than or greater than the Alfvén speed, or equivalently, by Eq. (6.5), the Keplerian speed, at the Alfvén radius. This implies that there is a stable equilibrium rotation rate Ω_{eq} (and period $P_{\text{eq}} = 2\pi/\Omega_{\text{eq}}$), with

$$\Omega_{\text{eq}} R_A = v_K = \sqrt{\frac{GM_2}{R_A}}, \tag{6.15}$$

and so from Eq. (6.14),

$$\frac{P_{\text{eq}}}{P_0} = \frac{\Omega_0}{\Omega_{\text{eq}}} = \left(\frac{GM_2}{c^2 R_2}\right)^{3/14} \left(\frac{\dot{M}_{\text{Edd}}}{\dot{M}_2}\right)^{3/7} \left(\frac{B_2}{B_0}\right)^{6/7}, \quad \Omega_0^2 \equiv \frac{GM_2}{R_2^3}. \tag{6.16}$$

Ω_0 is more or less the break-up angular velocity, corresponding to periods $P_0 \sim 12\ \text{s}$ for a white dwarf and $\sim 0.5\ \text{ms}$ for a neutron star. Table 6.1 lists $\dot{M}_{\text{Edd}}$ and P_{eq} for some cases.

6.2 Accretion discs

An accretion disc is likely to form within the Roche lobe of the gainer $(*2)$ if $*2$ is much smaller than its Roche lobe, and not so strongly magnetic that $R_A > R_D$. This disc would be a ring if it were not for the possibility of a torque that causes angular momentum to be transported in an outward direction, allowing material to spiral in to smaller orbits of lower angular momentum (Appendix F). The force whose azimuthal component provides this torque is commonly described as 'viscous', and modelled by a Navier–Stokes term of the form (in cylindrical polars) $R^{-2} \partial/\partial R\{\chi \rho R^3\ \partial\Omega/\partial R\}$, Ω being the Keplerian angular velocity within the disc and X the coefficient of viscosity. This viscosity coefficient is usually written as $X = \alpha p / \rho \Omega$, with α a dimensionless constant.

Recent work (Balbus and Hawley 1991, Stone *et al.* 1992, 1996, Tout 1997) makes it rather clear that the torque is in practice magnetic, and should be modelled with the Lorentz force $\mathbf{j} \times \mathbf{B}$. The magnetic field is expected to be chaotic, because any weak, systematic seed field in the presence of rotational shear is expected to be amplified strongly by a hydromagnetic instability (Chandrasekhar 1961). Since the amplification of the field, presumably until it reaches some quasi-steady amount dictated by the balance of magnetic pressure with gas pressure, depends on the shear $\partial \Omega / \partial R$, it is not unreasonable that the torque due to the azimuthal part of the Maxwell tensor might have much the same mathematical form as the Navier–Stokes term above, with the viscosity coefficient X replaced by an effective 'viscosity': $X \sim \alpha B^2 / \mu_0 \rho \Omega$ (Shakura and Sunyaev 1973; Appendix F). It is therefore possibly still reasonable to model disc accretion as an 'α-disc'.

Modelling of the chaotic magnetic field to be expected will no doubt be complicated by the fact that the saturation of the magnetic field at some mean value (assumed to be when magnetic pressure $\sim$ gas pressure) will probably not be achieved by simple ohmic diffusion, but rather by 'field-line reconnection', a highly non-equilibrium process (Syrovatskii 1981) such as is seen in Solar flares as well as in laboratory MHD. When regions of fluid containing frozen-in field of opposing sign collide with each other, a singularity develops, which leads to an explosive release of energy. In Appendix F we suppose for simplicity that there is an ohmic diffusion which is sufficiently large that the heat production is effectively the same as might be achieved by sporadic field-line reconnection.

The α-disc gives estimates – Appendix F, Eqs (F25) and (F26) – for, among other things, the optical depth $\tau_0(R)$, and the thickness of the disc as a fraction $\delta(R)$ of the radial coordinate R. For the sake of argument we adopt $\alpha =$ constant. The model is only valid if $\delta^2 \ll 1, \alpha^2 \delta^2 \ll 1$, but δ is indeed expected to be small $(0.01 - 0.1)$ for a fairly wide range of values of L_{acc}, M_2, R_2 and R. We can estimate α crudely from the observed timescale on which discs evolve, which should be $\sim R^2 / X$. The model gives this as a multiple $\alpha^{-1} \delta^{-2}$ of the timescale of Keplerian rotation Ω^{-1}. Accretion discs in cataclysmic variables seem to require $\alpha \sim 0.01 - 0.1$. Ordinary molecular or radiative viscosity would give a smaller α by many orders of magnitude, but magnetic processes seem capable of giving α in this range (Stone and Norman 1994).

An alternative source of viscosity that has often been invoked is turbulence, perhaps driven by convection. Turbulence is not expected to be generated by the shear, because the Rayleigh criterion for instability would require the angular momentum (per unit mass) to increase inward, and in a Keplerian disc it increases outward. But a torque capable of causing material to spiral inwards would release gravitational energy, and this energy has to be transported within the disc before it can be radiated from the surface (at least if the disc is optically thick). The temperature gradient required might well be unstable to convection for much the same reason that stellar envelopes can be unstable to convection, because of the rapid rise of opacity with temperature during partial hydrogen ionisation at $\sim 6-10$ kK (Section 2.2.3). However, numerical simulations (Balbus *et al.* 1996) suggest that this will not achieve an α in excess of $\sim 10^{-4}$. The basic reason for the relative insignificance of convective turbulence appears to be that it is not driven directly by the shear, whereas the hydromagnetic instability is. Nevertheless, at low temperatures of a few hundreds kelvin, such as is expected in accretion

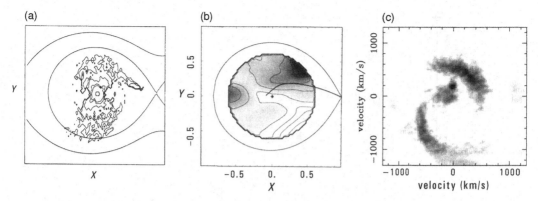

Figure 6.1 Models of accretion discs in cataclysmic variables, from (a) theory and (b) and (c) observation. In (a), after Lin and Pringle (1976), contours of energy-production rate are plotted, from a model where the gas is treated as collisionless particles, except that an artificial viscosity is included. In (b), after Wood *et al.* (1989), contours of surface brightness in the disc of the dwarf nova OY Car are plotted, reconstructed from eclipse mapping. The models are not directly comparable, having mass ratios (gainer/loser) of 2.5 and 10 respectively. (c) Spiral arms in the accretion disc of IP Peg during outburst (Harlaftis *et al.* 1999). Doppler tomography maps the gas in velocity space, large velocities (moduluswise) being associated with the innermost part of the disc, and vice versa.

discs around protostars, it may be necessary to have some other source of viscosity than frozen-in magnetic fields, since the gas will be almost completely un-ionised.

Figure 6.1 shows three views, one theoretical and two observational, of accretion discs. The left panel is a model where the gas was treated as a stream of particles coming from the L1 point and subject to the acceleration of the Roche potential (and Coriolis force). The effect of viscosity was simulated by dividing the area into small cells and replacing the velocities of all particles in a given cell by the average for the cell. This allows for an estimate of the local energy release, and hence of the local temperature. The centre panel is a reconstruction of the surface brightness in the disc of OY Car, a short-period (0.063 days) binary with a red dwarf loser and a white dwarf gainer. The track expected for particles falling freely from the L1 point is marked; perhaps the hot region in the top right is where the heat from the collision of the stream with the disc is released. The right panel shows the distribution of the gas in velocity space from Doppler tomography applied to a rather similar system, but with longer period (0.158 days). If one assumes a model of a steady disc with a Keplerian velocity field, one can map from velocity space to coordinate space; but such a model would not produce this two-armed spiral pattern.

As a star accretes it is liable to spin up, because the accreting gas acquires angular momentum from Coriolis force in the corotating frame. In the relatively simple case that accretion is via a disc, the newly accreted material has Keplerian velocity when it is added to $*2$. If the rotation is redistributed to uniformity within $*2$, we can write

$$\frac{d}{dt} M_2 (k_2 R_2)^2 \Omega_2 = \sqrt{GM_2 R_2}\, \frac{dM_2}{dt}, \tag{6.17}$$

where k_2 is the dimensionless radius of gyration – Eq. (3.8). If $R_2 \propto M_2^a$ and k_2 is a constant, we can easily integrate this to see that $*2$ is spun up from $\Omega_2 = 0$ at $M_2 = M_{20}$ to break up

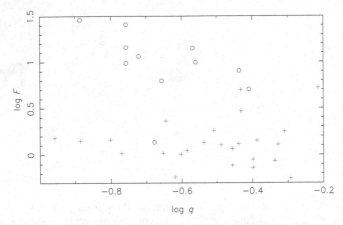

Figure 6.2 Over-rotation factor F against mass ratio q in semidetached binaries. F is the ratio of the gainer's angular velocity to the orbital angular velocity. Circles: long-period systems, $P > 4.5$ days. Pluses: shorter-period systems.

at $M_2 = M_{2f}$ when the mass has been increased by a factor

$$\frac{M_{2f}}{M_{20}} = \left(\frac{1}{1 - bk_2^2}\right)^{1/b}, \quad b = \frac{3 + a}{2}. \tag{6.18}$$

This is largely independent of the exponent a (or b) since k_2^2 is small. For a non-relativistic white dwarf $k_2^2 \sim 0.2$, and for a main-sequence star above $\sim 0.5\, M_\odot\ k_2^2 \lesssim 0.08$. Thus the mass can only be increased by ~ 8–20%. What happens next is not clear, but it is possible that the gainer develops a differentially rotating outer shell, which may allow it to be substantially bigger. There is evidence in some Algols that the gainer has up to twice the radius expected, and is also in very rapid rotation. We noted in Section 3.5 that RZ Sct has an anomalously large *2.

Although spin-up of the gainer drains angular momentum from the orbit, the effect should not normally be substantial, as the gainer is, by the hypothesis of an accretion disc, well inside its Roche lobe. Nevertheless, it can be allowed for fairly easily in computations by adding (or subtracting) an appropriate term in Eqs (4.74) and (4.76).

There is plenty of observational evidence to confirm that some gainers are indeed rotating substantially faster than synchronously. Van Hamme and Wilson (1990) determined rates of rotation relative to orbital rotation (a factor F, say) in the gainers of 36 Algols – Fig. 6.2. Eleven of them had $F > 5$, three of them $F = 2$–5, and the remainder $F < 2$. RY Per ($F = 10$), V356 Sgr ($F = 5$) and RZ Sct ($F = 6$) are examples of gainers rotating at several times the orbital rate; probably β Lyr is another. Some of the systems of Fig. 6.2 are what we identify as HMS; many are GMS and some are MMS, but it is sometimes hard to discriminate.

In a simplistic model, we would expect that (a) at the beginning the gainer accelerates only moderately during the thermal-timescale mass transfer, because it is fairly large relative to its Roche lobe, which both limits the amount of angular momentum the stream can pick up and allows tidal friction to work towards corotation and (b) as the mass ratio starts to drop well below the reciprocal of its intial value the system widens, and so in contrast to (a) the gainer can spin up strongly. There is some evidence for this: among the longer-period systems ($P > 4.5$ days) there is a substantial correlation of F with q. Among the shorter-period systems there is no such correlation; almost all the gainers are still rotating rather slowly.

However, the simple picture above does not take account of mass loss and angular momentum loss from the system. Four of the 36 systems, all represented by pluses, have $0.35 < X < 1$, where X is the parameter (Section 3.5) that describes how wide the system was at age zero. $X < 1$ indicates a substantial amount of orbital angular momentum loss; and it is likely that this is coupled with substantial mass loss. In fact most of the systems represented by pluses in Fig. 6.2 have $X < 2$, and in fully conservative evolution these would not be able to evolve to $q \lesssim 0.35$ before coming into contact in case AS. To avoid this probably requires a fairly specific amount of mass loss. If the gainer fails to gain *all* of the mass lost by the loser, the acceleration of its evolution, leading to contact, can be mitigated. But on the other hand if it fails to gain *any* of the mass lost, then it is difficult to see how values $\log q \lesssim -0.7$ can be reached. There is scope here for a rather considerable investigation, which would have to include all of mass loss, angular momentum loss, spin-up of the gainer and tidal friction, at least.

6.3 Partial accretion of stellar wind: mode PA

From the point of view of the long-term evolution of a binary system, the things that matter most are (a) the fraction of the material lost by ∗1 which is gained and retained by ∗2, and (b) the specific angular momentum carried off by the fraction of the material which is lost to the system from either ∗1 or ∗2; in other words the parameters $\zeta_1, \zeta_2, \xi, K_1$ and K_2 of Section 4.3. Modelling these parameters is necessarily very tentative. In a relatively simple case, with a radial wind of speed V_w from ∗1 and with no mass loss *from* ∗2, we might estimate ξ/ζ_1 as follows:

$$-\xi - \zeta_1 \equiv \dot{M}_1 \sim -4\pi d^2 \rho V_w, \quad \xi \equiv \dot{M}_2 \sim \pi R_{\text{acc}}^2 \rho |V_w \mathbf{d}/d - \dot{\mathbf{d}}|, \qquad (6.19)$$

$\mathbf{d}$ and $\dot{\mathbf{d}}$ being the separation and relative velocity of the two components, as usual. Hence, using Eq. (6.9) to estimate the accretion radius,

$$\frac{\xi}{\xi + \zeta_1} \sim \frac{1}{4} \left(\frac{GM_2}{d} \right)^2 \frac{1}{V_w |V_w \mathbf{d}/d - \dot{\mathbf{d}}|^3}. \qquad (6.20)$$

Both $\mathbf{d}$ and $\dot{\mathbf{d}}$ might be variable functions of orbital phase, if the orbit is eccentric – see next section. If V_w is small compared with $|\dot{\mathbf{d}}|$ the above formula must clearly be modified to prevent $\dot{M}_2 > |\dot{M}_1|$. However, we would usually not expect V_w to be so small, because, in the absence of the gainer, the wind would have to be expanding with a speed at least equal to the escape speed in order to leave the loser, and the escape speed is itself larger than the orbital speed.

An alternative estimate of $\dot{M}_2$ in Eq. (6.19), which probably represents an upper limit for winds whose velocities are comparable to the orbital velocity, comes from supposing that the accreted fraction of the outgoing wind is given by the fractional solid angle that ∗2's Roche lobe subtends at ∗1. This leads to the estimate

$$\frac{\xi}{\xi + \zeta_1} \sim \frac{1}{4} x_L^2 (M_2/M_1), \qquad (6.21)$$

with $x_L(q)$ coming from Eq. (3.5). It is difficult to be much more precise. Full three-dimensional modelling of the accretion problem should help, but it should be noted that

Table 6.2. *Some Ba stars and possibly related systems*

Name	Spectra	State	P	e	f_2	Reference
ζ Cap	WD + G5IbBa2	WHE	2380	0.28	0.0042	Böhm-Vitense 1980, McClure and Woodsworth 1990
ξ¹ Cet	WD + G7IIIBa0.4	WGD	1642	0	0.035	Böhm-Vitense and Johnson 1985, Griffin and Herbig 1981
ζ Cyg	WD + G8IIIBa0.5	WGE	6489	0.22	0.0227	Dominy and Lambert 1983, Griffin and Keenan 1992
HD31487	? + K1Ba5	wGD	1066	0.045	0.0379	Jorissen *et al.* 1998
105 Her	? + K3IIIBa0.5	wGE	486	0.36	0.135	Scarfe *et al.* 1983
HD 77247	? + G7IIIBa1	wGE	80.53	0.09	0.0050	McClure 1983
HD123949	? + Ba4	wGE	9200	0.97	0.105	Jorissen *et al.* 1998
DR Dra	WD + K0III	WGE	903.8	0.072	0.0035	Fekel *et al.* 1993
V832 Ara	WD + K0III/IIBa	WGE	5200:	0.18:	0.03:	Fekel *et al.* 1993
−43°14304	110kK + K5-M0	WsE	1450	0.2:	0.013	Schmidt *et al.* 1998
V2012 Cyg	S3,1 + ?	SmE	669	0.08	1.23	Jorissen *et al.* 1998
BD Cam	WD + S3.5/2	WSE	597	0.09	0.037	Jorissen *et al.* 1998
AG Dra	SDOe + K3pIIBa	WsE	549	0.13	0.006	Mikolajewska *et al.* 1995, Smith *et al.* 1996
T CrB	Be + M4III	wSS	227.7	0	0.30	Belczyński and Mikolajewska 1998

For V2012 Cyg alone, the mass function is f_1, not f_2.

some winds have an MHD origin, and so the problem may be dominated by MHD rather than just hydrodynamics.

Barium stars are a group of stars that have clearly been affected by mode PA. These are ∼3% of all G/K giants; they are fairly normal, but on close inspection of their spectra show an overabundance of Ba, and a few other elements such as C, Zr (Bidelman and Keenan 1951). The overabundant elements all appear to relate to the s-process (Section 2.3.2), and suggest that some of the material of the star has been subjected to a flux of low-energy neutrons. These neutrons can be generated during the thermal pulses of an AGB star, but it is odd that very few Ba stars (with the possible exception of ζ Cap) are of high enough luminosity, or low enough temperature, to be comprehended as such stars.

The answer (McClure 1983) appears to be that the nuclear processing took place in a *companion* star, formerly an AGB star but now a white dwarf. Most, and arguably all, Ba stars turn out to be spectroscopic binaries, and in a handful the companion can actually be recognised as a white dwarf (Fig. 1.1b). Table 6.2 lists a few. Almost all Ba star orbits have periods in the range 400−4000 days, and this is much like the range expected for stars that are able to reach the AGB. There does not appear to have been much orbital shrinkage, despite the fact that one might reasonably expect mode CE (Section 5.2) for binaries in this period range. Here we attribute the lack of mode CE in Ba-star precursors to the fact that these systems presumably had mass ratios in the range 1–2. Evidently ∗2 of a Ba binary is ≳1 $M_\odot$, since it is massive enough to have left the main sequence; whereas ∗2 in the post-CE binaries of Table 5.1 are more typically M dwarfs than F/G dwarfs. In Section 5.2 we suggested that a mass ratio of more than ∼4 is necessary for the drastic orbital shrinkage characteristic of mode CE. Hence we argue that a relatively massive ∗2 can shake loose the envelope of an AGB star without much orbital shrinkage, while a low-mass ∗2 cannot avoid being caught up in the expanding AGB envelope. This points to modes $\mathcal{C}$W, $\mathcal{C}$U or $\mathcal{C}$UN (Section 4.6), but not mode $\mathcal{C}$UD.

It is likely that Ba stars owe their characteristics to the combination of mode PA with modes NW or EW. Their orbits are commonly more circular than most normal G/K giants (Fig. 1.9b) in the same period range (which presumably have unevolved companions), although at least one has a markedly eccentric orbit. Within the period range of Ba-star binaries there are some, such as DR Dra, which do *not* show significant Ba enrichment, despite having a recognised white-dwarf companion. Possibly this is simply at the low edge of the distribution of Ba abundances that can be expected.

While Ba-rich red giants are reasonably well explained by mode PA, we would expect that there must also exist (a) some red supergiants, single and binary, that are Ba-enriched by virtue of their own intrinsic s-processing and (b) some similar red supergiants that *do* have white-dwarf companions, but that are now further enriching themselves by intrinsic s-processing. Examples can indeed be found of both kinds, and two (V2012 Cyg and BD Cam) are listed in Table 6.2. The former has a mass function too large to allow credibly for a white-dwarf companion.

When the barium-rich red giant evolves sufficiently, it is possible for reverse mass transfer to take place, initially with accretion from a wind but perhaps later (if the wind does not exhaust the envelope) by RLOF. This could make the system a 'symbiotic binary', i.e. one in which the spectrum shows evidence of both a cool component (usually MIII) and a hot component (usually SDOBe). AG Dra and T CrB are fairly typical of these. We might expect the final result to be either a wide or a close pair of white dwarfs. In the last section of Table 5.2, where the results of the two stages, forward and reverse, of evolution leading to WWD binaries are estimated, only EG52, with a long period, seems arguably to have had the parameters of a typical Ba star between the forward and reverse stages; but there is a great deal of guesswork in this table, as there is bound to be in any scenario involving mode CE. However, the mass ratio in a Ba star is not likely to be very extreme. With a white-dwarf mass of $0.55-0.65\ M_\odot$, say, and a mass for $*2$ of typically $1.5-2.5\ M_\odot$, most (according to our tentative criterion of $q \lesssim 4$) should avoid mode CE in its rather severe form.

6.4 Accretion: modes BP, IR

When material lost by $*1$ is accreted by $*2$, energy is released, and a part of this energy may be used to drive off a part of the mass that is trying to be accreted. Indeed, it is not entirely an exaggeration to say that 'wherever theorists talk of accretion, observers see an outflow'. This may happen in a number of ways:

(a) Many objects that involve accretion discs are also seen to be accompanied by bipolar jets apparently emerging from the central region normal to the disc. The disc need not be caused only by binary-star interaction, because the phenomenon is seen in young stellar objects (YSOs) where the disc is simply high-angular-momentum material left over during star formation. Jets are also seen in active Galactic nuclei (AGNi), where it is believed that a central massive black hole is accreting neighbouring material. I will discuss accretion discs very briefly in Appendix F. It may be that a very strong chaotic magnetic field is produced by differential rotation, and that near the centre magnetic pressure overcomes gas pressure, in a vertical direction, driving the jets.

(b)   Accretion discs often show P-Cyg type absorption lines that may be caused by coronal heating above (and below) the disc. A tenuous wind may be driven away from the disc, somewhat like a stellar wind.

(c)   Hydrogen-rich material accreted by a $*2$ which is a white dwarf or neutron star accumulates on the surface of $*2$, but when this layer is massive enough the pressure and temperature at its base, or at the base of the He shell below it, become enough to trigger nuclear burning in a highly unstable way – nova explosions, for a white dwarf (Truran *et al.* 1977), or Type I X-ray bursts for a neutron star (Taam 1981). In the former case, though not the latter, there is easily enough energy for much or all of the accreted layer to be ejected, roughly spherically, and even for some of the underlying white dwarf, if it has mixed to some extent with the accreted layer (MacDonald 1983); so that over long times the mass of $*2$ might actually *decrease* ($\zeta_2 > \xi$, Section 4.3). For a neutron-star gainer, however, the nuclear energy is well below the gravitational binding energy, and the material is unlikely to escape.

In all three cases, a first guess at the amount of angular momentum removed from the system is that it is the same as the orbital angular momentum of $*2$ ($K_2 = 1$). I have referred to this process (Section 4.3) as bi-polar reemission (mode BP).

For systems in which accretion takes place on to a compact gainer ($*1$ probably), the luminosity from the accretion process may dominate the accretion from either component, and may have an important feedback on the evolution of $*2$ since $*2$ may be irradiated by a fraction of the accretion energy – mode IR. The expectation is that $*2$ will be somewhat swollen by irradiation, and that this will increase the rate of transfer. If the mass transfer is on the thermal timescale of $*2$, then the ratio of irradiated luminosity to intrinsic luminosity is

$$\frac{\Delta L}{L_2} \lesssim \frac{\pi R_2^2}{4\pi a^2} \frac{M_1/R_1}{M_2/R_2}. \tag{6.22}$$

This is based on the solid angle subtended by the loser at the gainer, and is likely to be an upper limit since the loser may be partly or largely in the shadow of the accretion disc. R_2/a relates to the mass ratio by Eq. (3.5) if the loser is semidetached. The ratio R_2/R_1 can be large if $*1$ is compact, and this can easily outweigh the modest solid-angle factor. But the shadowing effect may be important, and is hard to assess without a reliable three-dimensional model of the disc, including its optical thickness to the accretion radiation incident on it.

If mass transfer is on a slower timescale than thermal, say mode NE or MB, the effect will be less, but still potentially significant for neutron-star gainers. However many systems with compact gainers – cataclysmic binaries and low-mass X-ray binaries – have M dwarf or even brown dwarf losers, for which mode MB and even mode GR may be on a thermal timescale.

I give here a very brief discussion of 'cataclysmic binaries', also known as 'cataclysmic variables' (CVs), which can be seen as the next stage of evolution of the short-period systems in Table 5.1. A whole book can be written – and has been, Warner (1995) – on this class, but here we must content ourselves with a few paragraphs. CVs include novae and dwarf novae, along with some other non-outbursting but otherwise similar systems. The large outbursts of novae are due to a thermonuclear explosion of hydrogen-rich material recently accreted by a white dwarf from a companion (Truran 1982). The smaller outbursts of dwarf novae, occuring every few weeks, may be due to instability in the accretion process, whereby material lost from the companion accumulates in a disc or ring around the white-dwarf gainer until some criterion is passed that increases the viscosity so that the accumulation is rapidly dumped into the deep potential well of the gainer (Bath and Pringle 1982). Additional 'superoutbursts', occurring at intervals of 6 months to 30 years in a fraction of the systems that usually also show normal dwarf outbursts, appear to be due to a dynamical 3:1 resonance between the orbit of the system and the orbit of material within the Roche lobe of the gainer (Whitehurst 1988). The interested reader should study Warner (1995) for a very comprehensive discussion of these phenomena; here however we shall treat even the rare outbursts of classical novae (perhaps one every $\gtrsim 10^4$ years) as minor perturbations (principally mode BP) on an otherwise fairly steady accretion at a rate estimated from the accretion luminosity as $\sim 10^{-3} - 10^{-5} \, M_\odot$/megayear. This quasi-steady evolution probably spans $10^2 - 10^4$ megayears.

Probably all CVs suffer thermonuclear outbursts every $\sim 10^4$ years, but most have not been observed to do so. When a white dwarf has acquired a thin hydrogen-rich shell of $\sim 10^{-4} \, M_\odot$, the density at the base of this shell is great enough to trigger hydrogen burning. The environment however is electron-degenerate, unlike in a normal red giant burning shell (Section 2.3.2) where the temperature is higher and the density lower. We therefore have an explosive runaway as in a core helium flash. The luminosity reaches the Eddington limit, and the shell is blown away. Thus over a long time span we expect mode BP, in addition to modes MB, GR or NE that presumably drive the long-term evolution. It is possible that mode BP blows away *more* than 100% of the accreted matter, since it is likely that shear instability at the interface may result in mixing of the recently-accreted envelope with deeper carbon-rich (or in some cases neon-rich, as in QU Vul) core. For unusually large white-dwarf masses, approaching the Chandrasekhar limit, the critical shell mass can be $\lesssim 10^{-6} \, M_\odot$, and thermonuclear outbursts correspondingly more frequent, as appears to be the case for recurrent novae.

Many of the shorter-period systems of Table 5.1, with $P \lesssim 3$ days and with a cool (G/K/M) main-sequence companion, may be capable of evolving by mode MB towards a semidetached state. Some with rather more massive companions might evolve by mode NE to a similar state; although the few systems in Table 5.1 with such massive companions (V651 Mon, FF Aqr) have substantially longer periods than one would expect as ancestors of those CVs in Table 6.3 (e.g. BV Cen, U Sco and GK Per) in which $*2$ is substantially evolved. A *wwd* system like 0957-666 (Table 5.1) can evolve by mode GR alone to something like AM CVn (Table 6.3).

Several hundred CVs have known orbital periods (Ritter and Kolb 1998), with a distribution over period showing (a) very strong and rather narrow peaks at ~ 0.075 days and 0.14 days, (b) a conspicuous shortage between 0.094 days and 0.125 days, the 'period gap', (c) a conspicuous shortage below ~ 0.058 days, the 'cutoff', and (d) a tail towards long periods with some but not many above 0.4 days. There is, however, a handful at ultrashort periods, e.g. AM CVn, in which the loser is probably a low-mass helium white dwarf rather than a low-mass main-sequence star. There are also about a dozen systems within the gap, e.g. QU Vul, compared with well over 100 in the peaks on either side. Those that are in the long-period tail tend to be fairly conspicuous because $*2$ is relatively massive and bright, but most recent discoveries have tended to be in the short-period peaks.

Figure 6.3a illustrates the evolution of $*2$ in the M_2, P plane, 'starting' from masses $0.6 + 0.4 \, M_\odot$ and period 0.2 days. Bipolar reemission was set at 90%, so that only 10% of the mass lost by $*2$ was permanently accreted by $*1$. The evolution is under three different assumptions: (a) angular momentum loss is given by the combination of mode GR (Section 4.1) and of mode MB with the specific model of Section 4.5, (b) it is given by mode GR alone and (c) the thermal perturbation to the radius is artificially suppressed, so that $*2$ simply slides down the ZAMS. Because, in the mass range plotted, $*2$ is largely or wholly convective, its response to mass loss is to expand, but because it is substantially less

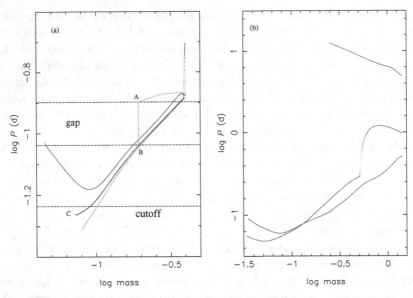

Figure 6.3 Theoretical evolution of *2 of a CV, in the log M_2–log P plane. (a) In all three tracks the system 'started' with parameters $(0.6 + 0.4\,M_\odot; 0.2\,\text{days})$. In the lowest curve, the thermal response of *2 to mass loss was ignored, so that *2 simply slid down the ZAMS. In the middle curve, GR alone drove the evolution; the thermal perturbation was slight except at the very shortest periods of $\lesssim 0.06$ days. In the top curve mode MB (Section 4.5) as well as mode GR drove the evolution. The thermal perturbation was substantial below 0.08 days, and caused the period to 'bounce' at ~ 0.065 days. The positions of the period gap and cutoff are indicated by horizontal lines. The dotted line is hypothetical evolution involving 'interrupted magnetic braking': see text. (b) Three tracks all start with masses $0.9 + 1.4\,M_\odot$, and periods of 0.5, 1.0 and 5 days. Roche-lobe overflow began at $M_2 \sim 0.5\,M_\odot$ in the intermediate-period system. The long-period system reached RLOF only briefly, shortly before *2 shrank away from its Roche lobe to become a well-detached white dwarf of low mass.

massive than *1 this does not lead to dynamic mode SR3 RLOF (Section 3.3). Instead it leads to steady mode SR2 RLOF – the timescale of modes MB plus GR on path (a) being somewhat coincidentally comparable to the thermal timescale, which is very long at these low masses. Path (c) terminates at about $0.08\,M_\odot$ because there are no ZAMS (i.e. thermal-equilibrium) stars below this mass. Path (b) terminates at a slightly lower mass. Such masses are allowed because the component is no longer in thermal equilibrium, but the approximate equation of state in our code becomes unreliable at these very low temperatures and high densities where non-ideal-gas effects dominate it. In fact, any equation of state is rather uncertain in this regime. On path (a) *2 was sufficiently expanded and heated that it avoided breakdown of the equation of state, but the models are nevertheless very uncertain.

Figure 6.3a suggests a reason for the observed cutoff at $P \sim 0.058$ days: it may be owing to the 'bounce' caused by increasing thermal disequilibrium, although only if mode GR is dominant and mode MB is not at a substantially greater rate. This – and another reason, see below – suggests that we should explore the possibility that mode MB is somehow switched off when *2 is below some critical mass or temperature. Let us suppose that above a certain mass mode MB is substantially *stronger* than the model used, by a factor of ~ 10. We sketch

Table 6.3. *Some cataclysmic and related binaries*

Name	Spectrum	State	Class	P	M_1	M_2	R_2	Reference
AM CVn	He em	WcS	USP	0.012		0.04:		Harvey *et al.* 1998
GP Com	He em	WcS	USP	0.032				Marsh *et al.* 1991
OY Car	SDBe + M7-M8	WMS	SU	0.063	0.685	0.07	0.127	Wood *et al.* 1989
ER UMa		WMS	ER	0.065				Thorstensen *et al.* 1997
GQ Mus		WMS	CN, MP, SSX	0.065				Shanley *et al.* 1995
HT Cas	SDBe + M5.4V	WMS	DN	0.074	0.61	0.09	0.154	Horne *et al.* 1991
T Pyx		WMS	RN	0.075:				Patterson *et al.* 1998
Z Cha	SDBe + M5.5V	WMS	SU	0.075	0.84	0.125	0.17	Robinson *et al.* 1995
ST LMi	SDBe + M5-6V	WMS	PL	0.079	0.76	0.17	0.20	Smith and Dhillon 1998
QU Vul		WMS	CN	0.112				Shafter *et al.* 1995
AM Her	SDBe + M4$^+$V	WMS	MP	0.129	0.44	0.29	0.33	Smith and Dhillon 1998
MV Lyr	SDBe + M5V	WMs	VY	0.134				Beuermann and Pakull 1984
UU Aqr	SDBe + ?	WMS	NL	0.164	0.67	0.20	0.34	Baptista *et al.* 1994
U Gem	SDBe + M4$^+$V	WMS	UG	0.177	1.26	0.57	0.51	Smak 1993
DQ Her	SDBe + M3$^+$V	WMS	CN, IP	0.194	0.60	0.40	0.49	Horne *et al.* 1993
UX UMa	SDBe + K M	WMS	UX	0.197	0.43	0.47		Shafter 1984
EM Cyg	SDBe + K3V	WMS	ZC	0.291	1.12	0.99	0.87	North *et al.* 2000
AC Cnc	SDBe + K0V	WMS	NL	0.300	0.82	1.02	0.92	Schlegel *et al.* 1984
BT Mon	SDBe + G8V	WMS	CN, SW	0.334	1.04	0.87	0.89	Smith *et al.* 1998
AE Aqr	SDBe + K4V	WgS	IP	0.412	0.79	0.50	0.86	Smith *et al.* 1998
V Sge	WN: + B8:	emc	SSX	0.514	0.9:	3.3:	2.1:	Herbig *et al.* 1965, Smak *et al.* 2001
BV Cen	SDBe + G6IV-V	WGS	DN	0.611				Williger *et al.* 1988
U Sco	SDBc + F8V	WhS	RN	1.23	1.55:	0.88	2.1	Thoroughgood *et al.* 2001
GK Per	SDBe + K1IV	WGS	CN, DN, IP	2.00	0.9:	0.5:	2.5:	Morales-Rueda *et al.* 2002

Some of the major classes are

CN – classical nova – one large outburst ($\Delta V \sim 10\text{--}15$) recorded
RN – recurrent nova – somewhat smaller outbursts, every ~ 30 years
DN – dwarf nova   moderate outbursts ($\Delta V \sim 3\text{--}5$) every few weeks

> UG – U Gem – repetitive lows last weeks, outbursts last days
> ZC – Z Cam – occasional long-lasting plateaus between maxima and minima
> SU – SU UMa – fairly regular, $\sim$ annual, 'superoutbursts' lasting 2 weeks, in addition to UG behaviour
> ER – ER UMa – as SU, but more frequent superoutbursts, $\sim$ monthly

NL – nova-like system – no outburst noted, but similar to CN, DN between outbursts

> UX – UX UMa – fairly steady accretion, no substantial outbursts
> VY – VY Scl – occasional cessation of mass transfer, lasting years
> SW – SW Sex – single-peaked emission; hot spot dominates disc

SSX – supersoft X-ray source – powerful source of soft X-rays, usually from rapid, steady accretion
USP – ultra-short period – hydrogen absent from spectra, $*2$ a He white dwarf
MP – magnetic polar – accretion column, not disc; flow dominated by magnetic field; white dwarf locked in corotation with binary
IP – intermediate polars – combination of column and disc; white dwarf rotation slow but not synchronous.

by hand a hypothetical track, the dotted curve in Fig. 6.3a. Starting from the top right, the star evolves well above the ZAMS to point A. Then mode MB is switched off. $*2$ shrinks on a thermal timescale inside its Roche lobe, mass transfer ceases, and the system becomes very faint (since $\sim$90% of the luminosity from such CVs is accretion luminosity). However, the Roche lobe also shrinks, more slowly, on the timescale of mode GR. When $*2$ stabilises on the ZAMS at point B, it will fill its Roche lobe again after a substantial wait, and evolution will continue along curve BC.

Such an assumption kills two birds with one stone. It explains the period gap (provided the critical mass is $\sim$0.19 $M_\odot$; see Fig. 6.3a), because stars on the portion AB are very inconspicuous, and it explains the period cutoff also. Note that we need the cessation of mode MB to be rather abrupt; a steady diminution would leave the star following a steady RLOF path, without getting substantially but temporarily fainter as required.

It is usually suggested that the sudden cessation of mode MB is due to the fact that low-mass stars change from having radiative cores to being fully convective at $\sim$0.35 $M_\odot$. However, the transition has to occur at about 0.19 $M_\odot$, to place the gap at the right period range. In fact, there are many later and lower-mass M dwarfs known that are conspicuously active, such as the prototype flare star UV Cet. But UV Cet rotates much more slowly. Perhaps rapid rotation alters (i.e. supresses) the dynamo, although most indications are that activity *increases*, or at least saturates, with increasing rotation.

Another difficulty may be that the model assumes that all CVs start somewhere near the upper right of Fig. 6.3a, perhaps even beyond the right-hand margin of the figure. But many of the precursor systems in Table 5.1 have a $*2$ that is already of quite low mass. If, say, KV Vel evolves in the fullness of time to a CV ($*2$ being above the critical mass at 0.23 $M_\odot$), it will become a CV squarely in the middle of the period gap. There are several other such precursor systems; they would have to avoid the mass range 0.17–0.35 $M_\odot$ in order not to do so. Although the statistics are very poor, the precursor detached $*2$s do not seem to do this, and there seems no reason why they should.

Patterson (1998) has discussed very carefully the implications for theoretical evolution that can be drawn from the short-period systems near the cutoff. He concludes that angular momentum loss in excess of GR by about 50% is desirable, and that after 'bouncing' the systems must dissipate themselves rather quickly, in $\sim$10% of the time that GR would allow. Evidently the distribution over period among CVs convolves both the current evolution mechanism and the distribution over mass (M_2) of precursor systems. The distribution of CVs over period has strong features in it that cry out for interpretation, but I feel that we are far from understanding them at present.

The modest subset of CVs that show substantial nuclear evolution in $*2$ – e.g. BV Cen to GK Per, Table 6.2 – will evolve in a very different manner to the low-mass systems that are incapable of mode NE. Whether the period increases or decreases will depend on the balance of mode MB to mode NE, but it is clear that mode MB cannot always dominate or something like GK Per would not exist. It is likely therefore that these systems evolve to longer, not shorter, periods, with a bifurcation at around $\sim$1 $M_\odot$ for the *original* mass of $*2$. The outcome in the long term would often be a wwD binary, with a period of a few days. $*1$ would be unusually massive, and $*2$ of unusually low mass.

Figure 6.3b shows possible evolution of systems with different initial periods, but the same initial masses ($0.9 + 1.4$ $M_\odot$). All of modes NE, EW, MB, PA and BP were included, along with (reverse) RLOF. Even the shortest-period system is able to do some significant nuclear evolution in its early stages, so that by the time that $*2$ is reduced to the mass where it becomes fully convective its uniform hydrogen abundance is reduced from 0.7 to 0.55. The next system shows modes NE and EW winning at first, so that the period increases, but then mode MB takes over and $*2$ fills its Roche lobe when it is reduced to $\sim 0.5\,M_\odot$. Its small, nearly-exhausted, core persists without being entirely mixed by convection, and so the period ultimately bounces at a somewhat shorter period than in the first case. In the widest system modes NE and EW dominate throughout, and the system widens from $P \sim 5$ days to $P \sim 13$ days, at which point $*2$ detaches from its Roche lobe and becomes a white dwarf of $\sim 0.25\,M_\odot$ with a rather thick hydrogen shell ($\sim 0.01\,M_\odot$).

It is possible for a white-dwarf gainer to grow fairly steadily in mass, if it is fed with fuel at a rate that is not much less than the rate at which a white dwarf is fed fuel if it is the core of a conventional red giant. In this case the white dwarf can be expected to retain most of its accreted mass, instead of blowing it away in intermittent CN outbursts. Some 'supersoft X-ray sources' (SSXs) may be in such situation. It is possible that the white dwarf will grow to the Chandrasekhar mass, and suffer 'accretion-induced collapse' (AIC), to a neutron star.

V Sge is a binary that has been known for a long time. It is not clear that it fits into any regular class of CVs, although it is usually grouped with them. The nature of *both* components has been arguable, but Smak *et al.* (2001) opt for the combination of a somewhat massive C/O white dwarf core surrounded by a helium-burning shell, and a late B main-sequence star; a possible product of case $\mathcal{C}$ evolution, which may have started with parameters not unlike that hypothesised for the precursor of IK Peg (Table 5.2). Both components of V Sge are of much greater luminosity than in normal CVs. They both appear to fill, or slightly overfill, their Roche lobes, but nevertheless to have very different temperatures (65 and 12 kK). It is unlikely that the outer layers are in a simple hydrostatic configuration, and there is evidence of a hot gaseous envelope around the system, fed by mass loss from the hot component. The luminosity varies erratically between lower and higher values, but not in the manner of dwarf nova outbursts.

There are observational indications of mass loss from the *system*, probably from $*1$ alone. It is not clear that there is any mass *transfer* at all; both components might be radiating intrinsic (nuclear) luminosity, while the Wolf–Rayet-like component may produce an intrinsic wind. But the fact that both components appear to be as large as their Roche lobes makes it rather unlikely that there is no mass transfer. If the system is simply the last, slightly detached, stage of case $\mathcal{C}$ RLOF that has suffered mode CE, we would expect $*1$ to shrink rapidly inside its lobe. But because of the mass ratio, it is quite likely that $*2$ loses mass on a thermal timescale (mode SR2) as a result of evolution that was originally (post-mode-CE) on a nuclear timescale. Some of this mass may be accreted, burn, and add to the core mass of $*1$, while some may be reemitted (mode BP). Since $L_1 \sim 400 L_2$, it is possible that $*1$ is burning a substantial fraction of the mass lost by $*2$, i.e. that $*1$'s nuclear timescale is comparable to $*2$'s thermal timescale. If indeed the process manages to average to a fairly steady rate of transfer and accretion, it does not seem unlikely that $*1$ will grow in mass to the Chandrasekhar mass.

However, it is something of a problem that we would expect rapid (thermal timescale) RLOF, because of the considerable mass ratio. The observed timescale of period decrease is $P/\dot{P} \sim -3$ megayears, and Smak *et al.* (2001) estimate the ratio of mass lost from the system to mass transferred (from $*2$ to $*1$) as 2/3. These correspond to our non-conservative parameters (Section 4.3) $\xi:\zeta_1:\zeta_2 = -3:2:0$. Then Eq. (4.42) – ignoring the terms in R_A and

Table 6.4. *Some low-mass X-ray and possibly related binaries*

Name	Alias	Spectrum	State	P	M_1	M_2	R_2	Reference
J1751-305		2.5ms + ?	NcS	0.029	0.0000013^a			Markwardt *et al.* 2002
UY Vol	J0748-6745	XR + ?	NMS	0.159	–	–		van Paradijs *et al.* 1988
V616 Mon	J0622-0020	XR + K5-7	BMS	0.323	10:	0.7:		Johnson *et al.* 1989
V818 Sco	Sco X-1	XR + A:	NMS	0.787	–	–		Priedhorsky and Holt 1987
HZ Her	Her X-1	1.2s+ A-FIII	NMS	1.70	1.3:	2.2:	3.9:	Deeter *et al.* 1991
V1033 Sco	J1654-3950	XR + F6IV	BMS	2.62	6.6	2.8	5.2	Shahbaz 2003
V404 Cyg	J2024 + 3352	XR + K0IV	BGS	6.47	6.8^a	0.4^a	5.0^b	Casares and Charles 1994
V1341 Cyg	Cyg X-2	XR + A9III	NHS	9.84	1.8:	0.6:	7.5	Orosz and Kuulkers 1999
J1012 + 5307		5.3ms + WD	NWD	0.605	1.6^a	0.12^a		van Kerkwijk *et al.* 1996a
J1857 + 0943	1855 + 09	5.4ms + WD	NWD	12.3	1.5	0.26		Kaspi *et al.* 1994b
J1640 + 2224		3.2ms + WD	NWD	175	0.0058^a			Lundgren *et al.* 1996

^{a} Mass function or, if two values, $M_i \sin^3 i$.
^{b} $R_2 \sin i$.

R – along with Eqs (4.36) gives $M_2/\dot{M}_2 \sim -21$ megayears. This is quite a lot slower than the expected thermal timescale. Perhaps the mass transfer is somehow stabilised by mode IR, which must be very important in this system.

I believe the system is still highly problematic. I have argued, regarding mode CE, that only systems with rather extreme initial mass ratios ($q_0 \gtrsim 4$) undergo conventional mode CE with a large period shrinkage; milder initial mass ratios may result in much mass loss but relatively little period shrinkage, as in IK Peg or V651 Mon (Table 5.2). Then add to this the difficulty of reconciling the timescales. But whatever the past evolution, future evolution seems quite likely to lead to AIC in $\lesssim 1$ megayear.

CVs are a class of binary where accretion energy is often the dominant contribution to the observed energy output, apart from occasional thermonuclear outbursts. This is also the case for low-mass X-ray binaries (LMXBs). Here the dominant energy is typically in X-rays, but even in the visual region much of the observed energy often comes from the accretion disc rather than either stellar component. The gainer is either a neutron star or a black hole. Mode IR is likely to be substantially more important than for CVs – see Eq. (6.22).

Some examples are given in Table 6.4. Although the origin of most CVs can be plausibly accounted for by mode CE, followed by mode MB or mode NE, or both, it is harder to see how LMXBs are formed. The transition $sMS;CE \rightarrow WMD$ seems reasonably natural, but it is not so clear how a neutron star would emerge from something similar. Currently there appear to be three main suggestions.

Firstly, the AIC process referred to above might convert a CV directly into an LMXB. A major problem is that this will not account for black holes, which are several times more massive than neutron stars. However, the existence of objects like V Sge, and other SSXs, does suggest that at least some LMXBs may form in this way.

Secondly, mode CE can occur when the core of the large star is still burning helium, or even somewhat earlier. Stars with $M_{10} \gtrsim 8\ M_{\odot}$, i.e. massive enough to form a neutron star, evolve rapidly across at least the first half of the Hertzsprung gap before igniting helium. Mode CE can be expected if q_0 is large, whether RLOF begins in case $\mathcal{B}$ or case $\mathcal{C}$. We obviously need a large q_0 to produce the combination of a neutron star (or black hole) and an A/F/G/K dwarf. Unfortunately, we do not have a clear idea of how common high-q_0 systems are. The largest q_0 ($\sim$8) measured fairly reliably in *MMD* systems is in EN Lac (Table 2.2), but such systems must be hard to detect.

Thirdly, supergiants with a neutron-star or black-hole core (TŻOs) *may* be the natural outcome of evolution of those HMXBs (Section 5.3) that have relatively short periods. If such a system lies within a wider triple, it is possible that the distant companion, presumably a low-mass star, is caught up in the envelope of the TŻO, and spirals in by mode CE. VV Ori (Table 4.10) might be such a triple, although we would prefer one with a longer outer period.

A substantial number of radio pulsars with pulse periods of a few milliseconds are found to be binary ('MSPBs'), and with orbital periods ranging from a few hours to a few years. The companions are very inconspicuous. They are conjectured to be white dwarfs; in a few cases a white dwarf is actually seen at the radio position. They could be the descendants of those LMXBs in which mode NE dominates over mode MB, e.g. HZ Her and V1341 Cyg. It is, however, surprising in that case that there are several with period under 1 day, such as J1751-305. We would expect a dichotomy, with shorter-period LMXBs evolving to shorter period still by mode MB with no mode NE, and remaining as LMXBs, while the longer-period LMXBs evolve by modes NE and SR1 to longer period still, until the envelope is exhausted and a white dwarf core is left. We seem to require that several systems 'start' in an intermediate regime where at first mode NE allows a small white-dwarf core to form while the binary widens, but then mode MB becomes dominant and shrinks the orbit while it exhausts the envelope. If this is so, it may be quite a powerful restriction on models of mode MB.

At the beginning of this section, I indicated three cases, (a)–(c), in which mode BP might be expected. In cases (a) and (b), since the energy being tapped is essentially energy from the accretion process itself, it is unlikely that more than a modest fraction of accreted mass is blown away. The third case does not apply to neutron stars, since even the nuclear energy from hydrogen-rich accretion is not enough to push material from the surface of the neutron star to infinity. Thus we appear to be arguing, for a neutron-star gainer, that in any circumstances it would be likely to accrete most of the material that falls into its potential well, rather than to reemit much of it. This is something of a problem, because (a) several MSPBs (three in Table 6.4), are seen with faint white-dwarf companions, presumably the relics of companions that were once $\gtrsim 1\ M_\odot$, and (b) neutron stars probably have an upper limit to their mass, because of their equation of state, and although this limit is not well known it would be surprising if it were over $\sim 2\ M_\odot$. If, for example, HZ Her evolves in a largely conservative way, with RLOF and with very little mode BP, one would expect it to evolve to a state with parameters say $(3.4 + 0.4\ M_\odot; 30\ \text{days})$. *1 would probably become a black hole in the course of this, though with substantially less mass than any currently-known black hole (as in V616 Mon).

A further reason for believing that neutron stars do not in fact accrete much mass comes from the fact that, as in Eq. (6.18), they would be spun up to breakup by the accretion of a rather modest amount of mass from the inner edge of a Keplerian disc. We cannot entirely discount the possibility that the neutron star is in a state of extreme *differential* rotation, in which case centrifugal support might allow it to be of substantially greater mass than any non-rotating limiting mass. But J1857 + 0943, the only system so far with an SB2 character, seems to argue against this. Although the inclination is unknown, eclipses being exceedingly improbable, the white-dwarf mass on the assumption that $i \sim 90°$ is about what is expected from the orbital period.

However, our non-conservative model of Section 4.5 predicts that LMXBs are quite likely to evolve mainly by modes EW, MB and PA, rather than RLOF. If the neutron star accretes only a modest proportion of the wind generated by dynamo activity in the companion, while the rest of the wind blows to infinity, then it may naturally increase its mass by only a modest amount, while accelerating its spin to the short periods observed.

Our canonical model of modes NE, EW and MB gives evolution in the $\log P$–$\log M_2$ plane which is, not surprisingly, little different from the CV evolution of Fig. 6.3. However, as mentioned above, it is not easy to see what starting conditions might produce the very short-period system J1751-305 in Table 6.4. The short-period (AM CVn) systems of Table 6.3 seem like legitimate descendants of *wwd* systems like 0957-666 in Table 5.1, which may themselves be legitimate descendants of post-Algols after mode CE in reverse RLOF. But it is not so easy to see how this kind of progression would occur if *1 were a neutron star rather than a low-mass white dwarf. Possibly the system 'started' as in Fig. 6.3b but somewhere intermediate between the initial periods of $\sim$1 day and $\sim$5 days; presumably very close to some critical period defining the dichotomy between late contraction and continued expansion. Such models were suggested by Nelson and Rappaport (2003), although as they did not model mode EW they favoured a somewhat shorter initial period.

Surprisingly many LMXBs, as well as MSPBs, are found in globular clusters. This suggests that mode DE is much more important there than in the bulk of the Galaxy. Neutron stars are, no doubt, formed from the massive stars that existed in very young globular clusters, but it is surprising that any have been retained, given the asymmetric kicks that seem to be necessary to produce the high proper motions of pulsars in the Solar neighbourhood. A bimodal distribution of kicks may be necessary. But given that apparently several are retained by their parent globular clusters, they presumably gravitate slowly to the centre once the mean mass of stars is reduced below $\sim$1.4 $M_\odot$, and there can interact by an exchange reaction (mode DE) with close primordial K/M dwarf pairs.

6.5   Accretion in eccentric orbits

Most binaries with $P \gtrsim 5$ days have eccentric orbits, at least initially. In many binaries with $P \sim 0.5$–100 years, we expect *1 to develop a wind, as a red giant or supergiant, and *2 is liable to accrete from this wind, even if the orbit is non-circular. Although tidal friction may well circularise the orbit if and when *1 becomes large enough, the wind may become significant before then. There are several binaries, such as some in Table 6.2, whose present evolutionary state suggests a previous interaction, and yet whose orbits are obstinately non-circular.

Although the concept of a Roche lobe only applies to circular orbits, we may expect that Eq. (3.10) can be loosely generalised to say that if at periastron the radius of *1 is a fraction $x_L(q)$ of the separation then some kind of overflow should take place. This means that P_{cr}, the period at which interaction first occurs as the star's radius expands, is in effect increased, or equivalently we should compare it to an 'effective orbital period' P' defined by

$$P' \sim P(1 - e)^{3/2}, \tag{6.23}$$

such that the semimajor axis of the effective circular orbit is the same as the periastron separation in the actual orbit. However, some interaction affecting the orbit might take place significantly before that, because of wind – either wind which leaves the system, or wind which is partially accreted by the companion. There is also the possibility of episodic accretion: this might take place mainly near periastron since the wind density will be highest there, but might alternatively be peaked to *apastron* since the relative velocity is slowest there and therefore the accretion radius is largest – Eqs (6.9), (6.22). Even without wind, and supposing that tidal friction has not in fact circularised the orbit, we expect some kind of episodic accretion once the star exceeds its 'Roche lobe' at periastron. A variant of this is that *1 might be a pulsator (Mira, Cepheid), which might transfer mass at some periastra, those which coincide, more or less, with a maximum stellar radius, but not at others.

It is difficult to model in detail the accretion process in such cases, but we can seek guidance in very simple models. In Appendix C(e) a model is put forward for dealing with the perturbation to an orbit due to mass loss with or without mass transfer. This model comes from considering the simplest generalisation of the usual momentum equations $M_1\ddot{\mathbf{d}}_1 = -M_2\ddot{\mathbf{d}}_2 = \mathbf{F}(\mathbf{d}_1 - \mathbf{d}_2)$ that

- (a) is invariant under Galilean transformations $\mathbf{d}_i \to \mathbf{d}_i + \mathbf{U}t$, where $\mathbf{U}$ is a constant,
- (b) is symmetrical with respect to suffices 1 and 2, so that it is unimportant whether one or the other star (or both) is losing mass,
- (c) leads, in the case of mass transfer with no wind mass loss, e.g. conservative RLOF in a circular orbit, to the familiar Eq. (3.13) and
- (d) leads, in the case of wind mass loss with no transfer to the familiar Eq. (4.35).

Replacing $M_1\ddot{\mathbf{d}}_1$ by $\mathrm{d}(M_1\dot{\mathbf{d}})/\mathrm{d}t = M_1\ddot{\mathbf{d}}_1 + \dot{M}_1\dot{\mathbf{d}}_1$ will not do, since the result is not Galilean invariant.

Let us write, as in Section 4.3, the following equations for the rates of change of mass:

$$\dot{M}_1 = -\zeta_1 - \xi, \quad \dot{M}_2 = -\zeta_2 + \xi, \quad \dot{M} = -\zeta_1 - \zeta_2; \qquad (6.24)$$

the ζ are the rates of loss to infinity, and ξ is the rate of transfer. Some or all of these we now imagine to be dependent on orbital phase. Then suitable equations that are manifestly Galilean invariant and symmetrical are

$$M_1\ddot{\mathbf{d}}_1 = -\frac{GM_1M_2\mathbf{d}}{d^3} + (\dot{\mathbf{d}}_1 - \mathbf{V})\xi, \quad M_2\ddot{\mathbf{d}}_2 = \frac{GM_1M_2\mathbf{d}}{d^3} - (\dot{\mathbf{d}}_2 - \mathbf{V})\xi, \qquad (6.25)$$

where

$$M\mathbf{V} \equiv M_1\dot{\mathbf{d}}_1 + M_2\dot{\mathbf{d}}_2 . \qquad (6.26)$$

Combining these into an equation for the relative motion,

$$\ddot{\mathbf{d}} = -\frac{GM\mathbf{d}}{d^3} + \mathbf{f}_{\mathrm{WT}}, \quad \mathbf{f}_{\mathrm{WT}} \equiv -\frac{\xi}{M_{\mathrm{WT}}}\mathbf{d}, \quad \frac{1}{M_{\mathrm{WT}}} \equiv \frac{1}{M_2} - \frac{1}{M_1}, \qquad (6.27)$$

and the label WT stands for wind transfer. This equation does have also the required property, as shown in Appendix C(e), of giving the two familiar results of conditions (c) and (d) above, in the appropriate limiting circumstances.

We can now determine the effect of the term $\mathbf{f}_{\mathrm{WT}}$ on a general Keplerian orbit, using the methodology of the LRL vector as described in Appendix C. In particular, we look for any general guidance on whether the eccentricity can be expected to decrease or increase, if the factor ξ in $\mathbf{f}_{\mathrm{WT}}$ is phase-dependent. Consider for example a situation where $\zeta_2 = 0$ (no wind out of *2) and where $\zeta_1 + \xi = -\dot{M}_1 \sim$ constant, but where ξ, and hence also ζ_1, is strongly phase-dependent. Using angular brackets for an average over an unperturbed Keplerian orbit, the rates of change of the specific angular momentum vector $\mathbf{h} \equiv \mathbf{d} \times \dot{\mathbf{d}}$, and of the LRL vector $\mathbf{e}$ which gives the direction of the semimajor axis and the magnitude of the eccentricity, are given – Eqs. (C106) – by

$$\dot{\mathbf{h}} = -\frac{1}{M_{\mathrm{WT}}}\mathbf{h}\langle\xi\rangle,$$

$$\dot{\mathbf{e}} = -\frac{2}{GMM_{\mathrm{WT}}}\langle\xi\dot{\mathbf{d}}\rangle\times\mathbf{h} - \frac{1}{M}\left(\left\langle\frac{\xi\mathbf{d}}{d}\right\rangle + \mathbf{e}\langle\xi\rangle\right). \qquad (6.28)$$

This is a special case of the more general Eq. (C100), which applies if any or all of the ζs and ξ are non-zero. We consider the following three models for ξ, which are loosely related to the discussion on accretion in Section 6.4, but are mainly chosen for their analytic convenience:

$$\xi = \frac{|\dot{M}_1|d_0^2}{d^2}\left(\frac{V_0}{|\dot{\mathbf{d}}|}\right)^{2j}, \qquad j = 0,\ 1\ \text{or}\ 2. \qquad (6.29)$$

The quantities d_0 and V_0 can be subsumed into one constant, but are given separate names to indicate their dimensionality: like $\dot{M}_1$, they are assumed constant on an orbital timescale, but might well vary on a longer timescale. The factor $1/d^2$ imitates the fact that the density of the wind can be expected to drop off with distance, and the $|\dot{\mathbf{d}}|$-dependence imitates an expected dependence on accretion radius. From the estimate of Eq. (6.23) we might expect that $j\sim0$ is relevant if the wind speed is high relative to the orbital speed, and $j\sim1.5$ is relevant if it is low. Equation (6.29) with $j=1.5$ leads to a tedious elliptic integral, but we hope to estimate the behaviour in this limit by looking at $j=1$ and 2.

The dependences on d and on $|\dot{\mathbf{d}}|$ in Eq. (6.29) work in opposite directions. The d-dependence alone gives most accretion at periastron, whereas the $|\dot{\mathbf{d}}|$-dependence alone gives most at apastron. The case $j=1$ is neutral: accretion has the same (local maximum) rate at both apses.

Performing the averages with respect to time over the unperturbed Keplerian orbit, with the help of the basic equations (C4)–(C9) of Appendix C, we obtain

$$j=0:\quad \dot{\mathbf{h}} = -\frac{\dot{M}_0}{M_{\mathrm{WT}}}\mathbf{h}, \qquad \dot{\mathbf{e}} = -\dot{M}_0\mathbf{e}\left(\frac{2}{M_{\mathrm{WT}}}+\frac{1}{M}\right); \qquad (6.30)$$

$$j=1:\quad \dot{\mathbf{h}} = -\frac{\dot{M}_0}{M_{\mathrm{WT}}}\mathbf{h}, \qquad \dot{\mathbf{e}} = 0; \qquad (6.31)$$

$$j=2:\quad \dot{\mathbf{h}} = -\frac{\dot{M}_0}{M_{\mathrm{WT}}}\mathbf{h}\frac{1+e^2}{1-e^2}, \qquad \dot{\mathbf{e}} = \dot{M}_0\mathbf{e}\left(\frac{2}{M_{\mathrm{WT}}}+\frac{1}{M}\right) \qquad (6.32)$$

where

$$\dot{M}_0 \equiv \frac{|\dot{M}_1|d_0^2}{a^2(1-e^2)^2}\left(\frac{V_0^2 a}{GM}\right)^j. \qquad (6.33)$$

Because of the symmetry of ξ about the major axis, there is no apsidal motion in these models, i.e. no term in $\dot{\mathbf{e}}$ proportional to $\mathbf{e}\times\mathbf{h}$. Thus $\mathbf{e}$ changes only in magnitude. We see that, in the $j=0$ case, the eccentricity decreases to a minimum at a specific mass ratio, which is $M_1/M_2 = 0.78$, and then increases again. In the $j=2$ case the eccentricity reaches a maximum at that value. One might wonder if this critical mass ratio is an artefact of the specific form of Eq. (6.29), but variants of this usually give a similar answer because of the more general character of Eq. (6.28). This typically leads to a critical value of M_1/M_2 that may depend slightly on **e**.

A different but related situation is one where, perhaps because $*1$ is a pulsator, say a Cepheid or Mira, mass transfer is very strongly peaked at periastron – perhaps not every periastron, if the pulsator happens to be at minimum radius, but nevertheless only near periastron if at all. A crude model of this can be obtained with a δ-function mass-exchange rate. To some extent the process might even be conservative in the sense that all the matter lost by the pulsator is gained by the companion. So let us consider $\zeta_1 = 0 = \zeta_2$, and

$$\xi = \dot{M}_0 P \sum_n \delta(t - t_n), \tag{6.34}$$

a delta-function pulse of mass transfer at periastron, when $t = t_n$ and $\dot{M}_0$ is a given (positive) constant. The equation for the evolution of **h** is the same as in Eq. (6.28), but the **e**-equation is simpler because of assumed mass conservation:

$$\dot{\mathbf{e}} = -\frac{2}{GMM_{\mathrm{WT}}} \langle \xi \dot{\mathbf{d}} \rangle \times \mathbf{h}. \tag{6.35}$$

Averaging over the Keplerian orbit, we obtain

$$\dot{\mathbf{h}} = \frac{\dot{M}_0}{M_{\mathrm{WT}}} \mathbf{h}, \quad \dot{\mathbf{e}} = -\frac{2\dot{M}_0}{M_{\mathrm{WT}}} \frac{1+e}{e} \mathbf{e}. \tag{6.36}$$

Thus, for $M_1 > M_2$ initially ($M_{\mathrm{WT}} > 0$), the eccentricity reduces to a minimum at $M_1 = M_2$, slightly earlier than in the previous case. Of course the model fails if e is reduced to zero before this: episodic accretion at periastron makes no sense once $e \sim 0$. Presumably if e reaches zero it stays zero, but if it does not reach zero before $M_1 = M_2$ then it increases again subsequently.

Tout (private communication) has suggested that a more valid model for $\mathbf{f}_{\mathrm{WT}}$ – Eq. (6.27) – would have $M_{\mathrm{WT}} = M_2$ rather than the definition (6.27). This is because it is unclear what happens to the angular momentum of the matter that is being accreted by $*2$. The matter from $*1$ gains angular momentum from Coriolis force as it flows towards $*2$; this is why an accretion disc is formed. As the matter flows in through the disc, its angular momentum flows out. If $*2$ is quite small (as is usually supposed) most of the angular momentum must be 'lost', but it is not clear where to. In principle, it might either escape from the system, carried by a very small amount of mass from the outer edge of the disc, or it might be reaccreted by $*1$ and thus get back into the orbit. Equation (6.27) indirectly assumes the latter: it assumes that all the matter getting to infinity carries only the specific orbital angular momentum of the star from which it came. The amended version – $M_{\mathrm{WT}} = M_2$ – assumes the former. Without a detailed hydrodynamic (or MHD) model, it is not clear which extreme is closer to the truth. Note that the amended version is no longer symmetrical as between $*1$ and $*2$.

6.6 Conclusions

There can be no doubt that RLOF is the most important way in which the evolution of a star can be affected by the presence of a binary companion. If the RLOF is largely conservative of both mass and orbital angular momentum there are several straightforward consequences that should follow. However, there are certainly some non-conservative possibilities that can arise in various situations; and the fact that even isolated stars are quite complicated entities means that we must always be cautious in applying 'standard' theoretical conceptions to real systems.

A number of fairly general problems are listed here. The first five are those considered to be most important, and the remainder are in no particular order.

(a) Even the evolution of *single* stars presents severe problems. I would like to emphasise the fact – Section 2.3.5 – that in several (arguably) non-interactive wide binaries with well-determined parameters where one component is a red giant and the other is a main-sequence dwarf, the dwarf's radius is *considerably* larger than the mass-ratio would imply in an alarming fraction of cases.

(b) Many of the closest systems (case A) must evolve into contact, and evolution beyond this point is very poorly understood. Some close binaries will probably merge, which means that a proportion of currently single stars may be former binaries, and of current binaries may be former triples.

(c) Common-envelope evolution is a particularly uncertain area, and it is not possible at the moment to make a clear a priori estimate of the period of a system that emerges from a common envelope. Nevertheless it is clear that some compact highly-evolved systems were created this way: we can only parametrise the situation, and hope that observed systems will give some sensible value for the parameters introduced.

(d) Many binaries containing cool stars show evidence of dynamo activity considerably in excess of similar single stars. Many also show some evidence of substantial loss of angular momentum or mass. These problems may be related; we need a non-conservative model of them. Such a model must also include tidal friction. Models of these processes have been described here, but are extremely tentative. Much work has to be done in this area.

(e) Some observed systems are found to disagree both with moderately conservative evolution (RLOF, but perhaps modified by winds) and the extremely non-conservative mechanism of common-envelope evolution. We need a compromise mode, which we shall call mode EJ, where much of an envelope is ejected rapidly but there is rather little change to the orbit.

(f) We have to understand better whether (i) all stars producing black holes are more massive than all stars producing neutron stars, (ii) there is something systematic but not necessarily monotonic in the production of compact remnants or (iii) it is a chaotic process, perhaps depending on the history of rotation and magnetic field in the core and envelope.

(g) The velocity field of material within even single stars is not well understood, and may have important effects on both mixing of composition and tidal dissipation.

(h) The heat energy deposited in the gainer as a result of mass transfer from the loser no doubt depends on the thermal history of the gas as it flows from one star to the other, either directly or through a disc. When, after much transfer, the material added to the gainer comes from near the core of the loser its composition will be different and might induce mixing, by the Rayleigh–Tayler instability. Probably these will only be small effects.

(i) Some of the accretion energy, particularly from a compact gainer, may be used to drive a bipolar outflow from near the central region of the accretion disc. Probably the material lost to the system carries off specific angular momentum equal to that of the gainer's orbital motion.

(j) The outer radius of the disc can be estimated by integrating the motion of particles falling inwards from the L_1 point, to the point where they have acquired an angular momentum (from Coriolis force) about the gainer equal to the angular momentum of a circular Keplerian orbit through the same point (Flannery 1975, Lubow and Shu 1975). However, this does not allow for the fact that angular momentum transport outwards in the disc, due to viscosity, will push out the outer edge of the disc; some discs in cataclysmic binaries appear to be substantially larger than the simple theory predicts. At the point where the stream of particles from L_1 impacts on the outer edge of the disc we expect a 'hot spot'.

(k) Averaged over a very long time, it is possible that the gainer might actually *lose* mass. If the gainer is a C/O white dwarf, and if accreted hydrogen-rich material mixes to some extent with the C/O material at the surface, a thermonuclear explosion will occur once the H-rich outer layer reaches a critical mass, and this explosion could eject not only the accreted material but also interior C/O material that it mixed with (MacDonald 1983). Thus the white dwarf, and hence the system, might *lose* mass in the long term.

(l) The gainer may acquire spin angular momentum, until it rotates at its own breakup speed ($\Omega_2^2 \sim GM_2/R_2^3$). This may act as a drain on the *orbital* angular momentum, though only a modest one. But it may prevent the gainer from actually accreting any more material, and further material may have to accumulate in some outer part of the Roche lobe until tidal friction transfers some of the gainer's spin angular momentum back into the orbit.

(m) If the gainer is not strongly magnetic, then the accretion disc should extend down to the stellar surface. But unless the gainer is rotating at breakup, there must be a boundary layer between the surface and the inner edge of the disc, in which there will be considerable shear. The boundary layer must liberate considerable energy; but the details of this region are not definitively modelled.

(n) Equation (6.22) assumes that the wind from the loser is spherically symmetrical, but real winds may well be confined towards the equatorial plane. This could increase significantly the fraction of mass transferred.

(o) The concept of RLOF is based on the notion that a star has a rather well-defined photospheric surface, but red supergiants probably have extremely poorly defined surfaces, which may be very asymmetric if dominated by a small number of large convective cells (Tuthill *et al.* 1999). Such stars may also be pulsating variables, either fairly regular (Miras) or, lower down the giant branch, semiregular or irregular. This might mean that Mode 3 RLOF (i.e. on a hydrodynamic timescale) could be irrelevant, or at least much modified.

(p) Roche-lobe overflow is in theory only meaningful in a circular orbit. It is usually assumed that an orbit will circularise in the interval prior to RLOF, thanks to tidal friction. But stars crossing the Hertzsprung gap evolve on a rather short timescale, and it is not so clear that tidal friction has time to circularise the orbit. Even when evolution slows down temporarily on the giant branch, the evolution is still rapid compared with main-sequence evolution, and tidal friction may not be fast enough for circularisation. Some kind of intermittent mass transfer may take place when a star approaches a radius comparable to the 'Roche lobe' radius at periastron.

(q) Triples, in particular fairly close triples where both periods are less than $\sim$30 years, while not *very* common, are also not *very* rare. Some of these will evolve in rather interesting ways that are not available to mere binaries. Triples are probably also particularly prone to dynamical encounters.

(r) Dynamical encounters, both in young star-forming regions and in older clusters that are still dense, may change binaries radically, probably causing collisions in some close primordial binaries, and also causing exchanges which mean that the current components may be non-coeval.

A number of the problems may only be approachable by three-dimensional modelling of entire stars, and indeed entire binaries. This subject is in its infancy, but rapid advances in computer technology may mean that a model with 10^{11} meshpoints in it will be available in perhaps ten years. Already the Djehuty project (named after the Egyptian god of astronomy) at the Lawrence Livermore National Laboratory can manipulate whole stellar models with $\sim 10^{8}$ meshpoints. This may be adequate for studying, say, convective motion in cores, but the larger number is minimal for resolving convection in surfaces.

Appendix A The equations of stellar structure

The equations of stellar evolution are presented here firstly in a traditional form and secondly in a form that I have found convenient for computation. Traditionally they are seen as four equations for the four structure variables p, T, r and L, with Lagrangian mass-cordinate m as the independent variable. Omitting a few refinements, they are:

$$\frac{\partial \log p}{\partial m} = -\frac{Gm}{4\pi r^4 p},$$
(A1)

$$\frac{\partial r}{\partial m} = \frac{1}{4\pi r^2 \rho},$$
(A2)

$$\frac{\partial \log T}{\partial m} = \frac{\partial \log p}{\partial m} \min(\nabla_r, \nabla_a), \quad \nabla_r \equiv \frac{3\kappa p L}{16\pi ac GmT^4}, \quad \nabla_a \equiv \left(\frac{\partial \log T}{\partial \log p}\right)_S,$$
(A3)

$$\frac{\partial L}{\partial m} = \epsilon - \epsilon_\nu - C_p T \left(\frac{\partial \log T}{\partial t} - \nabla_a \frac{\partial \log p}{\partial t}\right).$$
(A4)

Density ρ, opacity κ, specific heat C_p, adiabatic gradient ∇_a, nuclear energy generation rate ϵ and neutrino loss rate ϵ_ν are known functions of pressure p, temperature T and the abundances X_i ($\Sigma X_i = 1$) of the various nuclear species. Equations (A1)–(A4) are solved for a given distribution of the $X_i(m)$, and these abundances are then updated according to the prescription that, at a radiative meshpoint ($\nabla_r < \nabla_a$), is

$$\frac{\partial X_i}{\partial t} = A_i \sum_j \alpha_{ij} R_j, \quad \sum_i A_i \alpha_{ij} = 0,$$
(A5)

where A_i is the atomic number, R_j is the local rate of the jth nuclear reaction, and the α_{ij} are stoichiometric integers giving the number of particles created or destroyed per reaction. The number of compositions solved for can be quite large, although in practice only a modest number of composition variables have an important influence on the structure of the star. In a convection zone a more complicated recipe is needed, based on the concept that the composition is uniform in the zone as a result of convective mixing. In a semiconvection zone a still more complicated recipe is needed, usually based on the concept that the composition is determined by the neutral condition for convection: $\nabla_r = \nabla_a$. It is often unclear in published work what these recipes are, and how they are implemented numerically.

The rates R_j determine the nuclear energy generation rate ϵ as well as the composition changes:

$$\epsilon = \sum_i \sum_j E_i A_i \alpha_{ij} R_j = \sum_j Q_j R_j, \qquad Q_j = \sum_i E_i A_i \alpha_{ij}, \qquad \text{(A6)}$$

where the E_i are the binding energies (per baryon) of the various species and the Q_j are the energy yields of the nuclear reactions. We will ignore in this discussion the fine distinction between atomic weight and atomic number, although it is taken account of in the code.

Note that Eqs (A4) above and (A14) below, the latter a consequence of the former, are not correct in situations where the composition is changing. Thermodynamics tells us that

$$dU + p dV = T dS + T \sum \psi_i dX_i / A_i$$

$$= C_p T (d\log T - \nabla_a d\log p) + \sum \left(\frac{\partial H}{\partial X_i} \right)_{p,T} dX_i, \qquad \text{(A7)}$$

where V is the specific volume $(1/\rho)$, the ψ_i are chemical potentials, available from the equation of state with a little extra trouble, and H is the enthalpy. The error in missing out the enthalpy term would only be of consequence if the composition were changing on a thermal timescale, but of course it normally changes on a nuclear timescale that is ~ 1000 times longer. The error could be significant if, say, a convective zone expands rapidly, on a thermal timescale, into a region with a substantial composition gradient.

In the code used here, we suppose that only five nuclear species (^{1}H, ^{4}He, ^{12}C, ^{16}O and ^{20}Ne) are important for the changing structure of the star. Then we think of the structure and composition equations together as a set of ten partial differential equations to determine ten 'dependent' variables – ten rather than nine because there is an extra one to determine the distribution of meshpoints (see below). The dependent variables are defined at a set of discrete meshpoints, whose positions within the star are determined implicitly by the equations themselves. This means for example that the mesh is non-Lagrangian, and so when Lagrangian time-derivatives are required an advection term must be included. Such an implicit adaptive mesh turns out to be very stable, at least in the context of stellar evolution, and allows timesteps to be taken that can in some circumstances be much larger than those that can be taken with a Lagrangian distribution of meshpoints.

The ten dependent variables are r, m, L, T, ψ (the electron chemical potential) and X_i, for $i = 1$ to 5, the fractional abundances by mass of ^{1}H, ^{4}He, ^{12}C, ^{16}O and ^{20}Ne. In principle there are many more than five abundances that have to be determined, but in practice these five are the main ones determining the structure up to and including the late stage of carbon burning. There are two independent variables, t (time), and a space-like quantity, k, which is, in principle, a continuous function of position but in practice can be thought of as taking consecutive integer values at the meshpoints. Since the non-Lagrangian mesh is arranged to give meshpoints only where they are needed, a quite small number of meshpoints (say 200) is normally adequate for the whole star, from surface to photo-sphere, independent of the evolutionary stage of the star.

It is convenient to define a number of subsidiary variables, which are functions of the dependent variables. The following 15 are functions only of the state variables T, ψ and X_i: $p, \rho, S, \kappa, \nabla_a, C_p, \gamma, R_i, \epsilon, \epsilon_\nu$ and χ. These are respectively pressure, density, entropy, opacity, adiabatic gradient $(\partial \log T / \partial \log p)_S$, specific heat at constant pressure $(\partial S / \partial \log T)_p$, compressibility ($\partial \log p / \partial \log \rho)_S$, the destruction rate of the ith nuclear species (a negative quantity if the species is being produced rather than destroyed), the nuclear energy generation rate, i.e. $\Sigma Q_j R_j$ as in Eq. (A6), the neutrino energy loss rate, and a radiative diffusion coefficient

$$\chi \equiv \frac{4acT^3}{3\kappa\rho^2 C_p}. \tag{A8}$$

In addition to ∇_r, defined in Eq. (A3), we define three more quantities which are also explicit functions of the dependent variables at a point, but not just of the state variables. These are l, the convective mixing length, w, the mean velocity of turbulent eddies in a convective zone, and ∇, the actual temperature gradient – approximated in (A3) as $\min(\nabla_r, \nabla_a)$. The last three are only estimates, of course, which come from the standard version of the mixing-length model of turbulent convection. They are given by

$$l \equiv \alpha \, \min\left(\frac{pr^2}{Gm\rho}, \sqrt{\frac{p}{G\rho^2}}\right), \tag{A9}$$

$$\left(\frac{lw}{\chi}\right)^3 + 2\left(\frac{lw}{\chi}\right)^2 + 9\frac{lw}{\chi} = \frac{\alpha^2 l^2}{4\chi^2} C_p \nabla_a T \, \max(0, \nabla_r - \nabla_a) \tag{A10}$$

and

$$\nabla = \nabla_r - \frac{4lw^3}{\alpha^2 \chi \nabla_a C_p T}. \tag{A11}$$

Equation (A10) is a cubic equation for lw/χ which is readily solved algebraically for the unique positive root: if $x^3 + 2x^2 + 9x = a \geq 0$ then $3x = c - 23/c - 2$, where $c^3 = b + \sqrt{b^2 + 23^3}$ and $2b = 146 + 27a$. Obviously Eq. (A10) gives $w = 0$ in a convectively stable region ($\nabla_r < \nabla_a$), and hence $\nabla = \nabla_r$ from Eq. (A11). Thus we can determine l, w, ∇ in terms of other local variables.

The ten differential equations for the ten dependent variables can now be written:

$$\frac{\partial \log p}{\partial k} = -\frac{Gm}{4\pi r^4 p} \frac{\partial m}{\partial k}, \tag{A12}$$

$$\frac{\partial r}{\partial k} = \frac{1}{4\pi r^2 \rho} \frac{\partial m}{\partial k}, \tag{A13}$$

$$\frac{\partial \log T}{\partial k} = -\frac{Gm\nabla}{4\pi r^4 p} \frac{\partial m}{\partial k}, \tag{A14}$$

$$\frac{\partial L}{\partial k} = \left(\sum_j Q_j R_j - \epsilon_\nu\right) \frac{\partial m}{\partial k} - C_p T \left[\frac{\partial \log T}{\partial t} - \nabla_a \frac{\partial \log p}{\partial t}\right] \frac{\partial m}{\partial k}$$

$$+ C_p T \left[\frac{\partial \log T}{\partial k} - \nabla_a \frac{\partial \log p}{\partial k}\right] \frac{\partial m}{\partial t}, \tag{A15}$$

$$\frac{\partial}{\partial k}\left[\frac{4\pi r^2 \rho w l}{\partial r / \partial k} \frac{\partial X_i}{\partial k}\right] = A_i \sum_j \alpha_{ij} R_j \frac{\partial m}{\partial k} + \frac{\partial X_i}{\partial t} \frac{\partial m}{\partial k} - \frac{\partial X_i}{\partial k} \frac{\partial m}{\partial t}, \quad i = 1 \text{ to } 5, \tag{A16}$$

$$\frac{\partial m}{\partial k} = \frac{1}{C}\left[\frac{a_1 Gm}{4\pi r^4 p} + \frac{a_2}{m^{1/3}}\right]^{-1}. \tag{A17}$$

In the last equation, C is a kind of eigenvalue, i.e. a constant (in space, but not in time) whose value is not known until the equations are solved; a_1 and a_2 are constants whose values have to be chosen in order to give a reasonable dissection of the star into finite mesh intervals. By combining Eqs (A12) and (A17), it can be seen that the effect of these equations is to make

$$k = C(-a_1 \log p + 1.5 \, a_2 m^{2/3}) + \text{constant}, \tag{A18}$$

and so the meshpoints, which are at equal intervals of k, are therefore at equal intervals of this function of pressure and mass. In the surface layers, where $m \sim$ constant, this means that they are at approximately equal intervals of $\log p$; while near the centre, where $p \sim$ constant, they are at approximately equal intervals of $m^{2/3}$. In practice, a slightly more complicated function, involving extra terms in $\log T$ and r^2, is used. The significance of the particular powers $m^{2/3}$, r^2 is that both go linearly with $\log p$, $\log T$ near the centre.

C is determined by the fact that the first-order differential equation (A17) has *two* boundary conditions, viz.

$$m = 0 \quad \text{at} \quad k = 0, \tag{A19}$$

and

$$\frac{\partial m}{\partial t} = \dot{M}_{\text{wind}} + \dot{M}_{\text{RLOF}} \quad \text{at} \quad k = K, \tag{A20}$$

where K is the number of the outermost meshpoint. The right-hand side of Eq. (A20) can be zero, if there is no mass loss by stellar wind, and no RLOF; otherwise Eqs (2.76) or (3.70) might be used for the two terms on the right-hand side. The remaining boundary conditions are

$$r = L = 0 \quad \text{at} \quad k = 0, \tag{A21}$$

$$L = \pi a c r^2 T^4, \quad p\kappa = \frac{2}{3} \frac{Gm}{r^2} \quad \text{at} \quad k = K, \tag{A22}$$

and

$$w \frac{\partial X_i}{\partial k} = 0 \quad \text{at} \quad k = 0 \quad \text{and} \quad k = K, \quad i = 1 \text{ to } 5. \tag{A23}$$

Notwithstanding the fact that the diffusion coefficient – essentially the factor wl – in Eq. (A16) should, for consistency, be derived from Eq. (A10), in computations such as those presented in this book we have artificially 'weakened' the diffusion coefficient by typically two orders of magnitude, because otherwise the gradient of composition is so shallow in full convection zones that it is difficult to compute even to double precision. In Eq. (A16) we replace wl by something roughly equivalent to

$$wl = \text{constant} \, (l^2 C_p T \chi)^{1/3} \, (\nabla_r - \nabla_a)^2. \tag{A24}$$

The constant, expected from Eq. (A10) to be of order unity, is sometimes chosen to be smaller, on an ad hoc basis according to the performance of the code on a particular machine. This weakened coefficient typically gives a change of composition across a convective core of

$\sim 10^{-7}$, whereas one might expect it to be more like $\sim 10^{-9}$. According to the discussion of Section 2.2.4, this recipe is further modified to incorporate an ad hoc recipe for convective overshooting.

Equation (A10) is an implementation of the K. Schwarzschild criterion for convection in a homogeneous star, and also of the M. Schwarzschild criterion for convection and semi-convection in a star with a composition gradient. In the latter case it is *not* a statement that convection sets in where the entropy decreases outwards, since the entropy gradient involves the composition gradients as well as the pressure and temperature gradients. Equations (A10) and (A16) include, without further modification, the possibility of semiconvective mixing, as discussed in Section 2.2.4. It is not necessary to search for the boundaries of these zones and apply different algorithms within them.

The equation of state has an unusual form in that it gives such quantities as $p, \rho, C_p, \ldots$ as functions of the two independent variables f, a parameter related to electron degeneracy, and T, the temperature. In addition, of course, the equation of state depends on the abundances of the various elements, which are given constants for present purposes. The choice of f rather than, say, ρ or p as independent variable is based on the fact that several physical processes, notably electron degeneracy in cores and ionisation in envelopes, are *explicit* functions of f, but not of ρ or p. So also would be such extra processes as pair production and inverse β decay, although these have not actually been programmed. By using explicit formulae the computation is rendered very efficient; there is no need to invert a complicated highly non-linear relation $\rho = \rho(f, T)$ to determine the electron degeneracy parameter f, which is needed in the Fermi–Dirac integrals.

Electron degeneracy is normally represented as a quantity ψ, which appears in such Fermi–Dirac integrals as

$$N_e \rho = \text{constant} \int_0^\infty \frac{x^2 \mathrm{d}x}{e^{Emc^2/kT - \psi} + 1}. \tag{A25}$$

The quantities x, E are the dimensionless momentum and energy of an electron, related by $E = \sqrt{1 + x^2} - 1$; N_e is the number of free electrons per atomic mass unit, which itself depends on ψ via ionisation (see below). The quantity f gives ψ explicitly, by definition, as

$$\psi = \ln \frac{\sqrt{1+f} - 1}{\sqrt{1+f} + 1} + 2\sqrt{1+f}, \quad \text{so that} \quad \frac{\mathrm{d}\psi}{\mathrm{d}f} = \frac{\sqrt{1+f}}{f}. \tag{A26}$$

In terms of a further function $g(f, T)$, defined by

$$g = T\sqrt{1+f}, \tag{A27}$$

the Fermi–Dirac integral above can be approximated, to about three significant figures for *all* physical f, T, by

$$N_e \rho = \text{constant} \frac{f}{1+f} \{g(1+g)\}^{3/2} \frac{\sum_0^3 \sum_0^3 a_{ij} f^i g^j}{(1+f)^3(1+g)^3}, \tag{A28}$$

where the a_{ij} are a set of 16 constant coefficients found by the least-squares method. It is easy to see that in four limiting circumstances we have:

$$f \ll 1, g \ll 1: \quad \psi \sim \ln\frac{f}{4} + 2, \quad g \sim T, \quad N_e\rho = \text{constant } 4a_{00}\, T^{3/2}e^{\psi-2} \quad \text{(A.29a)}$$

$$f \ll 1, g \gg 1: \quad \psi \sim \ln\frac{f}{4} + 2, \quad g \sim T, \quad N_e\rho = \text{constant } 4a_{03}\, T^3 e^{\psi-2} \quad \text{(A.29b)}$$

$$f \gg 1, g \ll 1: \quad \psi \sim 2\sqrt{f}, \quad g \sim \frac{\psi T}{2}, \quad N_e\rho = \text{constant } a_{30}\,(\psi T)^{3/2} \quad \text{(A.29c)}$$

$$f \gg 1, g \gg 1: \quad \psi \sim 2\sqrt{f}, \quad g \sim \frac{\psi T}{2}, \quad N_e\rho = \text{constant } a_{33}\,(\psi T)^3. \quad \text{(A.29d)}$$

The functional forms of f, g as functions of ψ, T have been deliberately chosen so that the series expansions of of Eq. (A29) in each of the four limits matches exactly the series approximations of the integral (A25) in corresponding regimes. This means that considerable accuracy can be achieved with rather few coefficients. Similar approximations exist for the pressure and the internal energy U, or equivalently the entropy S, of the free electrons. The coefficients are to be found in Eggleton *et al.* (1973).

A second virtue of the above approximation is that, because it is closely based on the analytic expansions of the integral in its various limiting regimes, its partial derivatives, even up to third derivatives, are reasonably accurate, and can also be written down analytically. Since much of the physics we need involves derivatives of p, S wrt T, ρ, it is important that the derivatives are also accurate. We can, in fact, ensure that certain relations between derivatives of the state variables (Maxwell's relations) are satisfied *exactly*, and not just to the accuracy of the numerical approximation.

Ionisation is expressed by a number of sets of equations of the form

$$\frac{N_{H^+}}{N_H} = \frac{\omega_{H^+}}{\omega_H}\, e^{\psi+\chi_H/kT}, \quad N_{H^+} + N_H = X_H, \quad \text{(A30)}$$

where the ω are statistical weights and χ_H is the ionisation potential, and X_H is the given abundance of hydrogen by mass. Obviously these two equations give both N_H and N_{H^+} simply and explicitly in terms of ψ, T or equivalently f, T. Similar equations (three rather than two) give the helium ionisation equilibrium. A slight complication is that for hydrogen we also have to consider the *molecular* equilibrium, but it turns out that this only means solving a quadratic rather than a linear equation for N_H.

Two substantial problems remain, one of which is an artefact of the choice f of independent variable, and one of which is a problem however we choose the independent variables. In order, (a) f or ψ becomes indeterminate if the gas is fully non-ionised, so that there are no free electrons and (b) the ionisation equation above breaks down at high density ('pressure' ionisation), because the atomic structure of ions is strongly modified when the ions are so closely packed together that their Bohr radii are less than their separation. We do not have an answer to (a), but mercifully most *stellar* material is hot enough, dense enough or dilute enough, that at least some atoms are ionised. We help this along by approximating both Si and Fe as *wholly* ionised, even although they are not.

We deal with (b) approximately, by assuming that both $\omega(H)$ and χ_H are dependent on ρ, T, which we approximate as an explicit dependence on f, T. This is best done by adding to the Helmholtz free energy a term of the form $\Delta F \equiv N_e F_0(N_e\rho, T)$. Then in the ionisation equation we have to add to ψ the additional chemical potential term $(kT)^{-1}(\partial\Delta F/\partial N_e)_{\rho,T}$. This works because $(\partial\Delta F/\partial N_e)_{\rho,T}$, unlike ΔF itself, is a function of N_e and ρ only through the combination $N_e\rho$, and this in turn is a function only of the input variables f, T. We must also for thermodynamic consistency add terms $\rho^2(\partial\Delta F/\partial\rho)_{N_e,T}$ to the pressure and $-(\partial\Delta F/\partial T)_{N_e,\rho}$ to the entropy.

We conclude by proving an analytic result which is useful in considering the question, 'Why are some stars (such as evolved red giants) very centrally condensed?' (Section 2.3.1). We need only the equations of hydrostatic equilibrium, Eqs (A1) and (A2). Let us define

$$X \equiv \frac{3m}{4\pi r^3},\tag{A31}$$

a quantity that ranges from the central density at $r = 0$ to the mean density at $r = R$. Let us further define some 'homology invariants' s, U, V, W thus:

$$s \equiv \frac{d\log\rho}{d\log p}, \quad U \equiv -\frac{d\log m}{d\log p} = \frac{4\pi r^4 p}{Gm^2},$$

$$V \equiv -\frac{d\log r}{d\log p} = \frac{rp}{Gm\rho}, \quad W \equiv \frac{d\log X}{d\log p} = 3V - U.\tag{A32}$$

The variable s is the 'softness index', closely related to the local polytropic index n by $s = n/n + 1$. We do *not* assume that s is a constant; it will in general vary through the star, in a manner dictated usually by the temperature gradient and the molecular-weight gradient, but for the present we think of it as some general variable. The theorem we prove, however, reqires that $0 < s < 5/6$ everywhere.

A self-gravitating entity in which p is a given function of ρ is a 'barytrope', the special case of a power law being a polytrope. However, *every* hydrostatic stellar model is in principle a barytrope, since once computed it will follow some specific curve in the $(\log\rho, \log p)$ plane whose (variable) slope is the softness index s.

As is generally true of differential equations like (A1) and (A2) which are 'homologous', i.e. have right-hand sides that are products of powers of the variables, we can reduce the order of the system by one, from two to one. We can do this in several ways, but we choose a way which gives dW/dU as a function of W and U (and also s). Differentiating U, V logarithmically wrt $\log p$, we obtain

$$\frac{d\log U}{d\log p} = 1 + 2U - 4V = \frac{1}{3}(3 + 2U - 4W),\tag{A33}$$

$$\frac{d\log V}{d\log p} = 1 - s - V + U = \frac{1}{3}(3 - 3s + 2U - W).\tag{A34}$$

Hence

$$W' \equiv \frac{dW}{dU} = \frac{3V}{U}\frac{d\log p}{d\log U} - 1 = \frac{3W - 3s(W + U) + 5UW - W^2}{U(3 + 2U - 4W)},\tag{A35}$$

and we also see that

$$\frac{d\log X}{d\log U} = W\frac{d\log p}{d\log U} = \frac{3W}{3 + 2U - 4W}.\tag{A36}$$

Equation (A35) for W as a function of U – and of s, which we can now think of as a given $s(U)$ – may not look very promising for analysis, but we can prove a by-no-means trivial inequality from it. We begin by proving successively three preliminary results:

(a) $W \to \dfrac{3}{5}s$, $U \to \infty$ as $r \to 0$

(b) $0 < W < \dfrac{1}{2}$ everywhere if $0 < s < \dfrac{5}{6}$ everywhere

(c) $W < W_{\mathrm{E}}$, where $W_{\mathrm{E}}(U)$ is the Emden polytropic solution for $s = $ constant $= s_{\max} \equiv \max_r s(r) < 5/6$.

To prove (a), we expand equations (A1) and (A2) about the origin. We see that if

$$p = p_c \left(1 - \frac{r^2}{a^2} \right), \tag{A37}$$

neglecting terms of order r^4, and with subscript c meaning a central value, then

$$\rho = \rho_c \left(1 - s_c \frac{r^2}{a^2} \right), \quad m = \frac{4\pi \rho_c r^3}{3} \left(1 - \frac{3}{5} s_c \frac{r^2}{a^2} \right), \quad X = 1 - \frac{3}{5} s_c \frac{r^2}{a^2}, \tag{A38}$$

and hence

$$W = \frac{3}{5} s_c. \tag{A39}$$

Clearly $U \sim 1/r^2$ and so $U \to \infty$ as $r \to 0$; also $s \to s_c$, so that $W \to \dfrac{3}{5}s$ as $U \to \infty$.

To prove (b), consider the value of W' on the line $W = 1/2$. It is

$$W' = \frac{3}{2U} \left(\frac{5}{6} - s \right) > 0. \tag{A40}$$

If, therefore, a barytropic solution $W = W(U, s)$ crosses the horizontal line $W = 1/2$ in the (U, W) plane, it crosses it sloping upwards to the right, as U increases (U is always positive by definition). The barytropic solution can therefore never get back *below* $W = 1/2$, and yet it has to, since $W = 3s/5 < 1/2$ at the centre, according to (a). By *reductio ad absurdum*, W must remain below $1/2$ throughout. By a similar *reductio ad absurdum*, W must also remain above zero; this utilises our assumption that $s > 0$ everywhere. Note that $W = 1/2$ is, in fact, the Emden solution for the $s = 5/6$ ($n = 5$) polytrope, and $W = 0$ the Emden solution for $s = 0$ ($n = 0$), i.e. a uniform-density sphere.

We prove (c) similarly. Consider the value of W' as the barytropic curve crosses (if it can) the Emden solution $W = W_{\mathrm{E}}(U)$ corresponding to a polytrope of softness index $s_{\max}$. The value satisfies

$$W' - W'_{\mathrm{E}} = \frac{3(U + W)(s_{\max} - s)}{U(3 + 2U - 4W)} \geq 0, \tag{A41}$$

where we use the fact that $W < 1/2$ to confirm that the denominator is always positive, and the fact that $W > 0$ to confirm that the factor $U + W$ in the numerator is always positive, too. Thus the barytropic curve crosses the Emden curve from below left to above right, and cannot cross back at greater U. But the central value of W is below (or at) the central Emden value, and so we have another *reductio ad absurdum*.

Having established that $0 < W < W_E < 1/2$ everywhere (if $0 < s \le s_{max} < 5/6$ everywhere), we see from Eq. (A36) that

$$\left[\log X\right]_{U=0}^{U=\infty} = \int_0^{\infty} \frac{3W\, dU}{U(3+2U-4W)}$$

$$< \int_0^{\infty} \frac{3W_E\, dU}{U(3+2U-4W_E)} = \left[\log X_E\right]_{U=0}^{U=\infty}. \qquad (A42)$$

The central condensation parameter C is defined as

$$C \equiv \frac{4\pi \rho_c R^3}{3M} = \frac{X(\text{centre})}{X(\text{surface})} = \frac{X(U=\infty)}{X(U=0)}, \qquad (A43)$$

and so we have shown that a barytrope with $0 < s \le s_{max} < 5/6$ is less centrally condensed than the polytrope with $s = s_{max}$ throughout.

Appendix B Distortion and circulation in a non-spherical star

(i)   The hydrostatic-equilibrium model

Here, and in Section 3.2.1, we consider a star which has a binary companion, which is in uniform rotation at rate Ω (not necessarily the same as the orbital rate, which might be varying if the orbit is eccentric) and which has a uniform composition. These conditions, via hydrostatic equilibrium, give

$$\nabla p = -\rho \nabla \phi, \tag{B1}$$

$$\nabla^2 \phi = 4\pi G \rho - 2\Omega^2, \tag{B2}$$

and lead to the result that ρ, p, T, s are all constant on surfaces of constant ϕ. We can also define variables V, m, L, r_* – volume, mass, nuclear luminosity and 'volume radius' respectively – which are constant on equipotentials: V is the volume contained within an equipotential, m and L are the integrals of ρ and $\rho\epsilon$ over this volume, and r_* is given by

$$\frac{4\pi}{3} r_*^3 = V(\phi). \tag{B3}$$

The quantities m, L clearly satisfy

$$\frac{dm}{dV} = \rho, \quad \frac{dL}{dV} = \rho\epsilon. \tag{B4}$$

Let us define

$$K(\phi) \equiv 4\pi G m - 2\Omega^2 V = \int \nabla \phi \cdot d\mathbf{\Sigma} = \int |\nabla\phi| d\Sigma, \tag{B5}$$

and note that

$$\nabla^2 \phi = \frac{dK}{dV}. \tag{B6}$$

We also see that, since the distance between adjacent equipotentials along a normal is $\delta l = \delta\phi/|\nabla\phi|$,

$$\frac{dV}{d\phi} = \int \frac{\delta l}{\delta\phi} d\Sigma = \int \frac{d\Sigma}{|\nabla\phi|}. \tag{B7}$$

Hence hydrostatic equilibrium can be written as

$$-\frac{1}{\rho}\frac{\mathrm{d}p}{\mathrm{d}r_*} = \frac{\mathrm{d}\phi}{\mathrm{d}r_*} = \frac{\mathrm{d}\phi}{\mathrm{d}V}\frac{\mathrm{d}V}{\mathrm{d}r_*} = \frac{4\pi r_*^2}{\int \mathrm{d}\Sigma/|\nabla\phi|}$$

$$= \left(\frac{Gm}{r_*^2} - \frac{2\Omega^2 r_*}{3}\right)\frac{(4\pi r_*^2)^2}{\int |\nabla\phi|\,\mathrm{d}\Sigma \int \mathrm{d}\Sigma/|\nabla\phi|}. \tag{B8}$$

The factor in parentheses simply cancels the first factor in the denominator to its right, by Eq. (B5); we write it this way to show that the ratio to the right of the parentheses clearly differs from unity in the *second* order if ϕ differs from spherical in the *first* order, so that we can write

$$-\frac{1}{\rho}\frac{\mathrm{d}p}{\mathrm{d}r_*} = \frac{\mathrm{d}\phi}{\mathrm{d}r_*} \approx \frac{Gm}{r_*^2} - \frac{2\Omega^2 r_*}{3}. \tag{B9}$$

Now consider the energy flux **F**, which in general is a combination of radiative and convective flux. In spherical symmetry, we usually write this as

$$F = -\frac{4acT^3}{3\kappa\rho}\frac{\mathrm{d}T}{\mathrm{d}r} + \rho w T\delta S, \tag{B10}$$

where w is the mean velocity of convection and $T\delta S$ is the mean heat excess of an upward-rising eddy. The mixing-length approximations (Section 2.2.2 and Appendix A) for w and $T\delta s$ are

$$w^2 \sim T\delta S \sim Tl\left[-\frac{\mathrm{d}S}{\mathrm{d}r}\right] \sim T\left[\frac{\mathrm{d}S}{\mathrm{d}\log p}\right], \tag{B11}$$

where l is the mixing length, normally estimated by $l \sim -\mathrm{d}r/\mathrm{d}\log p$, and where the square brackets have the meaning $[X] \equiv \max(X, 0)$. In a non-spherical situation the generalisation of the radiative term in the energy flux is obvious; and of the many possible generalisations for the convective term we choose one which is

$$\mathbf{F} = -\frac{4acT^3}{3\kappa\rho}\nabla T + \rho\left[T\frac{\mathrm{d}S}{\mathrm{d}\log p}\right]^{3/2}\frac{\mathrm{d}r_*}{\mathrm{d}\phi}\nabla\phi, \tag{B12}$$

i.e.

$$\mathbf{F} = \chi(\phi)\nabla\phi, \qquad \chi(\phi) \equiv -\frac{4acT^3}{3\kappa\rho}\frac{\mathrm{d}T}{\mathrm{d}\phi} + \rho\left[T\frac{\mathrm{d}S}{\mathrm{d}\log p}\right]^{3/2}\frac{\mathrm{d}r_*}{\mathrm{d}\phi}. \tag{B13}$$

Thus the equation of energy production and transport is taken to be

$$\nabla\cdot\chi\nabla\phi = \rho\epsilon - \rho T\mathbf{v}\cdot\nabla S, \tag{B14}$$

where **v** is the meridional velocity field, satisfying

$$\nabla\cdot\rho\mathbf{v} = 0. \tag{B15}$$

We first establish that the circulation term carries no *net* energy across an equipotential surface, i.e. that $\int \rho T\mathbf{v}\cdot\nabla S\,\mathrm{d}V = 0$, where the integral is over the interior of an equipotential surface. From thermodynamics and hydrostatic equilibrium,

$$T\mathrm{d}S = \mathrm{d}U + p\mathrm{d}\frac{1}{\rho} = \mathrm{d}\left(U + \frac{p}{\rho}\right) - \frac{1}{\rho}\mathrm{d}p = \mathrm{d}\left(U + \frac{p}{\rho} + \phi\right). \tag{B16}$$

Hence, using (B15),

$$\int \rho T \mathbf{v} \cdot \nabla S \, dV = \int \rho \mathbf{v} \cdot \nabla \left(U + \frac{p}{\rho} + \phi \right) dV = \int \left(U + \frac{p}{\rho} + \phi \right) \rho \mathbf{v} \cdot d\mathbf{\Sigma}. \quad \text{(B17)}$$

Since the expression in parentheses is constant on an equipotential it can come outside the last integral, and since $\int \rho \mathbf{v} \cdot d\mathbf{\Sigma} = 0$ by Eq. (B15) we have the result. Hence

$$\int \mathbf{F} \cdot d\mathbf{\Sigma} = \int \chi \nabla \phi \cdot d\mathbf{\Sigma} = \int \rho \epsilon dV = L. \quad \text{(B18)}$$

Since χ is constant on equipotentials, we can write this, using Eq. (B5), as

$$L = \chi \int \nabla \phi \cdot d\mathbf{\Sigma} = \chi K, \quad \text{i.e.} \quad \chi = L/K. \quad \text{(B19)}$$

It follows that the only effects of the distortion on the structure equations (A1)–(A4) are that the factor Gm in Eqs (A1) and (A2) is to be replaced by $Gm - 2\Omega^2 r^3/3$.

Using Eqs (B4)–(B7) with (B19), Eq. (B14) becomes

$$\frac{L}{K} \frac{dK}{dV} + |\nabla \phi|^2 \frac{d}{d\phi} \frac{L}{K} = \frac{dL}{dV} - \rho T \frac{dS}{d\phi} \mathbf{v} \cdot \nabla \phi, \quad \text{(B20)}$$

i.e., after some manipulation,

$$\rho T \frac{dS}{d\phi} v_\perp = \left(\frac{1}{|\nabla \phi|} \int |\nabla \phi| d\mathbf{\Sigma} - |\nabla \phi| \int \frac{d\mathbf{\Sigma}}{|\nabla \phi|} \right) \frac{d}{dV} \frac{L}{4\pi Gm - 2\Omega^2 V}, \quad \text{(B21)}$$

where $v_\perp$ is the component of $\mathbf{v}$ in the direction of $\nabla \phi$.

(ii)   The degree of internal distortion

We now define the relative distortion parameter $\alpha(r)$ of an internal equipotential, and relate it to the quadrupole moment of a star distorted by either rotation or the effect of a companion.

An equipotential surface can be approximated by

$$r \approx r_*\{1 - \alpha(r_*) P_2(\cos\theta)\}, \quad r_* \approx r(1 + \alpha(r) P_2), \quad \text{(B22a,b)}$$

where θ is the angle from the axis of symmetry. Since α is first order, it can be thought of as a function of either r or r_*. The contribution to the quadrupole moment q from mass between surfaces ϕ and $\phi + d\phi$ is

$$\frac{dq}{d\phi} = \rho(\phi) \int \frac{2\pi r^2 \sin\theta d\theta}{|\nabla \phi|} r^2 P_2(\cos\theta), \quad \text{(B23)}$$

where we use Eq. (B7) to estimate the volume element. Now, ϕ is a function of r_* only, and r_* is a function of r, θ given by Eq. (B22b), so that

$$|\nabla \phi| = \frac{d\phi(r_*)}{dr_*} |\nabla r_*| \approx \frac{d\phi(r_*)}{dr_*} \left(1 + \frac{d \, r\alpha(r)}{dr} P_2 \right) \approx \frac{d\phi(r_*)}{dr_*} \left(1 + \frac{d r_* \alpha(r_*)}{dr_*} P_2 \right). \quad \text{(B24)}$$

Hence

$$\frac{dq}{dr_*} \approx \rho(r_*) \int \left(1 - \frac{d r_* \alpha(r_*)}{dr_*} P_2 \right) r_*^4 (1 - 4\alpha P_2) P_2 \, 2\pi \sin\theta d\theta = -\frac{4\pi}{5} \rho r_*^4 \left(4\alpha + \frac{d r_* \alpha}{dr_*} \right),$$
$$\text{(B25)}$$

and so, over the whole star,

$$q = -\frac{1}{5} \int_0^M (5\alpha + r_* \alpha') r_*^2 dm. \tag{B26}$$

(iii)   The effect of rotation

First consider rotation alone. In this case the potential distribution outside the star is

$$-\phi(\mathbf{r}) \approx \frac{GM_1}{r} + \frac{Gq^{\text{rot}}}{r^3} P_2(\cos\theta) + \frac{1}{2}\Omega^2 r^2 \sin^2\theta. \tag{B27}$$

On the stellar surface ($r_* = R_1$) this means, using Eq. (B22a), that

$$-\phi(R_1) \approx \frac{GM_1}{R_1} + \frac{1}{3}\Omega^2 R_1^2 + \left(\alpha_1 \frac{GM_1}{R_1} + \frac{Gq^{\text{rot}}}{R_1^3} - \frac{1}{3}\Omega^2 R_1^2\right) P_2(\cos\theta), \tag{B28}$$

where $\alpha_1 = \alpha(R_1)$. This must be independent of θ, so that

$$\alpha_1 = R_1^3 \left(\frac{\Omega^2}{3GM_1} - \frac{q^{\text{rot}}}{M_1 R_1^5}\right). \tag{B29}$$

Using Eq. (B26), this leads to

$$\alpha_1 = \frac{\Omega^2 R_1^3}{3GM_1} \frac{1}{1-Q}, \quad q^{\text{rot}} = -\frac{\Omega^2 R_1^5}{3G} \frac{Q}{1-Q}, \quad Q = \frac{1}{5} \frac{\int_0^M r^2 dm(5\alpha + r\alpha')}{M_1 R_1^2 \alpha_1}. \tag{B30a,b,c}$$

Since α is first order, we do not need to distinguish between $\alpha(r_*)$ and $\alpha(r)$ in the integral.

(iv)   The equation for $\alpha(r)$

From $r_*(r, \theta)$ as given by Eq. (B22b),

$$|\nabla r_*|^2 \approx 1 + 2(r\alpha)' P_2, \tag{B31}$$

and

$$\nabla^2 r_* \approx \frac{2}{r} + \left(\frac{1}{r^2}\frac{d}{dr}r^2\frac{dr\alpha}{dr} - \frac{6\alpha}{r}\right) P_2 \approx \frac{2}{r_*} + \left(r\alpha'' + 4\alpha' - \frac{2\alpha}{r}\right) P_2, \tag{B32}$$

so that

$$\nabla^2\phi = \phi''|\nabla r_*|^2 + \phi'\nabla^2 r_* \approx \phi'' + \frac{2}{r_*}\phi' + \left[r\alpha'' + 4\alpha' - \frac{2\alpha}{r} + 2\frac{\phi''}{\phi'}(r\alpha' + \alpha)\right]\phi' P_2$$

$$= 4\pi G\rho(r_*) - 2\Omega^2. \tag{B33}$$

For this to be true for all θ, we need

$$\phi'' + \frac{2}{r_*}\phi' = 4\pi G\rho(r_*) - 2\Omega^2, \quad \text{i.e.} \quad \phi' = \frac{K(r_*)}{4\pi r_*^2}, \tag{B34}$$

where K is the quantity defined in Eq. (B5). Equation (B34) is the same as (B9). We also need

$$\alpha'' - \frac{6\alpha}{r^2} + \frac{2rK'}{K}\left(\frac{\alpha'}{r} + \frac{\alpha}{r^2}\right) = 0. \tag{B35}$$

Since α is first order in Ω^2, we see that (a) the bracketed expression in Eq. (B33) can be thought of as a function of either r or r_* and (b) we can take $K = 4\pi Gm$ in (B35). Thus our equation for α is

$$\alpha'' - \frac{6\alpha}{r^2} + \frac{2rm'}{m} \left(\frac{\alpha'}{r} + \frac{\alpha}{r^2} \right) = 0. \tag{B36}$$

We therefore determine α by first solving the stellar structure equations to obtain $m(r)$, and then integrating (B36) with this $m(r)$, subject to α and α' being finite at the centre. These determine α up to a multiplicative constant, since at $r = 0$ we have $rm'/m = 3$ and hence from (B36) $\alpha \sim B + C/r^5$ there. The second term has to be excluded on account of its singularity. Then $\alpha(r)$ determines the constant Q unambiguously, since the definition (B30c) of Q is independent of a constant factor in α. A numerical treatment of Eqs (B36) and (B30c) for polytropes of index $n < 4.95$ leads to the interpolation formula

$$Q \approx \frac{3}{5} \left(1 - \frac{n}{5} \right)^{2.215} e^{0.0245n - 0.096n^2 - 0.0084n^3} \pm 1.5\% \text{ rms}. \tag{B37}$$

(v)   An approximation for $\alpha(r)$

At a lower level of approximation, suppose that (a) the mass is concentrated entirely at the centre and (b) the quadrupole is concentrated entirely at the surface, where the distortion is greatest. Then the potential *inside* the star ($r_* \leq R_1$) – cf. Eq. (B27) for *outside* – is given by

$$-\phi(\mathbf{r}) \approx \frac{GM_1}{r} + \frac{Gq^{\text{rot}}r^2 P_2}{R_1^5} + \frac{1}{2}\Omega^2 r^2 \sin^2\theta. \tag{B38}$$

The condition that ϕ is constant on any interior potential $r_* = \text{constant}$ implies that

$$-\phi(r_*) \approx \frac{GM_1}{r_*} + \frac{1}{3}\Omega^2 r_*^2 + \left(\alpha \frac{GM_1}{r_*} + \frac{Gq^{\text{rot}}r_*^2}{R_1^5} - \frac{1}{3}\Omega^2 r_*^2 \right) P_2(\cos\theta) \tag{B39}$$

is independent of θ and hence that

$$\alpha = r_*^3 \left(\frac{\Omega^2}{3GM_1} - \frac{q^{\text{rot}}}{M_1 R_1^5} \right) \approx r^3 \left(\frac{\Omega^2}{3GM_1} - \frac{q^{\text{rot}}}{M_1 R_1^5} \right). \tag{B40}$$

Equation (B29) is just Eq. (B40) evaluated at $r_* = R_1$. We confirm that $\alpha \propto r^3$ satisfies (B36) at least in the outer layers, for centrally condensed stars, since $m' \sim 0$ in that case. Putting $\alpha \propto r^3$ in (B30c), we obtain

$$Q \approx \frac{8}{5} \frac{\int r^5 dm}{R_1^5 M_1}. \tag{B41}$$

A numerical evaluation of Eq. (B41) for polytropes gives

$$Q \approx \frac{3}{5} \left(1 - \frac{n}{5} \right)^{2.205} e^{-0.437n + 0.066n^2 - 0.023n^3} \pm 2\% \text{ rms}, \tag{B42}$$

for polytropes of index $n < 4.95$.

If the star is so centrally condensed that $m' \sim 0$ for $r > 0$, Eq. (B36) has the solution $Br^3 + C/r^2$. This cannot persist all the way to the centre, of course, because the density, though large there, is finite. Taking $C = 0$ nevertheless, we have solution (B40). Numerical calculation of polytropes shows that $C \to 0$ as $n \to 5$, but unfortunately C is not really negligible in the important regime $n \sim 1.5$–3. In the opposite extreme that the star is of constant density, $rm'/m = 3$ throughout and Eq. (B36) has the solution $\alpha = B + C/r^5$. We must take $C = 0$ to exclude the singularity at the origin. This is just the well-known case of 'liquid' stars. For such stars all the equipotentials are similar ellipsoids with eccentricity e where

$$\frac{\Omega^2 R_1^3}{GM_1} = \frac{3\sqrt{1-e^2}}{2e^3}\{(3-2e^2)\sin^{-1}e - 3e\sqrt{(1-e^2)}\} \approx \frac{2}{5}e^2. \qquad (B43)$$

The quadrupole moment of a uniform ellipsoid is

$$q^{\text{rot}} = -\frac{1}{5}M_1 R_1^2 e^2 (1-e^2)^{1/3} \approx -\frac{\Omega^2 R_1^5}{2G}. \qquad (B44)$$

This 'agrees' with (B41) in the case that $\rho = $ constant and hence $Q = 3/5$. The agreement is providential, however, since we ought to use Eq. (B30c) with $\alpha = $ constant, whereas Eq. (B41) assumed $\alpha \propto r^3$; but both expressions give the same answer if $\rho = $ constant. The agreement, providential or not, at $n = 0$, coupled with agreement in the limit $n \to 5$, suggests that the approximation (B41) might in practice be good enough over the whole range of models from uniform density to centrally condensed. However, detailed comparison of approximations (B37) and (B42) shows that Q from Eq. (B42) can be in error by ~ 40–50%.

(vi) **The distortion due to the companion**
For a star distorted by the gravitational field of a companion, and *not* rotating, the calculation is the same as in Section (iii) except that the potential (B27) is replaced by

$$-\phi(\mathbf{r}) \approx \frac{GM_1}{r} + \frac{Gq^{\text{comp}}}{r^3}P_2(\cos\theta') + \frac{GM_2 r^2}{d^3}P_2(\cos\theta'), \qquad (B45)$$

θ' being measured from the line of centres rather than the rotation axis. This differs from (B27) mainly in the replacement of $\Omega^2/3$ by $-GM_2/d^3$ (apart from the orientation). Thus for the quadrupole moment we similarly obtain

$$\alpha_1 = -\frac{M_2 R_1^3}{M_1 d^3}\frac{1}{1-Q}, \qquad q^{\text{comp}} = \frac{M_2 R_1^5}{d^3}\frac{Q}{1-Q}, \qquad (B46a,b)$$

with the same Q as before – either the accurate (to first order) Eqs (B30c) and (B36) or the approximation (B41).

(vii) **Schwarzschild's derivation**
Schwarzschild (1958) obtained, by a slightly different route, a result which in our notation is

$$q^{\text{comp}} = \frac{M_2 R_1^5}{d^3}\left(\frac{3\alpha - r\alpha'}{2\alpha + r\alpha'}\right)_{r=R_1}. \qquad (B47)$$

Although superficially very different from (B46b) with Q given by (B30c) and (B36), it is in fact the same by virtue of the fact that

$$\frac{\text{d}}{\text{d}r}mr^2(3\alpha - r\alpha') = r^2(5\alpha + r\alpha')m', \qquad (B48)$$

as can be verified by using Eq. (B36) to eliminate α'' from the left-hand side of Eq. (B48). Thus

$$Q = \left(\frac{3\alpha - r\alpha'}{5\alpha}\right)_{r=R_1} = \frac{1}{5M_1 R_1^2 \alpha_1} \int (5\alpha + r\alpha') r^2 dm. \qquad (B49)$$

Note, however, that $\alpha \propto r^3$ gives zero in the differential form of Eq. (B47), as against approximation (B41) in the integral form of Eq. (B49). The differential form requires us to obtain a more accurate solution for α than the integral form, as is not unusual.

(viii)   The circulation velocity

The determination of $\alpha(r)$ allows us to estimate the angular-dependent term in the circulation velocity (B21). It is

$$\frac{1}{|\nabla\phi|} \int |\nabla\phi| \, d\Sigma - |\nabla\phi| \int \frac{d\Sigma}{|\nabla\phi|}$$

$$\approx -\frac{8\pi R_1^3}{M_1(1-Q)} \frac{r^2}{\alpha(R_1)} \frac{d(r\alpha)}{dr} \left[\frac{\Omega^2}{3G} P_2(\cos\theta) - \frac{M_2}{d^3} P_2(\cos\theta')\right], \qquad (B50)$$

where θ is latitude measured from the rotation axis and θ' is latitude measured from the line of centres.

In addition to the rotationally-driven circulation of Eq. (B50), there is in principle a circulation due to the part of the potential that comes from $*2$'s gravity. But unless and until $*1$ is brought into synchronism, and the orbit circularised, this contribution will fluctuate about zero with the period of $*1$'s relative rotation. It will therefore be insignificant until synchronism is reached.

(ix)   The quadrupole tensor

For a quadrupole moment q with symmetry axis $\mathbf{k}$, the quadrupole tensor q_{ij} is

$$q_{ij} = \frac{q}{2k^2}(3k_i k_j - k^2 \delta_{ij}). \qquad (B51)$$

Since the symmetry axes are Ω for rotation and $\mathbf{d}$ for the companion, results (B30b) and (B46b) tell us finally that the quadrupole tensor of a star in a binary is

$$q_{ij} = q_{ij}^{\text{rot}} + q_{ij}^{\text{comp}}, \qquad q_{ij}^{\text{rot}} = -\frac{A}{6G}(3\Omega_i\Omega_j - \Omega^2\delta_{ij}),$$

$$q_{ij}^{\text{comp}} = \frac{M_2 A}{2d^5}(3d_i d_j - d^2\delta_{ij}), \qquad A = \frac{R_1^5 Q}{1-Q}. \qquad (B52)$$

(x)   The force between the stars

When we allow for the quadrupole distortion above, we can write down the potential ϕ' at a general point $\mathbf{s}$ outside $*1$ in the inertial frame centred on the centre of gravity of the binary. With $*1$ at $\mathbf{d}_1$, $*2$ at $\mathbf{d}_2$, and remembering that $\mathbf{d} = \mathbf{d}_1 - \mathbf{d}_2$, ϕ' is the following:

$$\phi'(\mathbf{s}, \mathbf{d}_1, \mathbf{d}_2, \Omega) = -\frac{GM_1}{|\mathbf{s} - \mathbf{d}_1|} - \frac{GM_2}{|\mathbf{s} - \mathbf{d}_2|} - \frac{G(s_i - d_{1i})(s_j - d_{1j})\left(q_{ij}^{\text{rot}}(\Omega) + q_{ij}^{\text{comp}}(\mathbf{d})\right)}{|\mathbf{s} - \mathbf{d}_1|^5}$$

$$= -\frac{GM_1}{|\mathbf{s} - \mathbf{d}_1|} - \frac{GM_2}{|\mathbf{s} - \mathbf{d}_2|} - \frac{G}{3}l_{ij}(\mathbf{s} - \mathbf{d}_1)\left[q_{ij}^{\text{rot}}(\Omega) + \frac{M_2 A}{2}l_{ij}(\mathbf{d})\right], \qquad (B53)$$

where the tensor l_{ij} is defined, for a general vector $\mathbf{a}$, by

$$l_{ij}(\mathbf{a}) \equiv \frac{3a_i a_j - a^2 \delta_{ij}}{a^5}, \quad \text{so that} \quad l_{ii} = 0. \tag{B54}$$

Note that ϕ' is different from ϕ of Eq. (B2) or Eq. (B27) because it is in an inertial frame whereas ϕ is in a frame that rotates with $*1$.

The force $\mathbf{F}_1$ on $*1$, or equivalently its negative $\mathbf{F}_2$ on $*2$, is

$$\mathbf{F}_1 = -\mathbf{F}_2 = \int_{V_2} \rho(\mathbf{s}) \nabla_s \phi'(\mathbf{s}, \mathbf{d}_1, \mathbf{d}_2, \boldsymbol{\Omega}) \, \mathrm{d}^3 s, \quad \rho = \frac{1}{4\pi G} \nabla_s^2 \phi' \tag{B55}$$

integrated over the interior of $*2$. In this region ρ is just a delta function $M_2 \delta(\mathbf{s} - \mathbf{d}_2)$, since the other terms in ϕ' give zero density outside $*1$. Excluding the self-term of $*2$, we obtain

$$\mathbf{F}_1 = M_2 \left[\nabla_s \phi' \right]_{s=d_2} = -\frac{GM_1 M_2 \, \mathbf{d}}{d^3} + \frac{GM_2}{3} \left[q_{ij}^{\text{rot}}(\boldsymbol{\Omega}) + \frac{M_2 A}{2} l_{ij}(\mathbf{d}) \right] \nabla_{\mathbf{d}} \, l_{ij}(\mathbf{d}). \tag{B56}$$

Because the same function, $l_{ij}(\mathbf{d})$, appears before and after the gradient operator in the term that relates to the companion's distortion, the resulting $\mathbf{F}$ is seen to be derivable from a new potential $\Phi(\boldsymbol{\Omega}, \mathbf{d})$:

$$\mathbf{F} = -\nabla_{\mathbf{d}} \Phi, \quad \Phi = -\frac{GM_1 M_2}{d} - \frac{GM_2}{3} \left[q_{ij}^{\text{rot}}(\boldsymbol{\Omega}) + \frac{AM_2}{4} l_{ij}(\mathbf{d}) \right] l_{ij}(\mathbf{d}), \tag{B57}$$

which, on replacing $l_{ij}(\mathbf{d})$ in terms of $\mathbf{d}$ with Eq. (B54), is

$$\Phi = -\frac{GM_1 M_2}{d} - \frac{GM_2 A d_i d_j}{d^5} \left[-\frac{1}{6G} \{3\Omega_i \Omega_j - \Omega^2 \delta_{ij}\} + \frac{M_2}{4d^5} \{3 d_i d_j - d^2 \delta_{ij}\} \right]$$

$$= -\frac{GM_1 M_2}{d} + AM_2 \left[\frac{(\boldsymbol{\Omega} \cdot \mathbf{d})^2}{2d^5} - \frac{\Omega^2}{6d^3} - \frac{GM_2}{2d^6} \right]. \tag{B58}$$

(xi)   The tidal velocity field

In addition to the circulation velocity field there is, in the case of eccentric orbits or non-corotating stars, a tidal velocity field driven by the time-dependent character of the distortion. In the frame that rotates with $*1$, this field can be determined by using (a) the conservation statement

$$\frac{\partial \rho}{\partial t} + \nabla \cdot \rho \mathbf{v} = 0, \tag{B59}$$

and (b) the constancy of ρ on equipotential surfaces, i.e. the fact that

$$\rho = \rho(r_*), \quad r_* = r + r\alpha P_2(\cos\theta). \tag{B60a,b}$$

We are only concerned here with the companion-induced distortion, so that α_1, the surface value of α, is given by Eq. (B46a).

Let

$$F \equiv r^2 P_2 = \frac{3}{2}(\mathbf{k} \cdot \mathbf{r})^2 - \frac{1}{2}r^2, \quad \text{so that} \quad \nabla^2 F = 0, \quad \mathbf{r} \cdot \nabla F = 2F, \tag{B61}$$

where $\mathbf{k} \equiv \mathbf{d}/d$. With $\mathbf{d}$ time-varying, both α and $\mathbf{k}$ depend on t, the former because $\alpha \propto 1/d^3$ (Eq. B46a). Then we can differentiate ρ (Eqs B60a,b) wrt time to get

$$\frac{\partial \rho}{\partial t} = \frac{d\rho}{dr_*}\frac{\partial r_*}{\partial t} = \frac{d\rho}{dr_*}\left(\frac{\partial \alpha}{\partial t}\frac{F}{r} + \frac{3\alpha G}{r}\right) = \frac{3\alpha}{r}\frac{d\rho}{dr_*}\left(-\frac{1}{d}\frac{\partial d}{\partial t}F + G\right), \quad (B62)$$

where

$$G \equiv \frac{1}{3}\frac{\partial F}{\partial t} = \mathbf{k}\cdot\mathbf{r}\frac{\partial \mathbf{k}}{\partial t}\cdot\mathbf{r}, \quad \text{so that} \quad \nabla^2 G = 0, \quad \mathbf{r}\cdot\nabla G = 2G. \quad (B63)$$

Then it is easy to see that a velocity field given by

$$\mathbf{v} = \frac{3\alpha_1}{2}\beta(r)\left(\frac{1}{d}\frac{\partial d}{\partial t}\nabla F - \nabla G\right), \quad (B64)$$

satisfies Eq. (B59) to first order, provided that

$$\frac{d\rho\beta}{dr} = \frac{\alpha}{\alpha_1}\frac{d\rho}{dr}, \quad \text{so that} \quad \beta\rho = \rho_1 + \frac{1}{\alpha_1}\int_{R_1}^{r}\alpha\frac{d\rho}{dr}dr. \quad (B65)$$

The constant of integration comes from the fact that on the free surface $\beta = 1$. The function β is determined unambiguously by the structure of the star, via Eq. (B36) determining $\alpha(r)$, and is well behaved at the surface even for polytropic ($0 < n < 5$) surfaces as $\rho \to 0$, despite the apparent singularity there. For the special case $n = 0$, i.e. uniform density, we have $\beta = \alpha/\alpha_1 = 1$.

Using suffices, and putting $\mathbf{k} = \mathbf{d}/d$, Eq. (B64) becomes

$$v_i = \frac{3\alpha_1}{2}\beta(r)s_{ij}(t)x_j, \quad s_{ij} \equiv \frac{1}{d^3}\frac{\partial d}{\partial t}(5d_id_j - d^2\delta_{ij}) - \frac{1}{d^2}\left(d_i\frac{\partial d_j}{\partial t} + d_j\frac{\partial d_i}{\partial t}\right). \quad (B66a,b)$$

The tensor s_{ij} is symmetrical and traceless. Equation (B66) allows us to calculate the rate of dissipation of energy due to the action of turbulent viscosity (or any other viscosity) on this velocity field, and this dissipation in turn determines the amount of 'tidal friction'.

(xii)   The rate of dissipation

The rate-of-strain tensor is

$$t_{ij} \equiv \frac{\partial v_i}{\partial x_j} + \frac{\partial v_j}{\partial x_i} = \frac{3\alpha_1}{2}\left(2\beta s_{ij} + \frac{\beta'}{r}\{s_{ik}x_kx_j + s_{jk}x_kx_i\}\right). \quad (B67)$$

Squaring this, and averaging it over an equipotential (which at this level of approximation can be taken to be spherical), we use the standard results

$$\frac{1}{4\pi}\int x_ix_j\,d\Omega = \frac{r^2}{3}\delta_{ij}, \quad \frac{1}{4\pi}\int x_ix_jx_kx_l\,d\Omega = \frac{r^4}{15}(\delta_{ij}\delta_{kl} + \delta_{ik}\delta_{jl} + \delta_{il}\delta_{jk}) \quad (B68)$$

to obtain

$$\frac{1}{4\pi}\int t_{ij}^2\,d\Omega = 9\alpha_1^2 s_{ij}^2\gamma(r), \quad \gamma \equiv \beta^2 + \frac{2}{3}r\beta\beta' + \frac{7}{30}r^2\beta'^2. \quad (B69)$$

Now, after some manipulation of Eq. (B66b),

$$s_{ij}^2 = \frac{2}{d^2}\left[2\left(\frac{\partial d}{\partial t}\right)^2 + \left(\frac{\partial \mathbf{d}}{\partial t}\right)^2\right] = \frac{2}{d^4}\frac{\partial d}{\partial t}\cdot\left[2\mathbf{d}\,\mathbf{d}\cdot\frac{\partial \mathbf{d}}{\partial t} + d^2\frac{\partial \mathbf{d}}{\partial t}\right], \quad (B70)$$

and so, using Eq. (B46a) for α_1, the rate of dissipation of mechanical energy is

$$-\dot{\mathcal{E}} = \frac{1}{2} \int \rho w l \, t_{ij}^2 \, dV$$

$$= \frac{9M_2^2 R_1^6}{M_1^2 (1-Q)^2} \frac{1}{d^{10}} \frac{\partial \mathbf{d}}{\partial t} \cdot \left[2\mathbf{d}\,\mathbf{d} \cdot \frac{\partial \mathbf{d}}{\partial t} + d^2 \frac{\partial \mathbf{d}}{\partial t} \right] \int_0^{M_1} w l \, \gamma(r) \, dm, \qquad (B71)$$

where w, l are the mean velocity and mean free path of turbulent eddies, assuming that turbulent viscosity is the dominant dissipative agent. If we define a viscous timescale t_{visc} for *1 by

$$\frac{1}{t_{\mathrm{visc}}} \equiv \frac{1}{M_1 R_1^2} \int_0^{M_1} w l \, \gamma(r) \, dm, \qquad (B72)$$

then Eq. (B71) tells us that the energy loss is equivalent to the rate of working of a resistive force $\mathbf{F}$ where

$$\mathbf{F} = -\frac{9M_2^2 R_1^8}{M_1 (1-Q)^2 t_{\mathrm{visc}}} \frac{1}{d^{10}} \left[2\mathbf{d}\,\mathbf{d} \cdot \frac{\partial \mathbf{d}}{\partial t} + d^2 \frac{\partial \mathbf{d}}{\partial t} \right], \qquad (B73)$$

which we can identify with the force due to tidal friction – see Appendix C(c).

Appendix C Perturbations to Keplerian orbits

Using the relative position vector $\mathbf{d} \equiv \mathbf{d}_1 - \mathbf{d}_2$, and mass $M \equiv M_1 + M_2$, the relative motion of a binary subject to (a) Newtonian point-mass gravity and (b) an additional acceleration $\mathbf{f}$ (the perturbing force per unit reduced mass $\mu \equiv M_1 M_2 / M$) is given by

$$\ddot{\mathbf{d}} = -\frac{GM\mathbf{d}}{d^3} + \mathbf{f}. \tag{C1}$$

Define $\mathcal{E}$ (the Keplerian energy per unit reduced mass), $\mathbf{h}$ (the angular momentum, similarly) and $\mathbf{e}$ (the Laplace–Runge–Lenz vector) by

$$\mathcal{E} \equiv \frac{1}{2}\dot{\mathbf{d}} \cdot \dot{\mathbf{d}} - \frac{GM}{d}, \quad \mathbf{h} \equiv \mathbf{d} \times \dot{\mathbf{d}}, \quad GM\mathbf{e} \equiv \dot{\mathbf{d}} \times \mathbf{h} - \frac{GM\mathbf{d}}{d}. \tag{C2a–c}$$

Note that $\mathcal{E}$ is not the *total* energy, if $\mathbf{f} \neq 0$, but only the part that is kinetic plus Newtonian point-mass energy. We can see, after some manipulation in the case of $\mathbf{e}$, that

$$\dot{\mathcal{E}} = \dot{\mathbf{d}} \cdot \mathbf{f}, \quad \dot{\mathbf{h}} = \mathbf{d} \times \mathbf{f}, \quad GM\dot{\mathbf{e}} = \mathbf{f} \times \mathbf{h} + \dot{\mathbf{d}} \times (\mathbf{d} \times \mathbf{f}). \tag{C3a–c}$$

Hence $\mathcal{E}$, $\mathbf{h}$ and $\mathbf{e}$ are all constants of the motion if $\mathbf{f} = 0$. Using a standard parametrisation of the Keplerian orbit in this case, for example either of the two parametrisations below, Eqs (C6)–(C10), we find that $\mathbf{e}$ is a vector in the direction of periastron, and has magnitude equal to the eccentricity (thus justifying belatedly the choice of name $\mathbf{e}$). Even if $\mathbf{f} \neq 0$, auxiliary variables a, b, l, ω (mean angular velocity) and p (period) can be *defined* in terms of $\mathcal{E}, \mathbf{h}, \mathbf{e}$ in the usual way:

$$a = -\frac{GM}{2\mathcal{E}}, \quad b = a\sqrt{1 - e^2}, \quad l = a(1 - e^2), \quad \omega = \frac{h}{ab} = \frac{2\pi}{p}. \tag{C4a–d}$$

For general $\mathbf{f}$, and not just $\mathbf{f} = 0$, four standard relations can be shown to be satisfied:

$$2h^2\mathcal{E} + G^2M^2(1 - e^2) = 0, \quad \mathbf{e} \cdot \mathbf{h} = 0, \quad h^2 = GMl, \quad \omega^2 = GM/a^3. \tag{C5a–d}$$

Thus even when $\mathbf{h}$, l, ω and a are continuously changing because $\mathbf{f} \neq 0$, the orbit can be perceived as always 'instantaneously Keplerian'. For example, the instaneous period and semimajor axis always satisfy Kepler's third law.

If $\mathbf{f}$ is small, we can estimate its effect on $\mathcal{E}, \mathbf{h}, \mathbf{e}$ by averaging over time the right-hand sides of Eqs (C3) in a Keplerian orbit. This is done most easily by writing the Keplerian orbit in Cartesian coordinates (origin at focus, $\mathbf{e}$ in the 1-direction, $\mathbf{h}$ in the 3-direction) using one

276

or other of the following parametric forms:

$$\mathbf{d} = \frac{l}{1 + e \cos \theta}(\cos \theta, \sin \theta, 0) = (a \cos \phi - ae, b \sin \phi, 0) \tag{C6}$$

$$\dot{\mathbf{d}} = \frac{\omega ab}{l}(-\sin \theta, \cos \theta + e, 0) = \frac{\omega}{1 - e \cos \phi}(-a \sin \phi, b \cos \phi, 0) \tag{C7}$$

$$\omega dt = \frac{l^2}{ab}\frac{d\theta}{(1 + e \cos \theta)^2} = (1 - e \cos \phi)d\phi \tag{C8}$$

$$d = \frac{l}{1 + e \cos \theta} = a(1 - e \cos \phi) \tag{C9}$$

$$\dot{d} = \frac{\omega abe}{l}\sin \theta = \frac{\omega ae \sin \phi}{1 - e \cos \phi}. \tag{C10}$$

Various scalar, vector and tensor functions of $\mathbf{d}$ can be averaged over time using these parametrisations, θ being more useful if the function contains a substantial negative (≤ -2) power of d and ϕ otherwise (≥ -1). It is convenient to introduce an orthogonal right-handed basis of vectors $\mathbf{e}$, $\mathbf{q} \equiv \mathbf{h} \times \mathbf{e}$ and $\mathbf{h}$. These are not unit vectors: we use overlines to define the corresponding unit vectors $\bar{\mathbf{e}}$, $\bar{\mathbf{q}}$, $\bar{\mathbf{h}}$.

Some examples, expressed in terms of polynomials $I_{n,l}(e)$, $n, l \geq 0$, defined below, are as follows:

$$\left\langle \frac{1}{d^{n+2}} \right\rangle = \frac{1}{abl^n} I_{n,0} \tag{C11}$$

$$\left\langle d^{n-1} \right\rangle = a^{n-1} I_{n,0} \tag{C12}$$

$$\left\langle \frac{\dot{d}_i \dot{d}_i}{d^{n+2}} \right\rangle = \frac{\omega^2 ab}{l^{n+2}}\{(1 + e^2)I_{n,0} + 2eI_{n,1}\} \tag{C13}$$

$$\left\langle d^n \dot{d}_i \dot{d}_i \right\rangle = \omega^2 a^{n+2}(I_{n,0} - eI_{n,1}) \tag{C14}$$

$$\left\langle \frac{\dot{d}^2}{d^{n+2}} \right\rangle = \frac{\omega^2 abe^2}{l^{n+2}}(I_{n,0} - I_{n,2}) \tag{C15}$$

$$\left\langle d^{n-1}d_i \right\rangle = -a^n(I_{n,1} + eI_{n,0})\bar{e}_i \tag{C16}$$

$$\left\langle \frac{d_i}{d^{n+3}} \right\rangle = \frac{1}{abl^n} I_{n,1}\bar{e}_i \tag{C17}$$

$$\left\langle \frac{\dot{d}_i}{d^{n+2}} \right\rangle = \frac{\omega}{l^{n+1}}(eI_{n,0} + I_{n,1})\bar{q}_i \tag{C18}$$

$$\left\langle \frac{d\dot{d}_i}{d^{n+3}} \right\rangle = \frac{\omega e}{l^{n+1}}(I_{n,0} - I_{n,2})\bar{q}_i \tag{C19}$$

$$\left\langle \frac{d_i d_j}{d^{n+4}} \right\rangle = \frac{1}{abl^n}\{I_{n,2}\bar{e}_i\bar{e}_j + (I_{n,0} - I_{n,2})\bar{q}_i\bar{q}_j\} \tag{C20}$$

$$2\left\langle d_i d_j \right\rangle = a^2\{(1 + 4e^2)\bar{e}_i\bar{e}_j + (1 - e^2)\bar{q}_i\bar{q}_j\} \tag{C21}$$

$$2\left\langle d_i \dot{d}_j \right\rangle = \epsilon_{ijk}h_k \tag{C22}$$

$$2\langle \dot{d}_i \dot{d}_j \dot{d}_k \rangle = eah(\overline{e}_i \overline{q}_j \overline{e}_k + \overline{q}_i \overline{e}_j \overline{e}_k - 2\overline{e}_i \overline{e}_j \overline{q}_k) \tag{C23}$$

$$\left\langle \frac{\dot{d}_i \dot{d}_j \dot{d}_k}{d^{n+5}} \right\rangle = \frac{1}{abl^n} \{ (I_{n,1} - I_{n,3})(\overline{e}_i \overline{q}_j \overline{q}_k + \overline{q}_i \overline{e}_j \overline{q}_k + \overline{q}_i \overline{q}_j \overline{e}_k) + I_{n,3} \overline{e}_i \overline{e}_j \overline{e}_k \} \tag{C24}$$

$$\left\langle \frac{\dot{d}_i \dot{d}_j \dot{d}_k}{d^{n+4}} \right\rangle = \frac{\omega}{l^{n+1}} \{ (I_{n,3} - I_{n,1})(\overline{e}_i \overline{q}_j \overline{e}_k + \overline{q}_i \overline{e}_j \overline{e}_k) + (I_{n,1} + eI_{n,0})\overline{q}_i \overline{q}_j \overline{q}_k$$
$$+ (I_{n,3} + eI_{n,2})(\overline{e}_i \overline{e}_j \overline{q}_k - \overline{q}_i \overline{q}_j \overline{q}_k) \}. \tag{C25}$$

The polynomials $I_{n,m}$ are defined by

$$I_{n,m}(e) \equiv \int_0^{2\pi} (1 + e\cos\theta)^n \cos^m \theta \, \frac{d\theta}{2\pi}. \tag{C26}$$

They are easily evaluated from

$$I_{0,2m} = \int_0^{2\pi} \cos^{2m} \phi \, \frac{d\phi}{2\pi} = \frac{(2m)!}{2^{2m}(m!)^2},$$
$$I_{0,2m+1} = 0, \quad I_{n+1,m} = I_{n,m} + eI_{n,m+1}. \tag{C27}$$

Clearly

$$\int_0^{2\pi} (1 - e\cos\phi)^n \cos^m \phi \, \frac{d\phi}{2\pi} = I_{n,m}(-e) = (-1)^m I_{n,m}(e). \tag{C28}$$

We can now apply these to the following: (a) apsidal motion driven by general relativity, (b) apsidal motion and precession driven by quadrupolar distortion, (c) tidal friction, (d) gravitational radiation, (e) mass loss and mass transfer and (f) a third body.

(a)   Apsidal motion from general relativity

The motion of a particle in the Schwarzschild metric is given by

$$\delta \int L \, dt = \delta \int \sqrt{c^2 F - \dot{d}^2/F - d^2\dot{\theta}^2} \, dt = 0, \quad F(d) = 1 - \frac{2GM}{c^2 d}. \tag{C29}$$

The equations of motion are, therefore,

$$\frac{d}{dt}\left(\frac{d^2}{L}\dot{\theta}\right) = 0 \tag{C30}$$

$$\frac{d}{dt}\left(-\frac{1}{FL}\dot{d}\right) = \frac{1}{2L}\left(c^2 F' + \dot{d}^2 \frac{F'}{F^2} - 2d\dot{\theta}^2\right) \tag{C31}$$

where primes are derivatives wrt d and dots wrt t. The Euler–Lagrange integral is

$$\dot{d}\frac{\partial L}{\partial d} + \dot{\theta}\frac{\partial L}{\partial \dot{\theta}} - L = \frac{cF}{L} = \text{constant.} \tag{C32}$$

Changing to the proper-time coordinate $ds \equiv L \, dt/c$, and now using dots for derivatives wrt s, we obtain (after some manipulation)

$$d^2\dot{\theta} = h = \text{constant}, \quad \ddot{d} - d\dot{\theta}^2 = -\frac{GM}{d^2} - \frac{3GMh^2}{c^2 d^4}. \tag{C33}$$

Thus the effective perturbing force $\mathbf{f}$ is radial:

$$\mathbf{f} = -\frac{3GMh^2}{c^2}\frac{\mathbf{d}}{d^5}. \tag{C34}$$

So by Eq. (C3a) $\dot{\mathcal{E}}$ averages to zero, and by Eq. (C3b) $\dot{\mathbf{h}}$ is zero. Also, by Eqs (C3c), (C4), (C17), (C26) and (C27),

$$GM\dot{\mathbf{e}} = \langle \mathbf{f} \times \mathbf{h} \rangle = \langle \mathbf{f} \rangle \times \mathbf{h}$$

$$= -\frac{3GMh^2}{c^2}\frac{I_{2,1}}{eabl^2}\,\mathbf{e} \times \mathbf{h} = \frac{3G^2M^2\omega}{c^2l}\,\overline{\mathbf{h}} \times \mathbf{e}. \tag{C35}$$

Hence $\mathbf{e}$ rotates about $\overline{\mathbf{h}}$ (apsidal motion) with angular velocity Z_{GR}, where

$$Z_{GR} = \frac{3GM\omega}{c^2l} = \frac{3GM\omega}{ac^2(1-e^2)}. \tag{C36}$$

(b)   Apsidal motion and precession

We consider here the effect on the orbit of the quadrupole distortion or 'equilibrium tide' of $*1$ due to its rotation and to the presence of a companion (Appendix B). We ignore the distortion of $*2$, but this can easily be added into the result (C40) below. The force between the stars can be derived – Appendix B(x) – from a potential Φ:

$$\mathbf{F} = -\nabla\Phi, \quad -\Phi = \frac{GM_1M_2}{d} + AM_2\left[\frac{\Omega^2}{6d^3} - \frac{(\mathbf{\Omega}\cdot\mathbf{d})^2}{2d^5} + \frac{GM_2}{2d^6}\right], \tag{C37}$$

where A is the structure constant given by Eq. (B51). Hence the perturbing force $\mathbf{f}_{QD}$ in the orbital equation due to the quadrupole distortion is given by

$$\ddot{\mathbf{d}} = -\frac{1}{\mu}\nabla\Phi = -\frac{GM\mathbf{d}}{d^3} + \mathbf{f}_{QD},$$

$$\mathbf{f}_{QD} = \frac{AM_2}{\mu}\left[\frac{-\Omega^2\mathbf{d}}{2d^5} + \frac{5(\mathbf{\Omega}\cdot\mathbf{d})^2\mathbf{d}}{2d^7} - \frac{\mathbf{\Omega}\cdot\mathbf{d}\,\mathbf{\Omega}}{d^5} - \frac{3GM_2\mathbf{d}}{d^8}\right]. \tag{C38}$$

We assume first that the stellar rotation is parallel to the orbital rotation, and so $\mathbf{\Omega}\cdot\mathbf{d} = 0$. Then $\mathbf{f}_{QD}$ is purely radial, and $\dot{\mathcal{E}}$, $\dot{\mathbf{h}}$ average to zero, as for any $\mathbf{f}$ of the form $F(d)\mathbf{d}$. The only effect of the quadrupole moment on the otherwise Keplerian motion is to make the Laplace–Runge–Lenz vector $\mathbf{e}$ rotate about the angular momentum vector $\mathbf{h}$. Using Eqs (C3c) and (C17) to average over the orbit,

$$GM\dot{\mathbf{e}} = \left(\frac{M_2A\Omega^2}{2\mu abl^2}I_{2,1} + \frac{3GM_2^2A}{\mu abl^5}I_{5,1}\right)\mathbf{h} \times \overline{\mathbf{e}}. \tag{C39}$$

By Eqs (C4), (C26), and (C27), this can be written

$$\dot{\mathbf{e}}_{QD} \equiv Z\,\overline{\mathbf{h}} \times \mathbf{e}, \quad Z = \frac{M_2A\Omega^2}{2\mu\omega a^5(1-e^2)^2} + \frac{15GM_2^2A}{2\mu\omega a^8}\frac{1+\frac{3}{2}e^2+\frac{1}{8}e^4}{(1-e^2)^5}. \tag{C40}$$

Z is the rate of apsidal motion.

If $\mathbf{\Omega}\cdot\mathbf{d} \neq 0$, the extra terms in $\mathbf{f}_{QD}$ give some extra apsidal motion. It is necessary to use some of the tensorial averages, Eqs (C24) and (C25). The extra apsidal motion turns out to be in a negative sense: Eq. (C40) should be replaced by

$$Z = \frac{M_2 A \left(O_h^2 - \frac{1}{2}O_e^2 - \frac{1}{2}O_q^2\right)}{2\mu\omega a^5 (1-e^2)^2} + \frac{15 G M_2^2 A}{2\mu\omega a^8} \frac{1 + \frac{3}{2}e^2 + \frac{1}{8}e^4}{(1-e^2)^5}, \tag{C41}$$

where Ω_e, Ω_q, Ω_h are the components of $\mathbf{\Omega}$ in the triad of vectors defining the orientation of the orbit.

The extra terms in $\mathbf{f}_{QD}$ still make $\mathcal{E}$ average to zero, as for any conservative force, but they give a couple

$$\dot{\mathbf{h}} = \langle \mathbf{d} \times \mathbf{f}_{QD} \rangle = \frac{M_2 A}{\mu} \left\langle \frac{\mathbf{\Omega} \cdot \mathbf{d} \, \mathbf{\Omega} \times \mathbf{d}}{d^5} \right\rangle = -\frac{M_2 A}{2\mu a b l h^2} \, \mathbf{\Omega} \cdot \mathbf{h} \, \mathbf{\Omega} \times \mathbf{h}, \tag{C42}$$

where the average was evaluated using Eq. (C20). This means that $\mathbf{h}$ precesses about $\mathbf{\Omega}$. However $\mathbf{\Omega}$ is not a vector fixed in space. Rather, the total angular momentum vector $\mathbf{H} \equiv \mu\mathbf{h} + I\mathbf{\Omega}$ is fixed in space, since $*1$ experiences a couple which is the negative of the couple on the binary; I is the moment of inertia of $*1$. Thus we can write

$$\dot{\mathbf{h}} = -\frac{M_2 A H}{2\mu a b l h^2 I} \, \mathbf{\Omega} \cdot \mathbf{h} \, \overline{\mathbf{H}} \times \mathbf{h}, \tag{C43}$$

and so $\mathbf{h}$, $\mathbf{\Omega}$ both precess about $\overline{\mathbf{H}}$ at a rate $\dot{\chi}$ given, using Eq. (C4), by

$$\dot{\chi} = -\frac{M_2 A \Omega_h H}{2\mu a b l h I} = -\frac{M_2 A \Omega_h H}{2\mu\omega a^5 I (1-e^2)^2}. \tag{C44}$$

We have assumed so far that the star is dissipationless, and so there are no secular terms leading to orbital shrinkage or circularisation. We now consider a model of dissipation, i.e. tidal friction.

(c)   Tidal friction

We continue to suppose (for simplicity) that $*1$ is extended while $*2$ is a point mass. In the frame that rotates with $*1$ the quadrupolar tide will in general be time-dependent: $*1$ will be continually changing its shape. If the star is not perfectly elastic, we expect a loss of total mechanical energy but no loss of total angular momentum.

In Appendix C(b) we had an acceleration $\mathbf{f}_{QD}$ which is derivable from a potential Φ given by Eq. (C37). This conserves total energy, as will be verified shortly. But if there is in addition some slow dissipation of energy ('tidal friction') we will have an extra force $\mathbf{f}_{TF}$, say. Writing

$$\ddot{\mathbf{d}} = -\frac{G M \mathbf{d}}{d^3} - \frac{1}{\mu}\nabla\Phi + \mathbf{f}_{TF}, \tag{C45}$$

we can see that total angular momentum

$$\mathbf{H} \equiv \mu\mathbf{d} \times \dot{\mathbf{d}} + I\mathbf{\Omega} \tag{C46}$$

is conserved if the couple on $*1$ is given by

$$I\dot{\mathbf{\Omega}} = \mathbf{d} \times (\nabla\Phi - \mu\,\mathbf{f}_{TF}). \tag{C47}$$

The total energy

$$\mathcal{E}' \equiv \mu\mathcal{E} + \Phi(d, \mathbf{\Omega} \cdot \mathbf{d}) + \frac{1}{2}I\Omega^2$$

$$= \frac{\mu}{2}\dot{\mathbf{d}} \cdot \dot{\mathbf{d}} - \frac{GM\mu}{d} + \Phi(d, \mathbf{\Omega} \cdot \mathbf{d}) + \frac{1}{2}I\Omega^2, \tag{C48}$$

may change not only because $\mathbf{d}$ in $\mathcal{E}$ and Φ varies but also because $\mathbf{\Omega}$, which appears in Φ as well as in $I\Omega^2$, can vary, for instance as a result of precession. We find however that $\dot{\mathcal{E}}' = 0$ (in the absence of tidal friction) provided that Φ depends only on d and $\mathbf{\Omega} \cdot \mathbf{d}$:

$$\dot{\mathcal{E}}' = \mu\dot{\mathbf{d}} \cdot \left\{\ddot{\mathbf{d}} + \frac{GM\mathbf{d}}{d^3} + \frac{1}{\mu}\nabla\Phi\right\} + \mathbf{d} \cdot \dot{\mathbf{\Omega}}\,\Phi' + I\mathbf{\Omega} \cdot \dot{\mathbf{\Omega}}$$

$$= \mu\dot{\mathbf{d}} \cdot \mathbf{f}_{\text{TF}} + \frac{\mathbf{d}}{I} \times (\nabla\Phi - \mu\,\mathbf{f}_{\text{TF}}) \cdot (\mathbf{d}\,\Phi' + I\mathbf{\Omega}), \tag{C49}$$

where we use Eqs (C45) and (C47) for $\ddot{\mathbf{d}}$ and $\dot{\mathbf{\Omega}}$; Φ' means the partial derivative of Φ wrt $\mathbf{\Omega} \cdot \mathbf{d}$. For $\Phi = \Phi(d, \mathbf{\Omega} \cdot \mathbf{d})$, $\nabla\Phi$ is entirely in the plane of $\mathbf{d}$ and $\mathbf{\Omega}$. Hence all terms in Eq. (C49) vanish except those with $\mathbf{f}_{\text{TF}}$:

$$\dot{\mathcal{E}}' = \mu(\dot{\mathbf{d}} - \mathbf{\Omega} \times \mathbf{d}) \cdot \mathbf{f}_{\text{TF}} = \mu\frac{\partial\mathbf{d}}{\partial t} \cdot \mathbf{f}_{\text{TF}}, \quad \text{since} \quad \dot{\mathbf{d}} = \frac{\partial\mathbf{d}}{\partial t} + \mathbf{\Omega} \times \mathbf{d}, \tag{C50a,b}$$

where $\partial/\partial t$ is a derivative in the frame that rotates with $*1$, $\dot{\mathbf{d}}$ being the derivative in an inertial frame.

When there is dissipation, probably the simplest assumption is that the rate of loss of energy is some positive-definite function of the rate of change of $*1$'s shape, e.g. of its quadrupole tensor, since this determines its shape to lowest order. We therefore write

$$\dot{\mathcal{E}}' = -\sigma\frac{\partial q_{ij}}{\partial t}\frac{\partial q_{ij}}{\partial t}, \tag{C51}$$

where q_{ij} is given by Eq. (B51) and σ is a dissipative constant intrinsic to $*1$ (dimensions $m^{-1}l^{-2}t^{-1}$). In the frame that rotates with $*1$, $\mathbf{\Omega}$ is a constant while $\mathbf{d}$ varies at rate $\partial\mathbf{d}/\partial t$, so that

$$\frac{\partial q_{ij}}{\partial t} = \frac{\partial}{\partial t}\frac{M_2 A}{2d^5}(3d_id_j - d^2\delta_{ij})$$

$$= \frac{3M_2 A}{2d^5}\left[d_i\frac{\partial d_j}{\partial t} + d_j\frac{\partial d_i}{\partial t} + d\frac{\partial d}{\partial t}\delta_{ij} - \frac{5}{d}\frac{\partial d}{\partial t}d_id_j\right]. \tag{C52}$$

Hence, after some manipulation,

$$\dot{\mathcal{E}}' = -\frac{9\sigma M_2^2 A^2}{2d^{10}}\frac{\partial\mathbf{d}}{\partial t} \cdot \left[2\mathbf{d}\,\mathbf{d} \cdot \frac{\partial\mathbf{d}}{\partial t} + d^2\frac{\partial\mathbf{d}}{\partial t}\right]. \tag{C53}$$

Comparing this with (C50a), we see that a consistent expression for $\mathbf{f}_{\text{TF}}$ is

$$\mathbf{f}_{\text{TF}} = -\frac{9\sigma M_2^2 A^2}{2\mu d^{10}}\left[2\mathbf{d}\,\mathbf{d} \cdot \frac{\partial\mathbf{d}}{\partial t} + d^2\frac{\partial\mathbf{d}}{\partial t}\right]$$

$$= -\frac{9\sigma M_2^2 A^2}{2\mu d^{10}}\left[3\mathbf{d}\,\mathbf{d} \cdot \dot{\mathbf{d}} + (\mathbf{h} - \Omega d^2) \times \mathbf{d}\right], \tag{C54a,b}$$

using Eqs (C50b) and (C2b). Comparing the acceleration $\mathbf{f}_{\mathrm{TF}}$ of Eq. (C54a) with the resistive force $\mathbf{F}$ (and hence acceleration $\mathbf{F}/\mu$) of Eq. (B67), we obtain σ as a function of the internal viscous dissipation timescale of $*1$ as defined in Appendix B(xii):

$$\sigma = \frac{2}{M_1 R_1^2 Q^2 t_{\mathrm{visc}}}. \tag{C55}$$

We first specialise to the case $\boldsymbol{\Omega} \parallel \mathbf{h}$. Using Eqs (C3b,c) for $\dot{\mathbf{e}}$ and $\dot{\mathbf{h}}$, with acceleration $\mathbf{f}$ from Eq. (C54b), and averaging various functions of $\mathbf{d}$ according to Eqs (C17)–(C19), we obtain, after substantial manipulation

$$\dot{\mathbf{e}} \equiv -V\mathbf{e}$$

$$= -\frac{9\sigma M_2^2 A^2}{2\mu abl^6} \left[4e I_{6,0} + I_{6,1} - 3e I_{6,2} + I_{7,1} - \frac{\Omega_h}{\omega} \frac{l^2}{ab} (e I_{4,0} + 2I_{4,1} + e I_{4,2}) \right] \bar{\mathbf{e}}, \tag{C56}$$

$$\dot{\mathbf{h}} = -\frac{9\sigma M_2^2 A^2}{2\mu abl^6} \left[I_{6,0} - \frac{\Omega_h}{\omega} \frac{l^2}{ab} I_{4,0} \right] \mathbf{h} \equiv -W\mathbf{h}. \tag{C57}$$

Evaluating the polynomials $I_{m,n}$ according to Eqs (C27), V and W can be seen to be

$$V = \frac{9}{t_{\mathrm{TF}}} \left[\frac{1 + \frac{15}{4}e^2 + \frac{15}{8}e^4 + \frac{5}{64}e^6}{(1-e^2)^{13/2}} - \frac{11\Omega_h}{18\omega} \frac{1 + \frac{3}{2}e^2 + \frac{1}{8}e^4}{(1-e^2)^5} \right], \tag{C58}$$

$$W = \frac{1}{t_{\mathrm{TF}}} \left[\frac{1 + \frac{15}{2}e^2 + \frac{45}{8}e^4 + \frac{5}{16}e^6}{(1-e^2)^{13/2}} - \frac{\Omega_h}{\omega} \frac{1 + 3e^2 + \frac{3}{8}e^4}{(1-e^2)^5} \right], \tag{C59}$$

where

$$t_{\mathrm{TF}} \equiv \frac{2\mu a^8}{9\sigma M_2^2 A^2} = \frac{2}{9\sigma} \frac{a^8}{R_1^{10}} \frac{M_1}{M_2 M} \left(\frac{1-Q}{Q} \right)^2 = \frac{2t_{\mathrm{visc}}}{9} \frac{a^8}{R_1^8} \frac{M_1^2}{M_2 M} (1-Q)^2. \tag{C60}$$

We write Ω_h, the component of $\boldsymbol{\Omega}$ parallel to $\mathbf{h}$, rather than Ω – despite the fact that they are equal in the present case – because V, W remain useful even in the non-parallel case (see below).

Using the identities (C3) and (C4), we also obtain

$$-\frac{2}{3} \frac{\dot{P}}{P} = -\frac{\dot{a}}{a} = \frac{\dot{\mathcal{E}}}{\mathcal{E}} = -\frac{2\dot{h}}{h} - \frac{2e\dot{e}}{1-e^2}$$

$$= \frac{2}{t_{\mathrm{TF}}} \left[\frac{1 + \frac{31}{2}e^2 + \frac{255}{8}e^4 + \frac{185}{16}e^6 + \frac{25}{64}e^8}{(1-e^2)^{15/2}} - \frac{\Omega_h}{\omega} \frac{1 + \frac{15}{2}e^2 + \frac{45}{8}e^4 + \frac{5}{16}e^6}{(1-e^2)^6} \right]. \tag{C61}$$

We can obtain the rate of change of the intrinsic spin Ω, using the constancy of $\mathbf{H}$ in Eq. (C45):

$$\frac{\dot{\Omega}}{\Omega} = -\frac{\mu h}{I\Omega} \frac{\dot{h}}{h} = \lambda W, \quad \lambda \equiv \frac{\mu h}{I\Omega}. \tag{C62}$$

The factor λ, the ratio of orbital to spin angular momentum, is usually large.

For the general case $\boldsymbol{\Omega} \times \mathbf{h} \neq 0$ there are more terms, but the problem is tractable. Equations (C3) give $\dot{\mathbf{e}}$ and $\dot{\mathbf{h}}$ in terms of the perturbing force; we average these over a Keplerian orbit, using the combination of the forces $\mathbf{f}_{QD}$ and $\mathbf{f}_{TF}$ of Eqs (C38) and (C55). The recipes (C11)–(C25) allow us to express the results in the form

$$\dot{\mathbf{e}}/e = -V\bar{\mathbf{e}} + \mathbf{U} \wedge \bar{\mathbf{e}} \tag{C63}$$

$$-I\dot{\boldsymbol{\Omega}}/\mu h = \dot{\mathbf{h}}/h = -W\bar{\mathbf{h}} + \mathbf{U} \wedge \bar{\mathbf{h}}, \tag{C64}$$

where the vector $\mathbf{U} \equiv X\bar{\mathbf{e}} + Y\bar{\mathbf{q}} + Z\bar{\mathbf{h}}$ is the angular velocity of the $\mathbf{e}, \mathbf{q}, \mathbf{h}$ frame relative to an inertial frame. V, W are as above – Eqs (C58) and (C59). For X, Y we obtain, after some manipulation,

$$X = -\frac{M_2 A}{2\mu\omega a^5} \frac{\Omega_h \Omega_e}{(1-e^2)^2} - \frac{\Omega_q}{\omega t_{TF}} \frac{I_{4,2}}{(1-e^2)^5}, \tag{C65}$$

$$Y = -\frac{M_2 A}{2\mu\omega a^5} \frac{\Omega_h \Omega_q}{(1-e^2)^2} + \frac{\Omega_e}{\omega t_{TF}} \frac{I_{4,0} - I_{4,2}}{(1-e^2)^5}, \tag{C66}$$

while Z is exactly the same as Eq. (C41): tidal friction contributes nothing extra to apsidal motion.

It is instructive to use Euler angles η, χ, ψ, say, to determine the orientations of $\bar{\mathbf{e}}, \bar{\mathbf{q}}, \bar{\mathbf{h}}$ relative to an inertial frame, say $\bar{\mathbf{E}}, \bar{\mathbf{Q}}, \bar{\mathbf{H}}$, a suitable choice for $\mathbf{H}$ being the total angular momentum vector of Eq. (C46):

$$\mathbf{H} = \mu\mathbf{h} + I\boldsymbol{\Omega}. \tag{C67}$$

$\mathbf{E}$ and $\mathbf{Q}$ are arbitrary, provided they make a right-handed orthogonal set with $\mathbf{H}$. The transformation from $\bar{\mathbf{E}}, \bar{\mathbf{Q}}, \bar{\mathbf{H}}$ to $\bar{\mathbf{e}}, \bar{\mathbf{q}}, \bar{\mathbf{h}}$ is the product of three successive simple rotations: by χ about $\bar{\mathbf{H}}$, by η about (new) $\bar{\mathbf{E}}$, and by ψ about (newer still) $\bar{\mathbf{H}}$, which now coincides with $\bar{\mathbf{h}}$. This gives

$$\bar{\mathbf{e}} = (\cos\chi\cos\psi - \sin\chi\sin\psi\cos\eta, \ \sin\chi\cos\psi + \cos\chi\sin\psi\cos\eta, \ \sin\eta\sin\psi) \tag{C68}$$

$$\bar{\mathbf{q}} = (-\cos\chi\sin\psi - \sin\chi\cos\psi\cos\eta \ -\sin\chi\sin\psi + \cos\chi\cos\psi\cos\eta, \ \sin\eta\cos\psi)$$

$$\tag{C69}$$

$$\bar{\mathbf{h}} = (\sin\eta\sin\chi - \sin\eta\cos\chi, \ \cos\eta), \tag{C70}$$

where the 1, 2, 3 axes are in the directions of $\bar{\mathbf{E}}, \bar{\mathbf{Q}}, \bar{\mathbf{H}}$. Differentiating Eqs (C68)–(C70) wrt time, and dotting by $\bar{\mathbf{e}}, \bar{\mathbf{q}}$ or $\bar{\mathbf{h}}$, it is straightforward to show that the components X, Y, Z of the angular velocity $\mathbf{U}$ in Eqs (C63) and (C64) relate to $\dot{\eta}, \dot{\chi}, \dot{\psi}$ by

$$X = \dot{\eta}\cos\psi + \dot{\chi}\sin\psi\sin\eta, \qquad Y = -\dot{\eta}\sin\psi + \dot{\chi}\cos\psi\sin\eta,$$

$$Z = \dot{\psi} + \dot{\chi}\cos\eta. \tag{C71a,b,c}$$

Define a constant Ω_0 by

$$\Omega_0 \equiv H/I, \quad \text{so that} \quad \Omega_\perp \equiv |\boldsymbol{\Omega} \times \bar{\mathbf{h}}| = \sqrt{\Omega_e^2 + \Omega_q^2} = \Omega_0\sin\eta, \quad \text{(C72a,b)}$$

by the triangle of angular momenta (C67). Then

$$\Omega_e \equiv \boldsymbol{\Omega} \cdot \bar{\mathbf{e}} = \frac{1}{I}(\mathbf{H} - \mu\mathbf{h}) \cdot \bar{\mathbf{e}} = \Omega_0\sin\eta\sin\psi = \Omega_\perp\sin\psi, \tag{C73}$$

$$\Omega_q \equiv \boldsymbol{\Omega} \cdot \bar{\mathbf{q}} = \Omega_0\sin\eta\cos\psi = \Omega_\perp\cos\psi, \quad \Omega_h = \boldsymbol{\Omega} \cdot \bar{\mathbf{h}} = \Omega_0\cos\eta - \mu h/I, \text{(C74)}$$

and so, from X, Y, Z in Eqs (C65), (C66) and (C41), we obtain

$$\dot{\eta} = X \cos\psi - Y \sin\psi = -\frac{\Omega_\perp}{2\omega t_{TF}} \frac{1 + 3e^2 + \frac{3}{8}e^4 + \frac{3}{2}e^2\left(1 + \frac{1}{6}e^2\right)\cos 2\psi}{(1-e^2)^5}, \qquad (C75)$$

$$\dot{\chi} = \frac{X \sin\psi + Y \cos\psi}{\sin\eta} = -\frac{M_2 A}{2\mu\omega a^5} \frac{\Omega_0 \Omega_h}{(1-e^2)^2} - \frac{3\Omega_0}{4\omega t_{TF}} \frac{e^2\left(1 + \frac{1}{6}e^2\right)\sin 2\psi}{(1-e^2)^5}, \qquad (C76)$$

and

$$\dot{\psi} + \dot{\chi}\cos\eta = Z = \frac{M_2 A\left(\Omega_h{}^2 - \frac{1}{2}\Omega_\perp^2\right)}{2\mu\omega a^5(1-e^2)^2} + \frac{15GM_2^2 A}{2\mu\omega a^8} \frac{1 + \frac{3}{2}e^2 + \frac{1}{8}e^4}{(1-e^2)^5}. \qquad (C77)$$

If $\sigma = 0$ ($t_{TF} = \infty$) we have steady precession, $\eta = $ constant, and the same precession rate $\dot{\chi}$ as in Eq. (C44).

We now have a complete set of equations for $\dot{h}$, $\dot{e}$, $\dot{\eta}$, $\dot{\chi}$, $\dot{\psi}$: Eqs (C63) and (C64) – dotting through by $\mathbf{e}$, $\mathbf{h}$ respectively – and (C75)–(C77). Ancillary variables a, ω, t_{TF}, Ω_h, $\Omega_\perp$ are given in terms of h, e, η by Eqs (C4), (C5), (C60), (C72b) and (C74). A and Ω_0 are constants given by Eqs (B51) and (C72a); M_1, M_2, R_1, Q, σ, H, I are given constants.

When *both* stars are extended bodies the Euler angles are less helpful, but we can, nevertheless, follow the motion numerically using the following larger set of equations:

$$\dot{\mathbf{e}}/e = -(V_1 + V_2)\bar{\mathbf{e}} + (\mathbf{U}_1 + \mathbf{U}_2) \times \bar{\mathbf{e}}, \qquad (C78)$$

$$\frac{I_1\dot{\boldsymbol{\Omega}}_1}{\mu h} = W_1\bar{\mathbf{h}} - \mathbf{U}_1 \times \bar{\mathbf{h}}, \qquad (C79)$$

$$\frac{I_2\dot{\boldsymbol{\Omega}}_2}{\mu h} = W_2\bar{\mathbf{h}} - \mathbf{U}_2 \times \bar{\mathbf{h}}, \qquad (C80)$$

with $\mathbf{h}$ now given by

$$\mathbf{h} = \frac{1}{\mu}(\mathbf{H} - I_1\boldsymbol{\Omega}_1 - I_2\boldsymbol{\Omega}_2). \qquad (C81)$$

$\mathbf{U}_1$, $\mathbf{U}_2$ are the obvious generalisations of $\mathbf{U}$ to $*1$ and $*2$ separately. These equations, by updating $\mathbf{e}$, $\boldsymbol{\Omega}_1$, $\boldsymbol{\Omega}_2$ and $\mathbf{h}$, also update $\mathbf{q} = \mathbf{h} \times \mathbf{e}$ and so Ω_q as well as Ω_e, Ω_h, for each component. As in the case of the single extended star, a, ω are obtained from $\mathbf{e}$, $\mathbf{h}$ by using Eqs (C4) and (C5).

(d)   Gravitational radiation

In the weak-field approximation, gravitational radiation implies that a system loses total energy ($\mu\mathcal{E}$) by gravitational radiation at a rate

$$\mu\dot{\mathcal{E}} = -\frac{4G}{45c^5} \frac{d^3 q_{ij}}{dt^3} \frac{d^3 q_{ij}}{dt^3}, \qquad (C82)$$

where q_{ij} is the quadrupole tensor of the matter distribution. For a binary,

$$q_{ij} = \frac{\mu}{2}(3d_i d_j - d^2\delta_{ij}), \qquad (C83)$$

and hence, using Eq. (C1) with $\mathbf{f} = 0$,

$$\frac{d^2 q_{ij}}{dt^2} = -\frac{GM\mu}{d^3}(3d_i d_j - d^2 \delta_{ij}) + \mu(3\dot{d}_i \dot{d}_j - \dot{\mathbf{d}} \cdot \dot{\mathbf{d}}\, \delta_{ij}), \tag{C84}$$

$$\frac{d^3 q_{ij}}{dt^3} = -\frac{GM\mu}{d^3}(6d_i \dot{d}_j + 6\dot{d}_i d_j - 9d_i d_j \dot{d}/d - d\dot{d}\,\delta_{ij}). \tag{C85}$$

Squaring Eq. (C85), we obtain, after some manipulation,

$$\dot{\mathcal{E}} = -\frac{32G^3 M^2 \mu}{5c^5}\left(\frac{\dot{\mathbf{d}} \cdot \dot{\mathbf{d}}}{d^4} - \frac{11}{12}\frac{\dot{d}^2}{d^4}\right). \tag{C86}$$

Averaging over the Keplerian orbit, using Eqs (C4), (C13) and (C15), we find that

$$\dot{\mathcal{E}} = -\frac{32G^3 M^2 \mu}{5c^5}\frac{\omega^2 ab}{l^4}\{2I_{3,0} - (1 - e^2)I_{2,0} - \frac{11}{12}e^2(I_{2,0} - I_{2,2})\} \tag{C87}$$

$$= \frac{64G^3 M^2 \mu \mathcal{E}}{5c^5 a^4}\frac{1 + \frac{73}{24}e^2 + \frac{37}{96}e^4}{(1 - e^2)^{7/2}}. \tag{C88}$$

In the same weak-field approximation, the loss rate of total angular momentum $\mu\mathbf{h}$ is

$$\mu\dot{h}_i = -\frac{8G}{45c^5}\epsilon_{ijk}\frac{d^2 q_{jl}}{dt^2}\frac{d^3 q_{kl}}{dt^3}, \tag{C89}$$

which, by Eqs (C84) and (C85) gives

$$\dot{\mathbf{h}} = -\frac{32G}{5\mu c^5}\left[\left(\frac{GM\mu}{d^3}\right)^2\frac{d^2}{2} + \frac{GM\mu^2}{d^3}\left(\frac{\dot{\mathbf{d}} \cdot \dot{\mathbf{d}}}{2} - \frac{3}{4}\dot{d}^2\right)\right]\mathbf{d} \times \dot{\mathbf{d}}. \tag{C90}$$

Averaging with Eqs (C14), (C19), (C21), (C27) and (C28), we get

$$\dot{\mathbf{h}} = -\frac{32G^3 M^2 \mu}{5c^5}\frac{1}{abl^2}\left[\frac{1}{2}I_{2,0} + \{I_{2,0} - \frac{1}{2}(1 - e^2)I_{1,0}\} - \frac{3}{4}e^2(I_{1,0} - I_{1,2})\right]\mathbf{h}, \tag{C91}$$

i.e.

$$\frac{\dot{h}}{h} = -\frac{32G^3 M^2 \mu}{5c^5 a^4}\frac{1 + \frac{7}{8}e^2}{(1 - e^2)^{5/2}}. \tag{C92}$$

As in the case of tidal friction, Eq. (C60), we can find $\dot{e}$ from $\dot{\mathcal{E}}$ and $\dot{h}$, and obtain

$$\frac{\dot{e}}{e} = -\frac{32G^3 M^2 \mu}{5c^5 a^4}\frac{\frac{19}{6} + \frac{121}{96}e^2}{(1 - e^2)^{5/2}}. \tag{C93}$$

(e)   Mass loss and mass transfer

When one or both stars are losing mass (by isotropic stellar winds), and perhaps also one star is gaining mass from the other (by either RLOF or accretion from a wind), the equations of motion have to be modified to take account of varying mass. In more general circumstances we might have to model the process in difficult detail, but we suppose here that we can write

$$\dot{M}_1 = -\zeta_1 - \xi, \quad \dot{M}_2 = -\zeta_2 + \xi, \quad \dot{M} = -\zeta_1 - \zeta_2; \tag{C94}$$

the ζ_i are the the rates of mass loss to infinity, and ξ is the rate of mass transfer. For the reduced mass

$$\frac{\dot{\mu}}{\mu} = \xi \left(\frac{1}{M_2} - \frac{1}{M_1} \right) - \zeta_1 \frac{M_2}{M_1 M} - \zeta_2 \frac{M_1}{M_2 M}. \tag{C95}$$

The equations of motion of the individual components, allowing for momentum transport between them, are

$$M_1 \ddot{\mathbf{d}}_1 = -\frac{G M_1 M_2 \mathbf{d}}{d^3} + \xi(\dot{\mathbf{d}}_1 - \mathbf{V}), \quad M_2 \ddot{\mathbf{d}}_2 = \frac{G M_1 M_2 \mathbf{d}}{d^3} - \xi(\dot{\mathbf{d}}_2 - \mathbf{V}), \tag{C96}$$

where

$$M \mathbf{V} \equiv M_1 \dot{\mathbf{d}}_1 + M_2 \dot{\mathbf{d}}_2. \tag{C97}$$

$\mathbf{V}$ is *not* the velocity of the centre of mass, since the masses are varying. $\mathbf{V}$ is not necessarily constant, and neither is $\dot{\mathbf{D}}$, the velocity of the centre of mass ($M \mathbf{D} \equiv M_1 \mathbf{d}_1 + M_2 \mathbf{d}_2$).

It is easy to see that

$$\ddot{\mathbf{d}} = -\frac{G M \mathbf{d}}{d^3} + \xi \left(\frac{1}{M_1} - \frac{1}{M_2} \right) \dot{\mathbf{d}}, \quad \text{i.e.} \quad \mathbf{f} = \xi \left(\frac{1}{M_1} - \frac{1}{M_2} \right) \dot{\mathbf{d}}, \tag{C98}$$

and further that

$$\dot{\mathbf{V}} = \frac{\zeta_2 M_1 - \zeta_1 M_2}{M^2} \mathbf{d}, \quad \dot{\mathbf{D}} = \mathbf{V} + \left(\frac{\zeta_2 M_1 - \zeta_1 M_2}{M^2} - \frac{\xi}{M} \right) \mathbf{d}. \tag{C99}$$

We can now determine the rates of change of $\mathcal{E}$, $\mathbf{h}$ and $\mathbf{e}$ from Eqs (C3), except that Eqs (C3) were obtained from the definitions (C2) on the assumption that $M = $ constant. Correcting for this, we replace (C3) by

$$\dot{\mathcal{E}} = \dot{\mathbf{d}} \cdot \mathbf{f} - \frac{G \dot{M}}{d}, \quad \dot{\mathbf{h}} = \mathbf{d} \times \mathbf{f},$$

$$G M \dot{\mathbf{e}} + G \dot{M} \mathbf{e} = \mathbf{f} \times \mathbf{h} + \dot{\mathbf{d}} \times (\mathbf{d} \times \mathbf{f}) - \frac{G \dot{M} \mathbf{d}}{d}. \tag{C100}$$

We can now average over the Keplerian orbit. If we assume for the moment that ξ is independent of phase over one orbit, then using Eqs (C12) and (C14) we obtain, after some manipulation,

$$\frac{d}{dt} \frac{\mathcal{E}}{M^2} = -2\xi \left(\frac{1}{M_1} - \frac{1}{M_2} \right) \frac{\mathcal{E}}{M^2}, \quad \text{or} \tag{C101a}$$

$$\frac{d}{dt} \frac{\mathcal{E}}{M^2 \mu^2} = \frac{2\mathcal{E}}{M^2} \left(\zeta_1 \frac{M_2}{M_1 M} + \zeta_2 \frac{M_1}{M_2 M} \right), \tag{C101b}$$

$$\dot{\mathbf{h}} = \xi \left(\frac{1}{M_1} - \frac{1}{M_2} \right) \mathbf{h}, \quad \text{or} \tag{C102a}$$

$$\frac{d}{dt} \mu \mathbf{h} = \left(\zeta_1 \frac{M_2}{M_1 M} + \zeta_2 \frac{M_1}{M_2 M} \right) \mu \mathbf{h}, \tag{C102b}$$

$$\dot{\mathbf{e}} = 0. \tag{C103}$$

Consider the two limiting cases of (a) winds with no transfer ($\xi = 0$) and (b) transfer with no winds ($\zeta_1 = 0 = \zeta_2$). In the first case, we get

$$\mathcal{E} \propto -M^2, \quad \mathbf{h} = \text{constant}, \quad \mathbf{e} = \text{constant}, \quad a = -\frac{GM}{2\mathcal{E}} \propto \frac{1}{M} \tag{C104}$$

and in the second,

$$M = \text{constant}, \quad \xi\left(\frac{1}{M_2} - \frac{1}{M_1}\right) = \frac{\dot{\mu}}{\mu}, \quad \mathcal{E} \propto \mu^2, \quad \mu h = \text{constant},$$

$$\mathbf{e} = \text{constant}, \quad a \propto \frac{1}{\mu^2}. \tag{C105}$$

The ζ and ξ were not assumed either small or constant in obtaining Eqs (C98)–(C100), but they were assumed small and nearly constant (i.e. constant on an orbital timescale) to do the averaging for Eqs (C101)–(C103). If they depend on orbital phase, say according to some prescribed dependence of ξ on $\mathbf{d}, \dot{\mathbf{d}}$, an averaging that incorporates this can still be done. In this case $\dot{\mathbf{e}}$ will not necessarily be zero.

In a particular case where $*2$ has no wind and $*1$ loses mass at a constant rate, we could have $\zeta_2 = 0$, $\xi + \zeta_1 = \text{constant}$, but $\xi = \xi(\mathbf{d}, \dot{\mathbf{d}})$. Then Eqs (C100) for $\dot{\mathbf{h}}$ and $\dot{\mathbf{e}}$ become

$$\dot{\mathbf{h}} = \mathbf{h}\left(\frac{1}{M_1} - \frac{1}{M_2}\right)\langle\xi\rangle,$$

$$\dot{\mathbf{e}} = \frac{2}{GM}\left(\frac{1}{M_1} - \frac{1}{M_2}\right)\langle\xi\dot{\mathbf{d}}\rangle \times \mathbf{h} - \frac{1}{M}\left(\left\langle\frac{\xi\mathbf{d}}{d}\right\rangle + \mathbf{e}\langle\xi\rangle\right). \tag{C106}$$

(f) Third body

If $\mathbf{f}$ is due to a third body at position $\mathbf{D}(t)$ ($D \gg d$) relative to the inner binary's centre of mass, the effect on the binary can be obtained by 'doubly averaging' (Heggie, private communication 1995) over both inner and outer orbits. This is reasonable since the timescale on which either orbit is changed by the other turns out to be long compared to either orbit. Let M_1 and M_2 be the masses of the inner pair, and M ($\equiv M_1 + M_2$) and M_3 be the masses of the outer pair. We use the quadrupole approximation for the perturbative force $\mathbf{f}$ on the inner pair:

$$f_i = S_{ij}d_j, \quad S_{ij} = GM_3\left(\frac{3D_i D_j}{D^5} - \frac{\delta_{ij}}{D^3}\right). \tag{C107}$$

The S_{ij} are functions of time, but assumed as usual to be nearly constant over one Keplerian orbit of the inner binary. By Eqs (C3) and (C21)–(C23), we obtain the following averages:

$$\dot{\mathcal{E}} = 0, \quad \text{and so} \quad a = \text{constant}, \quad p = \text{constant} \tag{C108}$$

$$\dot{\mathbf{e}} = \frac{eal}{2h}\{-5S_{12}\bar{\mathbf{e}} + (4S_{11} - S_{22})\bar{\mathbf{q}} - S_{23}\bar{\mathbf{h}}\}, \tag{C109}$$

$$\dot{\mathbf{h}} = \frac{a^2}{2}\{(1 - e^2)S_{23}\bar{\mathbf{e}} - (1 + 4e^2)S_{13}\bar{\mathbf{q}} + 5e^2 S_{12}\bar{\mathbf{h}}\}, \tag{C110}$$

with the 1,2,3 axes in the directions of $\mathbf{e}, \mathbf{q}, \mathbf{h}$ respectively. From these, the angular velocity $\mathbf{U} = (X, Y, Z)$ of Eqs (C63) and (C64) can be read off, and also the rates V, W of the change of magnitude of eccentricity and angular momentum.

Let us assume that $H \gg h$, so that $\mathbf{H}$ rather than $\mathbf{H} + \mathbf{h}$ can be taken to be a constant vector in direction and magnitude; in fact at the level of the quadrupole approximation $\mathbf{H}$ is exactly constant though **h** is not. We specify the orientation of $\bar{\mathbf{e}}$, $\bar{\mathbf{h}}$ relative to $\overline{\mathbf{E}}$, $\overline{\mathbf{H}}$ by Euler angles η, χ, ψ as in Section (c), Eqs (C68)–(C70). The rates of change of these angles are found from (X, Y, Z) using Eqs (C71). We now average S_{ij} over an *outer* orbit, using Eqs (C11) and (C20) with $n = 1$. We use capitals $\mathbf{E}, \mathbf{Q}, \mathbf{H}, A, B, L, P, \ldots$ to mean the same quantities as $\mathbf{e}, \mathbf{q}, \mathbf{h}, a, b, l, p, \ldots$ for the inner orbit. Since S_{12}, for example, means $S_{ij}\bar{e}_i\bar{q}_j$,

$$\langle S_{12}\rangle_{\text{outer}} = -\frac{3GM_3}{2ABL}\,\overline{\mathbf{H}}\cdot\bar{\mathbf{e}}\,\overline{\mathbf{H}}\cdot\bar{\mathbf{q}} = -\frac{3GM_3}{2ABL}\sin^2\eta\sin\psi\cos\psi, \text{ etc.} \tag{C111}$$

Then, after considerable manipulation, we obtain the following average rates of change:

$$t_{\text{TB}}\sqrt{1-e^2}\,\dot{e} = 5e(1-e^2)\sin^2\eta\sin\psi\cos\psi \tag{C112}$$

$$t_{\text{TB}}\sqrt{1-e^2}\,\dot{\eta} = -5e^2\sin\eta\cos\eta\sin\psi\cos\psi \tag{C113}$$

$$t_{\text{TB}}\sqrt{1-e^2}\,\dot{\chi} = -\{1+e^2(5\sin^2\psi-1)\}\cos\eta \tag{C114}$$

$$t_{\text{TB}}\sqrt{1-e^2}\,\dot{\psi} = 2(1-e^2)+5(e^2-\sin^2\eta)\sin^2\psi. \tag{C115}$$

The constant t_{TB} determines the timescale:

$$t_{\text{TB}} = \frac{2P^2}{3\pi p}(1-E^2)^{3/2}\frac{M+M_3}{M_3} = \frac{2P^2}{3\pi p}(1-E^2)^{3/2}\frac{M_1+M_2+M_3}{M_3}. \tag{C116}$$

Eqs (C112)–(C115) combine to give two integrals:

$$(1-e^2)\cos^2\eta = \text{constant}, \quad e^2(2-5\sin^2\psi\sin^2\eta) = \text{constant}. \tag{C117}$$

If we identify the integration constants by taking $e = e_a$ at $\psi = 0$ and $e = e_b$ at $\psi = \pi/2$, and then eliminate η, ψ from Eq. (C112), we obtain the following equation for e:

$$t_{\text{TB}}\,e_b\,e\,\dot{e} = \pm\sqrt{2(e_b^2-e^2)(e^2-e_a^2)(2e_a^2+3e^2e_b^2)}. \tag{C118}$$

Appendix D Steady, axisymmetric magnetic winds

Most manifestations of magnetohydrodynamics in stars are non-steady, or non-axisymmetric, or both. However, much insight can be gained by considering steady axisymmetric configurations, which are relatively amenable to analysis. We are also helped by the assumption of high, in fact infinite, conductivity – even although some dissipation of magnetic energy via finite conductivity or field-line reconnection is probably what drives winds in many stars. We can further simplify matters by assuming that the wind is adiabatic. This leads, as we show below, to five equations in five unknowns, four of which can be integrated analytically.

Since the magnetic field $\mathbf{B}$ is solenoidal, and since $\rho\mathbf{v}$ is also solenoidal in a steady situation, both can be written in terms of a toroidal component combined with a poloidal component derivable from a stream function. Using cylindrical polars (R, ϕ, z), we can write

$$\rho\mathbf{v} = \rho v_\phi \overline{\phi} + \rho\mathbf{v}_P = \rho v_\phi \overline{\phi} - \frac{1}{R}\overline{\phi} \times \nabla P, \tag{D1}$$

$$\mathbf{B} = B_\phi \overline{\phi} + \mathbf{B}_P = B_\phi \overline{\phi} - \frac{1}{R}\overline{\phi} \times \nabla Q, \tag{D2}$$

with $\rho, v_\phi, B_\phi, P, Q$ all functions of R, z only, because of the assumed axisymmetry. The dynamo equation, for steady fields with infinite conductivity, is

$$\nabla \times (\mathbf{v} \times \mathbf{B}) = 0. \tag{D3}$$

This (with several other results below) is best written in terms of Jacobians:

$$J(X, Y) \equiv \frac{\partial X}{\partial R}\frac{\partial Y}{\partial z} - \frac{\partial X}{\partial z}\frac{\partial Y}{\partial R}, \tag{D4}$$

for any $X(R, z), Y(R, z)$. Using Eqs (D1) and (D2) we see for instance that

$$\rho\mathbf{v} \cdot \nabla X = \frac{1}{R}J(P, X), \quad \mathbf{B} \cdot \nabla Y = \frac{1}{R}J(Q, Y). \tag{D5a,b}$$

Hence Eq. (D3) can be written, after some manipulation, as

$$\overline{\phi}\left[J\left(Q, \frac{v_\phi}{R}\right) - J\left(P, \frac{B_\phi}{\rho R}\right)\right] + \frac{\overline{\phi}}{R} \times \nabla \frac{J(P, Q)}{\rho R} = 0. \tag{D6}$$

Generally, if $J(X, Y) = 0$ then Y is a function of X, and conversely; and for any further function $Z(X)$,

$$J(Z(X), Y) = \frac{\mathrm{d}Z}{\mathrm{d}X}J(X, Y) = J\left(X, \frac{\mathrm{d}Z}{\mathrm{d}X}Y\right). \tag{D7}$$

289

Consequently a sufficient, though not necessary, condition for the poloidal part of Eq. (D6) to vanish is that $J(P, Q) = 0$, so that $P = P(Q)$ and hence

$$\rho \mathbf{v_P} = f(Q)\mathbf{B_P}, \quad \text{where} \quad f(Q) \equiv \frac{dP}{dQ}. \tag{D8}$$

The toroidal part of Eq. (D6) will then also vanish provided that

$$J\left(Q, \frac{v_\phi}{R}\right) = J\left(P, \frac{B_\phi}{\rho R}\right) = J\left(Q, \frac{dP}{dQ}\frac{B_\phi}{\rho R}\right), \tag{D9}$$

using Eq. (D7), so that

$$\frac{v_\phi}{R} - \frac{f(Q)B_\phi}{\rho R} = g(Q). \tag{D10}$$

Putting Eqs (D1), (D2), (D8) and (D10) together, a very general condition for a frozen-in velocity field is that

$$\mathbf{v} = \frac{f(Q)\mathbf{B}}{\rho} + Rg(Q)\overline{\boldsymbol{\phi}}, \tag{D11}$$

where f and g are arbitrary functions constant on field lines of $\mathbf{B}$ and stream lines of $\mathbf{v}$, and determined by conditions at the base of these lines, where they leave the star. $P(Q)$ is the first integral of $f(Q)$.

The equation of motion also yields an integral, which comes from its toroidal component and expresses the gain of angular momentum under the action of the magnetic torque. For steady axisymmetric motion, with pressure $p(R, z)$ and gravitational potential $\Phi(R, z) = -GM/\sqrt{R^2 + z^2}$,

$$\rho \mathbf{v} \cdot \nabla \mathbf{v} + \nabla p + \rho \nabla \Phi = \mathbf{j} \times \mathbf{B} = \frac{1}{\mu_0}(\nabla \times \mathbf{B}) \times \mathbf{B}. \tag{D12}$$

Using Eq. (D2), we can write the toroidal component of the magnetic term as

$$\overline{\boldsymbol{\phi}} \cdot (\nabla \times \mathbf{B}) \times \mathbf{B} = \frac{1}{R^2}J(Q, RB_\phi). \tag{D13}$$

We can write the inertia term in Eq. (D12) as

$$\rho \mathbf{v} \cdot \nabla \mathbf{v} \equiv \frac{1}{2\rho}\nabla(\rho^2 v^2) - \mathbf{v}\mathbf{v} \cdot \nabla \rho + \frac{1}{\rho}(\nabla \times \rho \mathbf{v}) \times \rho \mathbf{v}. \tag{D14}$$

The toroidal term, using (D1), and by analogy with Eq. (D13), is

$$\overline{\boldsymbol{\phi}} \cdot (\rho \mathbf{v} \cdot \nabla \mathbf{v}) = -v_\phi \mathbf{v} \cdot \nabla \rho + \frac{1}{\rho R^2}J(P, R\rho v_\phi)$$

$$= -\frac{v_\phi}{\rho R}J(P, \rho) + \frac{1}{\rho R^2}J(P, R\rho v_\phi) = \frac{1}{R^2}J(P, Rv_\phi). \tag{D15}$$

Since the pressure and potential in Eq. (D12) do not contribute toroidal terms, the toroidal component of (D12) is, from (D13), (D15) and (D7),

$$\frac{1}{\mu_0} J(Q, RB_\phi) = J(P, Rv_\phi) = J(Q, f(Q)Rv_\phi), \tag{D16}$$

which integrates immediately to

$$\frac{RB_\phi}{\mu_0} - f(Q)Rv_\phi = -h(Q), \tag{D17}$$

where h is another arbitrary function of Q.

Combining the two integrals (D10) and (D17), we obtain

$$\frac{v_\phi}{R}\left(1 - \frac{\mu_0 f^2}{\rho}\right) = g\left(1 - \frac{\mu_0 f h}{g\rho R^2}\right), \tag{D18a}$$

$$RB_\phi\left(1 - \frac{\mu_0 f^2}{\rho}\right) = -\mu_0 h\left(1 - \frac{R^2 f g}{h}\right). \tag{D18b}$$

The quantities f, g, h are all constant on field lines, and can be thought of as determined by given conditions at the surface of the star where the field lines originate. Equations (D18) then determine B_ϕ and v_ϕ as functions of ρ and R on each field line. Clearly both B_ϕ and v_ϕ would have singularities at the point where $\rho = \mu_0 f^2$, *unless* at that point the terms in parentheses on the right-hand sides also vanish. This means that there is a critical (Alfvénic) surface in the flow, at $R = R_A(Q)$, $\rho = \rho_A(Q)$ say, where

$$R_A^2 = \frac{h}{fg}, \qquad \rho_A = \mu_0 f^2. \tag{D19a,b}$$

Note that for B_ϕ to be non-singular on the axis $R = 0$ we need to have $h(Q) \propto Q \propto R^2$ near the axis.

If $\mathbf{B}_P$ near the star is roughly dipolar, as we expect, then field lines emerging from the northern hemisphere can be expected to be of two types, as illustrated in Fig. 3.7:

(a) closed field lines, which emerge between the equator and a critical latitude, cross the equatorial plane, and return to the symmetrical point on the southern hemisphere

(b) open field lines, which emerge from a polar cap north of the critical latitude, and are dragged out roughly radially to infinity by the wind – and similar field lines in the southern hemisphere, which connect the symmetrical polar cap to infinity but with the field reversed.

In the region of closed field lines, the 'dead zone', we expect from symmetry that there is no poloidal flow:

$$\mathbf{v}_P = f = 0, \qquad RB_\phi = \mu_0 h, \qquad v_\phi = Rg. \tag{D20}$$

The last two results come from Eqs (D18) and (D19). In the regions of open field lines on the stellar surface, P increases from (say) zero to $|\dot{M}|/4\pi$ in the northern hemisphere, and further to $|\dot{M}|/2\pi$ in the southern hemisphere. Since Q will have the same value at the same (positive or negative) latitude on the two hemispheres, P and f must be two-valued functions of Q.

We see from Eq. (D18) that provided both ρ and ρR^2 decrease outwards, the angular velocity $\Omega \equiv v_\phi / R$ is roughly constant on field lines within the Alfvénic cylinder, and decreases roughly as $1/R^2$ outside it, i.e. the material is forced to corotate with the star out to R_A and then expands freely outside R_A conserving its angular momentum.

If we assume that the flow is approximately adiabatic, we can obtain a second integral, of Bernoulli type, from the equation of motion. The pressure term in Eq. (D12) is

$$\nabla p = \nabla K \rho^\gamma = \rho \, \nabla \frac{\gamma}{\gamma - 1} \frac{p}{\rho}. \tag{D21}$$

We dot Eq. (D12) through by $\mathbf{v}$, and on its left-hand side use Eq. (D5a). On the right-hand side we use Eqs (D11) and (D13) to obtain

$$\mathbf{v} \cdot \{ (\nabla \wedge \mathbf{B}) \wedge \mathbf{B} \} = \frac{g(Q)}{R} J(Q, RB_\phi). \tag{D22}$$

Consequently (D12) gives

$$\frac{1}{R} J \left(P, \, \frac{1}{2} v^2 + \frac{\gamma}{\gamma - 1} \frac{p}{\rho} + \Phi \right) = \frac{g(Q)}{\mu_0 R} J(Q, RB_\phi), \tag{D23}$$

which, with the help of Eqs (D8), (D1), (D7), (D16) and (D17), leads to

$$\frac{|\nabla P|^2}{2R^2 \rho^2} + \frac{1}{2} v_\phi{}^2 + \frac{\gamma}{\gamma - 1} \frac{p}{\rho} - \frac{GM}{\sqrt{R^2 + z^2}} - g(Q) R v_\phi = k(Q), \tag{D24}$$

where $k(Q)$ is yet another arbitrary function that is constant on field lines. Equation (D24) remains true in the dead zone, where from Eqs (D20) it simplifies to

$$\frac{\gamma}{\gamma - 1} \frac{p}{\rho} - \frac{1}{2} R^2 g^2 - \frac{GM}{\sqrt{R^2 + z^2}} = k(Q). \tag{D25}$$

Five equations, i.e. the dynamo equation (two components) and the equation of motion (three components), determine the five unknown functions $\rho, v_\phi, P, B_\phi, Q$. They have yielded four integrals, Eqs (D8), (D10), (D17) and (D24). We can use Eq. (D24), in principle, to obtain ρ as a function of R, z, Q and $|\nabla Q|$, since v_ϕ and B_ϕ are known functions of Q, R and ρ via Eqs (D18). The remaining component of the equation of motion is then a highly non-linear second-order partial differential equation for $Q(R, z)$. Since we have already taken components in the $\overline{\phi}$ and $\mathbf{v}$ directions, the remaining component can be taken in the direction of ∇P (or ∇Q) since this is perpendicular to both. Unfortunately this component is not particularly simple; it can be written in the form

$$\left(\frac{1}{\mu_0} - \frac{f^2}{\rho} \right) |\nabla Q|^2 \left(R \frac{\partial}{\partial R} \frac{1}{R} \frac{\partial Q}{\partial R} + \frac{\partial^2 Q}{\partial^2 z} \right) = |\nabla Q|^2 \left[\frac{f^2}{\rho} \nabla Q \cdot \nabla \left(\log \frac{f}{\rho} \right) - R^2 \rho \frac{dk}{dQ} \right]$$

$$+ \nabla Q \cdot \left[\rho \nabla \left(\frac{R^2 v_\phi{}^2}{2} \right) - R^2 \rho \nabla (g R v_\phi) - \nabla \left(\frac{R^2 B_\phi{}^2}{2\mu_0} \right) \right]. \tag{D26}$$

Let us now attempt a few simplifications, to obtain some order-of-magnitude estimates of, for example, the Alfvén radius and the braking torque. We think of the mass-loss rate as given, and also the strength of $\mathbf{B}_P$ on the surface, and then we hope that R_A, B_ϕ and other quantities will follow.

We first make a very bland estimate of the field that solves Eq. (D26). We take it to be dipole-like near the stellar surface, and a 'split monopole' further out: the wind drags out field lines until they are roughly radial away from the star in one hemisphere, and towards it in the other. Separating the two hemispheres at the equator (beyond the dead zone; see Fig. 3.7) must be a plane current sheet, to support the assumed discontinuity in the tangential field. Supposing that the dead zone does not extend anything like as far as the Alfvén radius, we simply approximate $\mathbf{B}_P$ near the equatorial plane by

$$|\mathbf{B}_P| \sim B_1 \left(\frac{R_1}{R}\right)^2, \quad \text{so that} \quad B_A \sim B_1 \left(\frac{R_1}{R_A}\right)^2, \tag{D27}$$

B_1 being the field on the stellar surface and B_A at the Alfvén radius. The velocity field beyond the dead zone is similarly approximated by a monopole (but 'unsplit').

We next estimate f, g, h of Eqs (D8), (D18) and (D19). The stream functions P and Q take the following values on the stellar equator:

$$P_{eq} = \frac{|\dot{M}_1|}{4\pi}, \quad Q_{eq} \sim B_1 R_1^2, \quad \text{hence} \quad f \sim \frac{|\dot{M}_1|}{4\pi R_1^2 B_1}. \tag{D28}$$

From (D18) and (D19), assuming that both ρ and ρR^2 increase inwards, we see that

$$g \sim \Omega_1 \sim \Omega_A, \quad h \sim \frac{R_1 |B_{\phi 1}|}{\mu_0} \sim \frac{R_A |B_{\phi A}|}{\mu_0}, \tag{D29}$$

and that $B_\phi \propto 1/R$ for $R \lesssim R_A$. Then by eliminating f, g, h in Eq. (D19a) we obtain

$$|\dot{M}_1| \Omega_1 R_A^2 \sim \frac{4\pi}{\mu_0} |B_{\phi 1}| B_1 R_1^3 \sim \frac{4\pi}{\mu_0} |B_{\phi A}| B_A R_A^3. \tag{D30}$$

Thus the torque is largely determined by the product of the poloidal and toroidal fields, as we expect on very general grounds.

The magnitude v_A of the poloidal velocity field at the Alfvén radius can be estimated thus:

$$v_A \equiv \left(\frac{|\nabla P|}{\rho R}\right)_A \sim \frac{|\dot{M}_1|}{4\pi R_A^2 \rho_A}. \tag{D31}$$

This, along with Eq. (D19b) for ρ_A, Eq. (D28) for f and Eq. (D27) for B gives

$$\rho_A v_A^2 \sim \frac{B_1^2}{\mu_0} \frac{R_1^4}{R_A^4} \sim \frac{B_A^2}{\mu_0}, \tag{D32}$$

a result that we might well have written down a priori.

We now use Eq. (D24) very crudely, ignoring the pressure term, to estimate R_A. Beyond R_A there is little further radial or tangential acceleration, and so the poloidal velocity field must reach escape speed there. On the equatorial plane $z = 0$, comparing $R = R_A$ with $R = \infty$, we obtain

$$v_A^2 \sim \frac{2GM}{R_A}. \tag{D33}$$

Then Eqs (D19b), (D28) and (D31) give

$$|\dot{M}_1| R_A^{3/2} \sqrt{2GM_1} = \frac{4\pi}{\mu_0} B_1^2 R_1^4. \tag{D34}$$

This gives R_A in terms of supposedly known quantities. Finally, the ratio of Eqs (D30) and (D34) gives us an estimate for the tangential magnetic field:

$$\frac{|B_{\phi 1}|}{B_1} \sim \Omega_1 R_1 \left(\frac{R_A}{2GM_1}\right)^{1/2}, \qquad \frac{|B_{\phi A}|}{B_A} \sim \Omega_1 \left(\frac{R_A^3}{2GM_1}\right)^{1/2}. \tag{D35}$$

Note that the B_ϕ in this appendix, which is external to the star, is *not* to be identified with the B_ϕ of the next appendix (the $\alpha\Omega$ dynamo), which is internal to the star. However, the global $\mathbf{B}_P$ field *is* perceived to be essentially the same field, and is continuous across the stellar surface.

Appendix E Stellar dynamos

Some stars, like the Sun, are active dynamos, producing magnetic energy out of rotational energy. It is not clear whether all magnetic stars (Bp stars, neutron stars, some white dwarfs, . . .) produce fields continuously, or whether some have 'fossil' fields generated during an earlier active phase; the timescale of magnetic diffusion in a large-scale field is $\sim 10^{10}$ years. To have an active dynamo it is thought necessary to have all three of (a) rotation, (b) *differential* rotation (the Ω mechanism) and (c) turbulent convection (the α mechanism). The first two alone might seem sufficient, but it is reasonable to suppose that they would be axisymmetric. Such motion can convert poloidal magnetic field into toroidal field, but not conversely – Eq. (E17), with $\alpha = 0$, below – and so the poloidal field is bound to decay by diffusion. Since the Sun reverses its poloidal field every 11 years, it must be making use of the turbulent convection in its surface layers to convert toroidal field back to poloidal field.

Even if we think of the velocity field as *given*, so that the induction equation

$$\frac{\partial \mathbf{B}}{\partial t} = \nabla \times (\mathbf{v} \times \mathbf{B}) + \lambda \nabla^2 \mathbf{B} \tag{E1}$$

is linear in the unknown $\mathbf{B}$, it is difficult to approach the fully three-dimensional problem; and in practice $\mathbf{B}$ might react back on $\mathbf{v}$ through the Lorentz force $\mathbf{j} \times \mathbf{B} = (\nabla \times \mathbf{B}) \times \mathbf{B}/\mu_0$. The usual approach (Steenbeck *et al.* 1966, Roberts and Stix 1972, Moffat 1978) uses a 'two-scale' model, with macroscopic ('M-scale') fields $\mathbf{v}_0$, $\mathbf{B}_0$ that are axisymmetric, and microscopic ('μ-scale') fields $\delta\mathbf{v}$, $\delta\mathbf{B}$ that are affected by turbulence. The essential result that emerges from the μ-scale analysis (crudely summarised below) is that it adds two extra terms to the equation for the M-scale field:

$$\frac{\partial \mathbf{B}_0}{\partial t} = \nabla \times (\mathbf{v}_0 \times \mathbf{B}_0 + \alpha \mathbf{B}_0) + (\lambda + \beta)\nabla^2 \mathbf{B}_0. \tag{E2}$$

The α term is the more important since, even if small, it 'closes' the system by allowing toroidal field to be converted into poloidal field. The β term does not change the nature of the system, but being of the same order as turbulent diffusion it does, in practice, dominate the magnetic diffusion. Crude orders of magnitude are $\alpha \sim 10^{-2}$ m/s, $\beta \sim 10^{11}$ m²/s and $\lambda \sim 10^5$ m²/s.

On the M scale the μ fields average to zero, except for their products, and so Eq. (E1) becomes

$$\frac{\partial \mathbf{B}_0}{\partial t} = \nabla \times (\mathbf{v}_0 \times \mathbf{B}_0 + \mathbf{E}) + \lambda \nabla^2 \mathbf{B}_0, \quad \text{where} \quad \mathbf{E} \equiv \langle \delta\mathbf{v} \times \delta\mathbf{B} \rangle. \tag{E3}$$

Subtracting this from Eq. (E1) we get an equation for the μ field:

$$\frac{\partial \delta \mathbf{B}}{\partial t} = \nabla \times (\delta \mathbf{v} \times \mathbf{B}_0 + \mathbf{v}_0 \times \delta \mathbf{B} + \mathbf{G}) + \lambda \nabla^2 \delta \mathbf{B},$$

$$\mathbf{G} \equiv \delta \mathbf{v} \times \delta \mathbf{B} - \langle \delta \mathbf{v} \times \delta \mathbf{B} \rangle. \tag{E4}$$

In an apparent inconsistency we drop $\mathbf{G}$ from Eq. (E4) while retaining $\mathbf{E}$ in Eq. (E3). $\mathbf{G}$ is second order, and so might reasonably be dropped. However, $\mathbf{E}$, although of the same order of smallness (at least by hypothesis), is crucial to Eq. (E3) because of its ability to regenerate poloidal field; it leads to the α term in Eq. (E2). A less obvious, but more important, inconsistency is that, as applied to turbulence, $\mathbf{G}$ is not small: we expect that $|\delta \mathbf{v}|$ is of the same order as the ratio of length scale to timescale of the μ field, so that $|\partial \delta \mathbf{B}/\partial t|$ is necessarily of the same order as $|\nabla \times (\delta \mathbf{v} \times \delta \mathbf{B})|$. However, we ignore that point, hoping to gain some insight anyway.

The μ fields are usefully represented by their Fourier transforms (FTs):

$$\delta \tilde{\mathbf{v}} = \int \delta \mathbf{v} \, e^{-i\mathbf{k} \cdot \mathbf{r} + i\sigma t} \, d'^4 r, \quad \delta \mathbf{v} = \int \delta \tilde{\mathbf{v}} \, e^{i\mathbf{k} \cdot \mathbf{r} - i\sigma t} \, d'^4 k, \tag{E5}$$

and similarly for $\delta \mathbf{B}$, where $d'^4 k$, $d'^4 r$ represent volume elements in $\mathbf{k}$, σ and $\mathbf{r}$, t space, both (somewhat unconventionally) divided by $(2\pi)^2$; the prime is a reminder. Strictly the integrals should be over all space, but we hope to get away with the concept that we can use a volume (V_4, say) that is large compared with the μ scale yet small compared with the M scale (and conversely in the Fourier space).

We assume, for the time being, that $\mathbf{B}_0$ and $\mathbf{v}_0$ are strictly constant on the μ scale. This allows us to estimate α, whereas to estimate β we need to allow $\mathbf{B}_0$ to have a slight (constant) gradient.

Following the above, Eq. (E4) has Fourier transform

$$-i\sigma \, \delta \tilde{\mathbf{B}} = i\mathbf{k} \cdot \mathbf{B}_0 \delta \tilde{\mathbf{v}} - i\mathbf{k} \cdot \mathbf{v}_0 \, \delta \tilde{\mathbf{B}} - \lambda k^2 \, \delta \tilde{\mathbf{B}}, \quad \delta \tilde{\mathbf{B}} = \frac{i\mathbf{k} \cdot \mathbf{B}_0 \delta \tilde{\mathbf{v}}}{\lambda k^2 - i\sigma + i\mathbf{k} \cdot \mathbf{v}_0}. \tag{E6}$$

The solenoidal character of $\mathbf{B}$ ensures that $\mathbf{k} \cdot \delta \tilde{\mathbf{B}} = 0$, and we have assumed incompressible motion so that $\mathbf{k} \cdot \delta \tilde{\mathbf{v}} = 0$ as well. Using a version of the convolution theorem,

$$V_4 \mathbf{E} = \int \delta \mathbf{v} \times \delta \mathbf{B} \, d'^4 r = \int \delta \tilde{\mathbf{v}}^* \times \delta \tilde{\mathbf{B}} \, d'^4 k = i \int \frac{\mathbf{k} \cdot \mathbf{B}_0 \delta \tilde{\mathbf{v}}^* \times \delta \tilde{\mathbf{v}} \, d'^4 k}{\lambda k^2 - i\sigma + i\mathbf{k} \cdot \mathbf{v}_0}. \tag{E7}$$

The asterisk indicates complex conjugation. The answer must be real, since the first integral is, and so we can replace the answer by the average of it with its complex conjugate:

$$V_4 \mathbf{E} \sim \mathbf{B}_0 \cdot \int i\lambda \mathbf{k} \frac{k^2 \delta \tilde{\mathbf{v}}^* \times \delta \tilde{\mathbf{v}} \, d'^4 k}{\lambda^2 k^4 + (\sigma - \mathbf{k} \cdot \mathbf{v}_0)^2}. \tag{E8}$$

The factor $\delta \tilde{\mathbf{v}}^* \times \delta \tilde{\mathbf{v}}$ in the integrand shows that the integral is related to the 'helicity' $\mathbf{v} \cdot \nabla \times \mathbf{v}$; for, by the same convolution theorem,

$$\int \delta \mathbf{v} \cdot \nabla \times \delta \mathbf{v} \, d'^4 r = \int \delta \tilde{\mathbf{v}} \cdot (i\mathbf{k} \times \delta \tilde{\mathbf{v}})^* \, d'^4 k = i \int \mathbf{k} \cdot \delta \tilde{\mathbf{v}} \times \delta \tilde{\mathbf{v}}^* \, d'^4 k. \tag{E9}$$

We also see that the magnetic diffusivity is, in fact, crucial to the dynamo process, since the answer in Eq. (E8) contains it as a factor. On the other hand, the term $\mathbf{k} \cdot \mathbf{v}_0$ in Eq. (E8) does not appear to be very significant, unlike the term in $\mathbf{k} \cdot \mathbf{B}_0$, and so we take $\mathbf{v}_0 = 0$ in future.

In a convective region helicity is generated by the Coriolis force. We can model this crudely by considering an approximate equation of motion including the Coriolis force, along with a buoyancy term that is a vertically upward force $-g\mathbf{n}\,\delta\rho/\rho$:

$$\frac{\partial \delta\mathbf{v}}{\partial t} = -\frac{g}{\rho}\mathbf{n}\,\delta\rho - 2\boldsymbol{\Omega}\times\delta\mathbf{v}, \quad -i\sigma\delta\tilde{\mathbf{v}} = -\frac{g}{\rho}\mathbf{n}\,\delta\tilde{\rho} - 2\boldsymbol{\Omega}\times\delta\tilde{\mathbf{v}},$$

$$\delta\tilde{\mathbf{v}} \approx -\frac{ig\,\delta\tilde{\rho}}{\rho\sigma}\left(\mathbf{n} - \frac{2i\boldsymbol{\Omega}\times\mathbf{n}}{\sigma}\right). \tag{E10}$$

We ignore quadratic terms in Ω/σ. Then

$$\delta\tilde{\mathbf{v}}^{*}\times\delta\tilde{\mathbf{v}} \approx \frac{4ig^2\,|\delta\tilde{\rho}|^2}{\rho^2\sigma^3}\mathbf{n}\times(\boldsymbol{\Omega}\times\mathbf{n}). \tag{E11}$$

Thus the helicity (Eq. E9), and also the turbulence-driven electromotive force (Eq. E8), is largely dictated by the Coriolis term, i.e. by the rotation of the star.

The integral on the right-hand side of Eq. (E8) is a tensor, say $V_4\alpha_{ij}$. It is common to assume in practice, for simplicity, that α_{ij} is isotropic, $\alpha_{ij} = \alpha\delta_{ij}$. This leads to the symmetrical result $\mathbf{E} = \alpha\mathbf{B}_0$ that was included in Eq. (E2). Given the inherent problems of the analysis one cannot be confident that this is a good approximation. But in fact we only need $\mathbf{E} \approx \alpha_{\phi\phi}B_{0\phi}\overline{\boldsymbol{\phi}}$ to accomplish the goal of turning toroidal field into poloidal field. It is probably not worthwhile to try and estimate $\alpha_{\phi\phi}$ directly from the integral, given the uncertainties. All we need to note is that it contains both λ and Ω as factors.

To estimate β in Eq. (E2), we replace the constant field $\mathbf{B}_0$ in Eq. (E4) by a magnetic field of constant gradient, $\mathbf{r}\cdot\mathcal{B}$, where $\mathcal{B}$ is a constant traceless tensor:

$$\mathcal{B}_{ij} = \partial B_{0j}/\partial x_i. \tag{E12}$$

Then the Fourier transform of Eq. (E4) leads to the following replacement for Eq. (E6):

$$-i\sigma\,\delta\tilde{B}_k = -\mathcal{B}_{lm}\left[\frac{\partial}{\partial k_l}(k_m\,\delta\tilde{v}_k) + \delta_{km}\,\delta\tilde{v}_l\right] - \lambda k^2\,\delta\tilde{B}_k, \tag{E13}$$

and the estimate for the mean turbulent electromotive force corresponding to Eq. (E7) is

$$V_4 E_i \sim \int (\delta\mathbf{v}\times\delta\mathbf{B})_i\,\mathrm{d}^{\prime 4}r = -\epsilon_{ijk}\mathcal{B}_{lm}\int \delta\tilde{v}_j^{*}\left[\frac{\partial}{\partial k_l}(k_m\,\delta\tilde{v}_k) + \delta_{km}\,\delta\tilde{v}_l\right]\frac{\mathrm{d}^{\prime 4}k}{\lambda k^2 - i\sigma}. \tag{E14}$$

Once again, we can do little better than hope that the integral, now a fourth-rank tensor, will be approximately isotropic, i.e. of the form $V_4(\beta'\delta_{jk}\delta_{lm} + \beta''\delta_{jl}\delta_{km} + \beta'''\delta_{jm}\delta_{kl})$. Defining $\beta'' - \beta''' \equiv \beta$, we obtain

$$E_i \sim -\beta\epsilon_{ijk}\mathcal{B}_{jk}, \quad \text{i.e.} \quad \mathbf{E} = -\beta\nabla\times\mathbf{B}_0, \tag{E15}$$

using the definition (E12) of $\mathcal{B}$. By these somewhat crude means, we smooth the turbulent-field Eq. (E4) to reduce the mean-field Eq. (E3) to its standard form, Eq. (E2).

Having shown qualitatively how the effect of the μ field can be incorporated into the equation for the M field, we now consider Eq. (E2) applied to an axisymmetric star. As in Appendix D, we describe $\mathbf{B}$ (dropping the suffix zero) terms of a toroidal component B_ϕ and a poloidal component with stream function Q; this can be done even when the magnetic field is time dependent. However, we simplify further by taking the velocity M field to be both steady and purely toroidal ($P = 0$), and write $v_\phi \equiv R\Omega(R, z)$. Then, with some manipulation, Eq. (E2) can be written

$$\frac{\partial B_\phi}{\partial t} = -J(\Omega, Q) - \frac{\alpha}{R} D^2 Q + \frac{\lambda + \beta}{R} D^2(RB_\phi), \tag{E16}$$

$$\frac{\partial Q}{\partial t} = \alpha R B_\phi + (\lambda + \beta) D^2 Q, \tag{E17}$$

J being the Jacobian operator of Eq. (D4) and D^2 a Laplacian-like operator

$$D^2 \equiv R \frac{\partial}{\partial R} \frac{1}{R} \frac{\partial}{\partial R} + \frac{\partial^2}{\partial z^2}. \tag{E18}$$

The Jacobian in Eq. (E16) is a source term that winds up the poloidal field to give a toroidal component, provided that surfaces of constant Ω do not coincide with surfaces of constant Q; and the α term in Eq. (E17) allows toroidal field to be converted back to poloidal field, although slowly, on the assumption that α is small.

To simplify still further, let us suppose that the poloidal field has the simplest possible structure, being uniform of strength $B_{P0}(t)$ within the star so that $Q = B_{P0}R^2/2$, $D^2 Q = 0$. Then the Jacobian source term involves only the z gradient of Ω, and we approximate this as $\Delta\Omega/l$, assuming that the gradient in Ω is confined to a shell of thickness l (as found by helioseismology at the base of the Solar convection zone). For consistency in Eq. (E17) we must take $B_\phi \propto R$, i.e. $B_\phi = B_{\phi 0}(t)R/R_0$, where R_0 is some characteristic radius and $B_{\phi 0}$ is independent of position. Then $D^2(RB_\phi) = 0$ as well, and we end up with

$$\dot{B}_{\phi 0} \sim \frac{R_0 \Delta\Omega}{l} B_{P0}, \qquad \dot{B}_{P0} = \frac{2\alpha}{R_0} B_{\phi 0}. \tag{E19}$$

These equations allow the kind of cyclic behaviour seen in the Sun and some other stars, provided that α is complex, as we reasonably expect from its derivation.

As emphasised at the end of Appendix D, the B_ϕ of Appendices D and E are not the same, even at the surface of the star; however, we do assume that the poloidal field $\mathbf{B}_{P0}$ of Eq. (E19) is basically the same in magnitude, at the stellar surface, as the uniform internal field B_1 of Eq. (D27).

Appendix F Steady, axisymmetric, cool accretion discs

Accretion discs are usually modelled in cylindrical polars, and assumed to be steady, axisymmetric, and thin in the z direction: $|z| \lesssim h(R)$, where

$$h/R \equiv \delta(R) \ll 1. \tag{F1}$$

The material has little velocity in the R and z directions, compared with its tangential velocity:

$$|v_z| \sim \delta|v_R| \ll |v_R| \ll v_\phi \equiv R\Omega(R). \tag{F2}$$

We take the temperature $T(R)$ to be largely independent of z, and low enough that the tangential velocity is strongly supersonic

$$\frac{\Re T(R)}{\mu} \equiv c_s^2, \qquad c_s \ll \Omega R. \tag{F3}$$

Of course c_s is not actually the sound speed, but is of the same magnitude. Pressure and density drop off rapidly with $|z|$, but T – as well as the radial velocity v_R, the viscosity χ_v (in dimensions of length²/time) and the opacity κ – are assumed to be nearly independent of z, though dependent on R. I will show below – Eq. (F10) – that $c_s \sim h\Omega$, so that the strong inequalities (F1) and (F3) are not independent. Most of the above assumptions are justified below a posteriori, from the basic ones: steady, axisymmetric and cool.

The equation of motion (EoM) includes four terms: the inertia term, the pressure gradient, gravity (potential $\Phi = GM_2/\sqrt{R^2 + z^2}$, ignoring the self-gravity of the disc), and Navier–Stokes viscosity. In the R direction, inertia and gravity dominate, their balance giving Keplerian motion. In the ϕ direction only inertia and viscosity contribute, and so they are in balance. In the z direction the pressure gradient and gravity (weak as it is in that direction) dominate, and balance. The equation of heat (EoH) has three terms: advected heat, radiative heat loss, and viscous heating. In the usual model, advected heat is assumed negligible, and the second and third terms balance.

In reality, it is likely that magnetic forces rather than viscous forces are important, at least when the disc material is hot enough to be substantially ionised, and so the Navier–Stokes term should be replaced by the Lorentz force $\mathbf{j} \times \mathbf{B}$, and the viscous heating by ohmic dissipation. We show below, after dealing with the standard viscous disc, that under some rather idealised circumstances the magnetic terms become surprisingly similar to the viscous terms. This does not really justify the viscous model, but it does leave it as a reasonable first approximation.

The continuity equation is

$$\frac{1}{R}\frac{\partial}{\partial R}R\rho v_R + \frac{\partial}{\partial z}\rho v_z = 0. \tag{F4}$$

Both terms are comparable if $|\partial/\partial z| \sim 1/h$ and $|\partial/\partial R| \sim 1/R$; but if we integrate in the z direction, taking $\rho \sim 0$ for $|z| \gg h$, and suppose that v_R is largely independent of z, we get

$$\frac{d}{dR}R\sigma(R)v_R(R) = 0, \quad R\sigma v_R = -\frac{\dot{M}_2}{2\pi} = \text{constant}, \tag{F5}$$

where σ is the surface density $\int \rho dz$, and $\dot{M}_2 > 0$ is the accretion rate.

The R component of the equation of motion is

$$-\Omega^2 R + \mathbf{v}\cdot\nabla v_R = -\frac{1}{\rho}\frac{\partial p}{\partial R} - \frac{GM_2 R}{(R^2 + z^2)^{3/2}} + \text{viscous terms involving } v_R, v_z. \tag{F6}$$

We ignore the terms involving v_R, v_z, and the radial pressure gradient, to obtain the usual Keplerian approximation

$$\Omega^2 = \frac{GM_2}{R^3} \quad (|z| \lesssim h \ll R). \tag{F7}$$

The z component of the equation of motion is

$$\mathbf{v}\cdot\nabla v_z = -\frac{1}{\rho}\frac{\partial p}{\partial z} - \frac{GM_2 z}{(R^2 + z^2)^{3/2}} + \text{viscous terms involving } v_R, v_z. \tag{F8}$$

Again ignoring v_R, v_z, but keeping the vertical pressure gradient, we get

$$\frac{\partial p}{\partial z} = -\Omega^2 \rho z \quad (|z| \lesssim h \ll R). \tag{F9}$$

With Eq. (F3) this implies that

$$\rho = \rho_0(R)e^{-z^2/2h^2}, \quad h = c_s/\Omega, \quad \sigma = \sqrt{2\pi}\,\rho_0 h, \tag{F10}$$

where ρ_0 is the density on the plane $z = 0$.

The ϕ component of the equation of motion has no terms from pressure or gravity, and so with the approximation $\Omega = \Omega(R)$ of Eq. (F7) it becomes

$$\frac{\rho v_R}{R}\frac{d}{dR}\Omega R^2 = \frac{1}{R^2}\frac{\partial}{\partial R}\chi_v \rho R^3 \frac{d\Omega}{dR}. \tag{F11}$$

Integrating over z, and then using Eqs (F5) and (F7), this gives

$$\dot{M}_2 = 3\pi\chi_v\sigma + \frac{\text{constant}}{R^{1/2}} \sim 3\pi\chi_v\sigma + \dot{M}_2\left(\frac{R_2}{R}\right)^{1/2}. \tag{F12}$$

The constant of integration represents the fact that the disc cannot be Keplerian all the way to the surface of $*2$. There must be a point near the inner edge of the disc where the viscous couple $\chi_v\sigma R^3\,\partial\Omega/\partial R$ goes through zero, which leads to the estimate above for the constant.

The equation of heat is

$$\rho T\mathbf{v}\cdot\nabla s = -\nabla\cdot\mathbf{F} + \rho\epsilon, \quad \epsilon = \chi_v R^2\left(\frac{\partial\Omega}{\partial R}\right)^2 = \frac{9}{4}\chi_v\frac{GM_2}{R^3}. \tag{F13}$$

The advection term on the left-hand side is usually expected to be small. The viscous dissipation rate ϵ neglects terms in v_R, v_z and $\partial\Omega/\partial z$, as usual. **F** is the radiative heat flux (though convection can be significant in some circumstances). $\mathbf{F}$ will be mainly in the z direction, with surface value of, say, $\pm F_0/2$ at $z \sim \pm h$. Integrating perpendicular to the disc,

$$F_0 = \int \rho \epsilon \, \mathrm{d}z = \frac{9}{4} \sigma \chi_{\mathrm{v}} \frac{GM_2}{R^3} = \frac{3}{4\pi} \frac{GM_2 \dot{M}_2}{R^3} \left[1 - \left(\frac{R_2}{R} \right)^{1/2} \right], \tag{F14}$$

where we have used Eq. (F12). We have assumed that χ_{v}, like T, varies only modestly in the z direction.

Radiative transport allows us to estimate F_0 in terms of the temperature and opacity κ, or more specifically the optical depth τ_0 at the plane $z = 0$:

$$2\tau_0 = \int \kappa \rho \mathrm{d}z = \kappa \sigma. \tag{F15}$$

We approximate Eq. (F13) as

$$\frac{\mathrm{d}F}{\mathrm{d}z} = \rho \epsilon, \quad \text{so that} \quad F = \frac{\epsilon}{\kappa} (\tau_0 - \tau) = \frac{F_0}{2\tau_0} (\tau_0 - \tau), \tag{F16}$$

where $\tau(z)$ is the optical depth of a general layer, measured from the upper surface. The usual equation of radiative transport in the grey-body approximation (Mihalas 1970) for the specific radiation intensity I, at depth τ and angle θ to the outward normal, is

$$\cos \theta \frac{\partial I}{\partial \tau} = I - \frac{acT^4}{4\pi}, \quad F = 2\pi \int_0^{\pi} I \cos \theta \sin \theta \, \mathrm{d}\theta. \tag{F17}$$

For $\tau_0 \gtrsim 1$, this has a solution similar in nature to the Milne–Eddington solution for a stellar atmosphere, but it differs slightly because, by Eq. (F16), the flux F is not independent of τ. The solution is

$$I = \frac{3F_0}{8\pi \tau_0} \left\{ (\tau_0 - \tau) \cos \theta - \cos^2 \theta - \frac{1}{2} (\tau_0 - \tau)^2 + A \right\},$$

$$acT^4 = \frac{3F_0}{2\tau_0} \left\{ A - \frac{1}{2} (\tau_0 - \tau)^2 \right\}. \tag{F18}$$

The constant $A(\tau_0)$ is determined by the condition that at $\tau = 0$ there is no net *inward* flux, i.e.

$$\int_{\pi/2}^{\pi} I \cos \theta \sin \theta \, \mathrm{d}\theta = 0, \quad \text{so that} \quad A(\tau_0) = \frac{1}{2} + \frac{2}{3} \tau_0 + \frac{1}{2} \tau_0^2. \tag{F19}$$

Thus, the temperature T_0 on the mid-plane, and the effective temperature T_{e} defined by $F_0/2 = $ surface flux $= acT_{\mathrm{e}}^4/4$, are given by

$$acT_0^4 = \frac{3\epsilon}{\kappa} A(\tau_0) = \frac{3F_0}{2\tau_0} A(\tau_0), \quad acT_{\mathrm{e}}^4 = \frac{4\tau_0 \epsilon}{\kappa} = 2F_0. \tag{F20}$$

The fact that T_0 differs from T_{e} by a factor $\sim (3\tau_0/8)^{1/4}$ for large optical depths apparently violates our assumption that T is roughly independent of z, but (a) empirical estimates suggest that τ_0 is usually not large and (b) the assumption is not very important, since it only influences in detail the estimate (F10) for the density distribution.

The optically thin case ($\tau_0 \ll 1$) is rather easier to estimate, directly from Kirchhoff's law:

$$acT_0^4 = acT_{\mathrm{e}}^4 = \frac{\epsilon}{\kappa} = \frac{F_0}{2\tau_0}. \tag{F21}$$

A single interpolation formula includes both the thick and the thin estimates:

$$\frac{ac T_0^4}{1 + 3\tau_0/8} \sim ac T_e^4 \sim \frac{1 + 4\tau_0}{2\tau_0} F_0, \tag{F22}$$

with F_0 given by Eq. (F14).

It is usual to introduce a dimensionless parameter

$$\alpha \equiv \frac{\chi_v}{c_s h}. \tag{F23}$$

In terms of α, along with δ of Eq. (F1), we can estimate all the neglected terms in the equations of motion and heat, relative to either of the leading two terms, and find that they are all of order δ^2, $\alpha^2 \delta^2$ or $\alpha^2 \delta^4$. Further, we can estimate δ, along with τ_0, from some of the above results, supposing that we know M_2, R_2, α, and either $\dot{M}_2$ or equivalently the accretion luminosity L_{acc} as a fraction of L_{Edd} (Eq. 3.5.11):

$$L_{acc} \sim \frac{GM_2 \dot{M}_2}{R_2}, \qquad \dot{M}_2 \sim \frac{4\pi c R_2}{\kappa_T} \frac{L_{acc}}{L_{Edd}}, \tag{F24}$$

where κ_T is the Thomson-scattering opacity. Then using the definitions (F1) and (F3) of δ and c_s, and using Eqs (F7), (F14) and (F22) for $\Omega(R)$, $F_0(R)$ and $T_0(R)$, we obtain

$$\delta^8 \sim C_1 \frac{L_{acc}}{L_{Edd}} \frac{R_2}{R_{ch}} \frac{R}{R_{ch}} \left(\frac{M_{ch}}{M_2}\right)^3 \frac{(8 + 3\tau_0)(1 + 4\tau_0)}{16\tau_0},$$

$$C_1 = \frac{3\mathfrak{R}^4 R_{ch}^2}{\mu^4 G^3 a \kappa_T M_{ch}^3} \sim 2 \times 10^{-14}, \tag{F25}$$

where we take $\mu \sim 0.6$; we refer radii and masses to the Chandrasekhar values of Eq. (2.51). From Eq. (F15) we estimate

$$\tau_0 \sim \frac{C_2}{\alpha \delta^2} \frac{L_{acc}}{L_{Edd}} \frac{R_2}{R_{ch}} \left(\frac{R_{ch}}{R}\right)^{1/2} \left(\frac{M_{ch}}{M_2}\right)^{1/2} \frac{\kappa}{\kappa_T},$$

$$C_2 = \frac{2}{3} \sqrt{\frac{c^2 R_{ch}}{G M_{ch}}} \sim 40. \tag{F26}$$

These linked estimates of δ and τ_0 can be solved simultaneously. Supposing α is independent of R (for which there is no basis), Eqs (F25) and (F26) give

$$\tau_0 \propto R^{-0.6}, \qquad \delta \propto R^{0.05}, \qquad \tau_0 \gtrsim 1 \quad (R \text{ small})$$

$$\tau_0 \propto R^{-1}, \qquad \delta \propto R^{0.25}, \qquad \tau_0 \lesssim 1 \quad (R \text{ large}). \tag{F27}$$

If we accept that in reality magnetic fields rather than viscosity are what really drives accretion, we should replace the viscous term in Eq. (F11) by

$$\left[\frac{1}{\mu_0} \nabla \cdot (\mathbf{BB})\right]_\phi = \frac{1}{\mu_0} \left[\frac{1}{R^2} \frac{\partial}{\partial R} R^2 B_R B_\phi + \frac{\partial}{\partial z} B_z B_\phi\right]. \tag{F28}$$

The field exists because any seed B_R or B_z will be turned rapidly by Keplerian differential rotation – and even more rapidly by MHD instability (Balbus and Hawley 1991) – into a large B_ϕ. We assume the field to be limited by magnetic diffusivity, although more realistically it is likely to be limited by field-line reconnection, a highly non-linear and non-equilibrium process such as happens in Solar flares. The ϕ component of the induction equation, assuming a quasi-steady balance, gives

$$[\nabla \times (\mathbf{v} \times \mathbf{B})]_\phi = RB_R \frac{d\Omega}{dR} + \text{terms involving } v_R, v_z$$

$$\sim [\nabla \times (\chi_{\mathrm{m}} \nabla \times \mathbf{B})]_\phi = -\frac{\partial}{\partial R} \frac{\chi_{\mathrm{m}}}{R} \frac{\partial}{\partial R} RB_\phi - \frac{\partial}{\partial z} \chi_{\mathrm{m}} \frac{\partial B_\phi}{\partial z}, \quad \text{(F29)}$$

where χ_{m} is the magnetic diffusivity, with same dimensions as χ_{v}. B_z will also be created, if it did not exist already, but we shall ignore B_z for mathematical simplicity; we continue also to ignore v_R, v_z. The B_R and B_ϕ fields are likely to be chaotic rather than systematic, so that we guess $\partial/\partial R \sim \partial/\partial z \sim 1/h$ and then approximate the balance of Eq. (F29) by

$$B_\phi \sim \frac{h^2}{\chi_{\mathrm{m}}} RB_R \frac{d\Omega}{dR}. \quad \text{(F30)}$$

Then the magnetic torque of Eq. (F28) is equivalent to the viscous torque of Eq. (F11), provided we *define* χ_{v} by

$$\chi_{\mathrm{v}} \equiv \frac{h^2}{\chi_{\mathrm{m}}} \frac{B_R^2}{\mu_0 \rho}. \quad \text{(F31)}$$

The magnetic field will also contribute a pressure term, which by magnetic confinement we can expect to be limited by balance with the gas pressure:

$$\frac{B_\phi^2}{\mu_0} \sim \rho c_{\mathrm{s}}^2. \quad \text{(F32)}$$

Using the estimate $h = c_{\mathrm{s}}/\Omega$ of Eq. (F10), and the definition (F23) of α, Eqs (F30)–(F32) tell us that

$$\chi_{\mathrm{v}} \sim \chi_{\mathrm{m}} \sim \alpha \frac{B_\phi^2}{\mu_0 \rho \Omega}, \qquad B_R \sim \alpha B_\phi \quad \text{(F33)}$$

The magnetic energy generation rate is

$$\epsilon = \frac{\chi_{\mathrm{m}}}{\mu_0 \rho} |\nabla \times \mathbf{B}|^2 \sim \frac{\chi_{\mathrm{m}} B_\phi^2}{\mu_0 \rho h^2} \sim \chi_{\mathrm{v}} \Omega^2, \quad \text{(F34)}$$

using estimate (F32) and Eqs (F10), and so is of the same order as the conventional viscous dissipation of Eq. (F13). This crude equivalence of magnetic stress and dissipation with viscous stress and dissipation suggests that we can continue to use estimates like (F25) and (F26) to model magnetically-driven accretion discs.

References

Aarseth, S. J. (1996) in *The Origins, Evolutions, and Destinies of Binary Stars in Clusters*, ed. Milone, E. F. and Mermilliod, J.-C. ASP Conf. **90**, p223

(2001) in *Evolution of Binary and Multiple Star Systems*, ed. Podsiadlowski, P., Rappaport, S., King, A. R., D'Antona, F. and Burderi, L. ASP Conf. **229**, p91

Abbott, D. C., Bieging, J. H. and Churchwell, E. (1981) *Ap. J.*, **250**, 645

Abt, H. A. (1983) *A. Rev. A. & A.*, **21**, 343

Ake, T. B. and Parsons, S. B. (1990) *Ap. J.*, **364**, L13

Alcock, C., Allsman, R. A., Alves, D. *et al.* (1997) *Astron. J.*, **114**, 326

Alexander, D. R. and Ferguson, J. W. (1994) *Ap. J.*, **437**, 879

Alexander, M. E. (1973) *Ap. Sp. Sci.*, **23**, 459

Andersen, J. (1991) *A. & A. Rev.*, **3**, 91

Andersen, J., Clausen, J. V. and Gimenez, A. (1993) *A. & A.*, **277**, 439

Andersen, J., Clausen, J. V., Nordström, B. and Popper, D. M. (1985) *A. & A.*, **151**, 329

Andersen, J., Clausen, J. V., Nordström, B., Tomkin, J. and Mayor, M. (1991) *A. & A.*, **246**, 99

Andersen, J., Nordstrom, B., Mayor, M. and Polidan, R. S. (1988) *A. & A.*, **207**, 37

Andersen, J., Pavlovski, K. and Piirola, V. (1989) *A. & A.*, **215**, 272

Anosova, J. P. (1986) *Ap. Sp. Sci.*, **124**, 217

Antonov, V. A. (1962) *Vest. Leningrad Univ.*, **7**, 135

Applegate, J. H. and Patterson, J. (1987) *Ap. J.*, **322**, 99

Arnett, D. W., Bahcall, J. N., Kirshner, R. P. and Woosley, S. E. (1989) *A. Rev. A. & A.*, **27**, 629

Aumann, H. H. (1985) *PASP*, **97**, 885

Babcock, H. W. (1953) *PASP*, **65**, 229

Backer, D. C. and Hellings, R. W. (1986) *A. Rev. A. & A.*, **24**, 537

Bagnuolo, W. G., Gies, D. R. and Wiggs, M. S. (1992) *Ap. J.*, **385**, 708

Bagnuolo, W. G., Riddle, R. L., Gies, D. R. and Barry, D. J. (2001) *Ap. J.*, **554**, 362

Bahcall, J. N. (1964) *Phys. Rev. Lett.*, **12**, 300

(2000) in *Unsolved Problems in Stellar Evolution*, ed. Livio, M., STScI Symp. Ser. **12**, Cambridge: Cambridge University Press, p126

Bailyn, C. D., Jain, R. K., Coppi, P. and Orosz, J. A. (1998) *Ap. J.*, **499**, 367

Balbus, S. A. and Hawley, J. F. (1991) *Ap. J.*, **376**, 214

Balbus, S. A, Hawley, J. F. and Stone, J. M. (1996) *Ap. J.*, **467**, 76

Balona, L. A. (1990) *MNRAS*, **245**, 92

Balona, L. A., Krisciunas, K. and Cousins, A. W. J. (1994) *MNRAS*, **270**, 905

Baptista, R., Steiner, J. E. and Cieslinski, D. (1994) *Ap. J.*, **433**, 332

Barlow, D. J., Fekel, F. C. and Scarfe, C. D. (1993) *PASP*, **105**, 476

Basu, S. and Rana, N. C. (1992) *Ap. J.*, **393**, 373

Bath, G. T. and Pringle, J. E. (1982) *MNRAS*, **199**, 267

Batten, A. H. (1973) *Binary and Multiple Systems of Stars*, Oxford: Pergamon Press

(1976) in *Structure and Evolution of Close Binary Systems*, ed. Eggleton, P. P., Mitton, S. and Whelan, J. A., J. Dordrecht: Reidel, IAU Symp. **73**, p303

Batten, A. H., Fletcher, J. M. and McCarthy, D. G. (1989) *PDAO*, **17**, 1

Beckers, J. M. (1993) *A. Rev. A. & A.*, **31**, 13

Belczyński, K. and Mikołajewska, J. (1998) *MNRAS*, **296**, 77

Bell, S. A., Pollacco, D. L. and Hilditch, R. W. (1994) *MNRAS*, **270**, 449
Bettwieser, E. and Sugimoto, D. (1984) *MNRAS*, **208**, 493
Beuerman, K. and Pakull, M. W. (1984) *A. & A.*, **136**, 250
Beust, H., Corporon, P., Siess, L., Forestini, M. and Lagrange, A. M. (1997) *A. & A.*, **320**, 478
Bidelman, W. P. and Keenan, P. C. (1951) *Ap. J.*, **114**, 473
Bisscheroux, B. C., Pols, O. R., Kahabka, P., Belloni, T. and van den Heuvel, E. P. J. (1997) *A. & A.*, **317**, 815
Blaauw, A. (1961) *BAIN*, **15**, 265
    (1964) *A. Rev. A. & A.*, **2**, 213
Blanco, V. M. and McCarthy, M. F. (1981) in *Physical Processes in Red Giants*, ed. Iben, I. and Renzini, A., Dordrecht: Reidel, p147
Boden, A. F. and Lane, B. F. (2001) *Ap. J.*, **547**, 1071
Bodenheimer, P. D. (1978) *Ap. J.*, **224**, 488
    (1991) in *Angular Momentum Evolution of Young Stars*, ed. Catalano, S. and Stauffer, J. R. Dordrecht: Kluwer, p1
Böhm-Vitense, E. (1958) *ZsAp*, **46**, 108
    (1980) *Ap. J.*, **239**, L79
Böhm-Vitense, E. and Johnson, H. R. (1985) *Ap. J.*, **293**, 288
Bolton, A. J. C. and Eggleton, P. P. (1973) *A. & A.*, **24**, 429
Bolton, C. T. (1975) *Ap. J.*, **200**, 269
Bolton, C. T. and Rogers, G. L. (1978) *Ap. J.*, **222**, 234
Bolton, C. T. and Thomson, J. R. (1980) *Ap. J.*, **241**, 1045
Bond, H. E., Liller, W. and Mannery, E. J. (1978) *Ap. J.*, **223**, 252
Borgman, E. R. and Lippincott, S. L. (1983) *Astron. J.*, **88**, 120
Boss, A. P. (1991) *Nature*, **351**, 298
    (1993) *Ap. J.*, **410**, 157
Bradstreet, D. H. (1985) *Ap. J. Suppl.*, **58**, 413
Brandenburg, A., Saar, S. H. and Turpin, C. R. (1998) *Ap. J.*, **498**, L51
Bratton, C. B., Casper, D., Ciocio, A. *et al.* (1988) *Phys. Rev. D*, **37**, 3361
Brosche, P. (1962) *ZsAp*, **56**, 181
Brown, A., Bennett, P. D., Baade, R. *et al.* (2001) *Astron. J.*, **122**, 392
Brown, G. E., Weingartner, J. C. and Wijers, R. A. M. J. (1996) *Ap. J.*, **463**, 297
Budding, E. (1984) *Bull. Inf. Cent. Don. Stellaires*, **27**, 91
Burki, G. and Mayor, M. (1983) *A. & A.*, **124**, 256
Burns, D., Baldwin, J. E., Boysen, R. C. *et al.* (1997) *MNRAS*, **290**, L11
Butler, R. P., Marcy, G. W., Williams, E., Hauser, H. and Shirts, P. (1997) *Ap. J.*, **474**, L115
Campbell, C. B. and Papaloizou, J. (1983) *MNRAS*, **204**, 433
Casares, J. and Charles, P. A. (1994) *MNRAS*, **271**, L5
Casey, B. W., Matthieu, R. D., Vaz, L. P. R., Andersen, J. and Suntzeff, N. B. (1998) *Astron. J.*, **115**, 1617
Catalán, M. S., Davey, S. C., Sarna, M. J., Smith, R. C. and Wood, J. H. (1994) *MNRAS*, **269**, 879
Caughlan, G. R. and Fowler, W. A. (1988) *At. Data Nucl. Data Tables*, **40**, 284
Caughlan, G. R., Fowler, W. A., Harris, M. J. and Zimmerman, B. A. (1985) *At. Data Nucl. Data Tables*, **35**, 198
Chandrasekhar, S. (1931) *MNRAS*, **91**, 446
    (1939) *An Introduction to the Study of Stellar Structure*, Chicago: University of Chicago Press
    (1961) *Hydrodynamic and Hydromagnetic Stability*, Oxford: Clarendon Press
Chapman, S. and Cowling, T. G. (1958) *The Mathematical Theory of Non-Uniform Gases*, Cambridge: Cambridge University Press
Chlebowski, T. and Garmany, C. D. (1991) *Ap. J.*, **368**, 24
Chochol, D. and Mayer, P. (2002) in *Exotic Stars as Challenges to Evolution*, ed. Tout, C. A. and Van Hamme, W., ASP Conf. **279**, p143
Christensen-Dalsgaard, J. and Däppen, W. (1996) *Science*, **272**, 1286
Claverie, A., Isaak, G. R., MacLeod, C. P., van der Raay, H. B., and Roca Cortés, T. (1979) *Nature*, **282**, 591
Clayton, D. D. (1968) *Principles of Stellar Evolution and Nucleosynthesis*, New York: McGraw-Hill
Cochran, W. D., Hatzes, A. P., Butler, R. P. and Marcy, G. W. (1997) *Ap. J.*, **483**, 457
Conti, P. S. (1982) in *Mass Loss from Astronomical Objects*, ed. Gondalekhar, P. M., RAL conf. pp45, 141, 617
Conti, P. S., Garmany, C. D., de Lore, C. and Vanbeveren, D. (1983) *Ap. J.*, **274**, 302

Counselman, C. C. (1973) *Ap. J.*, **180**, 307

Cox, A. N., Morgan, S. M., Rogers, F. J. and Iglesias, C. A. (1992) *Ap. J.*, **393**, 272

Crawford, J. A. (1955) *Ap. J.*, **121**, 71

Darwin, G. H. (1879) *Proc. Roy. Soc. London*, **29**, 168

   (1880) *Phil. Trans. Roy. Soc.*, **171**, 713

Davis, R., Jr, Harmer, D. S. and Hoffman, K. C. (1968) *Phys. Rev. Lett.*, **20**, 1205

d'Cruz, N. L., Dorman, B., Rood, R. T. and O'Connell, R. W. (1996) *Ap. J.*, **466**, 359

Dearborn, D. S. P., Liebert, J., Aaronson, M. *et al.* (1986) *Ap. J.*, **300**, 314

Deeter, J. E., Boynton, P. E., Miyamoto, S. *et al.* (1991) *Ap. J.*, **383**, 324

de Jager, C., Nieuwenhuizen, H. and van der Hucht, K. (1988) *A. & A. Suppl.*, **72**, 259

de Kool, M. (1990) *Ap. J.*, **358**, 189

de Koter, A., Heap, S. R. and Hubeny, I. (1997) *Ap. J.*, **477**, 792

Delfosse, X., Forveillet, T., Beuzit, J.-L. *et al.* (1999) *A. & A.*, **344**, 897

Delgado-Marti, H., Levine, A. M., Pfahl, E. and Rappaport, S. (2001) *Ap. J.*, **546**, 455

de May, K., Aerts, C., Waelkens, C. and Van Winckel, H. (1996) *A. & A.*, **310**, 164

De May, K., Daems, K. and Sterken, C. (1998) *A. & A.*, **336**, 527

Dewi, J. D. M. and Tauris, T. M. (2000) *A. & A.*, **360**, 1043

Dominy, J. F. and Lambert, D. L. (1983) *Ap. J.*, **270**, 180

Dudley, R. E. and Jeffery, C. S. (1990) *MNRAS*, **247**, 400

Duquennoy, A. and Mayor, M. (1991) *A. & A.*, **248**, 485

Dziembowski, W. A., Moskalik, P. and Pamyatnykh, A. A. (1993) *MNRAS*, **265**, 588

Dziembowski, W. A. and Pamyatnykh, A. A. (1993) *MNRAS*, **262**, 204

Eddington, A. S. (1926) *The Internal Constitution of the Stars*, Cambridge: Cambridge University Press

Eggleton, P. P. (1971) *MNRAS*, **151**, 351

   (1972) *MNRAS*, **156**, 361

   (1976) in *Structure and Evolution of Close Binary Systems*, ed. Eggleton, P. P., Mitton, S. and Whelan, J. A. J., Dordrecht: Reidel, p209

   (1983a) *Ap. J.*, **268**, 368

   (1983b) *MNRAS*, **204**, 449

Eggleton, P. P. and Cannon, R. C. (1991) *Ap. J.*, **383**, 757

Eggleton, P. P. and Faulkner, J. (1981) in *Physical Processes in Red Giants*, ed. Iben, I. and Renzini, A., Dordrecht: Reidel, p179

Eggleton, P. P., Faulkner, J. and Flannery, B. P. (1973) *A. & A.*, **23**, 325

Eggleton, P. P. and Kiseleva, L. G. (2001) *Ap. J.*, **562**, 1012

Eggleton, P. P., Kiseleva, L. G. and Hut, P. (1998) *Ap. J.*, **499**, 853

Ergma, E. V. and van den Heuvel, E. P. J. (1998) *A. & A.*, **331**, L29

Etzel, P. B. and Olson, E. C. (1995) *Astron. J.*, **110**, 1809

Evans, D. S. (1968) *QJRAS*, **9**, 388

   (1977) *Rev. Mex. A. & A.*, **3**, 13

Evans, N. R. and Bolton, C. T. (1990) *Ap. J.*, **356**, 630

Fabian, A. C., Pringle, J. E. and Rees, M. J. (1975) *MNRAS*, **172**, 15

Falk, S. W. and Arnett, W. D. (1973) *Ap. J.*, **180**, L65

Fan, X., Burstein, D., Chen, J.-S. *et al.* (1996) *Astron. J.*, **112**, 628

Faulkner, J. (1966) *Ap. J.*, **144**, 978

Fekel, F. C. (1981) *Ap. J.*, **246**, 879

Fekel, F. C., Henry, G. W., Busby, M. R. and Eitter, J. J. (1993) *Astron. J.*, **106**, 2370

Fekel, F. C. Jr., and Tomkin, J. (1982) *Ap. J.*, **263**, 289

Ferguson, D. R., Liebert, J., Haas, J., Napiwotzki, R. and James, T. A (1999) *Ap. J.*, **518**, 866

Fitzpatrick, E. L. and Garmany, C. D. (1990) *Ap. J.*, **363**, 119

Flannery, B. P. (1975) *MNRAS*, **170**, 325

Frank, J., King, A. R. and Raine, D. J. (2002) *Accretion Power in Astrophysics*, Cambridge: Cambridge University Press

Fricke, K. J. (1968) *ZsAp*, **68**, 317

Fulbright, M. S., Liebert, J., Bergeron, P. and Green, R. (1993) *Ap. J.*, **406**, 240

Gamow, G. (1943) *Ap. J.*, **98**, 500

Garmany, C. D., Olson, G. L., van Steenberg, M. E. and Conti, P. S. (1981) *Ap. J.*, **250**, 660

Garrido, R., Sareyan, J.-P., Gimenez, A. *et al.* (1983) *A. & A.*, **122**, 193

Georgakarakos, N. (2003) *MNRAS*, **345**, 340

Ghez, A. M., Neugebauer, G., Gorham, P. W. *et al.* (1991) *Astron. J.*, **102**, 2066

Gies, D. R., Bagnuolo, W. G. Jr, Ferrara, E. C., Kaye, A. B. and Thaller, L. T. (1998) *Astron. J.*, **115**, 2561

Gies, D. R., Huang, W. and McSwain, M. V. (2002) *Ap. J.*, **578**, 67

Girard, T. M., Grundy, W. M., López, C. E. and van Altena, W. F. (1989) *Astron. J.*, **98**, 227

Girard, T. M., Wu, H., Lee, J. T. *et al.* (2000) *Astron. J.*, **119**, 2428

Goldreich, P. and Schubert, G. (1967) *Ap. J.*, **150**, 571

Gordon, K. D., Clayton, G. C., Smith, T. L. *et al.* (1998) *Astron. J.*, **115**, 2561

Greenstein, J. L., Dolez, N. and Vauclair, G. (1983) *A. & A.*, **127**, 25

Griffin, R. E. M. and Griffin, R. F. (2000) *MNRAS*, **319**, 1094

    (2002) *MNRAS*, **330**, 288

    (2004) *MNRAS*, **350**, 685

Griffin, R. E. M., Hünsch, M., Marshall, K. P., Griffin, R. F. and Schröder, K.-P. (1993) *A. & A.*, **274**, 225

Griffin, R. E. M., Schröder, K.-P., Misch, A. and Griffin, R. F. (1992) *A. & A.*, **254**, 289

Griffin, R. F. (1983) *The Observatory*, **103**, 273

    (1985) in *Interacting Binaries*, ed. Eggleton, P. P. and Pringle, J. E., NATO ASI Ser. C, **150**, p1

    (1988) *The Observatory*, **108**, 49

Griffin, R. F. and Duquennoy, A. (1993) *The Observatory*, **113**, 53

Griffin, R. F., Gunn, J. E., Zimmerman, B. A. and Griffin, R. E. M. (1988) *Astron. J.*, **96**, 172

Griffin, R. F. and Herbig, G. H. (1981) *MNRAS*, **196**, 33

Griffin, R. F. and Keenan, P. C. (1992) *The Observatory*, **112**, 168

Grønbech, B., Gyldenkerne, K. and Jørgensen, H. E. (1977) *A. & A.*, **55**, 401

Gualandris, A., Portegies Zwart, S. and Eggleton, P. P. (2004) *MNRAS*, **350**, 615

Guinan, E. F. and Koch, R. H. (1977) *PASP*, **89**, 74

Guinan, E. F., Marshall, J. J. and Maloney, F. P. (1994) *IBVS*, **4101**, 1

Hakala, P., Ramsay, G., Wu, K. *et al.* (2003) *MNRAS*, **343**, L10

Halbwachs, J. L. (1983) *A. & A.*, **128**, 399

    (1986) *A. & A.*, **168**, 161

Halbwachs, J. L., Mayor, M., Udry, S. and Arenou, F. (2003) *A. & A.*, **397**, 159

Hale, A. (1994) *Astron. J.*, **107**, 306

Han, Z., Podsiadlowski, P. and Eggleton, P. P. (1994) *MNRAS*, **270**, 121

Hansen, B. M. S. and Phinney, E. S. (1997) *MNRAS*, **291**, 569

Harlaftis, E. T., Steeghs, D., Horne, K., Martín, E. and Magazzú, A. (1999) *MNRAS*, **306**, 348

Harmanec, P., Habuda, P., Stefl, S. *et al.* (2000) *A. & A.*, **364**, L85

Harries, T. J. and Hilditch, R. W. (1997) *MNRAS*, **291**, 544

Harries, T. J., Hilditch, R. W. and Hill, G. (1998) *MNRAS*, **285**, 277

Hartkopf, W. I., McAlister, H. A. and Franz, O. G. (1989) *Astron. J.*, **98**, 1014

Harvey, D. A., Skillman, D. R., Kemp, J. *et al.* (1998) *Ap. J.*, **493**, L105

Harvin, J. A., Gies, D. R., Bagnuolo, W. G., Jr, Penny, L. R. and Thaller, M. L. (2002) *Ap. J.*, **565**, 1216

Hata, N. and Langacker, P. (1995), *Phys. Rev. D*, **52**, 420

Haxton, W. C. (1995) *A. Rev. A. & A.*, **33**, 459

Hayashi, C., Hoshi, R. and Sugimoto, D. (1962) *Supp. Prog. Theor. Phy.*, **22**, 1

Heggie, D. C. (1975) *MNRAS*, **173**, 729

Heggie, D. C. and Aarseth, S. J. (1992) *MNRAS*, **257**, 513

Heggie, D. C. and Hut, P. (2003) *The Gravitational Million-Body Problem*, Cambridge: Cambridge University Press

Heintz, W. D. (1969) *JRASC*, **63**, 275

    (1974) *Astron. J.*, **79**, 819

Heintze, J. R. W. and van Gent, R. H. (1989) in *Algols*, ed. Batten, A. H. Dordrecht: Kluwer Academic, p264

Herbig, G. H. (1960) *Ap. J. Suppl.*, **4**, 337

Herbig, G. H., Preston, G. W., Smak, J. and Paczyński, B. (1965) *Ap. J.*, **141**, 617

Herbig, G. H. and Terndrup, D. M. (1986) *Ap. J.*, **307**, 609

Hesser, J. E., Harris, W. E., Vandenberg, D. A. *et al.* (1987) *PASP*, **99**, 739

Hilditch, R. W. (1981) *MNRAS*, **196**, 305

    (1984) *MNRAS*, **211**, 943

    (2001) *An Introduction to Close Binary Stars,* Cambridge: Cambridge University Press

Hilditch, R. W. and Bell, S. A. (1987) *MNRAS*, **229**, 529
   (1994) *MNRAS*, **267**, 1081
Hilditch, R. W., Cameron, A. C., Hill, G., Bell, S. A. and Harries, T. J. (1997) *MNRAS*, **291**, 749
Hilditch, R. W., Harries, T. J. and Hill, G. (1996) *MNRAS*, **279**, 1380
Hilditch, R. W., Hill, G. and Bell, S. A. (1992) *MNRAS*, **255**, 285
Hilditch, R. W., King, D. J. and MacFarlane, T. M. (1988) *MNRAS*, **231**, 341
   (1989) *MNRAS*, **237**, 447
Hilditch, R. W., Skillen, I., Carr, D. M. and Aikman, G. C. L. (1986) *MNRAS*, **222**, 167
Hill, G. and Holmgren, D. E. (1995) *A. & A.*, **297**, 127
Hill, L. C., Jr (1993) *QJRAS*, **34**, 73
Hirata, K. S., Kajita, T., Koshiba, M. *et al.* (1988) *Phys. Rev. D*, **38**, 448
Hjellming, M. S. and Webbink, R. F. (1987) *Ap. J.*, **318**, 794
Hoffleit, D. and Jaschek, C. (1983) *The Bright Star Catalogue*, 4th ed., New Haven: Yale University
   Observatory
Höflich, P., Wheeler, J. C. and Thielemann, F. K. (1998) *Ap. J.*, **495**, 617
Hogeveen, S. J. (1990) *Ap. Sp. Sci.*, **173**, 315
Hoogerwerf, R., de Bruijne, J. H. J. and de Zeeuw, P. T. (2000) *Ap. J.*, **544**, L136
Horne, K. (1985) *MNRAS*, **213**, 129
Horne, K., Welsh, W. F. and Wade, R. A. (1993) *Ap. J.*, **410**, 357
Horne, K., Wood, J. H. and Stiening, R. F. (1991) *Ap. J.*, **378**, 271
Howarth, I. D., Stickland, D. J., Prinja, R. K., Koch, R. H. and Pfeiffer, R. J. (1991) *The Observatory*, **111**,
   167
Hrivnak, B. J., Guinan, E. F. and Lu, W. (1995) *Ap. J.*, **455**, 300
Huang, S.-S. (1966) *Ann. d'Ast.*, **29**, 331
Hummel, C. A., Armstrong, J. T., Quirrenbach, A. *et al.* (1993) *Astron. J.*, **106**, 2486
Humphreys, R. M. and Davidson, K. (1979) *Ap. J.*, **232**, 409
Hut, P. (1981) *A. & A.*, **99**, 126
Imbert, M. (1987) *A. & A. Suppl.*, **71**, 69
Inagaki, S. (1984) *MNRAS*, **206**, 149
Isaak, G. R. (1986) in *Seismology of the Sun and Distant Stars*, ed. Gough, D. O., Dordrecht: Reidel, p223
Itoh, N., Adachi, T., Nakagawa, M., Kohyama, Y. and Munakata, H. (1989) *Ap. J.*, **339**, 354
Itoh, N., Mutoh, H., Hikita, A. and Kohtama, Y. (1992) *Ap. J.*, **395**, 622
Jameson, R. F., Dobbie, P. D., Hodgkin, S. T. and Pinfield, D. J. (2002) *MNRAS*, **335**, 853
Janot-Pacheco, E., Motch, C. and Mouchet, M. (1987) *A. & A.*, **177**, 91
Jasniewicz, G., Duquennoy, A. and Acker, A. (1987) *A. & A.*, **180**, 145
Jeffery, C. S. and Simon, T. (1997) *MNRAS*, **286**, 487
Jeffreys, H. (1959) *The Earth*, 4th ed., Cambridge: Cambridge University Press
Jha, S., Torres, G., Stefanik, R. P., Latham, D. W. and Mazeh, T. (2000) *MNRAS*, **317**, 375
Johnson, H. M., Kulkarni, S. R. and Oke, J. B. (1989) *Ap. J.*, **345**, 492
Jorissen, A., Van Eck, S., Mayor, M. and Udry, S. (1998) *A. & A.*, **332**, 877
Judge, P. G. and Stencel, R. E. (1991) *Ap. J.*, **371**, 357
Kambe, E., Ando, H. and Hirata, R. (1993) *A. & A.*, **273**, 435
Kaper, L., Lamers, H. J. G. L. M., Ruymaekers, E., van den Heuvel, E. P. J. and Zuiderwijk, E. J. (1995) *A. &*
   A., **300**, 446
Kaspi, V. M., Johnston, S., Bell, J. F. *et al.* (1994a) *Ap. J.*, **423**, L43
Kaspi, V. M., Taylor, J. H. and Ryba, M. F. (1994b) *Ap. J.*, **428**, 713
Kato, S. (1966) *PASJ*, **18**, 374
Kelley, R. L., Rappaport, S., Brodheim, M. J., Cominsky, L. and Stothers, R. (1981) *Ap. J.*, **251**, 630
Khallasseh, B. and Hill, G. (1992) *A. & A.*, **257**, 199
Kilkenny, D., Lynas-Gray, A. E. and Roberts, G. (1996) *MNRAS*, **283**, 1349
Kippenhahn, R. and Weigert, A. (1967) *ZsAp*, **65**, 251
Kleinman, S. J., Nather, R. E., Winget, D. E. *et al.* (1998) *Ap. J.*, **495**, 424
Knee, L. B. G., Scarfe, C. D., Mayor, M., Baldwin, B. W. and Meatheringham, S. J. (1986) *A. & A.*, **168**, 72
Koen, C., Orosz, J. A. and Wade, R. A. (1998) *MNRAS*, **300**, 695
Koester, D. and Chanmugan, G. (1990) *Rep. Prog. Phys.*, **53**, 837
Kopal, Z. (1959) *Close Binary Stars*, New York: Wiley
Kozai, Y. (1962) *Astron. J.*, **67**, 591

Kozyreva, V. S. and Khalliulin, K. F. (1999) *A. Rep.*, **43**, 679 (≡ *A. Zh.*, **76**, 775)
Kraft, R. P. (1967) *Ap. J.*, **150**, 551
Kroupa, P., Gilmore, G. and Tout, C. A. (1991) *MNRAS*, **251**, 293
Kruszewski, A. (1966) *Adv. Astr. Ap.*, **4**, 233
Kurtz, D. W. (1990) *A. Rev. A. & A.*, **28**, 607
Kuzmin, V. A. (1966) *Sov. Phys. JETP*, **22**, 1051
Labeyrie, A. (1970) *A. & A.*, **6**, 85
Lacy, C. H. S., Helt, B. E. and Vaz, L. P. R. (1999) *Astron. J.*, **117**, 541
Lada, C. J. (1985) *A. Rev. A. & A.*, **23**, 267
Lamb, F. K., Pethick, C. J. and Pines, D. (1973) *Ap. J.*, **184**, 271
Lamers, H. J. G. L. M. (1981) *Ap. J.*, **245**, 593
Landsman, W., Simon, T. and Bergeron, P. (1993) *PASP*, **105**, 841
LaSala, J., Charles, P. A., Smith, R. A. D., Bałucinska-Church, M. and Curch, M. J. (1998) *MNRAS*, **301**, 285
Lattimer, J. M. and Yahil, A. (1989) *Ap. J.*, **340**, 426
Leonard, P. J. T., Hills, J. G. and Dewey, R. J. (1994) *Ap. J.*, **423**, L19
Lestrade, J.-F., Phillips, R. B., Hodges, M. W. and Preston, R. A. (1993) *Ap. J.*, **410**, 808
Leung, K.-C. and Schneider, D. P. (1978) *Ap. J.*, **224**, 565
Leung, K.-C. and Wilson, R. E. (1977) *Ap. J.*, **211**, 853
Lewin, W. H. G. and van den Heuvel, E. P. J. (1983) *Accretion-Driven Stellar X-Ray Sources*, Cambridge: Cambridge University Press
Lin, D. C. and Pringle, J. E. (1976) in *Structure and Evolution of Close Binary Systems*, ed. Eggleton, P. P., Mitton, S. and Whelan, J. A. J., Dordrecht: Reidel, IAU Symp. **73**, p237
Linsky, J. L. and Haisch, B. M. (1979) *Ap. J.*, **229**, 27
Lipari, S. L. and Sisterò, R. F. (1986) *MNRAS*, **220**, 883
Lloyd, C. and Wonnacott, D. (1994) *MNRAS*, **266**, L13
Loinard, L., Rodríguez, L. F. and Rodríguez, M. I. (2003) *Ap. J.*, **587**, L47
Lowrance, P. J., Kirkpatrick, J. D. and Beichman, C. A. (2002) *Ap. J.*, **572**, L79
Lubow, S. H. and Shu, F. H. (1975) *Ap. J.*, **198**, 383
Lucy, L. B. (1968) *Ap. J.*, **153**, 877
Lucy, L. B. and Abbott, D. C. (1993) *Ap. J.*, **405**, 738
Lucy, L. B. and Ricco, E. (1979) *Astron. J.*, **84**, 401
Lucy, L. B. and Solomon, P. (1970) *Ap. J.*, **159**, 879
Lundgren, S. C., Cordes, J. M., Foster, R. S., Wolszczan, A. and Camilo, F. (1996) *Ap. J.*, **458**, L33
Lüst, R. (1952) *Z. Naturforsch.*, **7a**, 87
Lynden-Bell, D. (1968) *Bull. Astron.*, **3**, 305
   (1969) *Nature*, **223**, 690
Lynden-Bell, D. and Eggleton, P. P. (1980) *MNRAS*, **191**, 483
Lyne, A. G., Burgay, M., Kramer, M *et al.* (2004) *Science*, **303**, 1153
Lyne, A. G. and Lorimer, D. R. (1994) *Nature*, **369**, 127
McAlister, H. A. (1985) *A. Rev. A. & A.*, **23**, 59
McAlister, H. A. and Hartkopf, W. I. (1988) *2nd Catalog of Interferometric Measurements of Binary Stars*, Center for High Angular Resolution Astronomy Contrib. 2, CHARA: Georgia State University
McClure, R. D. (1983) *Ap. J.*, **268**, 264
McClure, R. D. and Woodsworth, A. W. (1990) *Ap. J.*, **352**, 709
MacDonald, J. (1983) *Ap. J.*, **273**, 289
McMillan, S. L. W., Hut, P. and Makino, J. (1990) *Ap. J.*, **362**, 522
Maeder, A. (1975) *A. & A.*, **40**, 303
   (1976) *A.& A.*, **47**, 389
Marchenko, S. V., Rauw, G., Antokhina, E. A. *et al.* (2000) *MNRAS*, **317**, 333
Marcy, G. W. and Butler, R. P. (1998) *A. Rev. A. & A.*, **36**, 57
Mardling, R. A. (1995) *Ap. J.*, **450**, 732
Margon, B. (1984) *A. Rev. A. & A.*, **22**, 507
Markwardt, C. B., Swank, J. H., Strohmeyer, T. E., in't Zand, J. J. M. and Marshal, F. E. (2002) *Ap. J.*, **575**, L21
Marsh, T. R. (1995) *MNRAS*, **275**, L1
Marsh, T. R., Dhillon, V. S. and Duck, S. R. (1995) *MNRAS*, **275**, 828
Marsh, T. R. and Horne, K. (1988) *MNRAS*, **235**, 269

Marsh, T. R., Horne, K. and Rosen, S. (1991) *Ap. J.*, **366**, 535
Mason, B. D., Gies, D. R., Hartkopf, W. I. *et al.* (1998) *Astron. J.*, **115**, 821
Massey, P. and Niemelä, V. S. (1981) *Ap. J.*, **245**, 195
Mathieu, R. D. (1994) *A. Rev. A. & A.*, **32**, 465
Mathieu, R. D., Latham, D. W., and Griffin, R. F. (1990) *Astron. J.*, **100**, 1859
Maxted, P. F. L., Heber, U., Marsh, T. R. and North, R. C. (2001) *MNRAS*, **326**, 1391
Maxted, P. F. L. and Hilditch, R. W. (1995) *A. & A.*, **301**, 149
Maxted, P. F. L., Hill, G. and Hilditch, R. W. (1994a) *A. & A.*, **282**, 821
    (1994b) *A. & A.*, **285**, 535
    (1995a) *A. & A.*, **301**, 135
    (1995b) *A. & A.*, **301**, 141
Maxted, P. F. L., Marsh, T. R. and Moran, C. (2002) *MNRAS*, **332**, 745
Maxted, P. F. L., Marsh, T. R., Moran, C., Dhillon, V. S. and Hilditch, R. W. (1998) *MNRAS*, **300**, 1225
Maxted, P. F. L., Marsh, T. R. and North, R. C. (2000) *MNRAS*, **317**, 41
Mayor M. and Queloz, D. (1995) *Nature*, **378**, 355
Mazeh, T., Goldberg, D., Duquennoy, A. and Mayor, M. (1992) *Ap. J.*, **401**, 265
Mazeh, T., Latham, D. W. and Stefanik, R. P. (1996) *Ap. J.*, **466**, 415
Mazurek, T. J. and Wheeler, J. C. (1980) *Fund. Cosm. Phys.*, **5**, 193
Méndez, R. H. and Niemelä, V. S. (1981) *Ap. J.*, **250**, 240
Mestel, L. (1952) *MNRAS*, **112**, 583
    (1968) *MNRAS*, **138**, 359
Mestel, L. and Spruit, H. (1987) *MNRAS*, **226**, 57
Michaud, G. (1970) *Ap. J.*, **160**, 641
Mihalas, D. (1970) *Stellar Atmospheres*, San Francisco: Freeman
Mikheyev, S. P. and Smirnov, A. Y. (1985) *Sov. J. Nucl. Phys.*, **42**, 913
Mikołajewska, J., Kenyon, S. J., Mikołajewski, M., Garcia, M. R. and Polidan, R. S. (1995) *Astron. J.*, **109**, 1289
Milano, L. and Russo, G. (1983) *MNRAS*, **203**, 235
Miller, G. E. and Scalo, J. M. (1979) *Ap. J. Suppl.*, **41**, 513
Moffat, H. K. (1978) *Magnetic Field Generation in Electrically Conducting Fluids*, Cambridge: Cambridge University Press
Morales-Rueda, L., Still, M. D., Roche, P., Wood, J. H. and Lockley, J. J. (2002) *MNRAS*, **329**, 597
Morgan, J. G. and Eggleton, P. P. (1979) *MNRAS*, **187**, 661
Morrison, N. D. and Conti, P. S. (1980) *Ap. J.*, **239**, 212
Morton, D. C. (1960) *Ap. J.*, **132**, 146
Nagataki, S., Shimizu, T. M. and Sato, K. (1998) *Ap. J.*, **495**, 413
Nakajima, T., Oppenheimer, B. R., Kulkarni, S. R. *et al.* (1995) *Nature*, **378**, 463
Napiwotzki, R., Koester, D., Nelemans, G. *et al.* (2002) *A. & A.*, **386**, 957
Nariai, K. (1967) *PASJ*, **19**, 564
Nauenberg, M. (1972) *Ap. J.*, **175**, 417
Naur, P. and Osterbrock, D. E. (1952) *Ap. J.*, **117**, 306
Nelson, C. A. and Eggleton, P. P. (2001) *Ap. J.*, **552**, 664
Nelson, L. A. and Rappaport, S. (2003) *Ap. J.*, **598**, 431
Niemelä, V. S. and Moffat, A. F. J. (1982) *Ap. J.*, **259**, 213
Nomoto, K. (1984) *Ap. J.*, **277**, 291
North, R. C., Marsh, T. R., Moran, C. K. J. *et al.* (2000) *MNRAS*, **313**, 383
O'Brien, M. S., Bond, H. E. and Sion, E. M. (2002) *Ap. J.*, **563**, 971
O'Brien, M. S., Vauclair, G., Kawaler, S. D. *et al.* (1998) *Ap. J.*, **495**, 458
O'Donoghue, D., Koen, C., Kilkenny, D. *et al.* (2003) *MNRAS*, **345**, 506
Olson, E. C. (1985) in *Interacting Binaries*, ed. Eggleton, P. P. and Pringle, J. E., NATO ASI Ser. C, **150**, p127
Olson, E. C. and Etzel, P. B. (1994) *Astron. J.*, **108**, 262
Olson, G. L. and Castor, J. I. (1981) *Ap. J.*, **244**, 179
Orosz, J. A., Wade, R. A., Harlow, J. J. B. *et al.* (1999) *Astron. J.*, **117**, 1598
Orosz, J. A. and Kuulkers, E. (1999) *MNRAS*, **305**, 132
Ostrov, P. G. (2001) *MNRAS*, **321**, L25
Paczyński, B. (1967) *Acta Astr.*, **17**, 193

(1971) *A. Rev. A. & A.*, **9**, 183

(1973) *Acta Astr.*, **23**, 192

(1976) in *Structure and Evolution of Close Binary Systems*, ed. Eggleton, P. P., Mitton, S. and Whelan J., IAU Symp. **73**, Dordrecht: Reidel, p75

(1986) *Ap. J.*, **304**, 1

Paczyński, B. and Sienkiewicz, R. (1972) *Acta Astr.*, **22**, 73

Paczyński, B. and Ziółkowski, J. (1968) *Acta Astr.*, **18**, 255

Patterson, J. (1998) *PASP*, **110**, 1132

Patterson, J., Kemp, J., Shambrook, A. *et al.* (1998) *PASP*, **110**, 380

Pauldrach, A. W. A., Hoffman, T. L. and Lennon, M. (2001) *A. & A.*, **375**, 161

Pennington, R. (1986) Unpublished Ph.D. thesis, Cambridge University

Peters, P. C. (1964) *Phys. Rev.*, **136**, B1224

Petrie, R. M. (1960) *Ann. Astrophys.*, **23**, 744

Pfahl, E., Rappaport, S., Podsiaplowski, P. and Spruit, H. (2002) *Ap. J.*, **574**, 364

Phinney, E. S. and Kulkarni, S. R. (1994) *A. Rev. A. & A.*, **32**, 639

Pizzo, V., Schwenn, R., Marsch, E. *et al.* (1983) *Ap. J.*, **271**, 335

Plavec, M. (1968) *Adv. Astr. & Af.*, **6**, 201

Plets, H., Waelkens, C. and Trams, N. R. (1995) *A. & A.*, **293**, 363

Pollacco, D. L. and Bell, S. A. (1994) *MNRAS*, **267**, 452

Pols, O. R. (1992) in *Evolutionary Processes in Interacting Binary Stars*, ed. Kondo, Y., Sisterò, R. and Polidan, R. S., IAU Symp. **151**, p527

(1994) *A. & A.*, **290**, 119

Pols, O. R., Schröder, K.-P., Hurley, J. R., Tout, C. A. and Eggleton, P. P. (1998) *MNRAS*, **298**, 525

Pols, O. R., Tout, C. A., Eggleton, P. P. and Han, Z. (1995) *MNRAS*, **274**, 964

Pols, O. R., Tout, C. A., Schröder, K.-P., Eggleton, P. P. and Manners, J. (1997) *MNRAS*, **289**, 869

Popper, D. M. (1980) *A. Rev. A. & A.*, **18**, 115

(1987) *Ap. J.*, **313**, L81

(1988a) *Astron. J.*, **95**, 1242

(1988b) *Astron. J.*, **96**, 1040

(1993) *PASP*, **105**, 721

(1994) *Astron. J.*, **108**, 1091

(1998) *Astron. J.*, **114**, 1195

Popper, D. M. and Hill (1991) *Astron. J.*, **101**, 600

Popper, D. M. and Plavec, M. J. (1976) *Ap. J.*, **205**, 462

Portegies Zwart, S. F. and McMillan, S. L. W. (2002) *Ap. J.*, **576**, 899

Pourbaix, D. (1999) *A. & A.*, **348**, 127

Poveda, A., Allen, C. and Parrao L. (1982) *Ap. J.*, **258**, 589

Preibisch, T., Weigelt, G. and Zinnecker, H. (2001) in *The Formation of Binary Stars*, ed. Zinnecker, H. and Matthieu, R. D., IAU Symp. **200**, p69

Priedhorsky, W. C. and Holt, S. S. (1987) *Ap. J.*, **312**, 743

Pringle, J. E. (1974) *MNRAS*, **168**, 13

(1981) *A. Rev. A. & A.*, **19**, 137

(1989) *MNRAS*, **239**, 361

Proffitt, C. R. and Michaud, G. (1991) *Ap. J.*, **380**, 238

Provencal, J. L., Shipman, H. L., Høg, E. and Thejll, P. (1998) *Ap. J.*, **494**, 759

Rappaport, S., Joss, P. C. and McClintock, J. E. (1976) *Ap. J.*, **206**, L103

Rappaport, S., Verbunt, F. and Joss, P. C. (1983) *Ap. J.*, **275**, 713

Rebolo, R., Zapatero-Osorio, M. R. and Martin, E. L. (1995) *Nature*, **377**, 129

Reimers, D. (1975) *Mem. Roy. Soc. Liège 6e Ser.*, **8**, 369

Reimers, D., Griffin, R. F. and Brown, A. (1988) *A. & A.*, **193**, 180

Reynolds, A. P., Bell, S. A. and Hilditch, R. W. (1992) *MNRAS*, **256**, 631

Rhie, S. H., Becker, A. C., Bennett, D. P. *et al.* (1999) *Ap. J.*, **522**, 1037

Richards, M. T., Albright, G. E. and Bowles, L. M. (1995) *Ap. J.*, **438**, L103

Richards, M. T., Jones, R. D. and Swain, M. E. (1996) *Ap. J.*, **459**, 249

Richer, J., Michaud, G. and Turcotte, S. (2000) *Ap. J.*, **529**, 338

Ritter, H. and Kolb, U. (1998) *A. & A. Suppl.*, **129**, 83

Roberts, P. H. and Stix, M. (1972) *A. & A.*, **18**, 453

Robertson, J. A. and Eggleton, P. P. (1977) *MNRAS*, **179**, 359

Robinson, E. L., Wood, J. H., Bless, R. C. *et al.* (1995) *Ap. J.*, **443**, 295

Roche, E. (1873) *Mem. Acad. Sci. Montpellier*, **8**, 235

Rogers, F. J. and Iglesias, C. A. (1992) *Ap. J.*, **401**, 361

Rucinski, S. M. (1969) *Acta Astr.*, **19**, 125

   (1973) *Acta Astr.*, **23**, 79

   (1983) *The Observatory*, **103**, 280

   (1998) *Astron. J.*, **116**, 2998

Rucinski, S. M. and Lu, W. (2000) *MNRAS*, **315**, 587

Rucinski, S. M. and Vandenberg, D. (1986) *PASP*, **98**, 669

Saffer, R. A., Wade, R. A., Liebert, J. *et al.* (1993) *Astron. J.*, **105**, 1945

Salpeter, E. E. (1955) *Ap. J.*, **121**, 161

Sandquist, E. L., Latham, D. W., Shetrone, M. D. and Milone, A. A. E. (2003) *Astron. J.*, **125**, 810

Sarma, M. B. K., Vivekananda Rao, P. and Abhyankar, K. D. (1996) *Ap. J.*, **458**, 371

Savonije, G. T. and Papaloizou, J. (1984) *MNRAS*, **207**, 685

Scalo, J. M. (1986) *Fund. Cosm. Phys.*, **11**, 1

Scarfe, C. D., Regan, J., Barlow, D. and Fekel, F. C. (1983) *MNRAS*, **203**, 103

Scharlemann, E. T. (1982) *Ap. J.*, **253**, 298

Schatzman, E. (1962) *Ann. d'Ast.*, **25**, 18

Schlegel, E. M., Kaitchuck, R. H. and Honeycutt, R. K. (1984) *Ap. J.*, **280**, 235

Schmidt, H. M., Dumm, T., Mürset, U., Nussbaumer, H., Schild, H. and Schmutz, W. (1998) *A. & A.*, **329**, 986

Schöffel, E. (1977) *A. & A.*, **61**, 107

Schönberner, D. (1978) *A. & A.*, **70**, 451

Schou, J., Antia, H. M., Basu, S. *et al.* (1999) *Ap. J.*, **505**, 390

Schröder, K.-P., Pols, O. R. and Eggleton, P. P. (1997) *MNRAS*, **285**, 696

Schröder, K.-P., Winters, J. M. and Sedlmayer, E. (2000) *A. & A.*, **349**, 898

Schwartz, R. D. (1983) *A. Rev. A. & A.*, **21**, 209

Schwarzschild, M. (1958) *Structure and Evolution of the Stars*, Princeton: Princeton University Press

Schwarzschild, M. and Härm, R. (1965) *Ap. J.*, **142**, 855

Schweikhardt, J., Schmutz, W., Stahl, O., Szeifert, T. and Wolf, B. (1999) *A. & A.*, **347**, 127

Shafter, A. W. (1984) *Astron. J.*, **89**, 1555

Shafter, A. W., Misselt, K. A., Szkody, P. and Politano, M. (1995) *Ap. J.*, **448**, 33

Shahbaz, T. (2003) *MNRAS*, **339**, 1031

Shakura, N. I. and Sunyaev, R. A. (1973) *A. & A.*, **24**, 337

Shanley, L., Ogelman, H., Gallagher, J. S., Orio, M. and Krautter, J. (1995) *Ap. J.*, **438**, 95

Shapiro, S. L. and Teukolsky, S. A. (1983) *Black Holes, White Dwarfs and Neutron Stars*, New York: Wiley

Shipman, H. L., Provencal, J. L., Hog, E. and Thejll, P. (1997) *Ap. J.*, **488**, L43

Shklovskii, I. (1970) *A. Zh.*, **46**, 715

Shu, F. H., Adams, F. C. and Lizano, S. (1987) *A. Rev. A. & A.*, **25**, 23

Simon, T., Fekel, F. C. and Gibson, G. M. (1985) *Ap. J.*, **295**, 153

Skilling, J. and Bryan, R. K. (1984) *MNRAS*, **211**, 111

Skinner, G. K., Bedford, D. K., Elsner, R. F. *et al.* (1982) *Nature*, **297**, 568

Skumanich, A. (1972) *Ap. J.*, **171**, 565

Smak, J. (1993) *Acta Astr.*, **43**, 121

Smak, J., Belczyński, K. and Zoła, S. (2001) *Acta Astr.*, **51**, 117

Smalley, B. (1997) *The Observatory*, **117**, 338

Smith, D. A. and Dhillon, V. S. (1998) *MNRAS*, **301**, 767

Smith, D. A., Dhillon, V. S. and Marsh, T. R. (1998) *MNRAS*, **296**, 465

Smith, M. A. (1977) *Ap. J.*, **215**, 574

Smith, V. V., Cunha, K., Jorissen, A. and Boffin, H. M. J. (1996) *A. & A.*, **315**, 179

Sneden, C., Kraft, R. P., Prosser, C. F. and Langer, G. E. (1991) *Astron. J.*, **102**, 2001

Soberman, G. E., Phinney, E. S. and van den Heuvel, E. P. J. (1997) *A. & A.*, **327**, 620

Söderhjelm, S. (1980) *A. & A.*, **89**, 100

Spruit, H. (1992) *A. & A.*, **253**, 131

   (1998) *A. & A.*, **333**, 603

St-Louis, N., Moffat, A. F. J., Lapointe, L. *et al.* (1993) *Ap. J.*, **410**, 342

Steenbeck, M., Krause, F. and Ädler, K. H. (1966) *Z. Nat. A.*, **21**, 369

Stępień, K. (1995) *MNRAS*, **274**, 1019

Stickland, D. J., Koch, R. H., Pachoulakis, I. and Pfeiffer, R. J. (1994) *The Observatory*, **114**, 107

Stickland, D., Lloyd, C. and Sweet, I. (1998) *The Observatory*, **118**, 7

Stone, J. M., Hawley, J. F., Evans, C. E. and Norman, M. L. (1992) *Ap. J.*, **388**, 415

Stone, J. M., Hawley, J. F., Gammie, C. F. and Balbus, S. A. (1996) *Ap. J.*, **463**, 656

Stone, J. M. and Norman, M. L. (1994) *Ap. J.*, **433**, 746

Syrovatskii, S. I. (1981) *A. Rev. A. & A.*, **19**, 163

Taam, R. E. (1981) *Ap. J.*, **247**, 257

Taam, R. E. and Sandquist, E. L. (2000) *A. Rev. A. & A.*, **38**, 113

Tapia, S. and Whelan, J. A. J. (1975) *Ap. J.*, **200**, 98

Tassoul, M. and Tassoul, J.-L. (1990) *Ap. J.*, **359**, 155

(1992) *Ap. J.*, **395**, 604

Tayler, R. J. (1952) *MNRAS*, **112**, 387

Taylor, G. I. (1919) *Phil. Trans. A*, **220**, 1

Taylor, J. H. and Weisberg, J. M. (1989) *Ap. J.*, **345**, 434

Terrell, D., Kaiser, D. H., Henden, A. A. *et al.* (2003) *Astron. J.*, **126**, 902

Terrell, D., Munari, U., Zwitter, T. and Nelson R. H. (2004) *Astron. J.*, **126**, 2988

Thorne, K. S. and Żytkow, A. N. (1977) *Ap. J.*, **212**, 832

Thoroughgood, T. D., Dhillon, V. S., Littlefair, S. P., Marsh, T. R. and Smith, D. A. (2001) *MNRAS*, **327**, 1323

Thorsett, S. E. and Chakrabarty, D. (1999) *Ap. J.*, **512**, 288

Thorstensen, J. R., Taylor, C. J., Becker, C. M. and Remillard, R. A. (1997) *PASP*, **109**, 477

Timmes, F. X., Woosley, S. E. and Weaver, T. A. (1996) *Ap. J.*, **457**, 834

Tokovinin, A. A. (1992) *A. & A.*, **256**, 121

(1997) *A. & A. Suppl.*, **124**, 75

Tokovinin, A. A., Balega, Y. Y., Pluzhnik, E. A. *et al.* (2003) *A. & A.*, **409**, 245

Tomkin, J. (1978) *Ap. J.*, **221**, 608

(1981) *Ap. J.*, **244**, 546

Torres, G. and Stefanik, R. P. (2000) *Astron. J.*, **119**, 1914

Tout, C. A. (1997) in *Viscosity and Large-Scale Magnetic Fields from Accretion Disc Dynamos*, ed. Wickramasinghe, D. T., Bicknell, G. V. and Ferrario, L., IAU Coll. **163** & ASP Conf. Ser. **121**, p190

Tout, C. A., Pols, O. R., Han, Z. and Eggleton, P. P. (1996) *MNRAS*, **281**, 257

Tout, C. A. and Pringle, J. E. (1992) *MNRAS*, **256**, 269

Trimble, V. (1987) *Astron. Nachr.*, **308**, 343

Truran, J. W. (1982) in *Essays in Nuclear Astrophysics*, ed. Barnes, C. A., Clayton, D. D. and Schramm, D. N., Cambridge: Cambridge University Press, p467

Truran, J. W., Starrfield, S. G., Strittmatter, P. A., Wyatt, S. P. and Sparks, W. M. (1977) *Ap. J.*, **211**, 539

Tuthill, P. G., Haniff, C. A. and Baldwin, J. E. (1999) *MNRAS*, **306**, 353

Udalski, A., Szymański, M., Kałużny, J. *et al.* (1995) *Acta Astr.*, **45**, 1

Underhill, A. B. (1983) *Ap. J.*, **265**, 933

(1984) *The Observatory*, **104**, 235

Vaccaro, T. R. and Wilson, R. E. (2002) *MNRAS*, **342**, 564

Van Hamme, W., Samec, R. G., Gothard, N. W. *et al.* (2001) *Astron. J.*, **122**, 3436

Van Hamme, W. and Wilson, R. E. (1990) *Astron. J.*, **100**, 1981

(1993) *MNRAS*, **262**, 220

van Kerkwijk, M. H., Bergeron, P. and Kulkarni, S. R. (1996a) *Ap. J.*, **467**, L89

van Kerkwijk, M. H., Geballe, T. R., King, D. L., van der Klis, M. and van Paradijs, J. (1996b) *A. & A.*, **314**, 521

van Kerkwijk, M. H., van Paradijs, J. and Zuiderwijk, E. J. (1995a) *A. & A.*, **303**, 497

van Kerkwijk, M. H., van Paradijs, J., Zuiderwijk, E. J. *et al.* (1995b) *A. & A.*, **303**, 483

van Leeuwen, F. and van Genderen, A. M. (1997) *A. & A.*, **327**, 1070

van Paradijs, J., van der Klis, M. and Pedersen, H. (1988) *A. & A. Suppl.*, **76**, 185

van Rensbergen, W. (2001) in *The Influence of Binaries on Stellar Population Studies*, ed. Vanbeveren, D., Dordrecht: Kluwer, Ap. Sp. Sci. Conf. Ser. **264**, p21

van 't Veer, F. (1976) in *Structure and Evolution of Close Binary Systems*, ed. Eggleton, P. P., Mitton, S. and Whelan, J. A. J., IAU Symp. **73**, Dordrecht: Reidel, p343

Vauclair, G., Pfeiffer, B., Grauer, A. D. *et al.* (1995) *A. & A.*, **299**, 707
Vennes, S. and Thorstensen, J. R. (1994) *Astron. J.*, **108**, 1881
Verbunt, F. and Zwaan, C. (1981) *A. & A.*, **100**, L7
Vilhu, O. (1981) *Ap. Sp. Sci.*, **78**, 401
Waelkens, C. (1991) *A. & A.*, **246**, 453
Waelkens, C. and Lampens, P. (1988) *A. & A.*, **194**, 143
Walborn, N. R. (1976) *Ap. J.*, **205**, 419
Walker, E. C. (1944) *JRASC*, **38**, 249
Wang, Y.-M. (1998) in *Cool Stars, Stellar Systems and the Sun*, ed. Donahue, R. A. and Bookbinder, J. A.,
    10th Cambridge Workshop, p131
Warner, B. (1995) *Cataclysmic Binary Stars*, Cambridge: Cambridge University Press
Webbink, R. F. (1976) *Ap. J.*, **209**, 829
Weidemann, V., and Koester, D. (1983) *A. & A.*, **121**, 77
Whitehurst, R. (1988) *MNRAS*, **232**, 35
Williger, G., Berriman, G., Wade, R. A. and Hassall, B. J. M. (1988) *Ap. J.*, **333**, 277
Willis, A. J. (1982) in *Mass Loss from Astronomical Objects*, ed. Gondalekhar, P. M., RAL Workshop, p1
Wilson, R. E. (1974) *Ap. J.*, **189**, 319
    (2001) *IBVS*, 5076, 1
Wilson, R. E. and Devinney, E. J. (1971) *Ap. J.*, **166**, 605
Wilson, R. E. and Rafert, J. B. (1981) *Ap. Sp. Sci.*, **76**, 23
Wilson, R. E. and Starr, T. C. (1976) *MNRAS*, **176**, 625
Witte, M. G. and Savonije, G. J. (1999) *A. & A.*, **350**, 129
Włodarczyk, K. (1984) *Acta Astr.*, **34**, 381
Wolfenstein, L. (1978) *Phys. Rev. D*, **17**, 2369
Wolszczan, A. and Frail, D. A. (1992) *Nature*, **355**, 145
Wood, F. B., Oliver, J. P., Florkowski, D. R. and Koch, R. H. (1980) *Penns. Univ. Pub.* **12**, 1
Wood, J., Horne, K., Berriman, G. *et al.* (1986) *MNRAS*, **219**, 629
Wood, J. H., Horne, K., Berriman, G. and Wade, R. A. (1989) *Ap. J.*, **341**, 974
Wood, J. H. and Saffer, R. A. (1999) *MNRAS*, **305**, 820
Woosley, S. E. and Weaver, T. A. (1995) *Ap. J. Suppl.*, **101**, 181
Worek, T. F. (2001) *PASP*, **113**, 964
Worley, C. E. and Douglas, C. G. (1984) *The Washington Visual Double Star Catalog*, Washington DC: US
    Naval Obs.
Yang, Y. and Liu, Q. (2002) *Astron. J.*, **124**, 3358
Yungelson, L. R. (1973) *Nauch. Inf. Akad. Nauk USSR*, **27**, 93
Zahn, J.-P. (1977) *A. & A.*, **57**, 383
Zahn, J.-P. and Bouchet, L. (1989) *A. & A.*, **223**, 112
Zapolsky, H. S. and Salpeter, E. E. (1969) *Ap. J.*, **158**, 809
Zhang, C. Y. and Kwok, S. (1993) *Ap. J. Suppl.*, **88**, 137
Zinnecker, H. (1984) *Ap. Sp. Sci.*, **99**, 41
Zoła, S. (1991) *Acta Astr.*, **41**, 213
Zuckerman, B., Webb, R. A., Becklin, E. E., McLean, I. S. and Malkan, M. A. (1997) *Astron. J.*, **114**, 805

Subject index

A case, 218
A-metallic stars, 56
AA sub-case, 199
AB Dor stars, 57
AB sub-case, 149, 151
absolute orbits, 3
accretion
    BP and IR modes, 241–250
    critical radii for, 231–235
    in eccentric orbits, 250–253
    and orbit perturbation, 126
    partial, 239–241
accretion discs
    bipolar output, 254
    modelling, 235–237, 299–303
    rotation, 238–239
    size of, 238, 255
accretion-induced collapse, 247, 248
accretion ring, 142–143
    in protostars, 61
AD sub-case, 149, 151, 153, 154
adaptive optics, 4
adiabatic gradient 43
AE sub-case, 149, 150, 151, 153
AG sub-case, 149, 150, 151, 153
AI Vel variables, 56
AL sub-case, 149, 150, 151, 176
Alfén radius, 178, 186, 187, 208, 232–233, 234, 235
Algol stars, 153, 154, 199–200, 227, 238
AM sub-case, 199
AN sub-case, 149, 151, 153, 156, 176, 225
angular momentum loss, 59–60, 135, 148, 154, 170
angular semi-major axis, 3, 5
anisotropic supernova explosions, 222–223
Ap stars, 56
aperture synthesis, 4
apsidal motion, 11–13, 121, 126, 127–128
AR sub-case, 149, 151, 153, 154, 156, 176
AS sub-case, 176
AS sub-case, 149, 150, 151, 153
astrometric binaries, 4
astrometric observations, 2–4
AT Peg, 152
atmospheric blurring, 4
AU sub-case, 175, 176
AUN sub-case, 175, 176
AUR sub-case, 175, 176
AUS sub-case, 175,176
AW sub-case, 175, 176

B case, 218
B-peculiar stars, 55–56
barium stars, 240–241
barytropes, 263–265
BB sub-case, 150, 156
BD sub-case, 150
Be stars, 55, 64
beginning giant branch (BGB), 32, 114
binaries, statistics, 65–69, 114
binary-binary encounters, 228–229
bipolar outflow, 254
bipolar re-emission (BP mode), 148, 168, 169, 173, 177, 242
BL sub-case, 150, 176
black holes
    evolutionary state notation, 147
    formation of, 89, 90, 225, 254
    mass of, 225
    Schwarzchild radius, 232
    and X-ray binaries, 248
blue loop, 85, 93
blue stragglers, 56, 227, 228
BN sub-case, 150, 156, 175, 176
bolometric vs. visual luminosity, 36–37
Bondi–Hoyle accretion radius, 233, 234
BP mode, 148
    see also bipolar re-emission
BR sub-case, 150
breakup times, 15, 66, 67
breathing, 80
brown dwarfs, 13, 42, 57–59
BU sub-case, 156, 175, 176
BUN sub-case, 175, 176, 218
BUR sub-case, 175
BW sub-case, 175, 176
BY Dra stars, 57, 191

carbon flash, 87
carbon-oxygen core, 87–88
case A evolution, 137, 149–154, 253, 254
case B evolution, 137, 150, 153, 154–157
case C evolution, 137, 150, 153, 194–195, 218
case D evolution, 176
cataclysmic binaries, 242–243
cataclysmic variables, 242–243
CD sub-case, 150
CE mode, 148, 215–218, 240, 248
    see also common envelope processes
central condensation, 70–73
Centre des Données astronomiques de Strasbourg, visual orbit data, 3

Stellar objects index

Printed in the United States
By Bookmasters